Computational Mathematics

This textbook is a comprehensive introduction to computational mathematics and scientific computing suitable for undergraduate and postgraduate courses. It presents both practical and theoretical aspects of the subject, as well as advantages and pitfalls of classical numerical methods alongside with computer code and experiments in Python. Each chapter closes with modern applications in physics, engineering, and computer science.

Features:

- No previous experience in Python is required.
- Includes simplified computer code for fast-paced learning and transferable skills development.
- Includes practical problems ideal for project assignments and distance learning.
- Presents both intuitive and rigorous faces of modern scientific computing.
- Provides an introduction to neural networks and machine learning.

Dimitrios Mitsotakis received a PhD in Mathematics in 2007 from the University of Athens. His experience with high-performance computing started while at the Edinburgh Parallel Computing Center at the University of Edinburgh. Dimitrios worked at the University Paris-Sud as a Marie Curie fellow, at the University of Minnesota as an associate postdoc and at the University of California, Merced as a Visiting Assistant Professor. Dimitrios is currently an associate professor/reader at the School of Mathematics and Statistics of Victoria University of Wellington. He has published his work in journals of numerical analysis and in more general audience journals in physics, coastal engineering, waves sciences, and in scientific computing. He develops numerical methods for the solution of equations for water waves, and he studies real-world applications such as the generation of tsunamis. Some of his main contributions are in the theory and numerical analysis of Boussinesq systems for nonlinear and dispersive water waves.

Advances in Applied Mathematics
Series Editors:
Daniel Zwillinger

Quadratic Programming with Computer Programs
Michael J. Best

Introduction to Radar Analysis
Bassem R. Mahafza

CRC Standard Mathematical Tables and Formulas, 33rd Edition
Edited by Daniel Zwillinger

The Second-Order Adjoint Sensitivity Analysis Methodology
Dan Gabriel Cacuci

Operations Research
A Practical Introduction, Second Edition
Michael Carter, Camille C. Price, Ghaith Rabadi

Handbook of Mellin Transforms
Yu. A. Brychkov, O. I. Marichev, N. V. Savischenko

Advanced Mathematical Modeling with Technology
William P. Fox, Robert E. Burks

Introduction to Quantum Control and Dynamics
Domenico D'Alessandro

Handbook of Radar Signal Analysis
Bassem R. Mahafza, Scott C. Winton, Atef Z. Elsherbeni

Separation of Variables and Exact Solutions to Nonlinear PDEs
Andrei D. Polyanin, Alexei I. Zhurov

Boundary Value Problems on Time Scales, Volume I
Svetlin Georgiev, Khaled Zennir

Boundary Value Problems on Time Scales, Volume II
Svetlin Georgiev, Khaled Zennir

Observability and Mathematics
Fluid Mechanics, Solutions of Navier-Stokes Equations, and Modeling
Boris Khots

Handbook of Differential Equations, Fourth Edition
Daniel Zwillinger, Vladimir Dobrushkin

Experimental Statistics and Data Analysis for Mechanical and Aerospace Engineers
James Middleton

Advanced Engineering Mathematics with MATLAB®, Fifth Edition
Dean G. Duffy

Handbook of Fractional Calculus for Engineering and Science
Harendra Singh, H. M. Srivastava, Juan J Nieto

Advanced Engineering Mathematics
A Second Course with MATLAB®
Dean G. Duffy

Quantum Computation
A Mathematical Foundation for Computer Scientists, Physicists, and Mathematicians
Helmut Bez and Tony Croft

Computational Mathematics
An Introduction to Numerical Analysis and Scientific Computing with Python
Dimitrios Mitsotakis

https://www.routledge.com/Advances-in-Applied-Mathematics/book-series/CRCADVAPPMTH?pd=published,forthcoming&pg=1&pp=12&so=pub&view=list

Computational Mathematics

An Introduction to Numerical Analysis and Scientific Computing with Python

Dimitrios Mitsotakis

CRC Press
Taylor & Francis Group
Boca Raton London New York

CRC Press is an imprint of the
Taylor & Francis Group, an **informa** business
A CHAPMAN & HALL BOOK

Designed cover image: https://exhibitions.lib.cam.ac.uk/linesofthought/artifacts/newton-by-kneller/
Sir Godfrey Kneller, Bart. Portrait of Sir Isaac Newton, 1689; oil on canvas

First edition published 2023
by CRC Press
6000 Broken Sound Parkway NW, Suite 300, Boca Raton, FL 33487-2742

and by CRC Press
4 Park Square, Milton Park, Abingdon, Oxon, OX14 4RN

CRC Press is an imprint of Taylor & Francis Group, LLC

ISBN: 978-1-032-26239-0 (hbk)
ISBN: 978-1-032-26240-6 (pbk)
ISBN: 978-1-003-28729-2 (ebk)

DOI: 10.1201/9781003287292

Typeset in CMR10
by KnowledgeWorks Global Ltd.

Publisher's note: This book has been prepared from camera-ready copy provided by the authors.
Access the Support Material: [https://www.routledge.com/9781032262390]

Dedicated to my family

"[...] he knew the difference between knowing the name of something and knowing something [...]"
by Ritchard P. Feynman talking about his father and the notion of deep understanding [40]

Contents

II Introduction to Computational Mathematics 125

Preface

This book is an introduction to some of the methods of computational mathematics. Computational mathematics (also known as numerical analysis or numerical mathematics) is a branch of applied mathematics dedicated to the development and justification of methods and algorithms for solving scientific problems using computers. Methods of computational mathematics have been designed to solve problems that can be described with mathematical equations. Many branches of science, including physics, engineering and finance rely on computational mathematics. The reason is simple: Scientific problems usually cannot be solved without computers. Alongside with the rest of applied mathematics, computational mathematics has become a key tool for understanding our complicated world. If we could describe applied mathematics as a circle starting with the statement of a scientific problem and its mathematical description, then computational mathematics as part of that circle would provide approximate solutions to the problem using scientifically sound numerical methods. A schematic representation of the processes in the circle of applied mathematics is depicted in Figure 1. Topics in the last three bubbles are introduced in this book and consist of the derivation and justification of numerical methods and numerical simulations. The main purpose of this book is to present how some numerical analysis methods work in theory and in practice.

The ultimate goal of this book though is to help students develop skills and learn classical and modern tools for solving mathematical problems using computers. However, in order to be able to understand the mathematical notions, numerical methods and proofs of this book, the reader should be equipped with basic knowledge in computer programming, basic calculus and linear algebra. Important methods and theories required for a complete and modern course in numerical analysis are to be found in this book. For the sake of completeness, I have included a few highly technical proofs among the basic proofs to justify the robustness of methods. The end of a proof is indicated by the symbol □. Technical proofs can be passed over by the reader who is not interested in the theoretical justification of numerical methods. In terms of typesetting, bold letters denote vectors and matrices, while regular letters have been used for scalar numbers. Most of the numerical methods are accompanied by their algorithms in pseudo-language and their implementation in Python. Therefore, it is our wish that this book will serve as both an introduction to Python and as a reference book to numerical analysis. Several applications can be found at the end of each chapter and in the exercises. These applications illustrate the use of numerical methods in practical problems of computer science, physics and machine learning.

The choice of programming language for scientific computations is a rather difficult task. One may choose among classical programming languages such as C/C++ and Fortran, or languages such as Python, Julia and MATLAB®. In this book, we present all the material in Python. Python is a free programming language with a wide variety of extensions. These include extensions to data science and machine learning, as well as more traditional areas, such as the numerical solution of differential equations. Python comes with functions for almost every method we present in this book, which makes Python a great pedagogical tool. The computer code you will find in this book is written in such a simple way that experts in Python will find my coding style very naive. However, I believe that this style helps students to learn programming. Each programmer can follow their own programming style or standards, like those suggested by the official "style guide for Python code" PEP. To ensure the validity of our code, we produced the output presented in this book with PythonTex [106]. In the text, we also test the majority of methods and codes using the so-called *method of manufactured solutions*. Specifically, we present experiments using simple problems with known solutions, and we verify experimentally the properties and accuracy of the numerical solutions. This is a common practice for experimental justification of numerical methods.

The book is organized into three parts. The first part contains a presentation of tools that we use to develop and study numerical methods. Chapters 1 and 2 briefly introduce Python, NumPy and SciPy and are dedicated to readers without previous experience in Python or those who want to refresh basic knowledge. Chapter 2 serves as a revision of linear algebra matters. It also introduces advanced methods, such as methods for special matrices. Chapter 3 is recommended to readers who are not familiar with floating point arithmetic and its consequences. The first part closes with a brief review of absolutely necessary mathematical notions and calculus tools to develop and justify numerical methods.

In the second part, we present the theory and implementation details of important and commonly used numerical methods in science. More precisely, Chapter 5 presents methods for the approximation of roots of nonlinear equations such as the bisection and Newton's method. Chapter 6 is dedicated to generating approximations of datasets, a procedure known as *interpolation* from known to unknown. It contains an introduction to the widely used method of regression (least squares approximation), which is one of the main ingredients of modern machine learning. Chapters 7 and 8 focus on two basic problems of infinitesimal calculus, namely differentiation and integration. Applications to ordinary differential equations are also presented. Chapter 9 is an introduction to numerical linear algebra and presents fundamental methods for linear systems.

The third part contains more advanced numerical methods. Some of them can be characterized as the pillars of scientific machine learning and neural networks. These include the Singular Value Decomposition and nonlinear optimization methods. We attempted to provide a practical point of view of the subject and an exposition of its theoretical background as well. This part usually cannot be covered in introductory courses and is usually found in more advanced classes. To emphasize the applicability of the field of numerical analysis to other scientific fields, such as the currently emerged field of machine learning, we present a brief introduction to artificial neural networks as well as eigenvalue problems and a very interesting application used by internet search engines.

I have to admit that this book cannot replace classical books of numerical analysis, and we encourage readers to read as many books as they can in the field. The phrase *hominem unius libri timeo*, meaning "I fear the person of a single book", applies to science as well. For this reason, in each chapter, we give references for further reading. Although I tried to include all the well-known classical books in each field, I am sure that I have forgotten to include some and apologize for that. My intention as always is to go back to the text and correct my mistakes, which can be from typographical to more serious mistakes such as failure to include important information. I close this paragraph by underlining the fact that whenever we write computer code, it unavoidably contains errors (bugs), and requires significant amount of time and effort to find them and correct them. I wish the readers good luck, and hope you enjoy learning computational mathematics. Finally, I would like to thank my friend, Ramsey Margolis, for his support, advice and help while I was writing this book. I would like also to express my gratitude to my teachers and colleagues for teaching me science over the years.

D. E. M.

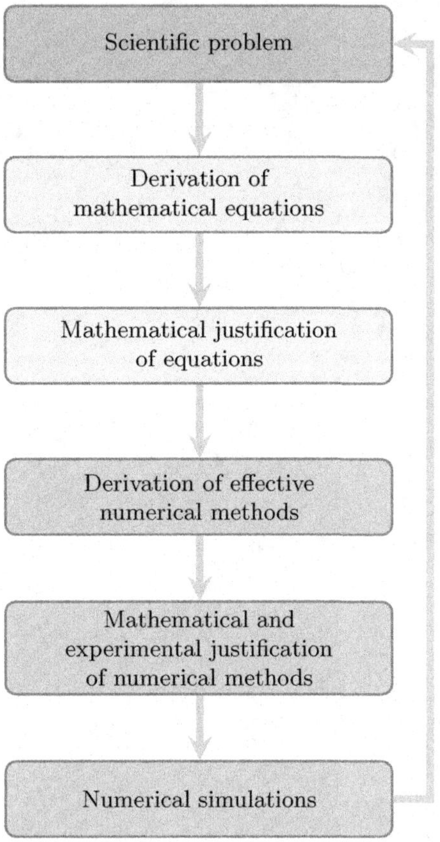

FIGURE 1
A schematic representation of applied mathematics as described by Prof. G. Birkhoff and transferred to me by two of his students (Professors J. Bona and V. Dougalis). See also [65] for more details; Computational mathematics deals with the last three topics.

Part I

Introduction to Scientific Computing with Python

1

Introduction to Python

Well, to begin with, we mention that G. Forsyth [42] pointed out that for the purposes of scientific computing, a *"math book isn't enough"*. In order to get the real essence of scientific computing and numerical analysis, some coding and experimentation are important. For this reason, we present the material of this book using Python as a modern and flexible programming language. This chapter is a brief introduction to Python and is dedicated to readers without previous experience. Special emphasis is given to mathematical applications of repetitive processes using `for` loops and to logical `if` statements. We also show how to write functions. We suggest to use Jupyter notebooks to test your code and run the examples of this book. For an extended description of Python, we refer to specialized books on the subject. A limited list of references can be found at the end of this chapter.

1.1 Basic Python

Python, as a modern programming language, is an interpreter that can be used as a calculator to perform instantly complicated operations, as opposed to compilers that require compilation of the computer code. We can perform the operation $1 + 1$ in a Jupyter notebook just by typing `1+1` and pressing `shift+return`. Notice that when we write a computer command or code, we use fonts that resemble a typewriter. In the following paragraphs, we explore some basic capabilities of Python.

1.1.1 Python, numbers and variables

0 Python plays the role of an instant interpreter between the programmer and the computer. The programmer gives commands to the computer via Python language. Python commands give directions to the computer system on how to perform operations. The most commonly used first example in programming is: Display on the screen of the computer the phrase `Hello world!`. This can be done easily by typing `print('Hello world!')` in our Jupyter notebook and then pressing `shift+return` (or in some keyboards is `shift+enter`). For example type the following command:

```
print('Hello world!')
```

```
Hello world!
```

Since we know how to display messages on the computer screen, we move to actual computations. We should not forget that computer systems were first developed to execute operations with numbers. For this reason, the fundamental element of any programming language, including Python, is numbers. Numbers in Python can belong to different sets, like the number sets we know from mathematics. Python practically has (almost) three main

DOI: 10.1201/9781003287292-1

sets of numbers. We used the word *almost* because there are many numbers missing from computer arithmetic. Python understands integer, real and complex numbers. We usually call the real numbers in Python `floats` or `floating point numbers` for reasons we will discuss later.

The main differences between integer, real and complex values are the following: a real number should contain a dot and/or a decimal part (for example `1.21`), and the complex number is defined from its real and imaginary parts, while the imaginary unit is denoted by `j`. For example, we write $2 + 3\mathrm{i}$ in Python as `2.0+3.0j`.

☞ Note that number `3` in Python is different from number `3.` or `3.0`. Although integers and floats (floating point numbers) are different numbers, we can mix them and perform operations with all of them, usually without problems. However, it is advisable to use the dot whenever you consider floating point numbers.

Unfortunately, computers have a finite memory and understanding. They only know some integer, real and complex numbers. Computer number systems are subsets of mathematical number systems. There is a maximum integer value that we can use, which usually depends on the computer system. We put this issue aside for the moment and continue with how to deal with these numbers in practice.

If we type `1+1` in Python, we will get 2. It is true that we will have the result, but we cannot use this result further. It is in fact forgotten by the computer immediately after it is computed. If we want to keep a number in computer memory for future reference, we need to ask computer for a storage place. The storage places in computer programming are called *variables*. We store numbers (and other values) in *variables*. A variable is an object that has a name, like variables in mathematics. We can have a variable with name x, and we can store values in this variable just by typing `x=2` or `x=1+1`.

We can name a variable with any name we want. Variable names should not contain spaces or symbols that are reserved for operations. We can use the underscore symbol `_`. Thus, we can use variables with names `x`, `x_`, `_x`, or we can use more descriptive names such as `number_of_students`. We can store in the variable `number_of_students` the number of students of a class. This can be done in Python by typing `number_of_students = 50` and pressing `shift+return`. If we want later on to see (access) the number of students in the specific class, we can write the command `print(number_of_students)`. Python will print the number `50` on the screen of the computer. We can even change the number of students to `40` by typing `number_of_students = 40`. The variable values can be changed just by assigning a new value to the name of the variable.

```python
number_of_students = 50
print( number_of_students )
number_of_students = 40
print( number_of_students )
```

```
50
40
```

Variables can store integer, real and complex numbers for future reference. We can also store other things such as matrices, or text (which we usually call string and are alpharithmetic sequences of letters, digits and symbols). To keep things simple, we consider only integer, real and complex numbers and variables for the moment. We can identify the type of a variable with the function `type`. For the variable `x = 1.9` we can type `type(x)` to find what is its kind. The type of this variable is `float`, which is equivalent to real.

```
x = 1.9
print( type(x) )
```

```
<class 'float'>
```

To provide more information related to a variable, we can use the print command to display the value of a variable combined with the appropriate explanatory text (string). Using the previous example, we can print the number of students using the more informative output command

```
print( 'The number of students =', number_of_students )
```

This will print next to the value of the variable `number_of_students` the text (string) `The number of students = `. This can make the communication between computer and human easier.

Another option is not to assign a value of a variable in the program, but instead to use the keyboard to enter the value we want. Python can read a value for a variable from our computer keyboard via the function `input`. The format of the function `input` is the following

```
variable = input( string )
```

This function will print the `string` on the screen, and the program will pause until we type a value for the variable `variable` and hit the return/enter key. This value will be stored in the variable `variable` in the computer memory. We rewrite the previous example using the new features for input and output:

```
1 number_of_students = input( 'Enter the number of students' )
2 print( 'The number of students = ', number_of_students )
```

In these lines of code, line 1 will pause the execution of the program, and the computer will wait for us to enter (type using the keyboard) the value we want to be stored in the variable `number_of_students`. Then line 2 prints the same value with the explanation of what that number is.

1.1.2 Basic mathematical operations

Now that we know about numbers and variables in Python, we continue with mathematical operations between them. The simplest example of an arithmetic operation is the addition of two numbers. We can perform very simple calculations such as $1+2$ by just typing `1+2` in a Jupyter notebook and pressing `shift+return`.

```
1 + 2
```

```
3
```

Obviously, the *arithmetic operator* $+$ indicates the sum of two numbers, and as we will see later, it can add other objects such as matrices. The complete list of Python arithmetic operators on numbers is presented in Table 1.1.

TABLE 1.1
Arithmetic operators in Python.

+	Addition	x+y
-	Subtraction	x-y
*	Multiplication	x*y
/	Division (float)	x/y
//	Division (floor)	x//y
%	Modulus (or remainder of division)	x%y
**	Raise to power	x**y

Float division results from the division between the numbers (or variables) x and y and is a real number. The result of the floor division is the integer quotient between two numbers or variables. We test the arithmetic operators using integers with the following code:

```

# Examples of Arithmetic Operators
x = 11
y = 2
# Addition of numbers
add = x + y
# Subtraction of numbers
sub = x - y
# Multiplication of numbers
mul = x * y
# Division (float) of numbers
div1 = x / y
# Division (floor) of numbers
div2 = x // y
# Modulo of both numbers
mod = x % y
# Raise to power
pow = x ** y
# print results
print( 'add = ', add )
print( 'sub = ', sub )
print( 'mul = ', mul )
print( 'div1 = ', div1 )
print( 'div2 = ', div2 )
print( 'mod = ', mod )
print( 'pow = ', pow )
```

```
add =  13
sub =  9
mul =  22
div1 =  5.5
div2 =  5
mod =  1
pow =  121
```

In the previous example, the commands and text following the symbol # are comments and are not taken into account by Python. We use comments to make our codes readable and also keep helpful notes for future reference. Observe that we can include helpful text in the print command using quotes ' or ". The text with the quotes is called alpharithmetic or string.

The priority of the operations is the same as the operations priority we use in mathematics. Python first executes operations in parentheses (square and curly brackets are reserved for other purposes), then exponents, then multiplications and divisions, and finally additions and subtractions. We test the operations priority using the following commands where we compute the variables a, b, c in one line of code using the separation operator ;

```
a = 1 + 2 * 3; b = (1 + 2) * 3; c = 2**2 * 3
print(a,b,c)
```

```
7 9 12
```

We can verify the result by doing the same thing with pen and paper:

$$a = 1 + 2 \cdot 3 = 1 + 6 = 7,\ b = (1 + 2) \cdot 3 = 3 \cdot 3 = 9,\ c = 2^2 \cdot 3 = 4 \cdot 3 = 12\ .$$

Another example of the arithmetic operations is the following: Suppose that there are 50 students registered for a course, number_of_students = 50. Later, 5 students drop the course. The updated number of students is the old number of students minus 5. This in python can be computed as easy as

```
number_of_students = 50
number_of_students = number_of_students - 5
print(number_of_students)
```

```
45
```

Initially number_of_students = 50. Then we assign to the variable with name number_of_students the result of the operation number_of_students - 5. Although we use the same name on both sides of the relation, the result stored in the variable with name number_of_students has been updated to 45.

☞ If we want to increase the value of the variable x by 1 the syntax is x = x + 1. If we want to decrease the value of the variable x by 3 the syntax is x = x - 3. We type x = 2 * x to double its value or x = x/3.0 to divide by 3. All these expressions are not equations but assignments. The operator = assigns the value of the right-hand side to the variable on left-hand side.

1.1.3 Mathematical functions and constants

It is common in scientific computations to use various mathematical functions and constants. Suppose that we need to compute the value of the expression $\sin(\pi/5)$. Fortunately, the most commonly used mathematical functions, including the sin function, are implemented in Python. These are part of specialized libraries (files containing computer code) called modules. There are many different Python modules. Mathematical functions can be found in the module math as well as in the module numpy. The module numpy will be a subject of later section, so for the moment we focus on the module math.

For efficiency reasons, modules are not loaded automatically. We can load an entire module (library of functions) using the command

```
import module_name
```

where `module_name` is the name of the module we want to load on computer's memory. As a result, all the names of functions, constants, and other objects, contained in the imported module will be loaded into the memory. For example, we can write `import math` to load the names of the functions included in the module math. If we need to use a specific function, then we call the name of the module "dot" the name of the function like in the following lines of code:

```
import math
print(math.sin(math.pi/5))
```

```
0.5877852522924731
```

In this example, `math.sin` is the sin function, and `math.pi` is the number $\pi = 3.14\cdots$. The result shows that $\sin(\pi/5) \approx 0.5877$.

You might think that it is a waste of time to write the "`math.`" every time you want to call a function, and this is reasonable. Sometimes, although it is not recommended, we can import a whole module (not just function names), in such a way that we do not need to specify the module again. This can be done by using the command `from math import *`. After that we can use the `sin` function without writing `math.sin`.

```
from math import *
print(sin(pi/5))
```

```
0.5877852522924731
```

The method we just mentioned is not recommended. The reason we should avoid using the last statement is because it makes the code slow, and in cases where we import more than one module, it can happen that different functions can be implemented under the same name in different modules. In such case, there will be a confusion on what function and from what module we are using. For this reason, we import modules by giving them short nicknames. This is also more convenient in terms of typing. For example, we can import the module `math` with the nickname `mt` in the following way:

```
import math as mt
print( mt.sin(mt.pi/5) )
```

```
0.5877852522924731
```

As another example of how to use the `math` module, we consider an approximation of π using the formula $\pi = 4\arctan(1)$:

```
import math as mt
print( 4.0*mt.atan(1.0) )
```

```
3.141592653589793
```

Since we only used the function `atan` from the `math` module we can import only that function by writing `from math import atan`. The complete code is the following:

```
from math import atan
print( 4.0*atan(1.0) )
```

```
3.141592653589793
```

Of course, in Python we do not really need to estimate the value of π as it is already in various modules. On the other hand, in classical programming languages such as FORTRAN and C we use the previous formula for the definition of π.

If you are curious to see how many different functions and constants you have at your disposal in the `math` module, try typing the command `dir(math)` in a Jupyter notebook.

```
import math
dir(math)
```

This will show you on your screen the contents of the math module.

1.1.4 Powers of 10 and scientific notation

Sometimes you will need to type a very small number or a very large number. For example, consider the number $0.0000015 = 1.5\times10^{-6}$. It is convenient to represent such small numbers using powers of 10. For this reason, we can use the so-called scientific notation. When we want to type a number $a\times10^b$ in scientific notation we write `a e b`, so the number 1.5×10^{-2} can be written as `1.5e-2`. Another example is 2×10^9. In this case, because there are no decimal digits we can write `2.e9` or just `2e9`.

It is worth mentioning that very small floating point numbers, such as `1.e-15` or less can be thought of good approximations of 0 when working with scales of order of 1 or greater. In order to see this, perform the following peculiar calculation in Python:

```
print(1.0+1.e-16)
```

```
1.0
```

In this case, the addition of a tiny number such as 10^{-16} to 1 returns 1. This is mathematically wrong, but it makes sense in computer arithmetic. More information about this phenomenon can be found in Chapter 3 along with a more detailed description of scientific notation.

1.2 Assignment Operators

We have already discussed that the sign `=` is used to assign a value to a variable. If we write `x = x + 1`, it means that the value of `x` will be replaced by the result of the operation `x + 1`. This can also be thought of as the operation that increases the value of `x` by `1`. In Python we can perform this operation by using the specialized operator `+=`. For example, we can write `x += 1` instead of `x = x + 1`. We can use this operator to increase the value of the variable `x` by any number. The command `x += 2` is equivalent to `x = x + 2` and will increase the value of `x` by 2.

TABLE 1.2
Assignment operators in Python.

Operator	Use	Equivalent to
=	x = y	x = y
+=	x += y	x = x + y
-=	x -= y	x = x - y
*=	x *= y	x = x * y
/=	x /= y	x = x / y
%=	x %= y	x = x % y
**=	x **= y	x = x ** y

We can perform all sorts of operations to assign new values to a variable in a similar way. If we want to multiply the value of variable x by 2, we can write `x *= 2`, which is the same as `x = x * 2`. The operators =, +=, *= are called *assignment operators*. A complete list of assignment operators can be found in Table 1.2.

1.3 Strings, Tuples, Dictionaries and Lists

Here, we introduce some commonly used data structures. These structures are the following: Strings that resemble text messages, tuples, which can combine more than one variable of different types, and lists, which can be seen as vectors or sequences of similar objects.

1.3.1 Strings

A string (or alpharithmetic) is a sequence of characters (letters, numbers and symbols) enclosed in single or double quotes. We can use strings when we want to display messages on the screen. We have already seen how to print on the screen the message `Hello world!` and we repeat it here to point out that the sentence `Hello world!` has been formed as a string in single quotes:

```
print( 'Hello world!' )
```

```
Hello world!
```

We can also store strings in variables. For example, we can set `my_name = 'James'` and `my_lastname = 'Bond'`. We can also add strings and concatenate them like

```
full_name = my_name + ' ' + my_lastname
```

where we add after the variable `my_name` a space and then the variable `my_lastname` forming the full name.

```
my_name = 'James'
my_lastname = 'Bond'
full_name = my_name + ' ' + my_lastname
print( full_name )
```

```
James Bond
```

We use the operator + to add strings although in this case the result is to merge the different strings.

The string stored in the variable `my_name` consists of 5 characters. Basically, it is stored character by character, and we can access any character we want as it is an ordered array of characters. For example, we can access the third letter of `my_name` by using the command `print(my_name[2])`. We use the index 2 in order to print the third letter because in Python counting starts from 0.

Starting counting from 0, we can write the full name using the slice operator (:) like `my_name[:]`. Similarly, we can access the first 5 characters with the command `my_name[0:5]`. This will access the entries of `my_name` with index from 0 to 5, excluding 5. This is because in Python the last index after the slice operator does not count. When we start from 0 we do not need to write it. So the previous command can be written as `my_name[:5]`. We can also start from any other entry in the particular sequence. For example, if we want to print the entries after the third one we can type `my_name[3:]`. This will print the fourth `my_name[3]` and fifth `my_name[4]` letters (since we start counting from 0).

```
my_name = 'James'
print( my_name[:] )
print( my_name[0:5] )
print( my_name[:5] )
print( my_name[3:] )
```

```
James
James
James
es
```

In the same way, we can access any character or part of the string we want. We will analyze further the slice operator when we discuss lists.

> ☞ In Python we start counting from 0. The slice operator : can be used as follows `start:stop` where `start` is the value from which we start counting and `stop-1` is where we want to stop. We say that Python is not inclusive. Python accesses the entire array when we use the slice operator : without specifying where to start and where to stop. If we don't provide the value `start` then this is taken 0. If we do not provide the value `stop`, then the counter will reach the last entry.

1.3.2 Tuples

A tuple is a sequence of (not necessarily same) objects separated with commas. For clarity, tuples are usually enclosed in parentheses. For example, a tuple can be the combination `2,'Hi'` or `(2,'Hi')`, where its first entry is the number 2 and its second entry is `'Hi'`. Tuples can be useful when we want to return more than one object in a function. We can access the entries of a tuple stored in a variable by using the location of the entry starting from location 0. If we store the previous tuple in the variable `x` typing `x = (2,'Hi')`, then we can access its first entry by typing `x[0]` and its second entry by typing `x[1]` without forgetting that we always start counting from 0.

```
x = ( 2, 'Hi' )
print( x[0] )
print( x[1] )
```

```
2
Hi
```

We can also have tuples that contain a single object. Tuples are unchangeable objects. Once we create a tuple we can either replace it with a new one or we can delete it using the command `del` but we cannot change its values.

We can access a range of entries in a tuple by using the slice operator :. For example we can access all its elements just by typing `x[:]`. To access the entries from the position 1 to 4 we write `x[1:5]`. We will discuss the slice operator in more detail below.

1.3.3 Dictionaries

A dictionary in Python is a data structure that creates references (keys) to some given values. It is pretty much the same as what we know as dictionary. We use curly brackets to define it and colon : to separate the key from its value.

For example, we can define a dictionary with name `variables` to store the variables $x = 0$, $y = 1$ and $z = 2$

```
variables = { \
    "x": 0, \
    "y": 1, \
    "z": 2 \
    }
```

The backslash operator \ in the previous example stands for line break. Although it is not required since the variables are contained in curly brackets, we still use line breaks in order to write the code in a clear way. The code continues in the next line avoiding any tabs or spaces.

We can access the key of a dictionary by using the name of the dictionary and the key name inside square brackets. For example we can access the value of the variable `x` by typing `variables["x"]`.

```
print( variables["x"] )
```

0

We can change the value of a key in a dictionary by setting the key of the dictionary variable equal to the new value:

```
variables["y"] = 10
print( variables["y"] )
```

10

Although dictionaries have many capabilities, we only mention the most fundamental, because in this book we will not make any use of them.

1.3.4 Lists

A list is a data structure that resembles a vector and contains data linked as in a chain. Lists can change size by adding entries anywhere we like. The list can contain only objects of the same type (i.e. numbers only, or strings only, etc.). The different entries are separated by commas and they are enclosed in square brackets. For example, we create the list (variable) with name `list` and entries `[5, 3, 7]` as simple as `list = [5, 3, 7]`. If we want to access an entry of a list, we ask for the specific entry (coordinate) starting counting from `0`. For example, the first entry of the list `list` is the entry `list[0]`. So we have the following example:

```
list = [5, 3, 7]
print(list[0])
print(list[1])
print(list[2])
```

```
5
3
7
```

A list is a periodic structure, in the sense that we can access the last entry of a list by calling its entry located before the first entry. This means that we can access the last entry of a list by writing `list[-1]`. In the previous example this will give

```
list = [5, 3, 7]
print(list[-1])
```

```
7
```

We can easily make changes to an entry of a list by using the right index in the list. For example, we can change the entry `list[1]` by just typing `list[1] = 9`

```
list[1] = 9
print(list)
```

```
[5, 9, 7]
```

We can add a new entry at the end of a list using the *member function* `append` which takes as an argument the new entry. We call this function a member because it is part of the structure list. For example, we can add the new value 10 at the end of the list `list` using the command `list.append(10)`.

```
list.append(10)
print(list)
```

```
[5, 9, 7, 10]
```

The length of a list can be found using the command `len(list)`. This command uses the function `len` with argument the name of a list, in our case `list`. Functions in Python execute a certain procedure given some input arguments and return a result (like in any other programming language). In the following example, we introduce a new list with the name `grades` that contains the grades of a student. We store also the length of the list in variable `n`.

```
grades = [9.0, 6.7, 8.5, 7.6, 9.8, 7.0]
n = len(grades)
print(n)
```

```
6
```

We can add new entries to the end of the list using the operator `+` avoiding the function `append` as follows

```
grades = grades + [6.2, 8.2, 5.9]
print(grades)
```

```
[9.0, 6.7, 8.5, 7.6, 9.8, 7.0, 6.2, 8.2, 5.9]
```

The slice operator : can be used again in order to extract certain parts of a list. For example, if we want to extract the entries 3, 4 and 5 (i.e. the entries `grades[2]`, `grades[3]` and `grades[4]`) we type

```
print(grades[2:5])
```

```
[8.5, 7.6, 9.8]
```

Observe that the slice operator includes the index value 5, indicating that we will go up to index 4. In order to extract all values starting from the beginning of list `grades` until a certain index (let's say index 3) we can write `grades[:4]` without specifying the starting index. Of course it is not wrong to write `grades[0:4]`. To extract all the values from index 4 until the end of the list we type `grades[4:]`. In general, the slice operator used to extract multiple entries from lists and matrices. For example, `start:stop:step` will return the values from `start` until `stop-1` with step `step`. The default value of `start` is 0, for `stop` is the end of the list, and for `step` is 1.

Copying one list variable to a new one needs some attention. If we use the command `list1 = list2`, then the new variable list1 will be practically the same with `list2`. Thus, when we make changes to any of these two lists, we modify the other one as well. In other words we create an *alias* of the `list2` with name `list1`. For example, consider the `list2 = [9, 6, 8, 7, 9, 7]` and we modify the entry `list2[1]`. This alters both lists `list1` and `list2`.

```
list2 = [9, 6, 8, 7, 9, 7]
list1 = list2
list2[1] = 10
print(list1)
print(list2)
```

```
[9, 10, 8, 7, 9, 7]
[9, 10, 8, 7, 9, 7]
```

If we want to create a new unrelated copy of `list2` with the name `list1`, then we need to type `list1 = list2[:]`.

```
list2 = [9, 6, 8, 7, 9, 7]
list1 = list2[:]
list2[1] = 10
print(list1)
print(list2)
```

```
[9, 6, 8, 7, 9, 7]
[9, 10, 8, 7, 9, 7]
```

Adding two lists results in a new list by concatenating the two lists. For example, if we want to concatenate `list1` and `list2` and create a new one with the name `list3`, we need to type `list3 = list1 + list2`.

```
list1 = [9, 6, 8, 7]
list2 = [1, 2, 3]
list3 = list1 + list2
print(list3)
```

```
[9, 6, 8, 7, 1, 2, 3]
```

Multiplying a list with a positive integer number will create a new list that will repeat the list as many times as is the value of the multiplier. For example

```
list1 = [1, 2, 3]
list2 = 3*list1
print(list2)
```

```
[1, 2, 3, 1, 2, 3, 1, 2, 3]
```

Next we present a way to represent matrices as we know them from mathematics, using lists of lists.

1.3.5 Lists of lists

Python does not have a built-in type for matrices. However, we can think of a matrix in Python as a list of lists. The entries of this list (matrix) represent the matrix rows and are lists. For example, the matrix

$$\boldsymbol{A} = \begin{pmatrix} 1 & 2 \\ 3 & 4 \end{pmatrix} ,$$

is represented in Python as `A = [[1,2],[3,4]]`.

```
A = [[1,2],[3,4]]
print(A)
```

```
[[1, 2], [3, 4]]
```

In linear algebra we usually refer to the entries a_{ij} of the matrix $\boldsymbol{A}$ using indices $i = 1, 2, \ldots$ and $j = 1, 2, \ldots$. In the previous example the entry $a_{11} = 1$ and the entry $a_{12} = 2$. The indices in Python start from 0 though. This means that in order to access the entry a_{12} of $\boldsymbol{A}$ we need to access the entry of the Pythonic matrix located at the 0th row and 1st column, indicating the row and the column in square brackets by typing `A[0][1]`.

```
a12 = A[0][1]
print(a12)
```

2

Since we deal with lists of lists, multiplication with a scalar will result in a new matrix (list) with different dimensions, which is not useful when we want to perform efficient matrix computations. For scientific computations with matrices we use a different data structure called *array*. This data structure is similar to the Pythonic list of lists but is equipped with the usual linear algebra properties and operations. For this reason, we focus on matrix computations using arrays. Arrays are provided by the module NumPy and will be subject of later section.

1.4 Flow Control

In the previous sections we studied data structures including lists and matrices that may contain thousands or millions of entries. Sometimes we need to access all of them one by one. This sounds impossible. On the other hand, these procedures can be automated by defining a repetitive process known as *loop*.

At other times we want to access an entry of a matrix when a specific condition holds. For example, assume we want to use an entry only if the entry is not 0. This logical comparison can be implemented within a logic control (boolean) statement.

1.4.1 `for` loops

Suppose we need to execute a specific command repeatedly for several times. For example, let's say that we want to print the text `Hello world!` three times. This can be done using a `for` loop. A `for` loop starts with the word `for` followed by a variable index that takes values in a specific order. The command ends with the slice operator ":". This indicates that the code beginning with a tab (four spaces) in lines that follow, will be executed every time the loop is executed. The variable index usually is declared to belong `in` a specific list. Then each command in the for loop will be executed once for each value of the variable in the specified list. The general for loop would look like this:

```
for variable in list:
    block of commands
```

☞ Python is space-sensitive language and different blocks of code can be distinguished using tabs (4 spaces).

So in our case, to print three times the text `Hello world!` we will use a variable `i` as a counter, which will take values from the list `[0, 1, 2]`. This can be implemented in the following lines:

```
for i in [0, 1, 2]:
    print('Hello world!')
```

```
Hello world!
Hello world!
Hello world!
```

It doesn't really matter what the values in the list are since we don't use the variable `i` in any calculation. This list could have been `[3, 5, 10]` or any other list with three entries. We can use the values of variable `i` by just using the variable in our code. For example, let's repeat the same loop but with printing the value of `i` instead of the string `Hello world!`.

```
for i in [0, 1, 2]:
    print(i)
```

```
0
1
2
```

Another example that demonstrates the same thing but instead of using a list with numbers we use a list with strings is the following

```
for animal in ['cat', 'dog', 'bird']:
    print(animal)
```

```
cat
dog
bird
```

If we want to repeat a process 1000 times, it will not be convenient to use the list `[0,1,...,999]` with values from 0 to 999. In this case, we can use the function `range`. The function `range` creates a sequence of values from a start value `start`, up to a value `stop-1` and they differ by `step`. The syntax of the function range is the following:

```
range([start ,] stop [,step])
```

The arguments in square brackets indicate optional arguments. When we omit the `start` value the default value is 0. When we omit the `step` value the default value is 1. For example, the command `range(100)` and `range(0,100,1)` will do the same thing: They will generate a list with entries from `0` to `99` with step `1`.

> ☞ In Python when we indicate a `stop` value in the `range` function, (like in slices), then this value is not used and the process ends at the value `stop-1`.

In our last example, if we want to print the values `i` from 0 to 2 we need to use the function `range(3)`, since this command will return the values 0, 1 and 2.

```
for i in range(3):
    print(i)
```

```
0
1
2
```

Python `for` loops can be combined with an `else` clause, offering more functionality compared to other programming languages. The `else` clause is executed when the `for` loop is completed normally, which means it was not interrupted by a `break` command. For more information on this Python `for/else` feature, we refer to more specialized texts on Python.

1.4.2 Useful applications

Here we discuss some useful applications that can be used in different circumstances. These applications require techniques and methods we have learned so far.

Computing the sum of several numbers

Suppose we want to compute the sum of the first 100 positive integers. Mathematically speaking this is the $sum = 1 + 2 + 3 + \cdots + 100 = \sum_{i=1}^{100} i$. In order to compute this sum we need a variable to store the result. We call this variable `sum`. Since we have to add a lot of numbers, it is better to use a `for` loop over the numbers we need to add. Initially, the variable `sum` has the value `0`. At every loop we add the corresponding number to the `sum` variable. This algorithm can be described using a pseudo-language like the one in Algorithm 1.

Algorithm 1 Summation of many numbers

$sum = 0$
for $i = 1 : n$ **do**
  $sum = sum + i$
end for

This is the first numerical algorithm we present in this book using a pseudo-language. When we present algorithms we tend to use natural indices instead of Python indices. So in this algorithm when we write $i = 1 : n$ it means that i takes all values from 1 to n with step 1. The general format of this iteration will be $i =$ start : stop : step. An implementation of this algorithm in Python can be the following:

```python
sum = 0
for i in range(1,101):
    sum = sum + i
print(sum)
```

```
5050
```

The command `sum = sum + i` can be replaced by the command `sum += i`. We could have also started with `sum = 1` instead of 0 and then add to this variable all the numbers from `2` to `100`. The result would have been the same using one addition less[1].

Creating a uniform grid

A typical example that we will consider many times in the future, especially when we deal with *discrete* problems (or formulations) instead of *continuous* is the following: Assume that we have an interval $[a, b]$ of the real line and that we want to divide it into N subintervals

[1]Every operation we perform in Python takes time and in lengthy calculations economy of operations can be crucial.

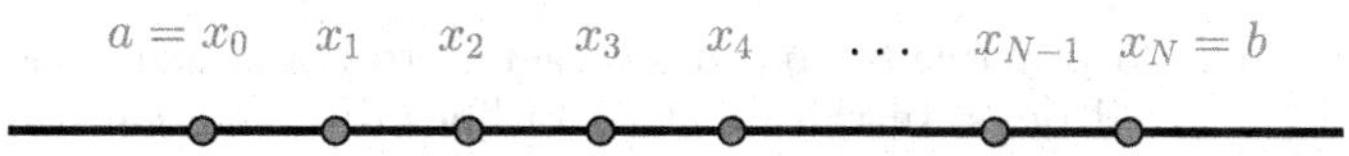

FIGURE 1.1
A uniform grid of an interval $[a, b]$.

$[x_i, x_{i+1}]$ such that $[a, b] = \cup_{i=0}^{N-1}[x_i, x_{i+1}]$. This requires to take $N + 1$ points x_i for $i = 0, 1, \ldots, N$ such that

$$a = x_0 < x_1 < x_2 < \cdots < x_{N-1} < x_N = b \ .$$

The points x_i are called *nodes* or *grid points* and they form a *partition* (usually called *grid* or *mesh*) of the interval $[a, b]$. The distance between two nodes is usually denoted by h_i or Δx_i and is defined as

$$h_i = x_{i+1} - x_i \ .$$

The maximum distance $h = \max_i\{h_i\}$ is called *gridsize* or *stepsize*. If $h_i = h$ for all values of i, then the grid is called *uniform*. If the grid is uniform then each subinterval $[x_i, x_{i+1}]$ of the interval $[a, b]$ has the same length $h = \frac{b-a}{N}$. In this case

$$x_{i+1} = x_i + h \quad \text{for} \quad i = 0, 1, \ldots, N \ .$$

A uniform grid is presented in Figure 1.1.

You may wonder why we need to generate partitions of intervals of real numbers. The answer is that when we perform scientific computations we cannot compute a solution or function everywhere in an interval. Instead, we compute the values of the function or the quantity only to a finite set of points $x_0, x_1, \ldots, x_N$ in the interval $[a, b]$. This set of points will be the grid.

Let us now use Python to create a uniform grid in the interval $[0, 1]$ with 5 nodes ($N = 4$ intervals). Each subinterval will have length $h = (1 - 0)/4 = 0.25$ and so the nodes will be

$$\{x_0, x_1, x_2, x_3, x_4\} = \{0, 0.25, 0.5, 0.75, 1\} \ .$$

In order to create the grid we need a `for` loop to create the end-point x_{i+1} of each interval $[x_i, x_{i+1}]$ and will store it in a list called `mesh`. We initialize the list `mesh` with the value of $x_0 = 0.0$.

```
1 a = 0.0; b = 1.0;
2 N = 4 # the number of subintervals
3 h = (b-a)/float(N) # the grid-size
4 x = 0.0 # the starting node
5 mesh = [x] # initialization of the mesh list
6 for i in range(N):
7     x = x + h
8     mesh.append(x)
9 print(mesh)
```

```
[0, 0.25, 0.5, 0.75, 1.0]
```

In the previous code, line 1 defines the endpoints of the interval $[a, b]$ and line 2 the number of subintervals. In line 3 we use the formula $h = (b - a)/N$ to define the uniform stepsize. Observe that we divide by the number `float(N)` instead of `N` to avoid the conversion of the result into integer, which can result to the number 0. Perhaps we didn't need to do this but it is a good practice to convert integers into floats using the function `float` whenever possible to avoid bugs. In the `for` loop of lines 6–8 we create the grid using the formula $x_{i+1} = x_i + h$ and we `append` our list `mesh` with the new value x_{i+1}.

It is to be noted here that the variable `x` initially had the value `0.0` (line 4). Then, during the first iteration of the `for` loop, the value `x` became `x+h`, which is $0 + 0.25 = 0.25$. This means that the initial value 0 has been replaced by 0.25 and entered at the end of the list `mesh`. During the second iteration of the `for` loop the value of `x` is replaced again by the result of the operation `x+h`, which is now $0.25 + 0.25 = 0.5$. And in the rest of the loops the code will compute the nodes 0.75 and 1.0.

Nested loops

Sometimes we need to use a `for` loop inside another `for` loop. For example let's try to access all the entries of the matrix

$$A = \begin{pmatrix} 11 & 12 & 13 \\ 21 & 22 & 23 \\ 31 & 32 & 33 \end{pmatrix} .$$

In this case, working with rows, we need first to access the entries 11, 12, 13 of the first row, then the entries 21, 22, 23 of the second row and finally the entries 31, 32, 33 of the third row. So we need a `for` loop for each row and while we are working with a specific row, we need a second loop for each column. This can be implemented using a nested loop as follows:

```
1 A = [[11, 12, 13], \
2      [21, 22, 23], \
3      [31, 32, 33]]
4 # The \ at the end of each row indicate a line break
5 # and the code continues in the next line
6 for i in range(3):
7     for j in range(3):
8         print( A[i][j] )
```

```
11
12
13
21
22
23
31
32
33
```

In the definition of the Pythonic matrix `A` we use the symbol "\" to indicate an explicit line break. The code continues in the next line avoiding possible tabs or spaces. This is not really necessary when introducing matrices or arrays. In general splitting brackets and parentheses do not require the explicit line break symbol "\", and we will try to avoid the use of the line break character in the rest of this book.

In this example the `for` loop of line 6 starts with the row `i=0`. Then, Python executes the `for` loop of line 7 with fixed row `i=0`, where `j` takes the values `0`, `1` and `2`. (Remember that the range function starts from 0 and never uses the value we set for upper bound).

So it will print the entries `A[0][0]`, `A[0][1]`, `A[0][2]`. When the second `for` loop finishes, Python returns back to the first loop which hasn't finished yet for `i` to take the value `1`. Now, for `i=1`, Python will execute again the second loop from the beginning, giving to `j` the values `0`, `1` and `2` and printing the entries `A[1][0]`, `A[1][1]`, `A[1][2]`. When the second `for` loop finishes again, Python will return back to first `for` loop which hasn't finished yet as it remains to give to the variable `i` the value 2. For `i=2`, Python will execute the second `for` for the last time (as `i` cannot take the value `3`) and will print the last row `A[2][0]`, `A[2][1]`, `A[2][2]`.

Nested loops are common in Python and the syntax is based again on tabs (or 4 spaces). If the first loop starts at some point of the code, the second loop needs to start 4 spaces (or a tab) after the first loop. Common mistakes in the syntax of loops are either to forget the slice operator (colon) or to use the wrong number of spaces under the `for` statement.

1.4.3 Boolean type of variables

So far we have encountered types of variables such as integers, floats, strings, lists, tuples etc., but there are even more types. A very useful type of variables is the boolean (logical). These variables can take the values `True` or `False` only. Assigning values to a boolean variable is like assigning the value 1 if it is `True` or 0 if it is `False`. Boolean variables can tell us if a statement is true or false. To understand how this can be done, consider the logical statement $1 < 2$. This statement is true since we know that 1 is less than 2. For Python, this is a boolean operation, and the result is true. In order to see that, we store the result in the variable a (boolean) by typing `a = 1<2` or `a = (1<2)` and we print it.

```
a = 1<2
print(a)
```

```
True
```

The result is `True` as expected.

There are several boolean operations. It is worth stressing again that the operator `=` doesn't mean equation or equality but assignment as it assigns to the variable on the left side whatever is the result on the right side of the operator `=`. In order to check if two variables are equal we use the operator `==`. To check whether they are not equal we use the operator `!=`. For example let us check if $1 = 1$. This can be done by typing `1==1`, or `1!=1`. The first operation gives the result `True` since $1 = 1$ indeed. The second operation gives the result `False` since the statement $1 \neq 1$ is false.

```
print(1==1)
print(1!=1)
```

```
True
False
```

A complete list of boolean operators in Python is presented in Table 1.3.

1.4.4 `while` loops

Now we are ready to discuss another kind of repetitive loop which can do a similar job to `for` loop. This is the `while` loop. A `while` loop executes the code block contained in its loop as long as a specified condition is `True`. The syntax of a `while` loop is the following:

TABLE 1.3
Boolean operators in Python

<	Less than	x < y
>	Greater than	x > y
==	Equal to	x == y
!=	Not equal to	x != y
<=	Less than or equal to	x <= y
>=	Greater than or equal to	x >= y

```
while condition:
    block of commands
```

where condition can be a boolean variable or a logical expression. In the following code, we create an infinite loop by keeping the condition always true.

```
cond = True
counter = 0
while cond:
    print(counter)
    counter +=1
```

In this example the variable cond will always remain true and the code will print the variable counter starting from 0. This is the simplest example of a while loop.

Another example is the following. Let's say that we want to print 1, 2, 3 on the screen. This can be done by using a variable i with initial value 1 and then print all values of i while they are less than or equal to 3. Inside the loop we need to increase the value of i every time we print its value.

```
i=1
while i <= 3:
    print(i)
    i = i + 1
```

```
1
2
3
```

In the previous code, initially we set the variable i to be 1. Then, Python checks if $i \leq 3$. If this is True, then Python executes the block of commands within the specific loop, and then it returns back to the while statement to check again whether $i \leq 3$. During the first loop Python prints the value of i, which is 1, and increases its value by 1. So the variable i becomes 2. Python then checks again if the value $i \leq 3$ is True or False. If it is True, Python executes the block of code in the loop again, and repeats the same cycle until the value of the variable i becomes $i > 3$ i.e. 4 since, here, i is an integer.

It should be noted that if the condition is always True then Python executes the specific loop forever until we interrupt the execution. This is a typical bug when we use while loops.

1.4.5 Decisions and the if statement

Sometimes, the flow of our code needs to change depending on the result of a boolean (logical) operation. For example, assume we want to divide 10 by a variable x. If $x = 0$ we know that this is impossible. For this reason, we need to use an `if` statement. This is formulated as follows

```
if condition:
    block of commands
```

And here is an example where we consider the division of 10 by a variable x

```
x = int ( input('Give me x = ') )
if x != 0:
    print(10/x)
```

In this case, our program will perform the division if $x \neq 0$ only, while x is given from the input. But, if we give $x = 0$, then our code does nothing. If we want to exhaust all possible scenarios, we need to use another `if` statement with the case $x = 0$. For example, we could tell Python to print an error message if $x = 0$. This can be done with the `elif` or/and the `else` statements. The command `elif` is an abbreviation of *else if* and we need to provide a new condition. On the other hand, with the `else` statement, there is no need to provide any additional condition as this statement will be executed if all the previous statements are `False`. In general, we can combine the `if` statement with `elif` and `else`. The general `if` statement can be formulated as:

```
if condition1:
    block of commands
elif condition2:
    block of commands
else:
    block of commands
```

where the blocks `elif` and `else` are optional. Observe that the `else` statement has no condition since it is satisfied if the `condition1` and `condition2` are both `False`.

The complete code for the example with the division would be:

```
x = int ( input('Give me x = ') )
if x != 0:
    print(10/x)
else:
    print('Division impossible')
```

The division can be performed if $x \neq 0$, otherwise the code prints the message `Division impossible`.

It should be noted that if Python encounters a division by zero the execution of the code will be terminated with an error message `ZeroDivisionError: division by zero`.

To illustrate the `if` clause better, consider the computation of the sign of a number without using the built-in function `sign`. In this example, the input to the code is the number stored in variable x. If this number is positive, $x > 0$, then the sign is $+1$, else if it is negative, $x < 0$, then the sign is -1, and if $x = 0$, then the sign is 0. The Python code should look like this:

```
x = int ( input('Give me x = ') )
if x > 0:
    print('sign(x) = +1')
elif x < 0:
    print('sign(x) = -1')
else:
    print('sign(x) = 0')
```

Sometimes, we need more than one condition to be satisfied at the same time. For example let's say that we want to check if the value of a variable x is in the interval $[0, 10]$, or in other words if $x \geq 0$ and $x \leq 10$ at the same time. This can be expressed in Python using the boolean operator **and**, and is as follows

```
x = int ( input('Give me x = ') )
if x >= 0 and x<=10 :
    print('x is in the interval [0,10]')
else:
    print('x is not in the interval [0,10]')
```

If we want to check whether the value x is outside the interval $[0, 10]$ we can check if $x < 0$ or $x > 10$ as these conditions cannot be true at the same time. This can be done using the boolean operator `or` as follows

```
x = int ( input('Give me x = ') )
if x < 0 or x > 10 :
    print('x is outside from the interval [0,10]')
else:
    print('x is in the interval [0,10]')
```

Another way to check whether the value of x is in $[0, 10]$ is to check if the value is not outside the interval $[0, 10]$ using the boolean operator `not` as follows

```
x = int ( input('Give me x = ') )
if not(x < 0 or x > 10) :
    print('x is in the interval [0,10]')
else:
    print('x is outside from the interval [0,10]')
```

The `not` operator acts like a function: it takes as an argument a boolean (`True` or `False`) and returns the opposite value.

1.4.6 Error control and exceptions

In the previous section we discussed the case where we wanted to divide a number by a variable `x`. We decided to use an `if` statement to avoid division by zero. This is because in the event of a division by zero, the execution of the code might stop with an error message. Problematic situations like the one we just mentioned can be avoided using the `try` and `except` statements, instead of `if` statements. The syntax of the `try` and `except` statements is the following

```
try:
    block of commands
except error:
    block of commands
```

The keyword error can be the name of the expected error, like `ZeroDivisionError`, or it can be omitted. The code with the division by variable `x` can be written without the use of the `if` statement in the following way:

```
x = 0
try:
    print(10/x)
except:
    print('The division is impossible')
x = 2
try:
    print(10/x)
except:
    print('The division is impossible')
```

```
The division is impossible
5.0
```

With the `try` and `except` statements we were able to perform the division 10/0 without terminating the execution of the code and continue with the division 10/2. If we hadn't used this technique the code would have been stopped when the division 10/0 was performed with the error `ZeroDivisionError`. We make use of the `try` and `except` statements in Chapter 9 in the implementation of Cholesky algorithm.

Similar to the `try` and `except` statements is the command `assert`. This command is formed as

```
assert expression
```

where `expression` can be a boolean statement. If this boolean statement is `False`, then the execution of the code will be interrupted with an error message. If the `expression` is `True`, then the assert statement has no effect in the execution of the code. The `assert` statement can be useful for debugging purposes. Next, we discuss how to interrupt the execution of a `for` loop.

1.4.7 The command `break`

The command `break` is used to terminate the execution of a `for` or `while` loop before the termination condition is satisfied. This is usually required when an exception during the loop process happens. The following example demonstrates the use of the command `break` through a simple example. We use a `for` loop to iterate the variable `i` from 0 to 9. We interrupt the `for` loop using a `break` statement when `i` becomes 5.

```
for i in range(10):
    print(i)
    if i == 5:
        break
```

```
0
1
2
3
4
5
```

In this example we observe that although the `for` loop indicates that the variable `i` can take values up to `9`, the loop is terminated when `i` becomes `5` because of the `break` command.

1.5 Functions

Functions are useful when we need to repeat a certain number of statements for several times. In particular, we include code that we want to use repeatedly in discrete blocks of code forming functions. These functions take some input and return some output. We can use them by calling their name and without caring about their content. Python functions can be thought of as factories where we give them some input variables, the function will process them, and finally will return the result in the output. The general notion of a Python function is not far from the general notion of a mathematical function. In mathematics we define a function as $y = f(x)$ where f is the name of the function, x is the input and y is the output. Similarly, in Python we define a function first by using the keyword `def` to indicate that we will define a function. Then we specify the name of the function and the input. If the input consists of more than one variable we separate them with commas. We close the line of the definition with the slice operator : and a block of code with the process follows. The result of the function is usually indicated at the very end of the function with the keyword `return`. We can return multiple variables or multiple values with the return statement. The general structure of a function is the following:

```
def function_name(variable1, variable2,...):
    block of commands
    return result_values
```

As a first example we define a function that we will call `double`. This function will take as input a variable x, and will return the product $2x$.

```
def double(x):
    y = 2 * x
    return y
```

We could have omitted the definition of the variable y by just putting the operation next to the `return` statement like:

```
def double(x):
    return 2 * x
```

We can call this function now anywhere in our code just by using its name and a valid argument `x`. The result can be stored in a new variable, or it can be printed on the screen:

```
x = 3
r = double(x)
print(double(x))
```

```
6
```

We now define a function that we call `div10`. This function will take as input a variable x and will return the result of the division $10/x$ if $x \neq 0$.

```
def div10(x):
    if x!=0:
        return 10/x
    else:
        print('Division impossible')
```

We can now define a `sign` function to return the sign of an input variable x

```
def sign(x):
    if x > 0:
        return +1
    elif x < 0:
        return -1
    else:
        return 0
```

Sometimes, we can give default value to one or all input variables. For example, if we want to compute the maximum between two variables without using Python's built-in function `max` we define the function:

```
def max(x, y = 0):
    if x >= y:
        return x
    else:
        return y
```

Here, the default value for the variable y is 0 but we can change it by providing a different input:

```
print( max( -1 ) )
print( max( 2, 5 ) )
print( max( x=2, y=5 ) )
a=2; b=5;
print( max( a, b ) )
```

```
0
5
5
5
```

Note that spaces between the different characters in the input don't matter. The first function call `max(-1)` assigns the value -1 to the variable x. Then it compares $x = -1$ with the default value $y = 0$ since there is no input for y. The rest of the function calls assign the value 2 to the variable x and the value 5 to the variable y with several different ways. We note that for the first two calls of the function `max`, the only variables being used are those included in the function. These are called *local* variables. For the last function calls we define two additional variables a and b. These new variables, for example the variable `a` and `b` are called *global* and are copied to the local functions variables `x` and `y` respectively. The local function variables x and y will be erased after the function call. Even if we make changes to the local variables this does not affect the actual *global* variables.

To see this we define a function that swaps the values between two variables. We employ a new variable `temp` to store the value of `x`. Then we replace the value of `x` to the value of `y` by setting `x=y`. At this stage, the value of `x` has been stored in the value of `temp` and the value of `y` has been stored in `x`. Then we set `y=temp` to get the value of `x`.

```
1 def swap(x, y):
2     temp = x
3     x = y
4     y = temp
5     print('local variables: x = ',x , 'y = ', y)
6     return x, y
7 x = 2; y = 3;
8 swap(x, y)
9 print('global variables: x = ',x , 'y = ', y)
```

```
local variables: x =  3 y =  2
global variables: x =  2 y =  3
```

After the call of the function `swap` (line 8) we observe that the global variables `x` and `y` are unchanged, while the function was expected to change their values. If in addition we print the local variables `x` and `y` inside the function `swap` we will see that these variables have been changed internally although externally they remain unchanged.

We can extend the scope of a local variable by using the commands `global` and `nonlocal`. The `global` statement can be used to indicate that particular variables can be used globally in the code even if they are defined in the scope of a function, while the `nonlocal` statement indicates that particular variables can be extended outside of the block of code that they are defined but they are bounded there. The command `nonlocal` is meaningful when we define a function inside a function.

☞ In Python, the variables defined inside a function are local variables. Local variables cannot be seen (or used) outside the function.

Finally, we discuss an alternative way to define functions in one line of code using the Python `lambda` statement. This can be useful for functions with simple formula. If we want to define a function with formula $f(x)$ this can be done by typing

```
f = lambda x: f(x)
```

Let's say that we want to define a function with formula $f(x) = x + 1$. This can be done by typing `f = lambda x : x + 1`. To compute, for example the value $f(1)$, we just type `f(1)`. This will return 2.

```
f = lambda x : x + 1
print(f(1))
```

2

The general description of a lambda function is the following

```
name of function = lambda arguments : formula
```

and is an alternative and fast way to introduce simple functions in our codes.

Recursive functions

A recursive function is a function that calls itself. One such example is the function that returns the Fibonacci sequence of numbers

$$0, 1, 1, 2, 3, 5, 8, 13, 21, 34, 55, 89, 144, 233, \ldots \ .$$

Let's denote $F(n)$ the function that returns the n-th Fibonacci number. Then obviously $F(0) = 0$ and $F(1) = 1$. The recursive function that returns all the rest of the Fibonacci numbers is

$$F(n) = F(n-1) + F(n-2) \quad \text{for} \quad n = 2, 3, \ldots \ .$$

Such a function can be implemented very easily with the following code.

```
def Fibonacci(n):
    if ( n==0 ):
        return 0
    elif ( n==1 ):
        return 1
    else:
        return Fibonacci(n-1)+Fibonacci(n-2)
# Print the first 14 Fibonacci numbers
for i in range(14):
    print( Fibonacci(i) )
```

This will print the numbers `0, 1, 1, 2, 3, 5, 8, 13, 21, 34, 55, 89, 144, 233`.

Recursive function calls are usually inefficient and its use is not recommended unless there are no alternatives.

1.6 Classes

Python is an object-oriented programming language. As such, it gives us the option to create our own objects. These objects are called `classes`. Objects can have their own new name. They can also have properties and methods. A new class can be defined as

```
class Name:
    block of code
```

where the `block of code` describes the various properties of the object with name `Name`. Let's take this simple example. Assume that we want to create a new object to describe vectors (x, y). Vectors are characterized by their x and y coordinates. Assume also that we would like to record their magnitude (norm). For example we can have a vector $\mathbf{a} = (3, 1)$ with magnitude $\|\mathbf{a}\| = \sqrt{3^2 + 1^2} = \sqrt{10}$ and we want all these information to be stored in a single variable. Similarly, we will need a second vector $\mathbf{b} = (2, 5)$. We will use a `class` object with name `vector`. This new class will contain three new variables such as the variables `x`, `y` and `norm`:

```
class vector:
    x = 0.0
    y = 0.0
    norm = 0.0
```

With the previous code we have created a class with three data members (`x`, `y` and `norm`). The three data members have all default value 0 (zero vector). If we want to consider our first vector we type

```
a = vector()
a.x = 3.0
a.y = 1.0
a.norm = np.sqrt(10.0)
```

This will create the variable `a` with the data of our first vector. Then we can create similarly a second variable with the second vector b:

```
b = vector()
b.x = 1.0
b.y = 5.0
b.norm = np.sqrt(26.0)
```

We can access the data of vector `a` by just printing its members:

```
print(b.x)
print(b.y)
print(b.norm)
```

```
1.0
5.0
5.0990195135927845
```

This was the simplest example of a class. Classes can also contain functions. The most common function is the function `__init__`. The function `__init__` initializes the class and can be seen as a constructor. We can create a new vector using the function init that can take as input all the members of the class. Our previous example can then be extended as follows

```
class vector:
    def __init__(self, a, b, c):
        self.x = a
        self.y = b
        self.norm = c
```

which means that we can now define a new vector such as

```
a = vector(3.0, 1.0, np.sqrt(10.0))
print(a.x)
print(a.y)
print(a.norm)
```

```
3.0
1.0
3.1622776601683795
```

There are other interesting properties of classes such as the inheritance but are out of the scope of this book and we omit them.

1.7 Accessing Data Files

When we deal with large amount of data we usually need to store them in data files and not in lists or arrays. This is due to memory limitations. Here we present how to access text (txt) files with data. We can access a txt file by first identifying and opening the file using the Python function `open`. This has the general function call

```
f = open(filename, mode)
```

where `filename` is a string with the name of the file, and `mode` is one of the following strings: `r` when we want to open a file just for reading, `w` to open a file by deleting its contents and writing new data, `a` to open the file by keeping existing data and append the file with writing new data at the end it. One can formulate combinations of these modes using the `+` symbol. For example, we can use the mode `r+` to read and write without creating a file. The function `open` returns a file object `f`. This object is equipped with the methods `read`, `readline` and `write` to read and write data. It also has the method `close` to close the file and interrupt any access to it.

Reading the first 5 characters from a file we just write `f.read(5)`. Text files have line breaks indicated by the invisible character '`\n`' at the end. If we want to read a whole line, then we can use the command `f.readline()`.

To write data into a text file is somewhat more complicated, especially if you want to write numbers because the method `write` accepts as input only strings. So if you want to store a variable, then you need to convert the variable into string using the command `str`. You can use the operator `+` if you want to add additional text.

As a simple example, we open a file with name `test.txt` for writing the numbers 0 to 4. To write the new data we use the mode `w` and the method `write` as follows

```
f = open("test.txt", "w")
for i in range(5):
    f.write(str(i)+'\n')
f.close()
```

Every time we want to write a new line we use the symbol '`\n`' at the end of the input of the command `write`. Otherwise, all input numbers will appear in one line. To write two columns of data `i` and `j`, we can use the command `f.write(stri(i)+' '+stri(j))` to merge the numbers i, j with a space between them.

1.8 Further Reading

There are many books and online resources which aim at teaching Python, including the official documentation of Python, NumPy, SciPy and MatPlotLib. We do not provide online references as these may change, but we do encourage the interested reader to explore these options. References [84, 87] consist of a complete introduction to programming in Python. This list is not restricted to those references though and among the online books there is a huge variety of books written for teaching Python. Books [88, 136] specialize in Python for data science and they also contain information about Jupyter. The books [78, 79, 50] serve as introductory textbooks to computational methods with Python and contain an introduction to Python and its numerical analysis modules.

Chapter Highlights

- Python is an interpreter of computer code, it translates and executes our code instantly.
- Python detects the type of a variable automatically. The basic types are `integer, float, string`.
- The variable `a = 1` is different from the variable `b = 1.0` as the first is integer and the second is float.
- Number of the form $a \times 10^b$ can be written using scientific notation as `a e b`. For example, we write `1.5e-4` instead of 1.5×10^{-4}.
- The priority of operations in Python is the same as the priority of operations in mathematics.
- The operator = indicates assignment and not equation. The command `x=2` assigns the value 2 in the variable `x`.
- The indices in matrices and lists start from `0`.
- There is no structure for matrices in Python and they can be represented as lists of lists. In this book we will use NumPy arrays to represent matrices.
- The slice operator is used to extract multiple entries from lists and matrices. For example, `start:stop:step` will return the values from `start` until `stop-1` with step `step`.
- The same with the function `range`, which is formed as `range(start, stop, step)` will return the values from `start` to `stop-1` with increment `step`. If we omit the `step` this implies that `step = 1`.
- For repetitive processes we use the commands `for` and `while`.
- We can control the execution of our code and take decisions using the command `if` followed by a boolean expression.
- Blocks of codes that perform certain procedures can be included in functions.
- Functions usually accept some input that is stored in local variables and return some output.
- Python has a wealth of functions for many applications. Libraries that contain functions and other objects are called modules.
- Mathematical functions and constants can be accessed from the `math` module. Later we will introduce more sophisticated modules for mathematical problems such as the modules NumPy and SciPy.
- We will not use the `math` module but the modules NumPy and SciPy instead.

Exercises

1. Write a Python program to take as input the user's first and last name and prints them in reverse order with a space in between.

2. Write a Python program to compute the volume of a sphere with radius 4.

3. Write a Python program to count the appearance of number 4 in a given list.

4. Write a Python program to check whether a specified value is contained in a list of integer values.

5. Write a Python program to compute an approximation of number π using the formula $\pi = 4\arctan(1)$ and floating point numbers. Then compute the difference between your π and the `pi` of the `math` module.

6. Write a Python program to find out whether a given number (given by the user) is even or odd, and print out an appropriate message.

7. Write a Python program to add three given integers. However, if two values are equal sum will be zero.

8. Write a Python program to add two given integers. However, if the sum is between 15 and 20, it will return 20.

9. Write a Python program that will return true if two given integer values are equal or their sum or difference is 5.

10. Write a Python program that will print all the even numbers from 1 to 100 and will compute their sum.

11. Write a Python program to find those numbers which are divisible by 7 and multiple of 5, between 1500 and 2700 (both included).

12. Write a Python program to count the number of even and odd numbers in a list of numbers.

13. Write a Python program to create the multiplication table (from 1 to 10) of a number.

14. Write a Python program to count the number of even and odd numbers from a series of numbers. Try the list `[1, 2, 3, 4, 5, 6, 7, 8, 9]`.

15. Write a Python program which takes two digits m (row) and n (column) as input and generates a two-dimensional array (list of lists). The value of the entry of the i-th row and j-th column of the array should be $i \cdot j$.

16. Write a Python program to check if a triangle is equilateral, isosceles or scalene.

17. Write a Python function to compute the circumference and the area of a circle of a given radius. The function must accept as input the radius of the circle and will return both the circumference and the area.

18. Write a Python function that accepts an integer n as input and returns the value of $n + n^2 + n^3$.

19. Write a Python function to compute the distance

$$d = \sqrt{(x_1 - x_2)^2 + (y_1 - y_2)^2} \ ,$$

between the points (x_1, y_1) and (x_2, y_2). All numbers including the result must be floating point numbers.

20. Write a Python function that will create a uniform grid of a given interval (a, b) using N subintervals. The function must accept the limits a and b of the interval (a, b) and the number of the subintervals N and will return a list with all the nodes of the uniform grid.

21. Write a Python program to construct the following pattern, using nested `for` loops.

```
*
* *
* * *
* * * *
* * * * *
* * * *
* * *
* *
*
```

2

Matrices and Python

Without exaggeration, most of the methods we use to solve approximately problems in physical sciences lead to linear algebra problems. Even approximations to solutions of non-linear equations can be found solving linear equations. Before we start exploring numerical methods, it will be useful to review some material from the field of linear algebra. We will introduce the modules NumPy and SciPy and we will start performing scientific computations with them. We will learn how to draw simple graphs on the screen of our computer using the module MatPlotLib. It is not by chance that we say "one picture is worth a thousand words". Graphs are used for debugging, to derive conclusions, make observations, and also for presentation purposes. We start with a revision of matrices and their properties.

2.1 Review of Matrices

An $n \times m$ (n by m) matrix is a rectangular array of elements with n rows and m columns. Not only are the values of the elements important, but their position in the array is as well. We usually denote an $n \times m$ matrix by a capital bold letter such as $\boldsymbol{A}$. The entries of the matrix are denoted by the lowercase letter of the matrix with two indices to indicate the number of row and the number of column (coordinates) where the entry is located in the array. A general entry of a matrix $\boldsymbol{A}$ is denoted by a_{ij}, and corresponds to the entry at the intersection of the i-th row and j-th column of matrix $\boldsymbol{A}$. Schematically, this can be seen as

$$\boldsymbol{A} = [a_{ij}] = \begin{pmatrix} a_{11} & a_{12} & \cdots & a_{1m} \\ a_{21} & a_{22} & \cdots & a_{2m} \\ \vdots & \vdots & & \vdots \\ a_{n1} & a_{n2} & \cdots & a_{nm} \end{pmatrix} .$$

If $m = n$ then the matrix is called *square* matrix. A matrix with only one column is called *column vector* and it is the $n \times 1$ matrix

$$\boldsymbol{b} = \begin{pmatrix} b_1 \\ b_2 \\ \vdots \\ b_n \end{pmatrix} ,$$

while a matrix with only one row is called *row vector* and it is the $1 \times n$ matrix

$$\boldsymbol{b} = \begin{pmatrix} b_1, b_2, \ldots, b_n \end{pmatrix} .$$

It is noted that $-\boldsymbol{A}$ denotes the matrix whose entries are $-a_{ij}$. Sometimes we write $\boldsymbol{A} \in \mathbb{R}^{n\times m}$ to indicate that $\boldsymbol{A}$ is $n \times m$ matrix with real coefficients. If the coefficients are allowed to be complex, then we write $\boldsymbol{A} \in \mathbb{C}^{n\times m}$.

DOI: 10.1201/9781003287292-2

2.1.1 Matrix properties

Matrices are useful. The main reason is because we can perform operations with them, such as addition, multiplication etc. Here we review basic matrix properties.

Matrix equality

Two matrices $\boldsymbol{A}$ and $\boldsymbol{B}$ are equal if they have the same number of rows and columns, say $n \times m$, and if $a_{ij} = b_{ij}$ for each $i = 1, 2, \ldots, n$ and $j = 1, 2, \ldots, m$.

Addition of matrices

If $\boldsymbol{A}$ and $\boldsymbol{B}$ are both $n \times m$ matrices, then the *sum* of $\boldsymbol{A}$ and $\boldsymbol{B}$, denoted by $\boldsymbol{A} + \boldsymbol{B}$, is an $n \times m$ matrix $\boldsymbol{C}$ whose entries are $a_{ij} + b_{ij}$, for each $i = 1, 2, \ldots, n$ and $j = 1, 2, \ldots, m$.

$$\boldsymbol{C} = \boldsymbol{A} + \boldsymbol{B} \Longleftrightarrow c_{ij} = a_{ij} + b_{ij} \ .$$

This means that for each i and j, the entry $c_{ij} = a_{ij} + b_{ij}$. If

$$\boldsymbol{A} = \begin{pmatrix} 1 & 2 & 0 \\ 3 & 2 & 2 \\ 0 & 3 & -4 \end{pmatrix} \quad \text{and} \quad \boldsymbol{B} = \begin{pmatrix} 2 & 4 & -1 \\ 0 & -4 & 1 \\ 1 & 2 & 6 \end{pmatrix} ,$$

then

$$\begin{aligned} \boldsymbol{A} + \boldsymbol{B} &= \begin{pmatrix} 1 & 2 & 0 \\ 3 & 2 & 2 \\ 0 & 3 & -4 \end{pmatrix} + \begin{pmatrix} 2 & 4 & -1 \\ 0 & -4 & 1 \\ 1 & 2 & 6 \end{pmatrix} = \begin{pmatrix} 1+2 & 2+4 & 0+(-1) \\ 3+0 & 2+(-4) & 2+1 \\ 0+1 & 3+2 & -4+6 \end{pmatrix} \\ &= \begin{pmatrix} 3 & 6 & -1 \\ 3 & -2 & 3 \\ 1 & 5 & 2 \end{pmatrix} . \end{aligned}$$

Scalar multiplication

We can multiply a matrix with a scalar (number). If we have a number λ (complex or real), the product $\lambda \boldsymbol{A}$ is a new matrix $\boldsymbol{B}$ with entries $b_{ij} = \lambda a_{ij}$ for all indices i, j. We can multiply the previous matrix $\boldsymbol{A}$ by 2 and this is

$$\boldsymbol{B} = 2\boldsymbol{A} = 2 \begin{pmatrix} 1 & 2 & 0 \\ 3 & 2 & 2 \\ 0 & 3 & -4 \end{pmatrix} = \begin{pmatrix} 2\cdot 1 & 2\cdot 2 & 2\cdot 0 \\ 2\cdot 3 & 2\cdot 2 & 2\cdot 2 \\ 2\cdot 0 & 2\cdot 3 & 2\cdot(-4) \end{pmatrix} = \begin{pmatrix} 2 & 4 & 0 \\ 6 & 4 & 4 \\ 0 & 6 & -8 \end{pmatrix} .$$

Matrix multiplication

Multiplication between two matrices is not as simple as the multiplication between a scalar and a matrix. An $n \times p$ matrix $\boldsymbol{A} = [a_{ij}]$ can be multiplied with a $p \times m$ matrix $\boldsymbol{B} = [b_{ij}]$. Note that the number of columns of $\boldsymbol{A}$ must agree with the number of rows of $\boldsymbol{B}$. Then the product $\boldsymbol{AB} = [(ab)_{ij}]$ is the $n \times m$ matrix with entries

$$(ab)_{ij} = \sum_{k=1}^{p} a_{ik} b_{kj}, \quad \text{for} \quad i = 1, 2, \ldots, n \quad \text{and} \quad j = 1, 2, \ldots, m \ .$$

Due to the restriction on the matrix dimensions, and also because of the definition of matrix multiplication, the commutative property is not true and in general

$$\boldsymbol{AB} \neq \boldsymbol{BA} \ .$$

For example, if $\boldsymbol{A}$ is a 2×2 matrix and $\boldsymbol{B}$ a 2×3 matrix

$$\boldsymbol{A} = \begin{pmatrix} 1 & 2 \\ 3 & 4 \end{pmatrix} \quad \text{and} \quad \boldsymbol{B} = \begin{pmatrix} 5 & 6 & 7 \\ 8 & 9 & 10 \end{pmatrix} ,$$

then the product $\boldsymbol{C} = \boldsymbol{AB}$ will be the 2×3 matrix

$$\begin{aligned} \boldsymbol{C} &= \begin{pmatrix} 1 & 2 \\ 3 & 4 \end{pmatrix} \begin{pmatrix} 5 & 6 & 7 \\ 8 & 9 & 10 \end{pmatrix} \\ &= \begin{pmatrix} 1 \cdot 5 + 2 \cdot 8 & 1 \cdot 6 + 2 \cdot 9 & 1 \cdot 7 + 2 \cdot 10 \\ 3 \cdot 5 + 4 \cdot 8 & 3 \cdot 6 + 4 \cdot 9 & 3 \cdot 7 + 4 \cdot 10 \end{pmatrix} \\ &= \begin{pmatrix} 5 + 16 & 6 + 18 & 7 + 20 \\ 15 + 32 & 18 + 36 & 21 + 40 \end{pmatrix} \\ &= \begin{pmatrix} 21 & 24 & 27 \\ 47 & 54 & 61 \end{pmatrix} . \end{aligned}$$

We cannot perform the multiplication $\boldsymbol{BA}$ for these matrices because the dimensions do not agree. This procedure is often called naive multiplication. Other more efficient algorithms for matrix multiplication exist, some of which we discuss in Section 2.2.8.

Zero and identity matrix

The zero matrix $\mathbf{0}$ is a matrix with all entries zero. For example, the 3×3 zero matrix is

$$\mathbf{0} = \begin{pmatrix} 0 & 0 & 0 \\ 0 & 0 & 0 \\ 0 & 0 & 0 \end{pmatrix} .$$

An identity matrix is a square $n \times n$ matrix with entries on its main (principal) diagonal to be 1 while all the off-diagonal entries are zero, i.e. $a_{ij} = 1$ if $i = j$ and $a_{ij} = 0$ if $i \neq j$. An identity matrix is usually denoted by $\boldsymbol{I}$. For example, the 3×3 identity matrix is the matrix

$$\boldsymbol{I} = \begin{pmatrix} 1 & 0 & 0 \\ 0 & 1 & 0 \\ 0 & 0 & 1 \end{pmatrix} .$$

The properties of matrix addition and scalar multiplication do not differ from those for real numbers. We summarize the most basic properties below.

- $\boldsymbol{A} + \boldsymbol{B} = \boldsymbol{B} + \boldsymbol{A}$ (commutative property)
- $(\boldsymbol{A} + \boldsymbol{B}) + \boldsymbol{C} = \boldsymbol{A} + (\boldsymbol{B} + \boldsymbol{C})$ (associative property)
- $\boldsymbol{A} + \mathbf{0} = \mathbf{0} + \boldsymbol{A} = \boldsymbol{A}$ (zero element)
- $\boldsymbol{A} + (-\boldsymbol{A}) = -\boldsymbol{A} + \boldsymbol{A} = \mathbf{0}$ (opposite matrices)
- $\lambda(\boldsymbol{A} + \boldsymbol{B}) = \lambda\boldsymbol{A} + \lambda\boldsymbol{B}$ (distributive property)
- $(\lambda + \mu)\boldsymbol{A} = \lambda\boldsymbol{A} + \mu\boldsymbol{A}$ (distributive property)
- $\lambda(\mu\boldsymbol{A}) = (\lambda\mu)\boldsymbol{A}$
- $1\,\boldsymbol{A} = \boldsymbol{A}\,1 = \boldsymbol{A}$

We will explore matrix multiplication later with some detail.

2.1.2 Special matrices

Matrices with special structure can lead to efficient algorithms. Matrices with many zero entries are special matrices called *sparse*. We tend to work only with the nonzero entries of sparse matrices to save memory and increase the speed of computations. Here we present some important examples.

Diagonal matrices

A *diagonal* matrix $\boldsymbol{D}$ is a square matrix with $d_{ij} = 0$ if $i \neq j$. For example, the identity matrix is a diagonal matrix. Another example is the matrix

$$\boldsymbol{D} = \begin{pmatrix} 1 & 0 & 0 \\ 0 & 2 & 0 \\ 0 & 0 & 3 \end{pmatrix} .$$

Banded matrices

There are cases where the main diagonal of a matrix, along with several other diagonals are not zero. Such matrices can be seen as a collection of non-zero diagonals and are usually referred to as *band* (or *banded*) matrices. A *band* matrix $\boldsymbol{A}$ is a matrix where $a_{ij} = 0$ if $j < i - l$ or $j > i + u$ for some $l, u \geq 0$. The quantities u and l are called *upper* and *lower* bandwidth; l is the number of non-zero diagonals below the main diagonal, and u the number of non-zero diagonals above the main diagonal. The bandwidth of a matrix is defined as the number $p = l + u + 1$, which is the number of its non-zero diagonals including the main diagonal.

According to the previous definition, an $n \times n$ full matrix can be treated as banded matrix with bandwidth equal to its $2n - 1$ diagonals. To make the notion of band matrix useful, we usually treat a matrix as band matrix if its bandwidth is reasonably small. For example, a band matrix with $l = 1$, $u = 2$ and total bandwidth $p = 4$ can be written as

$$\begin{pmatrix} a_{11} & a_{12} & a_{13} & 0 & 0 & 0 \\ a_{21} & a_{22} & a_{23} & a_{24} & 0 & 0 \\ 0 & a_{32} & a_{33} & a_{34} & a_{35} & 0 \\ 0 & 0 & a_{43} & a_{44} & a_{45} & a_{46} \\ 0 & 0 & 0 & a_{54} & a_{55} & a_{56} \\ 0 & 0 & 0 & 0 & a_{65} & a_{66} \end{pmatrix} .$$

It should be noted that:

- a band matrix with $l = u = 0$ is a *diagonal* matrix
- a band matrix with $l = u = 1$ is a *tridiagonal* matrix
- a band matrix with $l = u = 2$ is a *pentadiagonal* matrix, and so on.

Such matrices occur in many circumstances, and for this reason, special algorithms and storage schemes have been introduced.

Triangular matrices

Other special matrices are the following:

- *Upper triangular* $n \times n$ matrix $\boldsymbol{U}$, where for each $j = 1, 2, \dots, n$ the entries $u_{ij} = 0$, for all $i > j$.

- *Lower triangular* $n \times n$ matrix $\boldsymbol{L}$, where for each $j = 1, 2, \ldots, n$ the entries $l_{ij} = 0$, for all $i < j$.

A 3×3 lower triangular matrix has the form

$$\begin{pmatrix} a_{11} & 0 & 0 \\ a_{21} & a_{22} & 0 \\ a_{31} & a_{32} & a_{33} \end{pmatrix},$$

while a 3×3 upper triangular matrix has the form

$$\begin{pmatrix} a_{11} & a_{12} & a_{13} \\ 0 & a_{22} & a_{23} \\ 0 & 0 & a_{33} \end{pmatrix}.$$

Triangular matrices are useful because we don't need to store their zero entries.

Symmetric matrices

Consider a matrix $\boldsymbol{A} = [a_{ij}]$. We define its *transpose* matrix $\boldsymbol{A}^T$, the matrix which has rows the columns of $\boldsymbol{A}$. We can write $\boldsymbol{A}^T = [a_{ji}]$. For example, if

$$\boldsymbol{A} = \begin{pmatrix} 1 & 2 & 3 \\ 4 & 5 & 6 \\ 7 & 8 & 9 \end{pmatrix} \quad \text{then} \quad \boldsymbol{A}^T = \begin{pmatrix} 1 & 4 & 7 \\ 2 & 5 & 8 \\ 3 & 6 & 9 \end{pmatrix}.$$

In general if

$$\boldsymbol{A} = [a_{ij}] = \begin{pmatrix} a_{11} & a_{12} & \cdots & a_{1m} \\ a_{21} & a_{22} & \cdots & a_{2m} \\ \vdots & \vdots & & \vdots \\ a_{n1} & a_{n2} & \cdots & a_{nm} \end{pmatrix},$$

then

$$\boldsymbol{A}^T = [a_{ji}] = \begin{pmatrix} a_{11} & a_{21} & \cdots & a_{n1} \\ a_{12} & a_{22} & \cdots & a_{n2} \\ \vdots & \vdots & & \vdots \\ a_{1m} & a_{2m} & \cdots & a_{nm} \end{pmatrix}.$$

The following properties involving the transpose of a matrix hold whenever the operation is possible:

- $(\boldsymbol{A}^T)^T = \boldsymbol{A}$
- $(\boldsymbol{A} + \boldsymbol{B})^T = \boldsymbol{A}^T + \boldsymbol{B}^T$
- $(\boldsymbol{A}\boldsymbol{B})^T = \boldsymbol{B}^T \boldsymbol{A}^T$

A *square* matrix $\boldsymbol{A}$ is called *symmetric* if $\boldsymbol{A} = \boldsymbol{A}^T$. This means that the rows of matrix $\boldsymbol{A}$ are the same with its columns. Thus, square matrices can only be symmetric. For example,

$$\boldsymbol{A} = \begin{pmatrix} 1 & 2 & 3 \\ 2 & 5 & 6 \\ 3 & 6 & 9 \end{pmatrix} \quad \text{and} \quad \boldsymbol{A}^T = \begin{pmatrix} 1 & 2 & 3 \\ 2 & 5 & 6 \\ 3 & 6 & 9 \end{pmatrix}.$$

Symmetric matrices are also convenient for matrix computations since we need to store only their upper or lower triangular part.

In Python we can define and use matrices with the help of Python's modules NumPy and SciPy[1]. Both are libraries for general purpose scientific computations, [58, 138]. We start with a brief introduction to NumPy.

2.2 The NumPy Module

We have already discussed the **math** module and the lists of lists to represent matrices. In practice, though, we use other modules with more efficient data structures and functions. In particular, we use the **numpy** and **scipy** modules for all our scientific computations. To plot graphs we use the **matplotlib** module. We start with a description of the **numpy** module.

NumPy is a Python module that defines a number of data structures and functions specialized for scientific computing. Like every other module, we need to import its functions in the beginning of our code. This can be done with one of the following commands:

```
# Use only one of the following
import numpy
import numpy as np
from numpy import *
from numpy import function
```

If we use the first command **import numpy** then when we call a function, let's say the sine function **sin** evaluated for x, we need to write **numpy.sin(x)**. We can avoid writing the name of the module if we use the second command **import numpy as np** where we choose to use the alias **np** instead of the word **numpy**. In this case, we need to write **np.sin(x)**. We use the third command **from numpy import ***, when we do not want to specify the module in a function call as we import everything from the specific module. We do not recommend the third statement as Python imports every single function and method from NumPy. This can cause confusion. For example, if we import both **scipy** and **numpy**, then we will be unsure which version of sine function we use since both have the same function.

☞ We usually import the module **numpy** in the beginning of our code. Most commonly, NumPy is imported via the command **import numpy as np**. This is what we will use in this book.

One of NumPy's most important data structures is the n-dimensional array **ndarray**, which we introduce immediately.

2.2.1 Matrices and arrays

Since most scientific computations require matrices and vectors (one-dimensional matrices) NumPy provides with an efficient way of matrix representation. We can define n-dimensional matrices using the NumPy data structure (function) **ndarray**. If the matrix is a usual two-dimensional $n \times m$ matrix, then the data structure is just **array**. In this book we will use only one- and two-dimensional matrices (with small exceptions) so we will use only the function **array**.

[1]NumPy stands for "Numerical Python" while SciPy for "Scientific Python".

A NumPy array is very similar to a list of lists and is actually constructed by a list of lists. We can define the simple 3×2 matrix

$$\boldsymbol{A} = \begin{pmatrix} 1 & 2 \\ 3 & 4 \\ 5 & 6 \end{pmatrix}$$

using the function `np.array` and appropriate list of lists as input. The output will be a new array.

```
import numpy as np

A = np.array( [[1, 2], [3, 4], [5, 6]] )
print(A)
```

```
[[1 2]
 [3 4]
 [5 6]]
```

Each row has two entries (one for each column), which is why we typed the matrix as a list with three entries, the three rows of the matrix. Although, the `array A` in the previous example looks like a Python list of lists, it is not exactly the same. A NumPy array is a proper matrix with proper multiplication while the plain Python list of lists is not.

Natural vs Python indices

Usually, the row and column indices in the representation $\boldsymbol{A} = [a_{ij}]$ take the values $i = 1, 2, \ldots$ and $j = 1, 2, \ldots$. Using these indices, the entry a_{12} corresponds to the entry located at the first row and second column of the previous matrix and is 2. We will call indices that start from 1 and go up to n *natural indices*. In Python (and also in NumPy) indices start from 0. The entry a_{12} of the previous matrix $\boldsymbol{A}$ can be accessed through the $(0, 1)$ entry of the array `A` by using the command `A[0, 1]`. These indices are called *Python indices*. This distinction is important as numerical algorithms in the literature are usually presented using natural indices.

```
print("a(1,2) = ", A[0,1])
```

```
a(1,2) =  2
```

To modify the value of the entry a_{22} into 10 we type `A[1,1] = 10`.

```
A[1,1]=10
print(A)
```

```
[[ 1  2]
 [ 3 10]
 [ 5  6]]
```

Similarly, we modify the $(0,1)$ entry of our array by adding 1 just performing the operation

```
A[0,1] = A[0,1] + 1

print(A)
```

```
[[1 3]
 [3 4]
 [5 6]]
```

☞ Python indices start from 0 contrary to the natural indices that start from 1. This means that if a natural index takes values up to N, then the corresponding Python index will go up to $N-1$.

2.2.2 Slicing

The slice operator : can be used to extract parts of a NumPy array. We can extract the first column (column with Python index 0) of the previous array just by setting the row of A to be the operator ":" like `A[:,0]`:

```
A = np.array( [[1, 2], [3, 4], [5, 6]] )
print(A[:,0])
print(type(A[:,0]))
```

```
[1 3 5]
<class 'numpy.ndarray'>
```

In this example we also used the function `type` which returns the data structure (or object) type of the input variable (in our case the variable `A[:,0]`). We observe that although the result looks like a list, it was extracted as a NumPy array. Storing the first column into a new vector `a` we get

```
a = A[:,0]
print(a)
print(type(a))
```

```
[1 3 5]
<class 'numpy.ndarray'>
```

In the same way, we can extract rows instead of columns. For example, to extract the second row of $\boldsymbol{A}$ (row with Python index 1) we type

```
print(A[1,:])
```

```
[3 4]
```

We can even extract blocks of a matrix using the slice operator. This can be done by specifying the indices we want to extract. Note that the slice operator ":" by itself means that we extract everything, while in the form `start:stop` means that we want to extract the entries with index from the `start` value until `stop-1`. For example, consider the following 3×3 matrix

$$\boldsymbol{B} = \begin{pmatrix} 1 & 2 & 3 \\ 4 & 5 & 6 \\ 7 & 8 & 9 \end{pmatrix} ,$$

for which we extract the 2×2 block consisting of the first 2 columns and 2 rows

$$\boldsymbol{C} = \begin{pmatrix} 1 & 2 \\ 4 & 5 \end{pmatrix} ,$$

using the following lines of code

```
B = np.array( [[1, 2, 3],
               [4, 5, 6],
               [7, 8, 9]] )
C = B[0:1, 0:1]
print(C)
```

```
[[1 2]
 [4 5]]
```

The definition of the array $\boldsymbol{B}$ uses three lines and the command `np.array` has been split. In Python we can split into multiple rows a piece of code that is included between brackets (parentheses, square brackets etc.). Otherwise, we need to use the indication of line break "\" at the end of each line.

> ☞ Python is not inclusive language and as such the slice operator ":" when used like `start:stop` means that the index takes the value `start` until `stop-1`. The general use of the slice operator is `start:stop:step` where `step` is the index jump. When the `step` is 1, then it can be omitted.

An alternative way to define matrices in Python is with the use of the NumPy class `matrix`. `matrix` objects inherit all the properties of the `ndarray` and can be used exactly like the arrays, but with some advantages. Namely, we can perform matrix multiplication with `matrix` objects by using the `*` operator while we cannot with `array` objects. In this book we will refrain from using `matrix` objects, though, and we will stick with arrays.

Vectors and arrays

So far we have considered vectors and matrices using two-dimensional NumPy arrays. For example, we consider the vector $\boldsymbol{a} = (1, 2, 3)$ as a 1×3 matrix:

```
a = np.array( [[1, 2, 3]] )
print(a.shape)
```

```
(1, 3)
```

Note that the command `a.shape` returns the actual size of an array. On the other hand, sometimes we need to use (or a function will return) vectors instead of arrays. Here we call vectors (row or column) arrays with shape `(n,)` instead of `(n,1)` or `(1,n)`. This can be done by using single square brackets instead of double in the definition of an array:

```
a = np.array( [1, 2, 3] )
print(a.shape)
```

```
(3,)
```

We can retrieve the entries of a one-dimensional vector by using a single index:

```
print(a[1])
```

```
2
```

In the unfortunate event where the shape of an array creates a problem, especially when it involves calls of functions that require single or double index for arrays, we can use the command `reshape`. The new shape is provided as a single integer or a tuple in the arguments of this function. The tuple can have -1 as dimension indicating that this has the dimension inferred from the length of the original array:

```
b = a.reshape((1,3))
c = a.reshape((1,-1))
print('b=', b)
print(b.shape)
print('c=', c)
print(c.shape)
```

```
b= [[1 2 3]]
(1, 3)
c= [[1 2 3]]
(1, 3)
```

The reshape function practically doesn't change the data buffer. Instead, it creates a new view of the array.

2.2.3 Special matrices, methods and properties

We discussed special matrices in the first section of this chapter. Python offers functions and methods (attributes) to increase the efficiency of our programs, especially when we deal with special matrices. Here, we go over some special matrices, how to create them, and how to retrieve basic properties of NumPy arrays in general.

The zero matrix

It is sometimes useful to create automatically an $n \times m$ matrix with all entries zero. This can be done with the NumPy function `zeros( (n,m) )` that takes input a `tuple` with the matrix dimensions.

```
A = np.zeros( (3,4) )
print(A)
```

```
[[0. 0. 0. 0.]
 [0. 0. 0. 0.]
 [0. 0. 0. 0.]]
```

This matrix is called the 3×4 zero matrix (3 by 4 zero matrix). The dimensions of a matrix can be retrieved using the function `shape` of a NumPy array, which returns a `tuple` with the number of rows and columns of the array. Other useful methods that return useful information are the attributes `ndim` and `size`. The attribute `ndim` returns the dimensionality of the matrix, for example, whether the array is 2-dimensional or 3-dimensional array (we don't discuss 3-dimensional arrays in this book). The attribute `size` returns the number of entries of the array.

```
print(np.shape(A))
print(A.ndim)
print(A.size)
```

```
(3, 4)
2
12
```

The identity matrix

Another special matrix of interest is the identity matrix, which is the square matrix (has the same number of rows and columns) with all entries zero except for the entries of the main diagonal, which are $a_{ii} = 1$ for all i.

This matrix can be created with the NumPy function `eye(dimension)`. For example, we create a 3×3 identity matrix by typing

```
I = np.eye(3)
print(I)
```

```
[[1. 0. 0.]
 [0. 1. 0.]
 [0. 0. 1.]]
```

NumPy can also create an $n \times m$ matrix full of ones easily with the function `ones( (n,m) )`. The dimensions of the array are provided with a tuple.

```
A = np.ones( (3,2) )
print(A)
```

```
[[1. 1.]
 [1. 1.]
 [1. 1.]]
```

Functions like `zeros`, `eye` and `ones` are common array constructors that can reserve memory for matrix computations. Note that all the arrays created in the previous examples have real numbers as entries. We can specify the type of a matrix, for example whether the values are integers, real or complex, by using the keyword `dtype`. For example, in order to create a 3×3 identity matrix with integer entries we can write `ones(3, dtype=int)`. (In some versions of Python we can neglect the keyword `dtype=`.)

```
I = numpy.eye(3, dtype=int)
print(I)
```

```
[[1 0 0]
 [0 1 0]
 [0 0 1]]
```

Triangular matrices

A triangular matrix is a square matrix with zero all its entries above or below the main diagonal. For example a 3×3 lower triangular matrix has the form

$$\begin{pmatrix} a_{11} & 0 & 0 \\ a_{21} & a_{22} & 0 \\ a_{31} & a_{32} & a_{33} \end{pmatrix},$$

while an upper triangular matrix has the form

$$\begin{pmatrix} a_{11} & a_{12} & a_{13} \\ 0 & a_{22} & a_{23} \\ 0 & 0 & a_{33} \end{pmatrix}.$$

In order to extract the lower triangular part of a matrix $\boldsymbol{A}$ we can use the NumPy function `tril` and for the upper triangular part the function `triu`. For example, if

$$\boldsymbol{A} = \begin{pmatrix} 1 & 2 & 3 \\ 4 & 5 & 6 \\ 7 & 8 & 9 \end{pmatrix},$$

then we can extract the upper and lower triangular parts of this matrix by typing

```
A = np.array( [[1, 2, 3],
               [4, 5, 6],
               [7, 8, 9]] )
L = np.tril(A)
U = np.triu(A)
print('A = '); print(A)
print('L = '); print(L)
print('U = '); print(U)
```

```
A =
[[1 2 3]
 [4 5 6]
 [7 8 9]]
L =
```

```
[[1 0 0]
 [4 5 0]
 [7 8 9]]
U =
[[1 2 3]
 [0 5 6]
 [0 0 9]]
```

Extracting the upper and lower triangular parts is a very useful operation especially when we deal with matrix factorizations as we shall see later.

Matrix transpose

For the same matrix $\boldsymbol{A}$, it is easy to generate its transpose matrix $\boldsymbol{A}^T$ using the NumPy function `transpose`. For example, we can type

```
A = np.array( [[1, 2, 3],
               [4, 5, 6],
               [7, 8, 9]] )
B = np.transpose(A)
print(B)
```

```
[[1 4 7]
 [2 5 8]
 [3 6 9]]
```

We will discuss other special matrices, such as banded matrices later in this chapter.

2.2.4 Matrix assignment

Python provides efficient tools to perform matrix operations. A tricky subject is the matrix assignment (i.e. how to make two matrices equal). If we set two NumPy arrays `A` and `B` equal, `B = A`, then the two matrices become practically identical. When we say identical we mean that when we modify one of them then we modify the other one as well. For example,

```
A = np.array( [[1, 2, 3],
               [4, 5, 6],
               [7, 8, 9]] )
B = A
A[0,0]=10
B[1,1]=20
print(A)
print(B)
```

```
[[10  2  3]
 [ 4 20  6]
 [ 7  8  9]]
[[10  2  3]
 [ 4 20  6]
 [ 7  8  9]]
```

In this example we see that by modifying $\boldsymbol{A}$ we modify $\boldsymbol{B}$ at the same time, and vice versa. If we need to create a new unrelated copy of A, we need to use the attribute copy(). For example, we need to type B=A.copy().

```
A = np.array( [[1, 2, 3],
               [4, 5, 6],
               [7, 8, 9]] )
B = A.copy()
A[0,0]=10
B[1,1]=20
print(A)
print(B)
```

```
[[10  2  3]
 [ 4  5  6]
 [ 7  8  9]]
[[ 1  2  3]
 [ 4 20  6]
 [ 7  8  9]]
```

Now we see that modifying entries of $\boldsymbol{A}$ we don't modify the matrix $\boldsymbol{B}$. Also modifying the matrix $\boldsymbol{B}$ we leave $\boldsymbol{A}$ unchanged.

☞ Setting two NumPy arrays to be equal A = B is not just an assignment of the matrix B to A. Any changes applied to A will affect B and vice versa. To create an unrelated copy of a NumPy array we can use the copy() attribute A = B.copy(). This will create an independent copy of B and changes in A will not affect B.

2.2.5 Matrix addition and scalar multiplication

Matrix addition can be performed between two or more NumPy arrays (matrices) in a very simple and straightforward manner. If we have two matrices $\boldsymbol{A}$ and $\boldsymbol{B}$, then we can find their sum $\boldsymbol{C} = \boldsymbol{A} + \boldsymbol{B}$ just by typing C=A+B. If we want to add more than two matrices we can still write D=A+B+C.

```
A = np.array( [[1, 2, 3],
               [4, 5, 6],
               [7, 8, 9]] )
B = np.array( [[9, 8, 7],
               [6, 5, 4],
               [3, 2, 1]] )
C = A + B
print(C)
```

```
[[10 10 10]
 [10 10 10]
 [10 10 10]]
```

The matrix $\boldsymbol{C}$ is a new matrix and is unrelated to $\boldsymbol{A}$ and $\boldsymbol{B}$. In the same way we can subtract, multiply a matrix with a scalar etc. For example, we can write

```
1 D = A - B
2 E = 2.0 * A
3 F = C / 2.0
4 G = A - 3.0*F
5 print(D)
6 print(E)
7 print(F)
8 print(G)
```

```
[[-8 -6 -4]
 [-2  0  2]
 [ 4  6  8]]
[[ 2.  4.  6.]
 [ 8. 10. 12.]
 [14. 16. 18.]]
[[5. 5. 5.]
 [5. 5. 5.]
 [5. 5. 5.]]
[[-14. -13. -12.]
 [-11. -10.  -9.]
 [ -8.  -7.  -6.]]
```

The command on line 1 performs the subtraction $\boldsymbol{A}-\boldsymbol{B}$, the command on line 2 performs the scalar multiplication $2\boldsymbol{A}$, the command on line 3 the scalar multiplication $1/2\ \boldsymbol{C}$, and the command on line 4 performs the operation $\boldsymbol{A}-3\boldsymbol{F}$.

In order to practice our programming skills we can perform manually the addition of two matrices by writing a simple function `add` which will take as input the matrices $\boldsymbol{A}$ and $\boldsymbol{B}$ and will return the sum $\boldsymbol{C}=\boldsymbol{A}+\boldsymbol{B}$. Apart from practicing and demonstrating the definition of matrix addition, the reason for writing this simple function is to implement practically the algorithm $c_{ij}=a_{ij}+b_{ij}$ for all i, j (see Algorithm 2) in Python.

Algorithm 2 Matrix addition

for $i=1:n$ **do**
    for $j=1:m$ **do**
        $\boldsymbol{C}(i,j)=\boldsymbol{A}(i,j)+\boldsymbol{B}(i,j)$
    end for
end for

Remember that when we present algorithms in this book we usually use natural indices instead of Pythonic ones. For example, when we write $i=1:n$ this means that i takes all values from 1 to n with step 1. The general format of this iteration will be $i=\text{start}:\text{stop}:\text{step}$. In pseudo-language we denote matrix entries a_{ij} by $\boldsymbol{A}(i,j)$ to distinguish them from Pythonic arrays and to comply with standard notation for example introduced in [48]. The implementation of Algorithm 2 in Python can be the following:

```
def add(A,B):
    (n, m) = np.shape(A)
```

```
    C = np.empty((n,m))
    for i in range(n):
        for j in range(m):
            C[i,j] = A[i,j] + B[i,j]
    return C
```

In this function we first extract the shape of the $n \times m$ matrix $\boldsymbol{A}$. Suppose that $\boldsymbol{A}$ and $\boldsymbol{B}$ have the same shape. We reserve the memory for the matrix $\boldsymbol{C}$ using the function `empty` of NumPy because otherwise we will receive an error message. As an exercise try to modify the function to check that the shapes are the same. We go over the n rows and m columns using a nested `for` loop and perform the addition element by element. Testing our new function `add` using the previous matrices $\boldsymbol{A}$ and $\boldsymbol{B}$ we compute the exact same result with the simple operation `C=A+B`:

```
A = numpy.array( [[1, 2, 3],
                  [4, 5, 6],
                  [7, 8, 9]] )
B = numpy.array( [[9, 8, 7],
                  [6, 5, 4],
                  [3, 2, 1]] )
C = add(A,B)
print(C)
```

```
[[10. 10. 10.]
 [10. 10. 10.]
 [10. 10. 10.]]
```

2.2.6 Vectorized algorithms

Modern scientific computing suggests that we avoid the use of `for` loops over arrays' entries, especially in interpreter programming languages such as Python. The reason is that every time Python tries to access an entry of an array, it searches for the location of the entry in the computer memory (RAM), and this is time-consuming. Algorithms that perform operations with blocks of (or entire) matrices are faster than those performing element-wise operations. Such algorithms are called *vectorized* algorithms or algorithms implemented in vector format.

Algorithm 3 Matrix addition in vector format

for $i = 1 : n$ **do**
  $\boldsymbol{C}(i,:) = \boldsymbol{A}(i,:) + \boldsymbol{B}(i,:)$
end for

Modifying Algorithm 2 so as to perform the same operations using entire columns, and instead of adding entries we add columns in Algorithm 3, it turns out that the vectorized algorithm is much faster. The implementation of Algorithm 3 into a new function `addv` could be the following:

```
def addv(A,B):
    (n, m) = np.shape(A)
    C = np.empty((n,m))
```

```
    for i in range(n):
        C[i,:] = A[i,:] + B[i,:]
    return C
```

We can test our function to see that is working as before.

```
A = np.array( [[1, 2, 3],
               [4, 5, 6],
               [7, 8, 9]] )
B = np.array( [[9, 8, 7],
               [6, 5, 4],
               [3, 2, 1]] )
C = addv(A,B)
print(C)
```

```
[[10. 10. 10.]
 [10. 10. 10.]
 [10. 10. 10.]]
```

In the next section we measure and compare the efficacy of the vectorized and naive addition algorithms.

2.2.7 Performance of algorithms

We compare the performance of the previous methods to add matrices (NumPy addition, `add` and `addv` functions) by measuring the time required separately by each method. To measure the elapsed time we use the module `time` and the function `time()`. The first time we call the function `time()` initiates a "stopwatch" and we store the initial time in a variable, let's say `tstart`. The second call of the function `time()` stops the "stopwatch". We store the final time in a different variable, let's say `tend`. The elapsed time in seconds is then `tend-tstart`.

Because 3×3 matrices are small enough to show any possible differences between vectorized and naive algorithms, we measure the times for adding large 100×100 matrices generated randomly using the NymPy function `rand` of the submodule `random`. The command `rand(n,m)` of the module `numpy.random` generates an $n \times m$ random matrix with real values in the interval $[0, 1]$.

```
import time
A = np.random.rand(100,100)
B = np.random.rand(100,100)
t0 = time.time()
C = A+B
t1 = time.time()
print('Python', t0-t1, 'seconds')
t0 = time.time()
C = add(A,B)
t1 = time.time()
print('add', t0-t1, 'seconds')
t0 = time.time()
C = addv(A,B)
```

```
t1 = time.time()
print('addv', t0-t1, 'seconds')
```

```
Python 0.0021219253540039062 seconds
add 0.3293418884277344 seconds
addv 0.0029468536376953125 seconds
```

What we see is that the fastest method is the default NumPy array addition. The function `addv` which implements the addition in vector format is faster than the naive addition `add`. An alternative to the function `time` is the function `time_ns` that returns the time in nanoseconds. This can be useful for small problems or when accuracy of the elapsed time is very important.

Matrix addition can be implemented using vector format in rows instead of columns. The algorithm and its implementation are very similar to Algorithm 3 and the function `addv`, and we leave it as an exercise for the reader.

> ☞ For efficient implementation of matrix operations in Python it is better to use vector format of algorithms using entire columns, rows or blocks instead of individual entries of matrices. If there is a Python implementation of the same algorithm, then this is usually the fastest implementation as it is precompiled, optimized, and maybe written in other programming languages.

2.2.8 Matrix multiplication

Multiplication is perhaps the most difficult and expensive operation between matrices. Recall that we can multiply matrices $\boldsymbol{A}$ and $\boldsymbol{B}$ if the number of columns in $\boldsymbol{A}$ is the same with the number of rows in $\boldsymbol{B}$. For example, if $\boldsymbol{A}$ is $n \times m$, then $\boldsymbol{B}$ must be $m \times p$ matrix. The matrix product of $\boldsymbol{A}$ and $\boldsymbol{B}$, denoted by $\boldsymbol{AB}$, is an $n \times p$ matrix $\boldsymbol{C}$ whose entries c_{ij} are

$$c_{ij} = \sum_{k=1}^{m} a_{ik} b_{kj} = a_{i1} b_{1j} + a_{i2} b_{2j} + \cdots + a_{im} b_{mj} \ .$$

In order to compute the entry c_{ij} of the product matrix we use the i-th row of $\boldsymbol{A}$ and the j-th column of $\boldsymbol{B}$. These must have the same length. We multiply the entries a_{ik} with the corresponding entries b_{kj} for $k = 1, 2, \ldots, m$, and then we add all these products to form the entry c_{ij}. In other words, we compute the inner product of the i-th row of $\boldsymbol{A}$ with the j-th column of $\boldsymbol{B}$. See also the example on Section 2.1. We call this algorithm of matrix multiplication the naive multiplication since this is perhaps the least efficient implementation one can use. Practically, we need to implement a sum over the products $a_{ik} b_{kj}$. We have already discussed the algorithm for the computation of a sum in Algorithm 1. The algorithm for the naive matrix multiplication can be described in Algorithm 4. In this algorithm the entry $\boldsymbol{C}(i, j)$ of matrix $\boldsymbol{C}$ plays the role of the variable `sum` in Algorithm 1. The loop over the variable k is the loop for the computation of the sum. The outer nested loops over the variables i and j are for the computation of all entries c_{ij}.

We implement Algorithm 4 in a new function called `mult` and we test it using the matrices

$$\boldsymbol{A} = \begin{pmatrix} 1 & 2 \\ 3 & 4 \end{pmatrix} \quad \text{and} \quad \boldsymbol{B} = \begin{pmatrix} 5 & 6 & 7 \\ 8 & 9 & 10 \end{pmatrix} \ .$$

Algorithm 4 Naive matrix multiplication - ijk implementation

for $i = 1 : n$ **do**
  for $j = 1 : p$ **do**
    $\boldsymbol{C}(i,j) = 0$
    for $k = 1 : m$ **do**
      $\boldsymbol{C}(i,j) = \boldsymbol{C}(i,j) + \boldsymbol{A}(i,k) \cdot \boldsymbol{B}(k,j)$
    end for
  end for
end for

In this function, we set the matrix $\boldsymbol{C}$ initially equal to zero to reserve memory for the result.

```
def mult(A,B):
    (n, m) = np.shape(A)
    (m, p) = np.shape(B)
    C = np.zeros((n, p))
    for i in range(n):
        for j in range(p):
            for k in range(m):
                C[i,j] = C[i,j] + A[i,k]*B[k,j]
    return C

A = np.array( [[1, 2],
               [3, 4]] )
B = np.array( [[5, 6, 7],
               [8, 9, 10]] )
C = mult(A,B)
print(C)
```

```
[[21. 24. 27.]
 [47. 54. 61.]]
```

Algorithm 4 is the ijk implementation of matrix multiplication in the sense that the nested loops are formed in the natural order (first for i, then for j and then for k). This order is not important, though, and we can take the loops in any order we prefer. For example, we can take the loop first for i, then for k and then for j. This observation leads to a better implementation of matrix multiplication in Algorithm 5. Observe that the entry

Algorithm 5 Naive matrix multiplication - ikj implementation

for $i = 1 : n$ **do**
  for $k = 1 : m$ **do**
    $tmp = \boldsymbol{A}(i,k)$
    for $j = 1 : p$ **do**
      $\boldsymbol{C}(i,j) = \boldsymbol{C}(i,j) + tmp \cdot \boldsymbol{B}(k,j)$
    end for
  end for
end for

$\boldsymbol{A}(i,k)$ is always the same in the inner loop, which is a loop for j. Thus instead of letting Python search in memory to find the entry $\boldsymbol{A}(i,k)$ for each j we bring this value in the *cache memory* of the computer (which is closer to the processor and thus faster) by assigning it to a variable *tmp* as it is demonstrated in Algorithm 5.

The *ikj* implementation of the naive matrix multiplication is very similar to the *ijk* implementation and is presented here:

```
def mult(A,B):
    (n, m) = np.shape(A)
    (m, p) = np.shape(B)
    C = np.zeros((n, p))
    for i in range(n):
        for k in range(m):
            tmp = A[i,k]
            for j in range(p):
                C[i,j] = C[i,j] + tmp*B[k,j]
    return C
```

Although the *ikj* implementation is somewhat faster than the *ijk* implementation of the naive matrix multiplication, referring to the comments of the previous section, the algorithm still uses individual entries of the matrices. In order to improve the naive matrix multiplication we need to observe the deeper structure of this algorithm and think in terms of rows and columns instead of entries.

To do so, we come back to the first comment about matrix multiplication. In order to compute the entry c_{ij} of $\boldsymbol{C}$ we multiply all the entries of the i-th row of the matrix $\boldsymbol{A}$ with the corresponding entries of the j-th column of the matrix $\boldsymbol{B}$. This means that to perform a matrix-matrix multiplication we multiply row with column vectors.

Consider matrix $\boldsymbol{A}$ written as a column vector with entries row vectors, and matrix $\boldsymbol{B}$ as a row vector with entries column vectors. Denoting the rows of $\boldsymbol{A}$ by $\boldsymbol{a}_i^T$ for $i = 1, 2, \ldots, n$ and columns of $\boldsymbol{B}$ by $\boldsymbol{b}_j$ for $j = 1, 2, \ldots, p$, i.e.

$$\boldsymbol{a}_i^T = \begin{pmatrix} a_{i1} & a_{i2} & \cdots & a_{im} \end{pmatrix} \quad \text{and} \quad \boldsymbol{b}_j = \begin{pmatrix} b_{1j} \\ b_{2j} \\ \vdots \\ b_{mj} \end{pmatrix},$$

then

$$\boldsymbol{A} = \begin{pmatrix} \boldsymbol{a}_1^T \\ - \\ \boldsymbol{a}_2^T \\ - \\ \vdots \\ - \\ \boldsymbol{a}_n^T \end{pmatrix} \quad \text{and} \quad \boldsymbol{B} = \begin{pmatrix} \boldsymbol{b}_1 & | & \boldsymbol{b}_2 & | & \cdots & | & \boldsymbol{b}_n \end{pmatrix}.$$

In this format, the c_{ij} entry of the product $\boldsymbol{C}$ can be written as $c_{ij} = \boldsymbol{a}_i^T \cdot \boldsymbol{b}_j$; that is,

$$c_{ij} = \begin{pmatrix} a_{i1} & a_{i2} & \cdots & a_{im} \end{pmatrix} \cdot \begin{pmatrix} b_{1j} \\ b_{2j} \\ \vdots \\ b_{mj} \end{pmatrix}.$$

This is the usual inner product of vectors. This product practically replaces the sum over the variable k. To use the vector format, we go back to Algorithm 4 and replace the loop over the variable k with the product of the i-th row of $\boldsymbol{A}$ with the j-th column of $\boldsymbol{B}$ devising Algorithm 6.

Algorithm 6 Matrix multiplication in vector form

for $i = 1 : n$ **do**
  for $j = 1 : p$ **do**
    $\boldsymbol{C}(i,j) = \boldsymbol{A}(i,:) \cdot \boldsymbol{B}(:,j)$
  end for
end for

The inner product of two vectors is implemented in NumPy with the function `inner` and so the Python implementation of the Algorithm 6 can be the following

```
def multv(A,B):
    (n, m) = np.shape(A)
    (m, p) = np.shape(B)
    C = np.zeros((n, p))
    for i in range(n):
        for j in range(p):
            C[i,j] = np.inner(A[i,:],B[:,j])
    return C

A = np.array( [[1, 2],
               [3, 4]] )
B = np.array( [[5, 6, 7],
               [8, 9, 10]] )
C = multv(A,B)
print(C)
```

```
[[21. 24. 27.]
 [47. 54. 61.]]
```

Now that we have discussed basic algorithms for matrix multiplication, we will examine the NumPy implementation of matrix multiplication. The operator `*` with NumPy arrays does not perform an actual matrix multiplication. It performs an element by element multiplication between two arrays. On the other hand, the operator `@` performs the usual matrix multiplication. For example,

```
A = np.array( [[1, 2],
               [3, 4]] )
B = np.array( [[5, 6],
               [7, 8]] )
C = A * B
print('C='); print(C)
D = A @ B
print('D='); print(D)
```

```
C=
[[ 5 12]
 [21 32]]
```

```
D=
[[19 22]
 [43 50]]
```

Matrix multiplication with NumPy arrays can be also performed using the NumPy function `dot`. We can multiply two matrices $\boldsymbol{A}$ and $\boldsymbol{B}$ by typing `np.dot(A,B)`. In our previous example we can have:

```
A = np.array( [[1, 2],
               [3, 4]] )
B = np.array( [[5, 6, 7],
               [8, 9, 10]] )
C = np.dot(A,B)
print(C)
```

```
[[21 24 27]
 [47 54 61]]
```

The operator `*` can be used for matrix multiplication only with NumPy matrices. NumPy matrix is a data structure that inherits all the properties and methods from NumPy array and perhaps are more convenient, but there is always a possibility of problematic function calls. For this reason, we use mainly NumPy arrays instead.

```
A = np.matrix( [[1, 2],
                [3, 4]] )
B = np.matrix( [[5, 6, 7],
                [8, 9, 10]] )
C = A*B
print(C)
```

```
[[21 24 27]
 [47 54 61]]
```

For the sake of completeness, we briefly state here some of the properties of matrix multiplication. Let $\boldsymbol{A}$ be an $n \times m$ matrix, $\boldsymbol{B}$ an $m \times k$ matrix, $\boldsymbol{C}$ a $k \times p$ matrix, $\boldsymbol{D}$ be an $m \times k$ matrix and λ real number, then the following properties hold true:

- $\boldsymbol{A}(\boldsymbol{B}\boldsymbol{C}) = (\boldsymbol{A}\boldsymbol{B})\boldsymbol{C}$
- $\boldsymbol{A}(\boldsymbol{B} + \boldsymbol{D}) = \boldsymbol{A}\boldsymbol{B} + \boldsymbol{A}\boldsymbol{D}$
- $\lambda(\boldsymbol{A}\boldsymbol{B}) = (\lambda\boldsymbol{A})\boldsymbol{B} = \boldsymbol{A}(\lambda\boldsymbol{B})$
- $\boldsymbol{A} \cdot \boldsymbol{I} = \boldsymbol{A} \cdot \boldsymbol{I} = \boldsymbol{A}$

When the matrix products $\boldsymbol{A}\boldsymbol{B}$ and $\boldsymbol{B}\boldsymbol{A}$ are both defined, their results might be different. Sometimes they do not even have the same shape (see for example the previous multiplication where the product $\boldsymbol{B}\boldsymbol{A}$ is not defined at all).

Multiplying triangular matrices

Assume that $\boldsymbol{A}$ and $\boldsymbol{B}$ are two $n \times n$ upper triangular matrices. In such a case, $a_{ik} = 0$ when $i > k$ and $b_{kj} = 0$ when $k > j$. Recall that the multiplication $\boldsymbol{AB}$ is a matrix $\boldsymbol{C}$ with

$$c_{ij} = \sum_{k=1}^{m} a_{ik} b_{kj} \ . \tag{2.1}$$

Thus, multiplying the two upper triangular matrices, the sum in formula (2.1) is not zero only when $i \leq k$ and $k \leq j$ at the same time. These two conditions are simultaneously satisfied only when $i \leq j$, and thus for $i > j$ we get $c_{ij} = 0$, and the resulting matrix is upper triangular. Similar, the product of two lower triangular matrices is a lower triangular matrix. As we will see later, this property is very useful in matrix decomposition techniques.

☞ The product of two upper triangular matrices is an upper triangular matrix, and the product of two lower triangular matrices is a lower triangular matrix.

Powers

We saw that the result of the operator * used with NumPy arrays returns the element-by-element product of $\boldsymbol{A}$ and $\boldsymbol{B}$ and not their actual matrix product unless the matrices are defined as NumPy matrices. Because such operators are very useful, the module NumPy provides with a similar function for raising the entries of matrices to a specified power. This is the function `power` which is formulated as `power(A,p)` to compute the matrix with entries a_{ij}^p.

```
A = np.array( [[1, 2],
               [3, 4]] )
B = np.power(A,2)
print(B)
```

```
[[ 1  4]
 [ 9 16]]
```

To compute the actual power of a matrix $\boldsymbol{A}^p = \boldsymbol{A} \cdot \boldsymbol{A} \cdots \boldsymbol{A}$ NumPy is equipped with the additional function `matrix_power` in the submodule `linalg`. We will make use of the submodule `linalg` extensively in this book and we will use the standard alias `npl` instead of `numpy.linalg`. So, in order to compute the power $\boldsymbol{A}^2$ of the previous matrix we can type

```
import numpy.linalg as npl

A = np.array( [[1, 2],
               [3, 4]] )
B = npl.matrix_power(A,2)
print(B)
```

```
[[ 7 10]
 [15 22]]
```

which is the same with the product $\boldsymbol{A} \cdot \boldsymbol{A}$.

```
A = np.array( [[1, 2],
               [3, 4]] )
B = np.dot(A,A)
print(B)
```

```
[[ 7 10]
 [15 22]]
```

2.2.9 Inverse of a matrix

An $n \times n$ matrix $\boldsymbol{A}$ is said to be *nonsingular* (or *invertible*) if there exists an $n \times n$ matrix denoted by $\boldsymbol{A}^{-1}$ such that $\boldsymbol{A} \cdot \boldsymbol{A}^{-1} = \boldsymbol{A}^{-1} \cdot \boldsymbol{A} = \boldsymbol{I}$. The matrix $\boldsymbol{A}^{-1}$ is called the *inverse* of $\boldsymbol{A}$. A matrix that has no inverse is called *singular* (or *noninvertible*).

For any nonsingular $n \times n$ matrix $\boldsymbol{A}$ we have the following properties for the matrix itself and its inverse:

- $\boldsymbol{A}^{-1}$ is unique
- $\boldsymbol{A}^{-1}$ is nonsingular and $(\boldsymbol{A}^{-1})^{-1} = \boldsymbol{A}$
- If $\boldsymbol{B}$ is also a nonsingular $n \times n$ matrix, then $(\boldsymbol{AB})^{-1} = \boldsymbol{B}^{-1}\boldsymbol{A}^{-1}$
- If $\boldsymbol{A}^{-1}$ exists, then $(\boldsymbol{A}^{-1})^T = (\boldsymbol{A}^T)^{-1}$

Finding the inverse of a matrix $\boldsymbol{A}$ could give the solution to the linear system $\boldsymbol{Ax} = \boldsymbol{b}$, since after multiplying with the matrix $\boldsymbol{A}^{-1}$ on both sides of the linear system, we obtain

$$\begin{aligned}\boldsymbol{A}^{-1}\boldsymbol{Ax} &= \boldsymbol{A}^{-1}\boldsymbol{b}\\ \boldsymbol{Ix} &= \boldsymbol{A}^{-1}\boldsymbol{b}\end{aligned}$$

and thus the solution $\boldsymbol{x} = \boldsymbol{A}^{-1}\boldsymbol{b}$. Unfortunately, the computation of the inverse matrix is a difficult and slow procedure. As we shall see later in this book, we can solve this problem using efficient numerical algorithms. For the moment, we limit the discussion on Python's capabilities.

Computation of the inverse of a matrix $\boldsymbol{A}$ can be performed using the function `numpy.linalg.inv` that takes as input a NumPy array and returns its inverse. For example,

```
1 A = np.array( [[1, 2],
2                [3, 4]] )
3 Ainv = npl.inv(A)
4 print(Ainv)
```

```
[[-2.   1. ]
 [ 1.5 -0.5]]
```

Line 3 in the previous code is responsible for the computation of the inverse of $\boldsymbol{A}$. In order to verify that the result is correct perform the multiplications $\boldsymbol{A} \cdot \boldsymbol{A}^{-1}$ and $\boldsymbol{A}^{-1} \cdot \boldsymbol{A}$ and verify that the result is the identity matrix $\boldsymbol{I}$. Do not forget that bugs in computer codes are unavoidable!

Inverse of triangular matrix

As an application we explore the inverse of triangular matrices. Assume that $\boldsymbol{L}$ is an invertible lower triangular matrix. Write

$$\boldsymbol{L}^{-1} = \begin{pmatrix} \boldsymbol{y}_1 & \boldsymbol{y}_2 & \cdots & \boldsymbol{y}_n \end{pmatrix} ,$$

where $\boldsymbol{y}_i$ is $n \times 1$ column vector for all i. Then we have that

$$\boldsymbol{L}\boldsymbol{L}^{-1} = \boldsymbol{I} = \begin{pmatrix} \boldsymbol{e}_1 & \boldsymbol{e}_2 & \cdots & \boldsymbol{e}_n \end{pmatrix} ,$$

where $\boldsymbol{e}_i$ are the $n \times 1$ unit vectors with 1 on their i-th entry and 0 elsewhere. Then,

$$\begin{pmatrix} \boldsymbol{e}_1 & \cdots & \boldsymbol{e}_n \end{pmatrix} = \boldsymbol{L}\boldsymbol{L}^{-1} = \boldsymbol{L}\begin{pmatrix} \boldsymbol{y}_1 & \boldsymbol{y}_2 & \cdots & \boldsymbol{y}_n \end{pmatrix} = \begin{pmatrix} \boldsymbol{L}\boldsymbol{y}_1 & \boldsymbol{L}\boldsymbol{y}_2 & \cdots & \boldsymbol{L}\boldsymbol{y}_n \end{pmatrix} ,$$

which implies that $\boldsymbol{L}\boldsymbol{y}_i = \boldsymbol{e}_i$, for all $i = 1, 2, \ldots, n$. Note that $\boldsymbol{e}_i$ has zeros above its i-th row and because $\boldsymbol{L}$ is lower triangular, then $\boldsymbol{y}_i$ has only zeros above the i-th row. Since $\boldsymbol{L}^{-1} = \begin{pmatrix} \boldsymbol{y}_1 & \cdots & \boldsymbol{y}_n \end{pmatrix}$ we deduce that $\boldsymbol{L}^{-1}$ is lower triangular too. To understand better why this is the case, try a simple 3×3 matrix and assume $\boldsymbol{L}^{-1}$ is not a lower triangular matrix.

This is also true for upper triangular matrices but we leave the proof for exercise.

2.2.10 Determinant of a matrix

The determinant of a matrix is a useful quantity for matrix computations. For example, it is known that a matrix $\boldsymbol{A}$ is invertible if – and only if – its determinant is not zero. We usually denote the determinant of a square matrix $\boldsymbol{A}$ by $\det(\boldsymbol{A})$, but it is common to use also the notation $|\boldsymbol{A}|$.

If $\boldsymbol{A} = [a]$ is a 1×1 matrix, then $\det(\boldsymbol{A}) = a$. The determinant of a 2×2 matrix is

$$\det\begin{pmatrix} a & b \\ c & d \end{pmatrix} = ad - bc .$$

If $\boldsymbol{A}$ is an $n \times n$ matrix with $n > 2$, then the determinant can be computed by choosing a row or a column of $\boldsymbol{A}$ and computing the *cofactors* $\boldsymbol{A}_{ij} = (-1)^{i+j}\boldsymbol{M}_{ij}$, where $\boldsymbol{M}_{ij}$ denotes the determinant of the $(n-1) \times (n-1)$ submatrix of $\boldsymbol{A}$ that can be obtained by ignoring the i-th row and j-th column of the original matrix $\boldsymbol{A}$. Then the determinant of $\boldsymbol{A}$ can be computed as

$$\det(\boldsymbol{A}) = \sum_{j=1}^{n} a_{ij}\boldsymbol{A}_{ij} = \sum_{j=1}^{n} (-1)^{i+j} a_{ij}\boldsymbol{M}_{ij} ,$$

if we choose to analyze the determinant using the row i or by

$$\det(\boldsymbol{A}) = \sum_{i=1}^{n} a_{ij}\boldsymbol{A}_{ij} = \sum_{i=1}^{n} (-1)^{i+j} a_{ij}\boldsymbol{M}_{ij} ,$$

if we use the column j.

For example, if

$$\boldsymbol{A} = \begin{pmatrix} 1 & 4 & 2 & 3 \\ 0 & 1 & 4 & 4 \\ -1 & 0 & 1 & 0 \\ 2 & 0 & 4 & 1 \end{pmatrix} ,$$

in order to compute the det($\boldsymbol{A}$), it is easier to use the second column because it has two zeros:

$$\det(\boldsymbol{A}) = a_{12}\boldsymbol{A}_{12} + a_{22}\boldsymbol{A}_{22} + a_{32}\boldsymbol{A}_{32} + a_{42}\boldsymbol{A}_{42} = 4\boldsymbol{A}_{12} + 1\boldsymbol{A}_{22} \ .$$

Eliminating the first row and the second column and subsequently the second row and the second column we obtain the analysis in cofactors:

$$\begin{aligned}\det(\boldsymbol{A}) &= -4 \cdot \det\begin{pmatrix} 0 & 4 & 4 \\ -1 & 1 & 0 \\ 2 & 4 & 1 \end{pmatrix} + \det\begin{pmatrix} 1 & 2 & 3 \\ -1 & 1 & 0 \\ 2 & 4 & 1 \end{pmatrix} \\ &= -4\left\{-4\det\begin{pmatrix} -1 & 0 \\ 2 & 1 \end{pmatrix} + 4\det\begin{pmatrix} -1 & 1 \\ 2 & 4 \end{pmatrix}\right\} + \left\{3\det\begin{pmatrix} -1 & 1 \\ 2 & 4 \end{pmatrix} + \det\begin{pmatrix} 1 & 2 \\ -1 & 1 \end{pmatrix}\right\} \\ &= 80 - 15 = 65 \ .\end{aligned}$$

A determinant of a matrix can be computed approximately using the function `det` of the module `numpy.linalg`. For example,

```
import numpy as np
from numpy.linalg import det
A = np.array( [[ 1, 4, 2, 3],
               [ 0, 1, 4, 4],
               [-1, 0, 1, 0],
               [ 2, 0, 4, 1]] )
print(det(A))
```

```
64.99999999999999
```

where we verify our theoretical computations.

Unfortunately, the computation of a determinant is very slow, and the method discussed before with the cofactors is perhaps the slowest algorithm for such problems. There are efficient algorithms for the computation of determinants but we discuss them later in this book. Next, we proceed with the numerical solution of systems of linear equations.

2.2.11 Linear systems

The most commonly encountered problem in scientific computations is the solution of systems of linear equations (simultaneous equations). A general system with n unknowns and n linear equations can be written in the form

$$\begin{aligned} a_{11}\,x_1 + a_{12}\,x_2 + \cdots + a_{1n}\,x_n &= b_1 \ , \\ a_{21}\,x_1 + a_{22}\,x_2 + \cdots + a_{2n}\,x_n &= b_2 \ , \\ &\vdots \\ a_{n1}\,x_1 + a_{n2}\,x_2 + \cdots + a_{nn}\,x_n &= b_n \ . \end{aligned}$$

In this notation, the coefficients a_{ij}, for $i, j = 1, 2, \ldots, n$, and b_j for $j = 1, 2, \ldots, n$ are all given. On the other hand we need to determine the unknowns $x_1, x_2, \ldots, x_n$.

In computer systems we express a linear system using matrices and vectors. This can be done by writing all the coefficients of the system into an $n \times n$ matrix

$$\boldsymbol{A} = \begin{pmatrix} a_{11} & a_{12} & \cdots & a_{1n} \\ a_{21} & a_{22} & \cdots & a_{2n} \\ \vdots & \vdots & & \vdots \\ a_{n1} & a_{n2} & \cdots & a_{nn} \end{pmatrix},$$

and by expressing the right-hand side of the system as an n-dimensional column vector

$$\boldsymbol{b} = \begin{pmatrix} b_1 \\ b_2 \\ \vdots \\ b_n \end{pmatrix}.$$

Then the linear system can be written using matrix-vector product in the form

$$\boldsymbol{Ax} = \boldsymbol{b} ,$$

where $\boldsymbol{x}$ is the unknown vector with entries the unknowns $x_1, x_2, \ldots, x_n$. Usually we store the coefficients of $\boldsymbol{A}$ and $\boldsymbol{b}$ in appropriate NumPy arrays, let's say `A` and `b`. The easiest way to solve a linear system of equations defined by the arrays `A` and `b` is by using the function `solve(A,b)` of the module `numpy.linalg`. This function accepts as input the arrays `A` and `b` and returns the solution of the system in an array `x`. For example, consider the following system of linear equations

$$\left.\begin{array}{lllll} 2x_1 & -x_2 & & & =1 \\ -x_1 & +3x_2 & -x_3 & & =1 \\ & -x_2 & +3x_3 & -x_4 & =1 \\ & & -x_3 & +2x_4 & =1 \end{array}\right\},$$

which can be written in matrix form $\boldsymbol{Ax} = \boldsymbol{b}$ with

$$\boldsymbol{A} = \begin{pmatrix} 2 & -1 & 0 & 0 \\ -1 & 3 & -1 & 0 \\ 0 & -1 & 3 & -1 \\ 0 & 0 & -1 & 2 \end{pmatrix} \quad \text{and} \quad \boldsymbol{b} = \begin{pmatrix} 1 \\ 1 \\ 1 \\ 1 \end{pmatrix}.$$

It can be easily verified that the specific system has the exact solution $\boldsymbol{x} = (1, 1, 1, 1)^T$. This is, for example, because each entry of the right hand side $\boldsymbol{b}$ is the sum of the entries of the corresponding row of matrix $\boldsymbol{A}$

$$b_i = \sum_{j=1}^{n} a_{ij}, \quad i = 1, 2, \ldots, n .$$

Moreover, the matrix $\boldsymbol{A}$ is tridiagonal and symmetric. It has even more nice properties that we will explore in later chapters.

One of the first methods we learn in linear algebra courses in order to solve linear systems is Cramer's rule. Cramer's rule is an explicit formula for the solution of a linear system $\boldsymbol{Ax} = \boldsymbol{b}$. Let $\boldsymbol{A}_i$ be the matrix formed by replacing the i-th column of $\boldsymbol{A}$ by the right-hand side vector $\boldsymbol{b}$. Then the components x_i of the solution $\boldsymbol{x} = (x_1, x_2, \ldots, x_n)^T$ are given by the formulas

$$x_i = \frac{\det(\boldsymbol{A}_i)}{\det(\boldsymbol{A})}, \quad i = 1, 2, \ldots, n .$$

It is clear that if $\det(\boldsymbol{A}) = 0$, then we cannot form any solution $\boldsymbol{x}$, which is expected in the case singular matrix $\boldsymbol{A}$.

To solve numerically a system of linear equations $\boldsymbol{Ax} = \boldsymbol{b}$ we use NumPy and its function `solve`. For example, to solve the previous linear system we type

```
from numpy.linalg import solve
A = np.array( [[ 2.0, -1.0,  0.0 , 0.0],
              [-1.0,  3.0, -1.0,  0.0],
              [ 0.0, -1.0,  3.0, -1.0],
              [ 0.0,  0.0 ,-1.0,  2.0]] )
b = np.ones((4,1))
x = solve(A,b)
print(x)
```

```
[[1.]
 [1.]
 [1.]
 [1.]]
```

We were careful with the floating point numbers in the definition of our NumPy arrays to avoid confusion with integer entries that can cause problems with division. This might not be so important in this example, but in general using integers instead of floating point numbers could cause serious errors in our codes. So, it is advisable to use always floating point numbers when we expect the result to be real. Moreover, note that we could have used `b=np.ones(4)` resulting to a row vector solution instead of column vector.

To solve a linear system in the following sections, we use the `numpy.linalg.solve` function as black box. We explain the most popular numerical methods for solving linear systems in Chapter 9.

2.2.12 Eigenvalues and eigenvectors

An eigenvalue λ of a square matrix $\boldsymbol{A}$ is a number (real or complex) that makes the matrix $\boldsymbol{A} - \lambda \boldsymbol{I}$ singular (not invertible). This means that the eigenvalues of a matrix $\boldsymbol{A}$ are the roots of the equation $\det(\boldsymbol{A} - \lambda\boldsymbol{I}) = 0$. The determinant $\det(\boldsymbol{A} - \lambda\boldsymbol{I})$ is always a polynomial and it is called the *characteristic polynomial.*

Let us for example consider the matrix

$$\boldsymbol{A} = \begin{pmatrix} 1 & 1 \\ 2 & 3 \end{pmatrix} .$$

The eigenvalues of $\boldsymbol{A}$ are the roots of the equation

$$\begin{aligned} 0 = \det(\boldsymbol{A} - \lambda\boldsymbol{I}) &= \det\left(\begin{pmatrix} 0 & 1 \\ -2 & -3 \end{pmatrix} - \lambda \begin{pmatrix} 1 & 0 \\ 0 & 1 \end{pmatrix}\right) = \det\left(\begin{pmatrix} 0 & 1 \\ -2 & -3 \end{pmatrix} - \begin{pmatrix} \lambda & 0 \\ 0 & \lambda \end{pmatrix}\right) \\ &= \det\begin{pmatrix} -\lambda & 1 \\ -2 & -3-\lambda \end{pmatrix} = -\lambda(-3-\lambda) + 2 = \lambda^2 + 3\lambda + 2 \\ &= (\lambda+1)(\lambda+2) . \end{aligned}$$

Thus, $\boldsymbol{A}$ has two eigenvalues $\lambda_1 = -1$ and $\lambda_2 = -2$.

For the eigenvalues of matrix $\boldsymbol{A}$, the matrix $\boldsymbol{A} - \lambda\boldsymbol{I}$ is singular and the system of equations $(\boldsymbol{A} - \lambda\boldsymbol{I})\boldsymbol{x} = \boldsymbol{0}$ can have an infinite number of solutions. (Otherwise, the matrix

$\boldsymbol{A} - \lambda \boldsymbol{I}$ would have been invertible with unique solution the vector $\boldsymbol{x} = \boldsymbol{0}$.) The non-trivial solutions $\boldsymbol{x} \neq \boldsymbol{0}$ of $(\boldsymbol{A} - \lambda \boldsymbol{I})\boldsymbol{x} = \boldsymbol{0}$ are called eigenvectors (or sometimes right eigenvectors).

Getting back to our examples, for the eigenvalue $\lambda_1 = -1$, the corresponding eigenvectors will satisfy the system

$$(\boldsymbol{A} - \lambda_1 \boldsymbol{I})\boldsymbol{x} = \boldsymbol{0} \ .$$

To solve this system, we write it in the form

$$\begin{pmatrix} 1 & 1 \\ -2 & -2 \end{pmatrix} \begin{pmatrix} x_1 \\ x_2 \end{pmatrix} = \boldsymbol{0} \ ,$$

and then write explicitly the simultaneous equations, which in this case are

$$\begin{aligned} x_1 + x_2 &= 0 \ , \\ -2x_1 - 2x_2 &= 0 \ . \end{aligned}$$

We observe that the two equations are identical since we can multiply the first equation with -2 to obtain the second one. For this reason, we have only one equation with two unknowns, and thus infinite number of solutions. Here, we set $x_1 = k \in \mathbb{R}$ and so $x_2 = -k$. Thus, the solution is the vector

$$\boldsymbol{x} = k \begin{pmatrix} 1 \\ -1 \end{pmatrix} \ ,$$

for any $k \in \mathbb{R}$. The eigenvector $\boldsymbol{e}_1$ associated with the eigenvalue λ_1 is a vector for any particular choice of the parameter k. For example, taking $k = 1$, the eigenvector can be

$$\boldsymbol{e}_1 = \begin{pmatrix} 1 \\ -1 \end{pmatrix} \ .$$

The modulus of the eigenvector $\boldsymbol{e}_1$ is $\|\boldsymbol{e}_1\| = \sqrt{1^2 + (-1)^2} = \sqrt{2} \approx 1.4142$. Taking $k = 1/\|\boldsymbol{e}_1\|$ we obtain the eigenvector

$$\boldsymbol{e}_1 = \begin{pmatrix} 0.70710678 \\ -0.4472136 \end{pmatrix} \ .$$

Therefore, the new eigenvector $\boldsymbol{e}_1$ has modulus 1, and we say that is normalized. Similarly, the eigenvector $\boldsymbol{e}_2$ associated with the eigenvalue λ_2 is the vector

$$\boldsymbol{e}_2 = \begin{pmatrix} 1 \\ -2 \end{pmatrix} \ ,$$

which after normalization becomes

$$\boldsymbol{e}_2 = \begin{pmatrix} -0.70710678 \\ 0.89442719 \end{pmatrix} \ .$$

We can compute the eigenvalues and eigenvectors of a matrix $\boldsymbol{A}$ in Python using the NumPy function `numpy.linalg.eig`. The general call of this function is `eig(A)` for a NumPy array `A` and returns an array `w` with columns the eigenvalues of the matrix, each one repeated according to its multiplicity, and an array `v` representing the matrix $\boldsymbol{V} = \begin{pmatrix} \boldsymbol{e}_1 & \boldsymbol{e}_2 & \cdots & \boldsymbol{e}_n \end{pmatrix}$ with columns $\boldsymbol{e}_i$ the normalized eigenvectors (with modulus 1). So the eigenvector `i`, which corresponds to the eigenvalue `w[i]` is the column `v[:,i]`.

Back to our example, we can verify easily that our previous computations were all correct:

```
from numpy.linalg import eig
A = np.array([[0, 1],[-2, -3]]) # Define the matrix
# Compute eigenvalues and eigenvectors
(w, v) = eig(A)
print('w='); print(w)
print('v='); print(v)
```

```
w=
[-1. -2.]
v=
[[ 0.70710678 -0.4472136 ]
 [-0.70710678  0.89442719]]
```

Observe that the eigenvalues computed with the previous code are the same with the theoretically estimated eigenvalues.

When we do not need the eigenvectors of $\boldsymbol{A}$ but only its eigenvalues, then we can use the function `numpy.linalg.eigvals` by typing `eigvals(A)`. The function `eigvals` returns only an array `w` with the eigenvalues of `A`. Specialized functions for the computation of eigenvalues and eigenvectors of symmetric (Hermitian) matrices are also provided by the module `numpy.linalg` such as the `numpy.linalg.eigh` and `numpy.linalg.eigvalsh` functions. Specialized algorithms for special matrices are preferred over generic alternatives because of their advantages. More detailed description of eigenvalues and eigenvectors is presented in Chapter 12.

2.3 The SciPy Module

The SciPy module is a library of functions, methods and data structures for scientific computations that goes side-by-side with NumPy. We will learn how to use SciPy in optimization, integration, interpolation and other fields. In this chapter, we discuss some of the capabilities of SciPy for linear algebra problems. The SciPy module for linear algebra is the `scipy.linalg`. It has the same name with NumPy and practically contains all the functions of NumPy and some extensions. It is based on Atlas, Lapack and Blas libraries, which are all written in Fortran and C programming languages.

Since `scipy.linalg` is a superset of `numpy.linalg` we can use all the previous methods we learned using SciPy. In addition, we will learn how SciPy can handle banded matrices and other special matrices.

2.3.1 Systems with banded matrices

A banded system of equations is a linear system where its coefficients form a banded matrix. For example, the system $\boldsymbol{Ax} = \boldsymbol{b}$ (that we have already seen before) with matrices

$$\boldsymbol{A} = \begin{pmatrix} 2 & -1 & 0 & 0 \\ -1 & 3 & -1 & 0 \\ 0 & -1 & 3 & -1 \\ 0 & 0 & -1 & 2 \end{pmatrix} \quad \text{and} \quad \boldsymbol{b} = \begin{pmatrix} 1 \\ 1 \\ 1 \\ 1 \end{pmatrix},$$

is a banded system. In practical applications, the resulting linear systems are much bigger, with thousands or millions of entries. This suggest issues with storage capacity. Because these systems contain many zeros, we prefer to store only their non-zero entries. Banded matrices can even be considered as sparse matrices if let's say more than half of their entries are zero. There are many ways to store a sparse matrix efficiently but here we discuss only the case of a *band storage.*

Consider a 5×5 general banded matrix with four diagonals (with entries not necessarily non-zero):

$$\boldsymbol{A} = \begin{pmatrix} a_{11} & a_{12} & 0 & 0 & 0 \\ a_{21} & a_{22} & a_{23} & 0 & 0 \\ a_{31} & a_{32} & a_{33} & a_{34} & 0 \\ 0 & a_{42} & a_{43} & a_{44} & a_{45} \\ 0 & 0 & a_{53} & a_{54} & a_{55} \end{pmatrix} .$$

This matrix has the following non-zero diagonals: the main diagonal, one diagonal above the main diagonal, and two diagonals below the main diagonal. We want to store only these diagonals (even if they contain some zeros). This can be done by storing the diagonals in the rows of a new matrix $\boldsymbol{A}_b$. Since here we have four diagonals, the matrix $\boldsymbol{A}_b$ must be 4×5 (4 rows for the 4 diagonals). The number of columns will be the same as $\boldsymbol{A}$.

$$\boldsymbol{A}_b = \begin{pmatrix} * & a_{12} & a_{23} & a_{34} & a_{45} \\ a_{11} & a_{22} & a_{33} & a_{44} & a_{55} \\ a_{21} & a_{32} & a_{43} & a_{54} & * \\ a_{31} & a_{42} & a_{53} & * & * \end{pmatrix} .$$

The entries marked with $*$ in the upper left and lower right corners of $\boldsymbol{A}_b$ need not be set, or can be zero.

Let l denotes the number of non-zero diagonals below the main diagonal (lower bandwidth) and u the number of non-zero diagonals above the main diagonal (upper bandwidth). The total number of non-zero diagonals in $\boldsymbol{A}$ will be $p = l + u + 1$, since we need to take into account the main diagonal as well. In our example we have $l = 2$ and $u = 1$, so $p = 4$ as the number of rows in $\boldsymbol{A}_b$. Moreover, we have that

$$\boldsymbol{A}_b(u + 1 + i - j, j) = \boldsymbol{A}(i, j) \quad \text{for} \quad \begin{array}{l} j = 1, \ldots, n \ , \\ \max(1, j - u) \leq i \leq \min(n, j + l) \ . \end{array}$$

Note that the indices here are natural indices and not Pythonic indices. To implement such an algorithm we need to modify the indices appropriately. Since we have two indices to vary (`i` and `j`) we first start with `j in range(n)`, which assigns to `j` the numbers from `0` to `n-1` (1 less than the natural values).

Furthermore, we need to transform the i index into Python index. The natural values will be from $\max(1, j - u)$ to $\min(n, j + l)$. Since `j` is a Python index, to make it natural we need to add `1` and compute these two values in natural coordinates. So we compute the maximum $\max(1, j - u)$ as `M = max(1,j+1-u)` where we take `j+1` instead of `j`. We compute the minimum $\min(n, j + l)$ as `m = min(n,j+1+l)`, again taking `j+1` instead of `j`. In this way, we compute the maximum and minimum in natural coordinates. Then we need to iterate over `i`. This can be done with the command `for i in range(M-1,m):` in analogy of $i = \max(1, j - u), \ldots, \min(n, j + l)$. The reason why we take `M-1` is because we computed M in natural coordinates, so we need to transform it into Python index. We keep `m` unchanged since the `range` function returns values up to `m-1`, and so there is no need to change it. The Python code can be the following:

```
import numpy as np
A = np.array( [[ 11, 12,  0,  0,  0],
               [ 21, 22, 23,  0,  0],
```

```
                 [ 31, 32, 33, 34,  0],
                 [  0, 42, 43, 44, 45],
                 [  0,  0, 53, 54, 55]] )
(n,n) = A.shape    # Get the dimensions of A
Ab = np.zeros((4,5))  # Reserve the appropriate space
# Define the lower and upper bandwidth
l = 2
u = 1
for j in range(n):    # Convert the matrix into banded format
    M = max(1, j+1-u)
    m = min(n, j+1+l)
    for i in range(M-1,m):
        Ab[u+i-j,j] = A[i,j]
print(Ab)
```

```
[[ 0. 12. 23. 34. 45.]
 [11. 22. 33. 44. 55.]
 [21. 32. 43. 54.  0.]
 [31. 42. 53.  0.  0.]]
```

In order to make the correct assignment $\boldsymbol{A}_b(u+1+i-j, j) = \boldsymbol{A}(i, j)$, we need to keep in mind that the Python indices `i` and `j` differ from the natural indices i and j by `1`. So, in order to use the row index $u+1+i-j$ in Python instead of using `i` we use `i+1` and instead of `j` we use `j+1`. This results in the natural (not Pythonic) index `u+1+i+1-j-1`. Then we subtract `1` to convert the result into a Python index. So, we need to take `u+1+i+1-j-1-1`, which is equal to `u+i-j` as a Python index for the computation of the index $u+1+i-j$. While this is not the most efficient way to implement such transformation, it serves well the purpose of understanding the transformation between natural and Python indices. In our previous example we have `(l,u)=(2,1)`.

Consider now the linear system $\boldsymbol{A}\boldsymbol{x} = \boldsymbol{b}$ with the banded matrix $\boldsymbol{A}$ and right hand side

$$\boldsymbol{b} = (23, 66, 130, 174, 162)^T \ ,$$

which has the exact solution

$$\boldsymbol{x} = (1, 1, 1, 1, 1)^T \ .$$

To verify the validity of this statement observe that the entries of the right hand side $\boldsymbol{b}$ are the sums of the entries of the corresponding rows of $\boldsymbol{A}$. This can happen when you multiply $\boldsymbol{A}$ with the vector $\boldsymbol{x}$. We can solve such a system efficiently using the SciPy function `solve_banded` from the module `scipy.linalg`. This function can be used in the form `solve_banded((l,u),Ab,b)` where `(l,u)` are the lower and upper bandwidths, respectively, `Ab` the matrix $\boldsymbol{A}$ stored in banded format, and `b` the right-hand side.

```
import numpy as np
from scipy.linalg import solve_banded
Ab = np.array( [[ 0, 12, 23, 34, 45],
                [11, 22, 33, 44, 55],
                [21, 32, 43, 54,  0],
                [31, 42, 53,  0,  0]] )
b = np.array( [[23], [66], [130], [174], [162]] )
l = 2
```

```
u = 1
x = solve_banded((l,u), Ab, b)
print(x)
```

```
[[1.]
 [1.]
 [1.]
 [1.]
 [1.]]
```

A dense matrix (a matrix with mainly nonzero entries) can be considered as a band matrix with $l + u + 1 = n$. This means that the methods described in this section can be applied to dense matrices as well although it is not recommended.

2.3.2 Systems with positive definite, banded matrices

Positive definite matrices form another special category of matrices. An $n \times n$ matrix $\boldsymbol{A}$ is positive definite if

$$\boldsymbol{x}^T\,\boldsymbol{A}\,\boldsymbol{x} > 0\ ,$$

for all n-dimension vectors $\boldsymbol{x} \neq \boldsymbol{0}$. For example, consider the following matrix

$$\boldsymbol{A} = \begin{pmatrix} 2 & -1 & 0 \\ -1 & 2 & -1 \\ 0 & -1 & 2 \end{pmatrix}\ ,$$

that may usually be seen in applications with differential equations. Obviously, the particular matrix $\boldsymbol{A}$ is symmetric. In order to prove that $\boldsymbol{A}$ is positive definite, take $\boldsymbol{x} \in \mathbb{R}^3$ any three-dimensional vector. Then,

$$\begin{aligned} \boldsymbol{x}^T\,\boldsymbol{A}\,\boldsymbol{x} &= \begin{pmatrix} x_1 & x_2 & x_3 \end{pmatrix} \begin{pmatrix} 2 & -1 & 0 \\ -1 & 2 & -1 \\ 0 & -1 & 2 \end{pmatrix} \begin{pmatrix} x_1 \\ x_2 \\ x_3 \end{pmatrix} \\ &= \begin{pmatrix} x_1 & x_2 & x_3 \end{pmatrix} \begin{pmatrix} 2x_1 - x_2 \\ -x_1 + 2x_2 - x_3 \\ -x_2 + 2x_3 \end{pmatrix} \\ &= 2x_1^2 - 2x_1x_2 + 2x_2^2 - 2x_2x_3 + 2x_3^2\ . \end{aligned}$$

Rearranging the terms in the last equality we have

$$\boldsymbol{x}^T\,\boldsymbol{A}\,\boldsymbol{x} = x_1^2 + (x_1 - x_2)^2 + (x_2 - x_3)^2 + x_3^2 \geq 0\ .$$

More precisely, $\boldsymbol{x}^T\,\boldsymbol{A}\,\boldsymbol{x} > 0$, unless $x_1 = x_2 = x_3 = 0$. Therefore, the specific matrix $\boldsymbol{A}$ is positive definite.

Remark 2.1. *An $n \times n$ matrix $\boldsymbol{A}$ with $\boldsymbol{x}^T\,\boldsymbol{A}\,\boldsymbol{x} \geq 0$ for all $\boldsymbol{x} \in \mathbb{R}^n$ is called positive semidefinite.*

Positive definite matrices have some favorable properties. In general, if $\boldsymbol{A}$ is a symmetric $n \times n$ positive definite matrix, then:

- $\boldsymbol{A}$ is invertible,

- $a_{ii} > 0$, for each $i = 1, 2, \dots, n$,[2]
- $\max_{1\leq k,j\leq n} |a_{kj}| \leq \max_{1\leq i\leq n} |a_{ii}|$,
- $(a_{ij})^2 < a_{ii}a_{jj}$, for each $i \neq j$.

There are some theoretical tests to verify if an $n\times n$ symmetric matrix is positive definite based on the following simple properties:

(i) a symmetric matrix $\boldsymbol{A}$ is positive definite if and only if all of its eigenvalues are positive,

(ii) a symmetric matrix $\boldsymbol{A}$ is positive definite if its leading principal minors are all positive.

Recall that eigenvalues λ are the roots of the characteristic polynomial $\det(\boldsymbol{A} - \lambda \boldsymbol{I}) = 0$, while the k-th leading principal minor of a matrix is the determinant of its upper-left $k \times k$ block.

Consider again the matrix of the previous example

$$\boldsymbol{A} = \begin{pmatrix} 2 & -1 & 0 \\ -1 & 2 & -1 \\ 0 & -1 & 2 \end{pmatrix} .$$

The leading principal minors are

$$\det(\boldsymbol{A}_1) = \det(2) = 2 > 0 ,$$

$$\det(\boldsymbol{A}_2) = \det\begin{pmatrix} 2 & -1 \\ -1 & 2 \end{pmatrix} = 4 - 1 = 3 > 0 ,$$

and

$$\begin{aligned}\det(\boldsymbol{A}_3) = \det\begin{pmatrix} 2 & -1 & 0 \\ -1 & 2 & -1 \\ 0 & -1 & 2 \end{pmatrix} &= 2\det\begin{pmatrix} 2 & -1 \\ -1 & 2 \end{pmatrix} - (-1)\det\begin{pmatrix} -1 & -1 \\ 0 & 2 \end{pmatrix} \\ &= 2(4-1) + (-2+0) = 4 > 0 ,\end{aligned}$$

and thus the matrix $\boldsymbol{A}$ is positive definite.

Although applications usually involve symmetric, positive definite matrices, the notion of positive definiteness can be extended to non-symmetric matrices, but the situation is more complicated. A non-symmetric, positive definite matrix can have complex eigenvalues. For example, the matrix

$$\boldsymbol{A} = \begin{pmatrix} 1 & 1 \\ -1 & 1 \end{pmatrix} ,$$

has eigenvalues $1\pm$i. On the other hand, the eigenvalues of a non-symmetric, positive definite matrix have always positive real part. This ensures the invertibility of the non-symmetric matrix but the properties we mentioned previously are not valid anymore.

Testing whether a large matrix is positive definite or not can be very difficult. Luckily, there are special numerical methods to solve linear systems efficiently that rely heavily on the fact that the coefficient matrix is positive definite and symmetric. Failure of these methods to solve a system imply that the matrix is not positive definite. Such methods can serve as efficient testing tools for positive definite matrices. Perhaps the most commonly used

[2] To see that $a_{ii} > 0$ for positive definite matrices it suffices to take $\boldsymbol{x}$ the unit vector with $x_j = 1$ for $j = i$ and $x_j = 0$ for $j \neq i$.

method for solving systems with positive definite and symmetric matrices is the Cholesky method. We will analyze this method later. For the moment we consider a more general method from SciPy.

The module `scipy.linalg` is equipped with the function `solveh_banded` to solve systems with symmetric and positive definite matrices. The particular method is very efficient for banded systems but we can use it even for dense (full) systems by taking the bandwidth $l+u+1=n$. The matrices we deal with here are symmetric with $l=u$, so we set $p=l=u$.

Since this function is designed for positive definite matrices that are symmetric, we need only to store the upper or the lower triangular part of the matrix. By avoiding storing unnecessary data we can save space in computer memory. We will call this particular symmetric storage scheme *symmetric band storage* scheme. We demonstrate the symmetric band storage for the following symmetric matrix

$$\boldsymbol{A} = \begin{pmatrix} a_{11} & a_{12} & a_{13} & & \\ a_{21} & a_{22} & a_{23} & a_{24} & \\ a_{31} & a_{32} & a_{33} & a_{34} & a_{35} \\ & a_{42} & a_{43} & a_{44} & a_{45} \\ & & a_{53} & a_{54} & a_{55} \end{pmatrix},$$

with $a_{ij} = a_{ji}$. In this case, we store the lower triangular part of $\boldsymbol{A}$. In particular, we store the main diagonal and the nonzero diagonals bellow the main diagonal in the rows of a matrix $\boldsymbol{A}_b$ such as

$$\boldsymbol{A}_b = \begin{pmatrix} a_{11} & a_{22} & a_{33} & a_{44} & a_{55} \\ a_{21} & a_{32} & a_{43} & a_{54} & * \\ a_{31} & a_{42} & a_{53} & * & * \end{pmatrix}.$$

Then

$$\boldsymbol{A}_b(1+i-j, j) = \boldsymbol{A}(i,j), \quad \text{for} \quad \begin{array}{l} j = 1, \ldots, n \;, \\ j \leq i \leq \min(n, j+p) \;. \end{array}$$

In this notation $*$ symbolizes empty or zero entry. This band storage generalizes the band storage format we introduced in Section 2.3.1.

Similarly, if we decide to store the upper triangular part of $\boldsymbol{A}$, we store the main diagonal and all the nonzero diagonals above the main diagonal in the rows of a matrix $\boldsymbol{A}_b$ such that

$$\boldsymbol{A}_b = \begin{pmatrix} * & * & a_{13} & a_{24} & a_{35} \\ * & a_{12} & a_{23} & a_{34} & a_{45} \\ a_{11} & a_{22} & a_{33} & a_{44} & a_{55} \end{pmatrix},$$

and in this case

$$\boldsymbol{A}_b(p+1+i-j, j) = \boldsymbol{A}(i,j), \quad \text{for} \quad \begin{array}{l} j = 1, \ldots, n \;, \\ \max(1, j-p) \leq i \leq j \;. \end{array}$$

In order to implement this matrix transformation, we work in a similar way to the previous section. We consider only the case where we store the lower triangular part, and leave the other case as an exercise. We take the matrix

$$\boldsymbol{A} = \begin{pmatrix} 2 & -1 & 0 & 0 & 0 \\ -1 & 2 & -1 & 0 & 0 \\ 0 & -1 & 2 & -1 & 0 \\ 0 & 0 & -1 & 2 & -1 \\ 0 & 0 & 0 & -1 & 2 \end{pmatrix}.$$

This matrix consists of three diagonals in total. The lower part has one nonzero diagonal ($p = 1$) and taking into account the main diagonal, the matrix $\boldsymbol{A}_b$ will be a 2×5 matrix. The implementation of this case can be in analogy to the band storage scheme as follows:

```
A = np.array( [[  2, -1,  0,  0,  0],
               [ -1,  2, -1,  0,  0],
               [  0, -1,  2, -1,  0],
               [  0,  0, -1,  2, -1],
               [  0,  0,  0, -1,  2]] )
(n,n) = A.shape
Ab = np.zeros((2,5))
p = 1
for j in range(n):
    m = min(n, j+1+p)
    for i in range(j,m):
        Ab[i-j,j] = A[i,j]
print(Ab)
```

```
[[ 2.  2.  2.  2.  2.]
 [-1. -1. -1. -1.  0.]]
```

Again, here we perform transformations from natural to Python indices as described in Section 2.3.1. The minimum `min(n, j+1+p)` is the same as before, but here we changed the l into p, and the index $1 + i - j$ is transformed to Python index by first transforming the Python indices `i`, `j` into natural indices by taking `i+1` and `j+1` and then subtracting 1 from the result to transform the whole index into Python, like `1+(i+1)-(j+1)-1=i-j`.

We consider now the linear system $\boldsymbol{Ax} = \boldsymbol{b}$ with right-hand side $\boldsymbol{b} = (1, 0, 0, 0, 1)^T$ that has exact solution $\boldsymbol{x} = (1, 1, 1, 1, 1)^T$. If we use the upper triangular part of $\boldsymbol{A}$, in order to solve this system using the module `scipy.linalg` we need to call the function `solveh_banded(Ab, b)`. We have the option to use the lower triangular part (instead of the upper). In that case we can use the same function but with the option `lower=True`, and so the call of the function will be `solveh_banded(Ab, b, lower=True)`.

```
from scipy.linalg import solveh_banded
Ab = np.array( [[  2,  2,  2,  2,  2],
                [ -1, -1, -1, -1,  0]] )
b = np.array( [[1], [0], [0], [0], [1]] )
x = solveh_banded(Ab, b, lower=True)
print(x)
```

```
[[1.]
 [1.]
 [1.]
 [1.]
 [1.]]
```

The function `solveh_banded` works also for complex matrices (whose entries are complex numbers). For complex matrices, the symmetric matrix is not exactly what we usually need in applications. Instead of just symmetry we need the matrix to be equal to its conjugate transposed matrix, denoted by $\boldsymbol{A}^H = \overline{\boldsymbol{A}^T}$. Such a matrix is called Hermitian. So, a Hermitian matrix $\boldsymbol{A}$ is such that $\boldsymbol{A} = \boldsymbol{A}^H$, or element-wise $a_{ij} = \bar{a}_{ji}$ for all i, j, where the bar denotes the complex conjugate of complex numbers.

2.3.3 Systems with other special matrices

The module `scipy.linalg` provides also functions to solve linear systems with other special matrices. Here we mention some of these special cases.

Triangular systems

We frequently transform linear systems of equations into systems that have a lower or upper triangular form. For example, when we perform Gaussian elimination we obtain a linear system with upper triangular coefficient matrix. With SciPy we can solve such triangular systems efficiently using the function `solve_triangular` of `scipy.linalg`. The general syntax of this function is the following:

```
solve_triangular(a,b,trans,lower,unit_diagonal,ovewrite_b,check_finite)
```

where

`a`: Is a triangular matrix

`b`: Is the right hand side of the system $\boldsymbol{A}\boldsymbol{x} = \boldsymbol{b}$

`trans`: Can be one of the values `0,1,2`, with 0 for the solution of the system $\boldsymbol{A}\boldsymbol{x} = \boldsymbol{b}$, 1 for the system $\boldsymbol{A}^T\boldsymbol{x} = \boldsymbol{b}$ and 2 for the system $\boldsymbol{A}^H\boldsymbol{x} = \boldsymbol{b}$. (Optional with default value `trans=0`)

`lower`: Is a boolean variable which indicates the use of the upper triangular part of matrix `A`. If `lower` is `True` then the function uses the lower triangular part of matrix `A`. (Optional with default value `lower=False`)

`unit_diagonal`: Is a boolean variable. If it is `True` then the diagonal elements of `A` are taken to be 1. (Optional with default value `unit_diagonal=False`)

`overwrite_b`: Is a boolean variable. If it is `True` then the function allows to overwrite data in `b`. (Optional with default value `overwrite_b=False`)

`check_finite`: Is a boolean variable that allows to check if the input matrices contain only finite numbers. (Optional with default values `check_finite=True`)

Toeplitz systems

A Toeplitz matrix is a matrix in which each diagonal has the same value. For example,

$$\boldsymbol{A} = \begin{pmatrix} a & b & c & d & e \\ f & a & b & c & d \\ g & f & a & b & c \\ h & g & f & a & b \\ i & h & g & f & a \end{pmatrix} .$$

If we know the first row and first column of a Toeplitz matrix, then we can reconstruct the full matrix. A Toeplitz system is a system with a Toeplitz matrix. The module `scipy.linalg` contains the function `solve_toeplitz` that allows us to solve such systems, and we only need to provide the first column and/or the first row of the matrix. If we provide only the first column, then the first row is assumed to be the complex conjugate of the first column ($\boldsymbol{r} = \boldsymbol{c}^H$, with $\boldsymbol{c}$ the first column and $\boldsymbol{r}$ the first row). The general call of the function `solve_toeplitz` is the following

```
solve_toeplitz(c_or_cr, b, check_finite)
```

where

c_or_cr: Can be the first column c or the tuple (c,r) that contains the first column and the first row

b: Is the right-hand side as usual

check_finite: Is a boolean that allows to check if the input matrices contain only numbers and not NaN or Infinity. (Optional with default values True)

Next we look at the special case of circulant systems.

Circulant systems

A circulant system is a system with a circulant matrix. A circulant matrix is a special Toeplitz matrix where each row is a rotation of the previous row by one position to the right. An example of circulant matrix is the following

$$\boldsymbol{A} = \begin{pmatrix} a & b & c & d & e \\ e & a & b & c & d \\ d & e & a & b & c \\ c & d & e & a & b \\ b & c & d & e & a \end{pmatrix} .$$

For circulant matrices we only need the first column (or row) since the rest of the columns (or rows) are just simple periodic shifts of the first column (or row). In Python we can generate a circulant matrix easily using the first column of the matrix and the function scipy.linalg.circulant. For example, to construct the matrix

$$\boldsymbol{A} = \begin{pmatrix} 2 & 3 & 4 \\ 4 & 2 & 3 \\ 3 & 4 & 2 \end{pmatrix} ,$$

we can simply type the following commands

```
from scipy.linalg import circulant
c = np.array([2, 4, 3])
A = circulant(c)
print(A)
```

```
[[2 3 4]
 [4 2 3]
 [3 4 2]]
```

We can solve circulant systems without generating the whole matrix but just using its first column with the help of the scipy.linalg function

```
solve_circulant(c, b, singular, tol, caxis, baxis, outaxis)
```

where

c: Is the first column c of the circulant matrix

`b`: Is the right hand side as usual

`singular`: Is a string variable that controls how a near singular circulant matrix should be handled. If `singular` is `"raise"` and the circulant matrix is near singular, then an error message will be returned. If `singular` is `"lstsq"`, then the least squares solution is returned. We will discuss the least squares later in this book. (Optional with default value `singular="raise"`)

`tol`: Is a real number, and serves as a tolerance related to the singularity of the matrix. If the circulant matrix has an eigenvalue less than this tolerance, then the matrix is consider to be near singular. (Optional with default value `tol=None`)

`caxis`: When `c` has dimension greater than 1, it is viewed as a collection of circulant vectors. In this case, `caxis` is the axis of `c` that holds the vectors of circulant coefficients. (Optional with default value `caxis=-1`)

`baxis`: When `b` has dimension greater than 1, it is viewed as a collection of vectors. In this case, `baxis` is the axis of `b` that holds the right-hand side vectors. (Optional with default value `baxis=0`)

`outaxis`: When `c` or `b` are multidimensional, the returned value is also multidimensional. In this case, `outaxis` is the axis of the result that holds the solution vectors. (Optional with default value `outaxis=0`)

The function `solve_circulant` returns only the solution of the circulant system in an array `x`. For example, if we want to solve the system with the matrix $\boldsymbol{A}$ generated using $\boldsymbol{c} = (2, 4, 3)^T$ and right-hand side $\boldsymbol{b} = (10, 10, 10)^T$ we write

```
from scipy.linalg import solve_circulant
c = np.array([2, 4, 3])
b = np.array([10, 10, 10])
x = solve_circulant(c, b)
print(x)
```

```
[1, 1, 1]
```

2.4 Drawing Graphs with MatPlotLib

It is not by chance that we use the phrase *one picture is worth a thousand words*; and it is very often where thousands of lines of code, theories, and many hours of hard work can all be hidden behind a graph. The most common tool for sketching graphs (plotting) with Python is the MatPlotLib package. In this section, we cover a few of the basic approaches to plotting figures. If you are interested in learning more about MatPlotLib or to see how you might create a particular plot check out the documentation of MatPlotLib for more information.

2.4.1 Plotting one-dimensional arrays

When we want to sketch a graph we define an x-axis and a y-axis. In Python we can sketch graphs in the same way but with the difference that the data cannot be continuous. Also the graph will consist of pixels. The main idea is to use a set x values and the corresponding set of y values. Python draws the points (x, y) on the screen in appropriate coordinate system. Then it joins them with straight lines and create a "continuous line". If we provide many points (x_i, y_i) and the distances between them are small, then the impression given is that of a smooth continuous graph.

We first explore the simplest case where we need to represent graphically a set of points (x_i, y_i), for $i = 1, 2, \ldots, N$. These points can be experimental data of some kind. To plot such points on the screen we need to import the module `matplotlib.pyplot`, which provides with the functions `plot` and `show` among other commands. The last function is used at the end of our code when we are ready to plot the graph on the screen. The function `plot` takes as input the arrays `x` and `y` to plot the points (x_i, y_i) and other arguments, such as the color of the graph, the size etc. We usually import the module `pyplot` with the alias texttttplt. Here, we will explore only some simple plotting options, while the entire spectrum of choices can be found online.

Let's say that we want to plot the points $(-1, -1)$, $(0, 0)$ and $(1, 1)$. We define our arrays `x`, `y`, and plot the results using the following commands:

```
import matplotlib.pyplot as plt

x = np.array([-1, 0, 1])
y = x.copy()
plt.plot(x,y,'o')
plt.show()
```

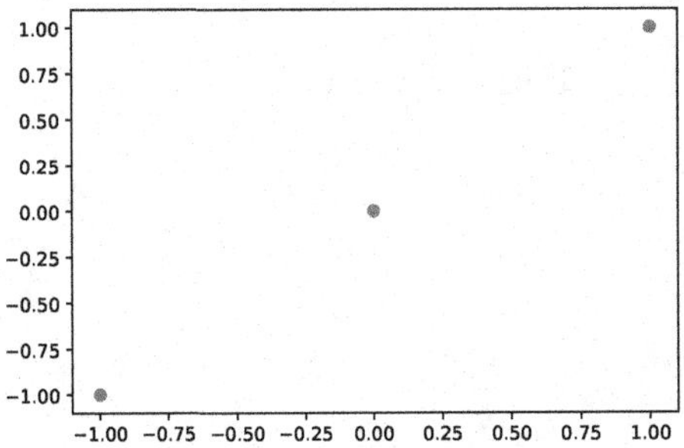

In this code we use the command `plot(x,y,'o')`. The symbol `o` is used to depict the points with filled circles. If we use the command `plot(x,y)`, then the points will be connected with straight line segments and the result will be a piecewise continuous line. It is also noted that in order for Python to display the graph on the screen we need to type the command `show()`. There are exceptions, though, in some environments where the `show()` command is not required.

Instead of using circles, we can use triangles and we can also use a different color instead of the default one. For example, if we want to use red triangles, then we can use the command `plot(x,y,'rv')`, where the letter `r` specifies the red color and the letter `v` the triangular shape. The most commonly used markers and colors are presented in Tables 2.1 and 2.2.

Assume now that we want to sketch the graph of a function $f(x)$. Python, like any other programming language, can plot only a finite set of points that can be joined with straight lines to give an impression of a curved line. For the same reason computers can only handle bounded intervals. We consider the interval $[a, b]$. We also consider a finite number of nodes $x_0 < x_1 < \cdots < x_N$ in $[a, b]$. After computing the values of the function $f(x)$ at the points

TABLE 2.1
Commonly used markers

Symbol	**Marker**	**Symbol**	**Marker**
`.`	point	`s`	square
`o`	circle	`p`	pentagon
`v`	triangle_down	`*`	star
`^`	triangle_up	`x`	x
`<`	triangle_left	`D`	diamond
`>`	triangle_right	`d`	thin_diamond

TABLE 2.2
Commonly used colors

Symbol	**Color**	**Symbol**	**Color**
`b`	blue	`m`	magenta
`g`	green	`y`	yellow
`r`	red	`k`	black
`c`	cyan	`w`	white

x_i, i.e. we compute the values $y_i = f(x_i)$, we can plot the points $(x_i, f(x_i))$, for $i = 0, \ldots, N$ on the screen using MatPlotLib as we did in the previous example.

In order to generate the points x_i we create a uniform grid for the interval $[a, b]$, using, for example, the methodology presented in Section 1.4.2. In particular, we create N equidistributed points x_i in $[a, b]$ such that $x_{i+1} = x_i + h$ with $h = (b - a)/N$. If the values a, b and N are given then we can use the NumPy function `linspace(a,b,N)`. This generates a one-dimensional array of N nodes x_i such that $x_{i+1} = x_i + h$ with $h = (b - a)/(N - 1)$. For example, the command `linspace(0.0, 1.0, 5)` will create the one-dimension NumPy array with 5 entries $(0, 0.25, 0.5, 0.75, 1)$ and $h = 0.25$. Note that the grid will contain the end-points a, b. If we need to exclude the end-points from our grid, we can use the optional argument `endpoint=False` in the `linspace` function. If the values a, b and h are given, then we can use the NumPy function `arange(a,b+h,h)` to generate similarly a uniform grid in the interval $[a, b]$ with stepsize h. Observe that in the function `arange` we use the right endpoint increased by h. This is because the function `arange` is not inclusive and will not include the value b in the grid unless we extend the endpoint by h.

```
a = 0.0
b = 1.0
N = 5 #Number of points
h = (b-a)/(N-1)
x = np.linspace( a, b, N )
print(x)
x = np.arange( a, b + h, h )
print(x)
```

```
[0.   0.25 0.5  0.75 1.  ]
[0.   0.25 0.5  0.75 1.  ]
```

After we generate the points x_i, we need to compute the corresponding points $f(x_i)$ and plot the points on the screen. The computation of the points $f(x_i)$ can be done instantaneously by typing `f(x)` with `x` being the array containing the mesh-points x_i. To plot the points on the screen we need to import the module `matplotlib.pyplot` which provides with the functions `plot` and `show`.

In the following example, we plot the function $f(x) = \sin(x)$ in the interval $[0, 2\pi]$ using 20 points.

```
x = np.linspace( 0, 2*np.pi, 20 )
y = np.sin(x)
plt.plot(x, y)
plt.xlabel('$x$')
plt.ylabel('$y$')
plt.title('$y=sin(x)$')
plt.xlim([0, 2*np.pi])
plt.ylim([-1.1, 1.1])
plt.show()
```

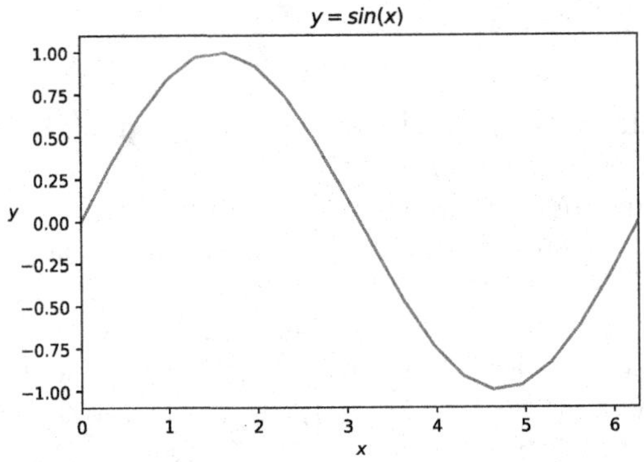

In this example we don't specify the symbol of the points and Python connects all the points with straight line segments forming on the screen the graph of a piecewise linear function. We also specify the titles of the x and y axes using the functions `xlabel` and `ylabel`, the title of the graph with the function `title` and the limits of the plotted axes using the function `xlim` and `ylim`. Observe that in the title and for the axes labels we use the symbol `$`. By doing this Python interprets the text with LaTeX and can handle mathematical formulas better. We also use the option `rotation=0` for the label of the y-axis in order to align the label vertically and not horizontally.

We can use different styles for lines such as broken or dotted line. A list with most commonly used line styles is presented in Table 2.3.

TABLE 2.3
Commonly used line styles

Symbol	Line style
-	solid
--	dashed
-.	dash-dot
:	dotted

We can also combine points with lines. For example, we can plot the points with circles connected with lines using the command `plot(x,y,'o-')`. In principle, we can use combinations of color, symbol and line from Tables 2.1–2.3. In the following example, we demonstrate the use of some different combinations of line styles to plot various plots in the same graph.

```
x = np.linspace( 0, 2*np.pi, 20 )
y1 = np.sin(x)
y2 = np.cos(x)
y3 = np.sin(x/2.0)
plt.plot(x,y1,'o-',x,y2,'rd:',x,y3,'k-.')
plt.xlabel('x')
plt.ylabel('y', rotation = 0)
plt.title('y=sin(x)')
plt.xlim([0, 2*np.pi])
plt.ylim([-1.1, 1.1])
plt.show()
```

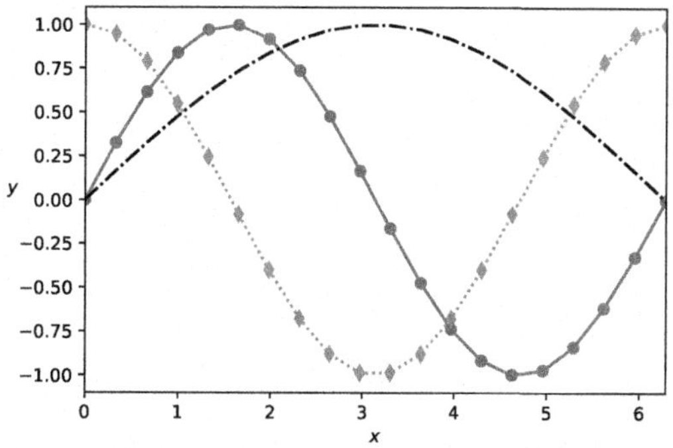

It should be noted that MatPlotLib has many more capabilities for sophisticated and complicated graphs. Alternative to `pyplot` is the module `pylab` of MatPlotLib which results in high quality figures, but we will stop here as we will not need anything more for the purposes of this book.

We close this chapter with a summary of the most commonly used import abbreviations for Python modules in this book (Table 2.4).

TABLE 2.4
Commonly used import abbreviations for various Python modules

Module	Abbreviation	Import
`numpy`	`np`	`import numpy as np`
`numpy.linalg`	`npl`	`import numpy.linalg as npl`
`scipy`	`sp`	`import scipy as sp`
`scipy.linalg`	`spl`	`import scipy.linalg as spl`
`scipy.sparse`	`sps`	`import scipy.linalg as sps`
`scipy.optimize`	`spo`	`import scipy.optimize as spo`
`scipy.interpolate`	`spi`	`import scipy.interpolate as spi`
`matplotlib.pyplot`	`plt`	`import matplotlib.pyplot as plt`

2.5 Further Reading

For a complete reference on matrices and linear algebra refer to the introductory books to linear algebra written by Gilbert Strang [128, 129]. A modern introduction to linear algebra and its tools to data science can be found in [130]. Another textbook that we suggest as an introduction to linear algebra is the book by David Poole [105], which also includes interesting applications of linear algebra in various fields. A complete and quite advanced exploration in linear algebra can be found in [67]. The book [48] consist of a complete guide for numerical methods in linear algebra. Specialized books on numerical algorithms for linear algebra include the classics [48, 47]. As NumPy, SciPy and MatPlotLib are the core modules for scientific computing and computational mathematics there are several books that explain the use of these modules. The book [72] is an extensive introduction to these modules for scientific computations. For a quick introduction to SciPy and NumPy we also refer to [16].

Chapter Highlights

- We can represent matrices using NumPy arrays. A NumPy array is created using the command `numpy.array( [[1, 2], [3, 4]] )` where the matrix is given as a list with entries the rows of the matrix. All entries and all rows are separated with commas.
- The (i, j) entry of an array can be accessed by writing `A[i,j]`.
- A mathematical index i that takes values $i = 1, 2, \ldots, n$ is called natural index, while the index `i` that takes values $\mathtt{i} = 0, 1, \ldots, \mathtt{n} - 1$ is called Python index.
- Mathematical algorithms are usually given with natural indices and we need to convert them to algorithms with Python indices.
- We can perform any matrix operation using the usual operators but the matrix multiplication can be computed using the function `dot` or the operator `@`. The `*` operator is used to perform entry by entry multiplication.
- We can access blocks of a matrix using the slice : operator. For example the first row of a matrix let's say `A` can be extracted as `A[0,:]`.
- The identity matrix is `numpy.eye(n)`. The zero $n \times m$ matrix is `numpy.zeros((n,m))`.
- To create an unrelated copy B of a matrix A we use the function `B=A.copy()`.
- The module `numpy.linalg` contains useful functions for the computation of determinants, inverses, eigenvalues and eigenvectors, and for the solution of linear systems.
- To raise each entry of a matrix in a power p we can use the NumPy function `power`.
- To raise a matrix in a power of p we can use the function `matrix_power` of the module `linalg` of NumPy.
- We can solve a linear system $\boldsymbol{Ax} = \boldsymbol{b}$ defined by the matrices `A` and `b` just using the function `solve` of the module `numpy.linalg` simply by typing `x=solve(A,b)`. The last command returns the approximation of the solution $\boldsymbol{x}$ in the NumPy array `x`.
- Banded matrices and other special matrices can be handled efficiently using the module SciPy.
- We can measure the performance of our code using the function `time` of module `time`.
- The product of lower(upper)-triangular matrices is lower(upper)-triangular matrix.
- The inverse of a lower(upper)-triangular matrix is lower(upper)-triangular matrix.
- We can plot graphs of functions using the module `matplotlib.pyplot` and the function `plot`.

Exercises

1. Write a Pyhton program using NumPy to:

   (a) Create a 3×3 matrix with values ranging from 1 to 9.

   (b) Create a null vector of size 5 and update third entry to 1.

   (c) Create a 7×7 array with 1 on the first and last rows and columns and 0 elsewhere.

   (d) Create a 8×8 matrix and fill it with a checkerboard pattern.

   (e) Create a 10×10 empty and a 10×10 full matrix.

   (f) Find and return the number of entries of the last created array.

   (g) Create an 8×8 array with ones on a diagonal and zeros elsewhere.

   (h) Create a 5×5 matrix with row values ranging from 0 to 4.

   (i) Compute the multiplication of two random 10×10 matrices.

   (j) Compute the determinant of a random 100×100 square array.

2. Consider the matrix

$$\boldsymbol{A} = \begin{pmatrix} 1 & 2 & 3 \\ 0 & 1 & 4 \\ 5 & 6 & 0 \end{pmatrix}.$$

   Using NumPy module

   (a) Compute the determinant of $\boldsymbol{A}$

   (b) Compute the inverse matrix of $\boldsymbol{A}$ if it exists

   (c) Work out the results without using computers

   (d) Compute the eigenvalues and the respective eigenvectors

3. Consider the vectors $\boldsymbol{x} = (0.1, 1.3, -1.5, 0, 12.3)$. Try to write your own implementations using for loops in Python (use Python functions only for question (v)) for the following problems:

   (a) Compute the sum $s_1 = \sum_{i=1}^{5} x_i$

   (b) Compute the sum $s_2 = \sum_{i=1}^{5} x_i^i$

   (c) Compute the maximum absolute value $M = \max\{|x_i|, i = 1, 2, \ldots, 5\}$

   (d) Compute the minimum absolute value $m = \min\{|x_i|, i = 1, 2, \ldots, 5\}$

   (e) Compare with the functions `sum`, `max` and `min` of NumPy.

4. Write the commands for each of the following operations:

   (a) Create a row vector `x` of 5 equally spaced elements between 2 and 3.

   (b) Add 1 to the second element

   (c) Create a second row vector `y` of the same dimension with elements equal to the successive even integers starting with 4.

   (d) Create the matrix `A`, whose first row is equal to `x`, whose second row is a line of ones, and whose third row is equal to `y`.

   (e) Define a row vector `z`, whose elements are equal to the mean value of the columns of `A`.

5. Create two matrices $\boldsymbol{A}$ and $\boldsymbol{B}$:

$$\boldsymbol{A} = \begin{pmatrix} 1 & 2 \\ 4 & -1 \end{pmatrix} \quad \text{and} \quad \boldsymbol{B} = \begin{pmatrix} 4 & -2 \\ -6 & 3 \end{pmatrix}$$

   (a) Compute the matrices $\boldsymbol{C}_1 = \boldsymbol{A} + \boldsymbol{B}$ and $\boldsymbol{C}_2 = \boldsymbol{A} - \boldsymbol{B}$.
   (b) Compute the matrix products $\boldsymbol{D}_1 = \boldsymbol{A} \cdot \boldsymbol{B}$ and $\boldsymbol{D}_2 = \boldsymbol{B} \cdot \boldsymbol{A}$.
   (c) Using element by element operations, compute the matrix F whose elements are obtained as follows: $F_{ij} = B_{ij} + A_{ij} B_{ij}^{1/3}$.
   (d) Are $\boldsymbol{A}$ and $\boldsymbol{B}$ singular? If not, then compute their inverse.
   (e) Compute the eigenvalues of $\boldsymbol{B}$. Comment in light of your previous answer.

6. Create two column vectors `x` and `y` of 100 elements with random values between -10 and 10. To create a random vector of N random values between a and b you can use the following command `y = a + (b-a)*rand(N)`.

   (a) Write a function `mydot(x,y)` computing the dot (inner) product between `x` and `y`.
   (b) Compute the dot product using your function. Compare the results with the results of the functions `inner(x,y)` and `dot(x,y)`.

7. Consider two vector $\boldsymbol{x} = (x_1, x_2, \ldots, x_N)^T$ and $\boldsymbol{y} = (y_0, y_1, \ldots, y_N)^T$.

   (a) Write a Python function that returns the following double sum

$$s = \sum_{i=1}^{N} \sum_{j=1}^{i} x_i y_j \ .$$

   (b) Calculate the number of multiplications and additions required in the computation for general N.
   (c) Modify the previous formula appropriately to reduce the number of floating point operations.

8. Consider the matrix

$$\boldsymbol{A} = \begin{pmatrix} 1 & 2 & 3 & \cdots & n-1 & n \\ 2 & 3 & 4 & \cdots & n & 1 \\ 3 & 4 & 5 & \cdots & 1 & 2 \\ \vdots & \vdots & \vdots & \ddots & \vdots & \vdots \\ n & 1 & 2 & \cdots & n-2 & n-1 \end{pmatrix} .$$

   (a) Show that

$$\det(\boldsymbol{A}) = (-1)^{n(n-1)/2} \frac{(n+1)n^{n-1}}{2} \ .$$

   (b) Write a Python function that generates matrix $\boldsymbol{A}$ for any value n.
   (c) Verify the formula for its determinant using the Python function `det(A)`.

9. Consider the linear system $\boldsymbol{A}\boldsymbol{x} = \boldsymbol{b}$ with

$$\boldsymbol{A} = \begin{pmatrix} 3 & 2 & 1 & 0 \\ -3 & -2 & 7 & 1 \\ 3 & 2 & -1 & 5 \\ 0 & 1 & 2 & 3 \end{pmatrix} \quad \text{and} \quad \boldsymbol{b} = \begin{pmatrix} 6 \\ 3 \\ 9 \\ 6 \end{pmatrix} .$$

(a) Solve the system above using the NumPy function `solve`.

(b) Write a Python function that stores its matrix $\boldsymbol{A}$ in a banded form.

(c) Using the SciPy function `solve_banded` solve the system $\boldsymbol{Ax} = \boldsymbol{b}$.

(d) Which of the two methods is faster?

10. Consider the linear system $\boldsymbol{Ax} = \boldsymbol{b}$ with

$$\boldsymbol{A} = \begin{pmatrix} 3 & -1 & 0 & 0 & -1 \\ -1 & 3 & -1 & 0 & 0 \\ 0 & -1 & 3 & -1 & 0 \\ 0 & 0 & -1 & 3 & -1 \\ -1 & 0 & 0 & -1 & 3 \end{pmatrix}, \quad \text{and} \quad \boldsymbol{b} = \begin{pmatrix} 1 \\ 1 \\ 1 \\ 1 \\ 1 \end{pmatrix}.$$

Such systems occur in numerical integration of differential equations with periodic boundary conditions. Taking advantage of their special structure we construct a fast and memory efficient numerical method for its numerical solution.

(a) Solve the system $\boldsymbol{Ax} = \boldsymbol{b}$ using the NumPy functions `solve` and `solve_circulant`.

(b) Consider the matrix $\bar{\boldsymbol{A}} = [\bar{a}_{ij}]$ with entries all the entries of the main band of $\boldsymbol{A}$. This means that $\bar{a}_{ij} = a_{ij}$ for $i = 1, \ldots, 5$ and j such that $|i - j| \leq 1$ otherwise $\bar{a}_{ij} = 0$. Find vectors $\boldsymbol{u}, \boldsymbol{v} \in \mathbb{R}^5$ such that $\bar{\boldsymbol{A}} = \boldsymbol{A} + \boldsymbol{uv}^T$. Choose $\boldsymbol{u}$ and $\boldsymbol{v}$ such that $\boldsymbol{v}^T \bar{\boldsymbol{A}}^{-1} \boldsymbol{u} \neq 1$.

(c) Prove that the matrix $\boldsymbol{A} = \bar{\boldsymbol{A}} - \boldsymbol{uv}^T$ is invertible, with inverse

$$\boldsymbol{A}^{-1} = \bar{\boldsymbol{A}}^{-1} + \alpha(\bar{\boldsymbol{A}}^{-1}\boldsymbol{u})(\boldsymbol{v}^T \bar{\boldsymbol{A}}^{-1}) ,$$

where

$$\alpha = \frac{1}{1 - \boldsymbol{v}^T \bar{\boldsymbol{A}}^{-1} \boldsymbol{u}} .$$

(d) Implement the following algorithm to solve the linear system $\boldsymbol{Ax} = \boldsymbol{b}$

- Solve the banded linear system $\bar{\boldsymbol{A}}\bar{\boldsymbol{x}} = \boldsymbol{b}$ using the function `solve_banded`
- Solve the banded linear system $\bar{\boldsymbol{A}}\boldsymbol{y} = \boldsymbol{u}$ using the function `solve_banded`
- Solve the banded linear system $\bar{\boldsymbol{A}}^T \boldsymbol{z} = \boldsymbol{v}$ using the function `solve_banded`
- Compute the numbers

$$\alpha = \frac{1}{1 - \boldsymbol{v}^T \boldsymbol{y}} \quad \text{and} \quad \beta = \boldsymbol{z}^T \boldsymbol{b} .$$

- Compute the solution $\boldsymbol{x} = \bar{\boldsymbol{x}} + \alpha\beta\boldsymbol{y}$.
- Verify that $\boldsymbol{Ax} - \boldsymbol{b} \approx \boldsymbol{0}$.

(e) Compare the efficiency of all the methods you used.

[The previous algorithm is known as Sherman-Morrison-Woodbury algorithm. For more information please see Chapter 9 and [73]]

11. Consider the matrix

$$\boldsymbol{A} = \begin{pmatrix} 1 & 0 & 0 \\ 2 & 3 & 0 \\ 4 & 5 & 6 \end{pmatrix} .$$

With the help of NumPy and SciPy:

(a) compute the determinant of $\boldsymbol{A}$. Recall that the determinant of a triangular matrix is the product of the main diagonal entries.

(b) compute the inverse matrix of $\boldsymbol{A}$ if it exists.

(c) solve the linear system $\boldsymbol{A}\boldsymbol{x} = \boldsymbol{b}$ with $\boldsymbol{b} = (1, 5, 15)^T$.

12. Consider the 100×100 circulant matrix $\boldsymbol{A}$ with first row

$$\boldsymbol{c} = \begin{pmatrix} 3 & -1 & 0 & \cdots & 0 & -1 \end{pmatrix} .$$

(a) Using SciPy generate the appropriate matrix $\boldsymbol{A}$.

(b) Using NumPy functions compute the eigenvalues of $\boldsymbol{A}$.

(c) Why $\boldsymbol{A}$ is invertible?

(d) Using Python generate appropriate right-hand side $\boldsymbol{b}$ such that the solution of the corresponding system $\boldsymbol{A}\boldsymbol{x} = \boldsymbol{b}$ has $x_i = 1$ for $i = 1, 2, \ldots, 100$.

(e) Solve the linear system $\boldsymbol{A}\boldsymbol{x} = \boldsymbol{b}$ using the function `solve_circulant` of SciPy.

(f) Solve the same linear system using the function `solve` of NumPy.

(g) Compare the efficiency of the two methods using the function `time` of the module `time`, and explain what are the advantages of each method.

13. Consider the vector $\boldsymbol{x} = (0, 0.01, 0.02, \ldots, 1)^T$ with values from 0 to 1 with distance 0.01, i.e. $x_i = x_{i-1} + 0.01$.

(a) Write down the Python commands that will plot the graphs of the following functions:

i. $\log(1 + \sqrt{x})$,

ii. e^{x+x^2},

iii. $\arccos(x)$,

iv. $\sqrt{1 + \ln^2(x)}$,

v. $\tan^2(x) - 1$.

(We denote by log the logarithm with base 10 and ln the one with base e.)

(b) Explain why in some cases we receive error messages. Try to correct the graphs so that no error messages occur.

3

Scientific Computing

Most algorithms in this book aim to approximate solutions of mathematical problems. The word approximation automatically infers to the introduction of errors. This is because any approximate value may differ from the actual solution of the problem. Another source of error is the finite precision arithmetic of computers. This is an unavoidable form of error in scientific computing. In this chapter, we will focus on those errors that occur from the representation of real numbers in a computer system.

3.1 Computer Arithmetic and Sources of Error

One example of approximation is the computation of $\pi = 3.14\cdots$, which is an irrational number with infinite many digits. Such a number cannot be stored in the memory of a computer due to its infinite size. But this is not the only case. There are many real numbers missing from computer arithmetic, and this is one of the main sources of errors in computer simulations. Because of these errors (in addition to other errors we will discuss later) the approximations are not always very accurate.

A disturbing example of finite precision arithmetic is the result of the operation $1.1+0.1$. We all know that this is equal to 1.2. By typing in Python `1.1+0.1==1.2` we receive the unexpected answer `False`.

```python
1.1 + 0.1 == 1.2
```

```
False
```

The question arising from this example is what could the result of that operation be? What we get in this case is not far away from being 1.2 but definitely is not exactly that.

```python
print(1.1 + 0.1)
```

```
1.2000000000000002
```

Although this is unbelievable, in reality the computer is missing one of the exact numbers involved in the previous computation or the operation itself can lead to small errors. If we count the number of digits in this result, these are sixteen. This is not by chance. It is true that we are usually confident to about the first 15 digits of the representation of a real number of $O(1)$ in a standard computer system (with the currently standard double precision arithmetic). There are cases where better precision can be achieved but this requires more resources. The errors occurring because of floating-point arithmetic are known as *floating-point errors.* The first fact related to this problem is that there is no continuity in computer number systems. This means that between floating-point numbers there are gaps.

DOI: 10.1201/9781003287292-3

Another source of error is the approximation of the original problem like the approximation of mathematical formulas. Sometimes the approximation of a problem is called discretization, especially if it involves transition from continuum to discrete level. These errors are known to as *discretization* or *truncation errors*. For example, suppose that we want to approximate the base of natural logarithms e. We know that the exponential function e^x can be expressed as the infinite sum (Taylor series[1]):

$$e^x = \sum_{k=0}^{\infty} \frac{x^k}{k!} ,$$

where $k! = 1 \cdot 2 \cdot 3 \cdots k$ is the factorial of k. In order to compute an approximation of e we take $x = 1$, and we approximate the infinite sum by the sum of the first five terms. Because the rest of the terms become smaller and smaller as k grows, discarding them will lead to a reasonably accurate approximation. After dropping the terms with $k > 4$ we obtain the approximation

$$e \approx 1 + 1 + \frac{1}{2} + \frac{1}{6} + \frac{1}{24} \approx 2.7083 .$$

The actual value of e with 5 digits is 2.7183, and thus the discretization error has the magnitude $O(0.01)$. Taking more terms into account we can compute the same number with better accuracy and thus smaller error.

We will discuss this kind of errors in later chapters for various numerical methods. In this chapter, we focus only on floating-point errors due to the finite precision arithmetic of computer systems. We continue with another example of floating-point error known as catastrophic cancelation.

> ☞ There are two main sources of error in scientific computing: Floating-point errors that are introduced by the computer arithmetic; and discretization errors that occur when we approximate the exact problem by a problem that computers can solve. Both errors are important, but usually discretization errors are larger in magnitude than floating-point errors.

3.1.1 Catastrophic cancelation

Catastrophic cancelation is the phenomenon where the subtraction of two almost equal numbers leads to a large error due to the finite precision arithmetic. In order to demonstrate this phenomenon we consider the function

$$f(x) = \frac{1 - \cos(x)}{x^2} .$$

This function is not defined at $x = 0$ but using de L'Hospital's rule we obtain the limit

$$\lim_{x \to 0} \frac{1 - \cos(x)}{x^2} = \lim_{x \to 0} \frac{(1 - \cos(x))'}{(x^2)'} = \lim_{x \to 0} \frac{\sin(x)}{2x} = \lim_{x \to 0} \frac{\cos(x)}{2} = \frac{1}{2} .$$

Let's pretend that we don't know the answer and we want to use Python to estimate the limit. In order to do so we consider the values $x = x_i = 10^{-i} \approx 0$ for $i = 5, \ldots, 11$, and we record the corresponding values $f(x_i)$. According to the theoretical estimation of the limit, these values should converge to 0.5.

[1] A revision of this subject can be found in the next chapter

```
import numpy
f = lambda x:  (1 - numpy.cos(x))/x**2
x = numpy.linspace(-5, 5, 100)
y = f(x)
xi = (1.e-5, 1.e-6, 1.e-7, 1.e-8, 1.e-9, 1.e-10, 1.e-11)
for x in xi:
    print('f(',x,')','=',f(x))
```

```
f( 1e-05 ) = 0.5000000413701854
f( 1e-06 ) = 0.5000444502911705
f( 1e-07 ) = 0.4996003610813205
f( 1e-08 ) = 0.0
f( 1e-09 ) = 0.0
f( 1e-10 ) = 0.0
f( 1e-11 ) = 0.0
```

The solution should have been

$$\lim_{x\to 0} f(x) = 0.5 \ .$$

In the beginning, the convergence looks correct, but after a while we get the impression that the limit is 0. We cannot get an accurate computation due to the phenomenon of *catastrophic cancelation* (also known as *loss of significance*). The explanation is the following: The difference $1 - \cos(x)$ for very small values of x is very close to 0. This difference will become very fast so small that the computer will approximate it by 0. The division of this 0 by the small value of x^2 will result again in a 0 and not to the expected value of 0.5.

To understand the cause of this problem, first keep in mind that computers can only store a finite number of decimal digits. Assume that our machine can store in its memory only 10 floating-point digits (decimal digits) and consider the number:

$$a = 0.12345678901234567890 \ .$$

A floating-point representation of this number in our machine would be

$$b = 0.1234567890 \ .$$

Now perform the calculation

$$a - b = 0.00000000001234567890 \ .$$

However, on the 10-digit floating-point machine, this last number can be considered by the machine as 0. This leads to loss of significance, because if we divide $a - b$ by any small number the result will be zero anyway.

3.1.2 The machine precision

Another unexpected phenomenon in computer arithmetic is the non-uniqueness of 0. We all know that $0 + 1 = 1$. In computer systems there are numbers other than 0 where their addition with the number 1 will result in 1. The smallest number x such that $x + 1 \neq 1$ is called the *machine epsilon* (also known as *unit roundoff*) and is denoted here by `EPS`. All numbers x between 0 and `EPS`, ($0 \leq x <$ `EPS`) satisfy $x + 1 = 1$.

In order to approximate the machine epsilon we usually start with $x = 1$ and we divide continuously by 2 until we find $x > 0$ such that $x + 1 = 1$. Then we choose the number $2x$ as an approximation of the machine epsilon. Algorithm 7 describes this process.

Algorithm 7 Computation of machine epsilon

Set $x = 1.0$
while $x + 1.0 \neq 1.0$ **do**
  $x = x/2.0$
end while
Set EPS $= 2.0 \cdot x$

A Python implementation of the previous algorithm can be the following:

```
x = 1.0
while (1.0 + x != 1.0):
    x = x / 2.0
EPS = 2.0*x
print('EPS = ', EPS)
print('x = ', x)
```

```
EPS =  2.220446049250313e-16
x =  1.1102230246251565e-16
```

Practically, we computed the number `1.1102230246251565e-16`, so that if we add it to `1.0` we get `1.0`. For this reason, when we work with scaled problems, we expect our results to have an error of order 10^{-16} at least.

Between any two consecutive floating-point numbers there is always an empty gap (without any floating-point number)[2]. The machine epsilon approximates the distance between `1.0` and the next floating point number. The gap between consecutive floating-point numbers is increasing with their magnitude. You can verify this by modifying the previous code to find the number x that added to 1000 equals 1000. We computed $x \approx 6 \times 10^{-14}$. This is another reason why we use scaled equations in mathematical modeling. For example, it is more convenient (and safer) to use scaled distances instead of meters for the equations of planetary motion[3].

☞ When we perform operations with numbers close to `1.0`, any other floating point number less than the machine epsilon `2.22e-16` can be considered to be 0. This introduces to our estimates a floating-point error of the order 10^{-16}. The accuracy is reduced even more when we use numbers larger than `1.0`

3.2 Normalized Scientific Notation

Normalized scientific notation is the notation we use to represent real numbers in a computer system. In normalized scientific notation the first digit is always zero while we shift the

[2]Contrary to computer arithmetic, in the real arithmetic there is always a real number between any two real numbers.

[3]To minimize floating point errors we scale the equations of planetary motion using the distance between the Sun and the Earth. This distance is 1 Astronomical Unit (AU).

decimal point. Appropriate powers of 10 are supplied so that all digits are located to the right of the decimal point. For example, the number -123.456 is written in normalized scientific notation as -0.123456×10^3.

More precisely, a number in scientific notation is written in the form

$$x = \pm r \times b^n ,$$

where

$b =$ the base of the system (for example 10 or 2)

$r =$ the mantissa, a number in the range $\frac{1}{b} \leq r \leq 1$

$n =$ the exponent, an integer

Although most computer systems use $b = 2$ as the base of their number system, in the examples used in this book we assume $b = 10$ for simplicity. We denote the lower and upper limits of exponent n by L and U, respectively. That is $L \leq n \leq U$. When the exponent n exceeds the limits, we say that an overflow (or underflow) occurs. The underflow region practically is a gap between the largest negative floating-point number and the smallest positive floating-point number.

As an example, consider a system with $b = 10$, $L = -1$, $U = 2$ and the numbers $a = 12.3$ and $b = 34.5$, then

$$a \times b = 424.35 = 0.42435 \times 10^3 .$$

In this case, the exponent is $3 > U$ and we have overflow. An overflowed number in various programming languages is represented by `NaN` (or `Inf`). The abbreviation `NaN` stands for the expression `Not a Number` while the `Inf` means infinity. On the other hand, an underflowed number is represented by a signed zero, i.e. `+0.0` or `-0.0` depending on whether it is positive or negative.

In finite precision arithmetic, the mantissa consists of a finite number of digits (known as significant digits or figures). Therefore, we write the mantissa

$$r = 0.d_1 d_2 \cdots d_t ,$$

where t is the precision. For example, consider the number $x = 3.141596$. In this scientific notation we can write $x = 0.3141596 \times 10^1$ with $t = 7$.

Modern computers usually use the binary system to store numbers in memory with binary digits (bits) and are characterized by the following floating-point arithmetic systems:

- **Single-precision floating-point arithmetic** consists of 32-bit long numbers stored in a bit pattern $b_1 b_2 \cdots b_9 b_{10} \cdots b_{32}$, which is interpreted as the real number

$$(-1)^{b_1} \times 2^{(b_2 b_3 \cdots b_9)_2} \times 2^{-127} \times (1.b_{10} b_{11} \cdots b_{32})_2 .$$

- **Double-precision floating-point arithmetic** consists of 64-bit long numbers stored in a bit pattern $b_1 b_2 \cdots b_9 b_{10} \cdots b_{64}$, which is interpreted as the real number

$$(-1)^{b_1} \times 2^{(b_2 b_3 \cdots b_{12})_2} \times 2^{-1023} \times (1.b_{13} b_{14} \cdots b_{64})_2 .$$

We prefer to use double-precision instead of single-precision floating-point arithmetic because we can store decimal digits, and the results are more accurate; catastrophic cancelation can be avoided, while an overflow or an underflow is less likely to happen. Python by default considers double-precision floating-point numbers unless otherwise specified.

3.2.1 Floats in Python

The module NumPy provides the function `numpy.finfo` which contains all the extreme values for floating point numbers (double-precision numbers). For example, in my computer system I got the following results:

```python
from numpy import finfo
print(finfo(float))
```

```
Machine parameters for float64
---------------------------------------------------------------
precision =  15   resolution = 1.0000000000000001e-15
machep =    -52   eps =        2.2204460492503131e-16
negep =     -53   epsneg =     1.1102230246251565e-16
minexp =  -1022   tiny =       2.2250738585072014e-308
maxexp =   1024   max =        1.7976931348623157e+308
nexp =       11   min =        -max
---------------------------------------------------------------
```

This means the following:

- There is no unique `0.0` stored in our machine. The smallest floating point number is `tiny = 2.2250738585072014e-308`.
- When we see in our display the number `0.0` or `-0.0` this can be any number between the `-tiny` and `tiny` numbers. This phenomenon is called *underflow.*
- If we compute any number greater than `max = 1.7976931348623157e+308` or less than `min = -1.7976931348623157e+308` we say that we get an *overflow* and modern programming languages represent these numbers by `inf` or `-inf` respectively.

3.3 Rounding and Chopping

If we enter in our code a number with more digits than the respective computer precision t, then the computer will approximate this number with a new one. This new number will be stored in the memory of the computer for future reference. There are two ways to approximate a number. One is by *chopping* where only the first t digits are stored (if the precision of the specific computer is t) ignoring the rest. The other one is by *rounding* where the last digit d_t is rounded up if $d_{t+1} \geq b/2$, while is rounded down if $d_{t+1} < b/2$, where b is the base of the number system we use. This rounding process in the decimal system happens when $d_{t+1} \geq 5$ or $d_{t+1} < 5$.

We denote the stored number (the floating-point representation of a real number x) with $fl(x)$. For example, consider the number $x = 3.141596$ in a system with

$$b = 10, t = 3, L = -10, U = 10 \ .$$

In this particular case the result is $fl(x) = 3.14$ with both ways of representation. If we take precision $t = 5$, then using *rounding* gives

$$fl(x) = 3.1416 \ ,$$

while using *chopping* results in

$$fl(x) = 3.1415 \ .$$

This is related with the floating-point errors that we have discussed so far. In order to quantify errors in computations we introduce specific formulas.

3.3.1 Absolute and relative errors

Whenever approximations occur there is an implied error in the computations. In order to estimate the errors in the various computations we introduce the notion of absolute and relative error defined by the following formulas

$$\text{Absolute Error} = |x - fl(x)| \ ,$$

and

$$\text{Relative Error} = \frac{|x - fl(x)|}{|x|}, \quad x \neq 0 \ ,$$

where $fl(x)$ stands for the floating point approximation of x. The absolute error depends on the magnitude of x. This means that larger numbers will have different error than smaller numbers, while the relative error is scaled and it does not depend on the magnitude.

For example, suppose that the numbers $x = 0.51$, $y = 1.73$ have the floating-point representations $fl(x) = 0.50$ and $fl(y) = 1.72$, respectively. The absolute error is

$$|x - fl(x)| = |y - fl(y)| = 0.01 \ ,$$

while the relative errors are

$$\frac{|x - fl(x)|}{|x|} = 0.0196 \qquad \text{and} \qquad \frac{|y - fl(y)|}{|y|} = 0.0058 \ .$$

This shows that y can be represented more accurately in the floating-point representation compared to x.

In general, given a true value of a quantity f and an approximation $\bar{f}$ we define the absolute error as

$$e = |f - \bar{f}| \ ,$$

and the relative error as

$$r = \frac{e}{|f|} = \frac{|f - \bar{f}|}{|f|} \ .$$

The relative error is the absolute error scaled by the exact quantity f. In computations where the actual quantity is unknown, we can scale by the numerical approximation if we are sure that the latter is accurate enough.

> ☞ The most common error representations between an actual quantity and its approximation are the absolute and relative errors. The relative error is the absolute error scaled by the exact value (or sometimes by the numerical approximation) and it is usually more useful compared to the absolute error.

3.3.2 Loss of significance (again)

The subtraction between two numbers that are close together (or between very large numbers) can lead to large errors. Consider for example a computer with numerical base $b = 10$ and with five-digit mantissa. Also consider the numbers

$$x = 0.543616731 \quad \text{and} \quad y = 0.543504564\ ,$$

and their difference

$$x - y = 0.000112167\ .$$

Because of the precision $t = 5$ using chopping we have

$$fl(x) = 0.54361 \quad \text{and} \quad fl(y) = 0.54350\ ,$$

and

$$fl(x) - fl(y) = 0.00011\ .$$

The relative error is then

$$\left|\frac{x - y - (fl(x) - fl(y))}{x - y}\right| = \left|\frac{0.000112167 - 0.00011}{0.000112167}\right| \approx 2 \times 10^{-2}\ .$$

This error is very large, especially if we consider that the numbers have 3 digits correct. This unexpected reduction of the significant digits in a computation is known as *loss of significance.*

3.3.3 Algorithms and stability

In mathematics and computer science an algorithm is a step-by-step procedure for calculations. More precisely, an algorithm is an effective method expressed as a finite list of well-defined instructions. Algorithms are characterized as *stable* or *unstable.* An algorithm is *stable* if small changes in the initial data lead to small changes in the final results. If an algorithm is not stable, then it is called *unstable.* Some algorithms are stable only for certain choices of data and are called *conditionally stable.*

Floating-point arithmetic will always induce small errors in computations and in data, thus stable algorithms are preferred. An unstable algorithm with these small errors will lead to untrustworthy results.

In the example of Section 3.1.1 we attempted to compute the limit

$$\lim_{x\to 0} \frac{1-\cos(x)}{x^2}\ ,$$

and we observed that the computation of the function $f(x)$ for x close to 0 lead to wrong results. The problem was caused by the subtraction of two almost equal numbers. Specifically, the difference $1 - \cos(x)$ in the formula of $f(x)$ for tiny values of x was approximated by 0.

We can overcome this difficulty by modifying the formula of the function to avoid the subtraction between 1 and $\cos(x)$. For example, using some trigonometry we can write

$$f(x) = \frac{1-\cos(x)}{x^2} = \frac{(1-\cos(x))(1+\cos(x))}{x^2(1+\cos(x))} = \frac{1-\cos^2(x)}{x^2(1+\cos(x))} = \frac{\sin^2(x)}{x^2(1+\cos(x))}\ .$$

The new formula for the function $f(x)$ can be used in our previous Python code to obtain a more accurate result.

```
f = lambda x:  np.sin(x)**2/(x**2*(1+np.cos(x)))
x = np.linspace(-5, 5, 100)
y = f(x)
xi = (1.e-5, 1.e-6, 1.e-7, 1.e-8, 1.e-9, 1.e-10,1.e-11)
for x in xi:
    print('f(',x,')','=',f(x))
```

```
f( 1e-05 ) = 0.4999999999958333
f( 1e-06 ) = 0.4999999999999583
f( 1e-07 ) = 0.4999999999999995
f( 1e-08 ) = 0.5
f( 1e-09 ) = 0.5
f( 1e-10 ) = 0.5
f( 1e-11 ) = 0.5
```

The technique we used to make the previous computation stable can be used also whenever we encounter differences of almost equal numbers. For example, if $x > y \geq 1$ and we want to compute the difference $\sqrt{x} - \sqrt{y}$ accurately, especially when we suspect that $x \approx y \approx 1$, we can compute the fraction

$$\sqrt{x} - \sqrt{y} = \frac{x - y}{\sqrt{x} + \sqrt{y}} ,$$

which can be more stable. This is because $x - y$ is larger than $\sqrt{x} - \sqrt{y}$. We will use such a trick in the following section to find stable approximations of π.

3.3.4 Approximations of π

We have seen that we can have accurate approximations of π by using the NumPy number `pi` or the formula $\pi = 4\arctan(1)$. In this section we present two different algorithms for the approximation of π one of which is Archimedes algorithm from his work 'On the Measurement of a Circle'. Archimedes algorithm is based on the approximation of a unit circle by inscribed regular polygons. Recall, that a unit circle has radius 1 and circumference 2π.

First, consider a regular polygon with 2^n sides inscribed in a unit circle. Each side of this polygon has length $s = 2\sin(\phi_n/2)$, where $\phi_n = \pi/2^{n-1}$ is the central angle. This can be seen in Figure 3.1. The circumference of such polygon is $C_n = 2^n s$ or more precisely

$$C_n = 2^{n+1} \sin \frac{\pi}{2^n}, \quad n = 1, 2, \dots .$$

We set $y_n = C_n/2$ to be half of the polygon circumference. This is

$$y_n = 2^n \sin \frac{\pi}{2^n}, \quad n = 1, 2, \dots . \tag{3.1}$$

Obviously, $y_1 = 2$. We also note that

$$y_{n+1} = 2^{n+1} \sin \frac{\pi}{2^{n+1}} .$$

Using the trigonometric identity $\sin x = [(1 - \cos(2x))/2]^{1/2}$, we get

$$y_{n+1} = 2^{n+1} \left[\frac{1 - \cos(\pi/2^n)}{2} \right]^{1/2} .$$

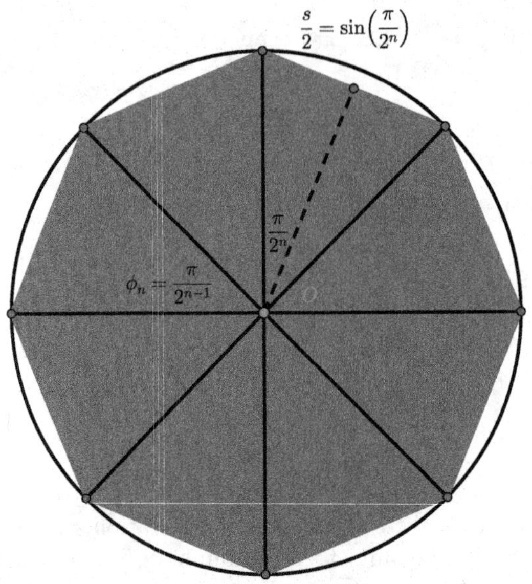

FIGURE 3.1
Approximation of a unit circle by regular polygons.

Using the identity $\sin^2 x + \cos^2 x = 1$ into the last identity leads to the formula

$$y_{n+1} = 2^{n+1}\left[\frac{1-\sqrt{1-\sin^2(\pi/2^n)}}{2}\right]^{1/2}.$$

Moreover, using the formula (3.1) for y_n we obtain

$$y_{n+1} = 2^{n+1}\sqrt{\frac{1}{2}\left(1-\sqrt{1-(2^{-n}y_n)^2}\right)},$$

which accompanied by the initial approximation $y_1 = 2$ forms a recursive sequence of approximations of π. This sequence approximates π since y_n is half of the perimeter of the polygon. The recursive sequence can be implemented in an algorithm such as Algorithm 8.

Algorithm 8 Unstable algorithm for the approximation of π

Set $y = 2.0$
for $n = 1, 2, \ldots, 30$ **do**
Compute

$$y = 2^{n+1}\sqrt{\frac{1}{2}\left(1-\sqrt{1-(2^{-n}y)^2}\right)}$$

end for
Return y as an approximation of π

The recursive sequence contains two subtractions indicating instabilities. Implementing Algorithm 8 our fears become reality after the 27th iteration.

```
yn = 2.0
for n in range(1,31):
    yn=2.0**(n+1) * np.sqrt(0.5*(1.0-np.sqrt(1.0-(2.0**(-n)*yn)**2)))
    print(n, yn)
```

```
1 2.8284271247461903
2 3.0614674589207187
3 3.121445152258053
4 3.1365484905459406
5 3.140331156954739
6 3.141277250932757
7 3.1415138011441455
8 3.1415729403678827
9 3.141587725279961
10 3.141591421504635
11 3.141592345611077
12 3.1415925765450043
13 3.1415926334632482
14 3.141592654807589
15 3.1415926453212153
16 3.1415926073757197
17 3.1415929109396727
18 3.141594125195191
19 3.1415965537048196
20 3.1415965537048196
21 3.1416742650217575
22 3.1418296818892015
23 3.142451272494134
24 3.142451272494134
25 3.1622776601683795
26 3.1622776601683795
27 3.4641016151377544
28 4.0
29 0.0
30 0.0
```

The sequence seems to approximate π up to y_{24}. Then the computation becomes unstable and returns 0 due to loss of significance.

To avoid the subtraction of almost equal numbers $a, b \geq 1$ we can use the identity $(a-b)(a+b) = a^2 - b^2$ with $a = 1$ and $b = \sqrt{1-(2^{-n}y_n)^2}$ to obtain

$$1 - \sqrt{1-(2^{-n}y_n)^2} = \frac{2^{-2n}y_n^2}{1+\sqrt{1-(2^{-n}y_n)^2}} .$$

This transforms the recursive sequence to

$$y_{n+1} = \sqrt{\frac{2}{1+\sqrt{1-(2^{-n}y_n)^2}}}\,y_n ,$$

where again $y_1 = 2$. Implementing the new recursive iteration we obtain a stable algorithm.

```
import numpy as np
yn = 2.0
for n in range(1,31):
    yn=np.sqrt(2.0/(1.0+np.sqrt(1.0-(2.0**(-n)*yn)**2)))*yn
    print(n, yn)
```

```
1 2.8284271247461903
2 3.0614674589207187
3 3.121445152258053
4 3.1365484905459398
5 3.1403311569547534
6 3.1412772509327733
7 3.1415138011443013
8 3.1415729403670913
9 3.1415877252771596
10 3.1415914215111997
11 3.1415923455701176
12 3.1415925765848725
13 3.141592634338563
14 3.141592648776985
15 3.1415926523865907
16 3.1415926532889924
17 3.141592653514593
18 3.141592653570993
19 3.1415926535850933
20 3.141592653588618
21 3.141592653589499
22 3.1415926535897194
23 3.1415926535897745
24 3.1415926535897882
25 3.141592653589792
26 3.1415926535897927
27 3.1415926535897927
28 3.1415926535897927
29 3.1415926535897927
30 3.1415926535897927
```

This new approximation converges to π and appears to be stable. The original Archimedes' algorithm for the approximation of π is stable and very elegant as well. Archimedes' algorithm consists of the following steps:

$$\begin{aligned} x_0 &= \sqrt{3}, & y_0 &= 2\ , \\ x_{n+1} &= x_n + y_n, & y_{n+1} &= \sqrt{1+(x_{n+1})^2} \quad \text{for } n = 0, 1, \cdots\ , \end{aligned}$$

where y_n converges to π as $n \to \infty$. Algorithm 9 presents Archimedes' algorithm.

In the previous implementations we computed 30 iterations. On the other hand, we can use different stopping criteria. If convergence is our goal, then we can perform as many iterations as it is required until the difference between two results is very small. The distance between the result y_{n+1} and y_n can be computed via the difference $e_n = |y_{n+1} - y_n|$. For a convergent sequence y_n this difference becomes small as n increases. To stop the execution of a code we choose a tolerance TOL (usually something like 10^{-10} depending

Algorithm 9 Archimedes algorithm for the approximation of π

Set $x = \sqrt{3}$ and $y = 2.0$
for $n = 1, \ldots, 30$ **do**
  Compute $x = x + y$ and $y = \sqrt{1 + x^2}$
end for
Return y as an approximation of π

on the problem) to compare with the difference e_n. Then we can execute the algorithm until $|y_{n+1} - y_n| < TOL$. When this difference becomes less than TOL indicates that the sequence is converging and it is unlikely to obtain better results with more iterations.

This stopping criterion though has a risk. If the sequence is not convergent, then the difference $|y_{n+1} - y_n|$ might never become less than TOL and our code will never stop. For this reason, we usually count the number of iterations using a counter, let's say *iter* and we never allow them to exceed a limit of $MAXIT$ iterations. Taking into account the new stopping criteria, we present Archimedes algorithm in Algorithm 10.

Algorithm 10 Archimedes algorithm for the approximation of π (Again)

Set the tolerances $TOL = 10^{-10}$ and $MAXIT = 100$
Set $x = \sqrt{3}$ and $y = 2.0$
Set $e = y$, and $iter = 0$
while $|e| \geq TOL$ and $iter < MAXIT$ **do**
  Set $e = y$
  Compute $x = x + y$ and $y = \sqrt{1 + x^2}$
  $iter = iter + 1$
  $e = e - y$
end while
Return y as an approximation of π

Note that the variable e takes initially the old value of y in the beginning of the `while` loop, and then we subtract from it the new value of y at the end of the loop. In the end we use absolute value to compute the distance between old and new approximations. We leave the implementation of Archimedes' algorithm as an exercise for the reader. We also recommend to test and change the given code as much as you can.

3.4 Further Reading

A classical and influential work in scientific computing and its pitfalls is a paper by Forsythe with title "Pitfalls in computation, or why a math book isn't enough", [42]. This also justifies why we present computer experiments side by side with proofs and theorems in this book. For a complete discussion of the theory of floating-point numbers, we refer to the classical works of Goldberg [46], Higham [64] and Wilkinson [141]. All classical, introductory books on numerical analysis and scientific computing contain extensive introductions to floating-point arithmetic and sources of errors. We indicatively refer to some classical books on numerical analysis [43, 25, 26, 27, 70]. An excelent book dedicated on scientific computations with finite precision arithmetic is [20]. Other modern textbooks on numerical analysis with similar material include but not limited to [57, 18, 22, 76, 89, 90, 108, 132].

Chapter Highlights

- Real numbers are represented in computer systems by floating-point numbers.
- Floating-point numbers are stored in memory in the form $\pm r \times 10^n$ where $r = 0.d_1 d_2 \cdots d_t$ is the mantissa, d_i the significant digits and t the precision.
- In scientific notation, a number $\pm r \times 10^n$ is written as `±r e n` where `e` represents the base 10. For example, `1.23456=0.123456e1`.
- Floating-point numbers consist of finite number of digits and are isolated in the sense that there are gaps between them.
- The distance between `1` and the next floating point number is called machine epsilon `EPS` and it is approximately `2.22e-16`. Usually, this is the minimum expected error in computations for scaled problems.
- Operations involving small numbers of magnitude less than the machine epsilon need extra care. Sometimes numbers smaller than the machine epsilon can be considered as approximations to 0.
- Any number in the interval between the smallest positive and the largest negative floating-point numbers is represented by `0.0` or `-0.0` depending if it is positive or negative. This problem is known as underflow.
- Any number larger than the largest possible floating-point number is represented by `inf` or `NaN`. This problem is known as overflow.
- The absolute error between f and its approximation $\bar{f}$ is the absolute distance $e = |f - \bar{f}|$.
- The relative error between f and its approximation $\bar{f}$ is its absolute error normalized by the exact value f, $r = |f - \bar{f}|/|f|$.
- The difference between two numbers of the same size can cause large errors. This phenomenon is known as catastrophic cancelation.
- To avoid catastrophic cancelation, we modify formulas and algorithms to avoid subtraction of two almost equal numbers.
- The results of a stable algorithm will not contain large amount of error given small errors in the input data. This is important because input data will always contain small errors due to finite precision arithmetic.
- Small changes in the input of unstable algorithms can result in large errors.
- We can improve unstable algorithms using algebraic or trigonometric identities.
- Other sources of error include the approximation of mathematical formulas. These errors are known as discretization errors.

Exercises

1. Construct representations of the following expressions that are stable when the computations are performed with finite precision arithmetic:

   (a) $1 - \cos(x)$, for $|x| \ll 1$.

   (b) e^{x-y}, x, y large numbers.

   (c) $\log(x) - \log(y)$, for $x, y > 0$.

   (d) $\sin(1 + x) - \sin(1)$, for $|x| \ll 1$.

   [The symbol $\ll$ means "much smaller than".]

2. Consider the functions

$$f(x) = \frac{1 - \cos(x)}{\sin^2(x)} \quad \text{and} \quad g(x) = \frac{1}{1 + \cos(x)} .$$

   (a) Show that $f(x) = g(x)$.

   (b) Show that

$$\lim_{x\to 0} f(x) = \lim_{x\to 0} g(x) = \frac{1}{2} .$$

   (c) Write a Python code to compute the values of $f(x)$ and $g(x)$ for

$$x = 10^0, 10^{-1}, 10^{-2}, \ldots, 10^{-11} .$$

   (d) Explain why the values are not the same.

3. Consider the quadratic equation

$$ax^2 + bx + c = 0 ,$$

   with $a = 1, b = 9^{12}$ and $c = -3$.

   (a) Show this quadratic equations has two roots and compute the roots using the formula

$$x_\pm = \frac{-b \pm \sqrt{b^2 - 4ac}}{2a} .$$

   (b) Observe none of the roots is equal to 0.

   (c) Write a Python code to compute the roots of this equation.

   (d) Explain why the x_+ computed by python is 0.

   (e) Show that we can rewrite the solution x_+ in the form

$$x_+ = \frac{-2c}{b + \sqrt{b^2 - 4ac}} .$$

   (f) Compute again the roots of the quadratic equation using the new formula and observe that the results are better. Explain why this is the case.

4. Using Python compute the absolute and relative errors between the exact value of x and its approximation $\tilde{x}$:

(a) $x = \pi$, $\tilde{x} = 355/113$.

(b) $x = \sqrt{2}$, $\tilde{x} = 1.41421$.

(c) $x = e$, $\tilde{x} = 2.718$.

(d) $x = \pi$, $\tilde{x} = 4\arctan(1)$.

(e) $x = \pi/4$, $\tilde{x} = 1 - 1/3 + 1/5 - 1/7 + 1/9$.

5. Consider the formula for the binomial coefficient

$$\binom{a}{b} = \frac{a!}{b!(a-b)!} ,$$

where the factorial $n! = 1 \cdot 2 \cdots n$.

(a) Write a Python function that returns the binomial coefficient of two integers a, b.

(b) Find the maximum value of a a you can use in your code without getting an overflow?

(c) Show that the binomial coefficient can be computed also using the formula

$$\binom{a}{b} = \frac{a}{b} \cdot \frac{a-1}{b-1} \cdots \frac{a-b+1}{1} .$$

(d) Write a new Python function to compute the binomial coefficient using the second formula.

(e) Find again the maximum value a you can use in your new function without getting an overflow.

6. Consider the function

$$f(x) = \frac{\ln(1-x)}{x} .$$

(a) Show that limit

$$\lim_{x \to 0} f(x) = 1 .$$

(b) Plot the function f in the interval $[-0.5, 0.5]$ using the module MatPlotLib.

(c) Magnify the plot around 0 using the interval $[-5\times 10^{-10}, 5\times 10^{-10}]$. Can you extract any safe conclusion related to the limit of the function f for $x \to 0$ using the specific magnification?

(d) Explain why there is such an inconsistency between the theory and the graph.

(e) Try to modify the function appropriately so as to make the computation of $f(0)$ stable.

7. Consider x and y two floating-point numbers with $x \approx y$.

(a) What is the relative error of the operation $x - y$.

(b) What formula would you use for the computation of the quantity $(x-y)^2$: $(x-y) \cdot (x-y)$ or $x^2 + y^2 - 2xy$?

(c) What formula would you use for the computation of the quantity $x^2 - y^2$: $(x-y) \cdot (x+y)$ or $x \cdot x - y \cdot y$?

(d) Compute the relative error of the expressions of the last two questions.

8. In Section 1.5 we introduced the Fibonacci sequence using a recursive function. An alternative formula for the Fibonacci sequence is

$$F(n) = \frac{1}{\sqrt{5}}\left[\left(\frac{1+\sqrt{5}}{2}\right)^n - \left(\frac{1-\sqrt{5}}{2}\right)^n\right], \tag{3.2}$$

where the fraction $(1+\sqrt{5})/2$ is known as the golden ratio.

(a) Write a function to compute any term of the Fibonacci sequence using the formula (3.2).

(b) Use your function to compute the term $F(100)$ and $F(1000)$.

(c) Compare the previous results with the results obtained with the `Fibonacci` function of Section 1.5.

(d) Describe the advantages and disadvantages of each method?

4

Calculus Facts

We have already used Python to solve calculus problems such as the estimation of limits of functions. The notions and theory of calculus are of fundamental importance for computational mathematics. Calculus provided with basic problems and tools for the development of numerical algorithms. For example, finding the value of a definite integral can be extremely difficult or even impossible task. Meanwhile tools like Taylor polynomials used extensively to analyze basic numerical methods. Calculus is interconnected with computational mathematics, and in this chapter, we review briefly some of its most important notions and theorems.

4.1 Sequences and Convergence

Part of this book deals with functions and numbers. Numbers in principle form sets (collections of numbers). If a number a is in a set A we write $a \in A$. On the other hand, if a number is not in A, then we write $a \notin A$. If a number belongs to two sets A and B at the same time, then we write $a \in A \cap B$. If a belongs to the set A or B or to both but we aren't sure, we write $a \in A \cup B$. The basic sets of numbers are namely the following:

- The set of natural numbers: $\mathbb{N} = \{0, 1, 2, 3, \dots\}$
- The set of integer numbers: $\mathbb{Z} = \{\cdots, -3, -2, -1, 0, 1, 2, 3, \dots\}$
- The set of rational numbers: $\mathbb{Q} = \{p/q : \ p, q \in \mathbb{Z}, \ q \neq 0\}$
- The set of irrational numbers: $\mathbb{I} = \{x : \ x \notin \mathbb{Q}\}$
- The set of real numbers: $\mathbb{R} = \mathbb{Q} \cup \mathbb{I}$

Usually the indices lie in the set of natural numbers $\mathbb{N}$ while the rest of the numbers, including floating point numbers, will be considered in $\mathbb{R}$ with the exception of complex numbers. Complex numbers are usually denoted by $z = a + ib$ where $a, b \in \mathbb{R}$ and i is the imaginary unit (with the property $\mathrm{i}^2 = -1$). If z is a complex number, then we write $z \in \mathbb{C}$. The complex conjugate of $z = a + ib$ will be denoted by $\bar{z} = a - ib$.

Sometimes, we will not consider a whole set A but a subset B of A. Such an inclusion is denoted as $B \subset A$ and is a set again. If $B \subset A$, then B does not necessarily contain all members of A. In some sense, B is included in A and perhaps is smaller. Useful examples of subsets of $\mathbb{R}$ are the intervals. For example, the interval $[a, b]$ contains all the real numbers x between a and b including a and b. Such an interval is called closed interval. In other words, all the real numbers x such that $a \leq x \leq b$ are included in the set $[a, b]$. Notice that we used square brackets since the endpoints a and b are included. If we don't want to include the endpoints of an interval, then we write (a, b), which means that if $x \in (a, b)$, then $a < x < b$, and the interval is called open. We can also have combinations of open and closed intervals, such as the interval $[a, b)$ with numbers x such that $a \leq x < b$.

DOI: 10.1201/9781003287292-4

Very often in computational mathematics we construct sequences of approximations a_n. A sequence is an enumerated collection of numbers $a_0, a_1, a_2, \dots$, and is denoted as $\{a_n\}_{n\in\mathbb{N}}$. The values of sequence may be real $a_n \in \mathbb{R}$, while the indices are natural numbers. Sometimes, in some computations of a_{n+1} we will use approximations of a_n. Methods that generate sequences of approximations are called iterative methods.

To give an example of an iterative method we will consider again the ancient method of Archimedes for the computation of π. Archimedes approximated π by the area of regular polygons inscribed in a unit circle (see also Section 3.3.4). The area of a regular polygon inscribed in a unit circle is

$$a_n = \frac{n}{2}\sin\left(\frac{\pi}{2n}\right) ,$$

where n is the number of equal sides of the regular polygon. As we increase the number of sides we get a larger polygon that covers more of the area of the unit circle. By taking $n = 3, 4, \dots$ we find approximations a_n of π, and we say that we generated a sequence of approximations. If we continue this process by increasing the index n for ever we finally construct the unit circle. This infinite process describes the notion of *convergence*.

Suppose that we have a method to generate a sequence of numbers $\{a_n\}_{n\in\mathbb{N}}$ that converges to a. This means that the values of a_n approximate a for large values of n, and if we take the index n to be huge (close to infinity) then we practically get the exact value of a. We write this as

$$\lim_{n\to\infty} a_n = a ,$$

and we say that the sequence a_n converges to a. Sometimes we just write $a_n \to a$ as $n \to \infty$.

In practice, we can compute only a finite number of iterations/approximations. The sequence $\{a_n\}_{n\in\mathbb{N}}$ approximates the exact solution, let's say a, and for some $n < \infty$ the distance $|a_n - a|$ defines the error in the approximation. Specifically, we defined the absolute error

$$|e_n| = |a_n - a| ,$$

and is the distance between the exact value a and the approximation a_n. Since $a_n \to a$, then the distance between a_n and a is decreasing until it is eliminated at infinity. In the end, we have that $|a_n - a| \to 0$ as $n \to \infty$, and therefore $|e_n| \to 0$. Ideally, in our iterative methods we want the error to go to zero as n grows so we can trust the approximations.

As an example, consider the sequence of numbers

$$a_n = 1 + \frac{1}{n} .$$

When $n \to \infty$ we have that $1/n \to 0$ becomes very small and thus $a_n \to 1$ or else $|a_n - 1| \to 0$ as $n \to \infty$. Taking $n = 10$ we observe that $a_{10} = 1 + 10^{-1} = 1.1 \approx 1$, while the error $|e_{10}| = |1.1 - 1| = 0.1$. Taking n even larger we can have smaller error $|e_n|$.

The formal definition of convergence deals with the error $|e_n|$. In formal mathematical language we say that for (any) small number $\varepsilon > 0$, there is an index (that might depend on ε) say $N = N(\varepsilon)$, such that the error $|a_n - a| < \varepsilon$ for all values of $n > N$. This means that we can make the error as small as we want (of order ε) by choosing appropriately large values of the index $n > N$.

☞ Usually iterative methods generate a sequence a_k, $k = 0, 1, 2, \ldots$ of approximations of the exact solution a. We prefer this sequence to converge to the exact solution. Then for any reasonably large value of k we will have $a_k \approx a$. We will use similar notation to refer to sequences of vectors. We will denote a sequence of n-dimensional vector by $\boldsymbol{a}^{(k)} = (a_1^{(k)}, a_2^{(k)}, \ldots, a_n^{(k)})^T$, with $k = 0, 1, 2, \ldots$. In this notation, the bold letter indicates a vector, while the superscript index indicates the k-th term of an infinite sequence of vectors.

4.2 Continuous Functions

In applications we tend to use functions to describe quantities and unknowns. Functions are mappings from one set to another. A function of a single variable $f(x)$ takes values x from one set A known to as the domain of f, and returns values $y = f(x)$ in a set B, which is called range of f. The set A can be an interval or a combination of intervals. To emphasize that f is a map from A to B we write $f : A \to B$. For example, the function $f(x) = \log(x)$ is defined for $x > 0$. So the domain of $\log(x)$ is the interval $(0, +\infty)$. The range of the logarithm is the whole real line $\mathbb{R}$. Another interesting example with a very similar domain is the function $f(x) = \sqrt{x}$. The domain of the $\sqrt{x}$ is the set $[0, +\infty)$ as it is defined for $x = 0$ but not for negative numbers (and we exclude complex numbers). Its range, though, is the same set as its domain $[0, +\infty)$.

A root of a function $f(x)$ is a number x^* in the domain of f such that $f(x^*) = 0$. For example, the function $f(x) = x^2 - 1$ has two roots $x^* = 1$ and $x^* = -1$. Assume that we have an iterative method that generates a sequence x_n of approximations of the root $x^* = 1$ and that this sequence eventually converges to x^*. This means that $x_n \to x^*$. Since $f(x^*) = 0$ we prefer for the approximations $f(x_n) \approx 0$, or in other words $f(x_n) \to f(x^*) = 0$ as $x_n \to x^*$. This property is a consequence of the continuity of the function and generalizes the notion of limit from sequences to functions.

We generalize the notion of limit to a variable x. We say that x approaches $x_0 \in \mathbb{R}$ if the distance between x and x_0 becomes very small or is eliminated, and we write $x \to x_0$. There are also two ways to approach x_0: From the left $x \to x_0-$, which means the $x < x_0$ or from the right $x \to x_0+$, which means that $x > x_0$. We can consider a general approach when the values x can approach x_0 from both sides. This is the meaning of $x \to x_0$.

Consider a function $f(x)$ with domain A and a value $x_0 \in A$. The function $f(x)$ is called continuous at x_0 if

$$\lim_{x \to x_0} f(x) = f(x_0) ,$$

which means that if we take values of x close to x_0, then the values of $f(x)$ are close to the values of $f(x_0)$. This guarantees that the graph of f is a continuous line and there is no discontinuity at x_0. We can think of discontinuities as jumps in the graph of a discontinuous function. The function

$$f(x) = \begin{cases} 1, & x = 0 \\ 0, & x \neq 0 \end{cases} ,$$

has a jump at $x = 0$. In this trivial case $f(0) = 1$ but for $x \neq 0$, $f(x) = 0$ and so no matter how we approach 0 we will always have $f(x) = 0$ but not the value $1 = f(0)$. This can be expressed as

$$\lim_{x \to 0} f(x) = \lim_{x \to 0} 0 = 0 \neq 1 = f(0) .$$

A function which is continuous at every point in an interval (a, b) is said to be continuous in (a, b). All continuous functions form a set. The set of continuous functions in an interval I is denoted by $C(I)$. For example, the set of continuous functions on (a, b) is denoted by $C(a, b)$. If the interval I is a closed interval $[a, b]$ then we will write $C[a, b]$.

Going back to our example with the root x^* of $f(x)$ we mention an important consequence of continuity that serves the purposes of computational mathematics. This is that instead of x we can choose a sequence x_n that converges to x^*, and if f is continuous, then $\lim_{n\to\infty} f(x_n) = f(x^*)$, which means that $f(x_n) \approx f(x^*) = 0$ for large values of n.

The formal definition of a continuous function is based on the notion of limit (and approximation). Specifically, a function $f(x)$ is continuous at a point x_0 if for every number $\varepsilon > 0$, we can find a number (that might depend on ε) $\delta > 0$ such that for all values of x δ-close to x_0, i.e. $|x - x_0| < \delta$, the values of $f(x)$ are ε-close to $f(x_0)$, i.e. $|f(x) - f(x_0)| < \varepsilon$.

This formal definition of continuity guarantees that the values of the function $f(x)$ become arbitrarily close to $f(x_0)$ for values of x very close to x_0, but says nothing on how much the values of $f(x)$ vary.

A Lipschitz continuous function is a continuous function that has a specific way to vary: A function $f(x)$ is *Lipschitz continuous* if for all x, y in the domain of f, there exists a constant $L > 0$ such that

$$|f(x) - f(y)| < L|x - y| \ .$$

This means that the distance between two values $f(x)$ and $f(y)$ cannot be very different than the difference between x and y.

Obviously, a Lipschitz continuous function is a continuous function, which in addition cannot vary arbitrarily. If in addition $0 < L < 1$, then the function is called *contraction* and has more favorable properties. For example, we can find x^* such that $x^* = f(x^*)$. The point x^* is called fixed point of f and as we will see later it can be very important for numerical computations.

> ☞ Lipschitz continuous functions are continuous functions that do not vary rapidly. The smoothness of their variations depends on the Lipschitz constant $L > 0$. If $L < 1$ this means that the function varies very smoothly and is called contraction.

To see why the notion of Lipschitz continuity is related with the variations of the function f consider the ratio

$$\frac{|f(x) - f(y)|}{|x - y|} < L \ .$$

Taking y very close to x, we practically set the slope of the secant line to the graph of f at the point x to be less than L (see Figure 4.1).

As an example of Lipschitz continuous function consider the function $f(x) = x^2$ in the interval $I = [-1, 1]$. For $x, y \in [-1, 1]$ we have $|x + y| \leq 2$ since in the worst case scenario $x = y = 1$. Therefore,

$$\begin{aligned} |f(x) - f(y)| &= |x^2 - y^2| \\ &= |(x - y)(x + y)| \\ &\leq 2|x - y| \ . \end{aligned}$$

Thus, $f(x) = x^2$ is Lipschitz continuous function in $[-1, 1]$ with Lipschitz constant $L = 2$[1]. The Lipschitz constant is not unique since one can take any constant $L \geq 2$ as a Lipschitz

[1]Functions that are Lipschitz continuous only in a subinterval of its domain are called locally Lipschitz continuous functions.

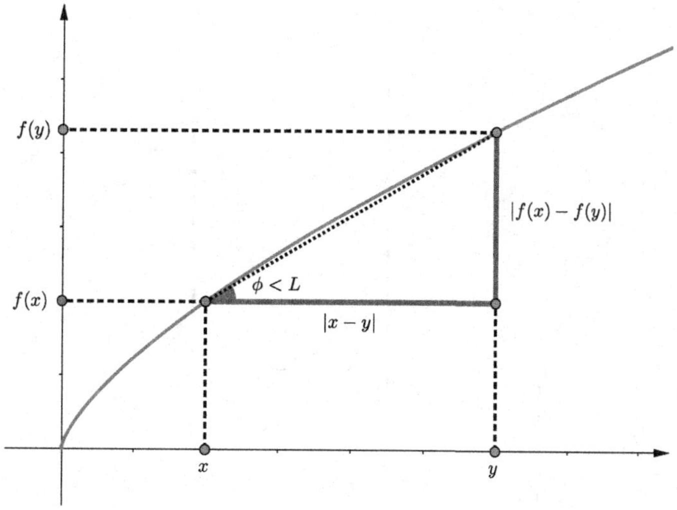

FIGURE 4.1
Slope of a secant of a Lipschitz continuous function.

constant. Also the bounds of the Lipschitz constant depends on the interval I. Finally we note that the particular function is not a contraction in I.

4.2.1 Bolzano's intermediate value theorem

Before start solving a problem it is very useful to know that the problem has a solution. In computational mathematics we wouldn't care about the exact solution of a problem since we compute approximations. We do care enough though to know that there is a solution and ideally that the solution is unique. Let's say that we want to compute the root of the function $f(x) = x^2+1$. Obviously there is no real root since the function f is always positive $f(x) \geq 1 > 0$ for all values of $x \in \mathbb{R}$. So there is no point to search for real solutions even numerically.

Another observation to root finding methods is that if $f(x)$ is continuous and changes sign, then it must have a root. Of course there are cases like the function $f(x) = x^2$ where it is positive and never becomes negative. Nevertheless, we will consider for the moment only functions that change sign. A necessary condition for a function that changes sign to have a root, is the function to be continuous. Otherwise the function can move from negative values to positive by jumping the x-axis without having a root. Bolzano's intermediate value theorem describes this situation.

Theorem 4.1 (Bolzano's theorem). *If a function $f(x)$ satisfies the following conditions:*

- *f is continuous in $[a, b]$*
- *$f(a) \cdot f(b) < 0$*

then, there exists at least one solution $x^* \in (a, b)$ of the equation $f(x) = 0$.

The condition $f(a) \cdot f(b) < 0$ means that $\text{sign}(f(a)) \neq \text{sign}(f(b))$, and thus the graph of the continuous function $f(x)$ crosses the x-axis necessarily at a point between a and b. The specific situation is illustrated in Figure 4.2.

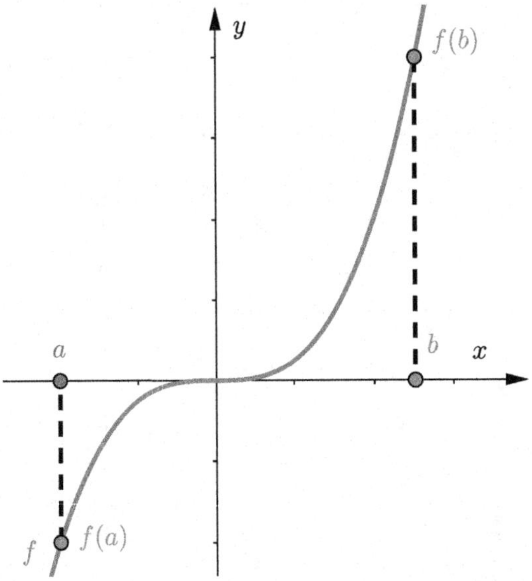

FIGURE 4.2
Illustration of Bolzano's theorem.

It is worth mentioning that Bolzano's theorem requires f to be continuous in a closed interval $[a,b]$ while the root is in the open interval (a,b). The last observation is a consequence of the fact that $f(a)\cdot f(b)<0$ strictly. If one of the values $f(a)$ or $f(b)$ is 0, then we don't need to search for a root as the root would be either a or b and $f(a)\cdot f(b)=0$.

Bolzano's theorem is a special case of the so-called intermediate value theorem for continuous functions. The intermediate value theorem states that if a function $f(x)$ is continuous in the interval $[a,b]$ and s is a value between $f(a)$ and $f(b)$, then there is a value $x_0\in(a,b)$ such that $f(x_0)=s$. Bolzano's theorem is actually a special case of the intermediate value theorem with $s=0$.

As an example consider the function $f(x)=x^3$. The particular function is continuous in $[-1,1]$. Moreover $f(-1)=-1$ and $f(1)=1$. Therefore, $f(-1)\cdot f(1)<0$ and thus there is at least one root $x^*\in[-1,1]$. The obvious root is $x^*=0$.

4.3 Differentiation

Continuity is an important property of functions, but quite often we require the functions to be smooth. By saying smooth we mean that a function will be continuous while its graphs has no corners. If the graph of f has no corners, then we will be able to draw a unique tangent line at any point of its graph. Smoothness is equivalent to the existence of derivatives.

Let f be a function defined in an open interval (a,b) containing a point x_0. The function f is called differentiable at x_0 if the limit

$$f'(x_0)=\lim_{x\to x_0}\frac{f(x)-f(x_0)}{x-x_0}\ ,$$

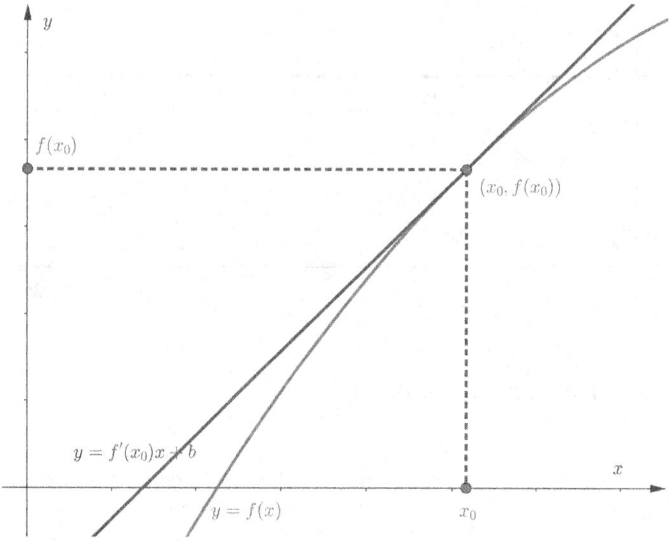

FIGURE 4.3
The tangent line of a differentiable function.

exists. The number $f'(x_0)$ is called the derivative of f at x_0. A function that has derivative at each point in an interval (a, b) is called differentiable on (a, b). Differentiable functions in the interval I form the set $C^1(I)$. If a function is differentiable in (a, b) we write $f \in C^1(a, b)$.

If a function is differentiable at x_0, then the function is smooth enough to define a unique tangent line passing through the point $(x_0, f(x_0))$. The derivative of a function $f'(x_0)$ is the slope of the tangent line at x_0. Therefore, the tangent line passing through the point $(x_0, f(x_0))$ can be described by the formula $y = f'(x_0)x + b$, where b is the point of y-axis that the line is passing through (see Figure 4.3).

The tangent line in Figure 4.3 coincides with the function f at $(x_0, f(x_0))$. Around that point $f(x)$ is very close to the tangent line. This gives us an idea of approximation. In particular, we could use this tangent line to approximate $f(x)$ around that point. As we go away from that point, this approximation may not be as good, but the limit in the definition of the derivative insures a good approximation at least close to $(x_0, f(x_0))$.

Higher order derivatives of a function f such as the second derivative $f''(x)$ are defined as the derivatives of the derivatives. This means, for example, that a twice differentiable function has a smooth derivative as well. If a function is twice differentiable in the interval (a, b), then we write $f \in C^2(a, b)$. The more times a function is differentiable the smoother it is. Ideally, we would prefer the functions to have derivatives of all orders. If a function f has the derivatives of all orders defined in the interval (a, b), then we write $f \in C^\infty(a, b)$. In computational mathematics, when we try to solve problems, we assume certain smoothness that will allow us to find unique solutions and estimate approximation errors. However, we can use our methods for less smooth functions but with no guarantee of the quality of the results.

Derivatives are also important for other reasons. Many scientific and engineering problems are related with derivatives. The reason is that the practical meaning of a derivative is the same with that of the rate of change. For example, in physics the acceleration $a(t)$ of an object is the rate of change of its velocity $v(t)$, and mathematically speaking $a(t) = v'(t)$. Here, the independent variable t represents time. Sometimes, in order to show clearly the variable in which we differentiate, we write the derivative using the Leibniz notation $v'(t) = \frac{d}{dt}v(t)$.

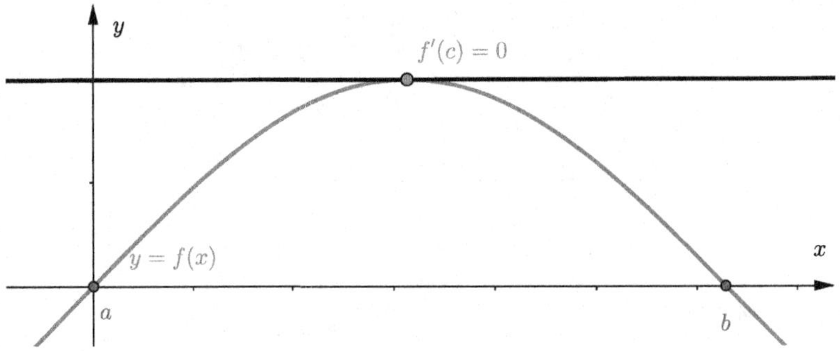

FIGURE 4.4
The tangent line at a maximum point of a function.

Positive acceleration implies increase in speed while negative acceleration implies decrease in speed. In general if $f(x)$ is differentiable and $f'(x) > 0$, then $f(x)$ is increasing (i.e. if $x_1 < x_2$ then $f(x_1) < f(x_2)$) while if $f'(x) < 0$, then $f(x)$ is decreasing (i.e. if $x_1 < x_2$ then $f(x_1) > f(x_2)$).

4.3.1 Important theorems from differential calculus

Optimization is a branch of computational mathematics that deals mainly with the computations of extreme values (extrema) of functions. Extrema of a function $f(x)$ are, for example, the maximum and the minimum values of $f(x)$. A (local) maximum of a function is a point $(c, f(c))$ such that $f(x) \leq f(c)$ for all values of x in a neighborhood of c. A (local) minimum of a function is a point $(c, f(c))$ such that $f(x) \geq f(c)$ for every value x in a neighborhood of c. If $f(x)$ is differentiable, then the derivative is zero $f'(c) = 0$ at the point c of extremum (maximum or minimum). Zero derivative indicates that the tangent line is parallel to the x-axis as its slope is zero. In Figure 4.4, we show the geometric interpretation of this fact for a point of maximum at $(c, f(c))$.

A very useful theorem for theoretical and practical situations is Rolle's theorem. Rolle's theorem guarantees the existence of a maximum or minimum for a smooth function.

Theorem 4.2 (Rolle's theorem). *Suppose that f is continuous in $[a, b]$ and differentiable on (a, b). If $f(a) = f(b)$, then there exists a number $c \in (a, b)$ such that $f'(c) = 0$.*

Note that Rolle's theorem doesn't specify if the point $(c, f(c))$ is a maximum or minimum. It is only guaranteed an extremum point such that $f(x) \leq f(c)$ or $f(x) \geq f(c)$ locally around the point c. The situation of Rolle's theorem is depicted in Figure 4.4.

Another important theorem is the Mean Value Theorem of differential calculus:

Theorem 4.3 (Mean Value Theorem). *If f is continuous in $[a, b]$ and differentiable in (a, b), then there exists a number $c \in (a, b)$ such that*

$$f'(c) = \frac{f(b) - f(a)}{b - a} .$$

The mean value theorem of differential calculus can be viewed in two different ways, an algebraic and a geometric: It relates the difference quotient $(f(b) - f(a))/(b - a)$ with the derivative of a function, and on the other hand states that the slope of the secant passing through the points $(a, f(a))$ and $(b, f(b))$ is the same as the slope of some tangent line to the graph of the function $f(x)$. This geometric interpretation is presented in Figure 4.5.

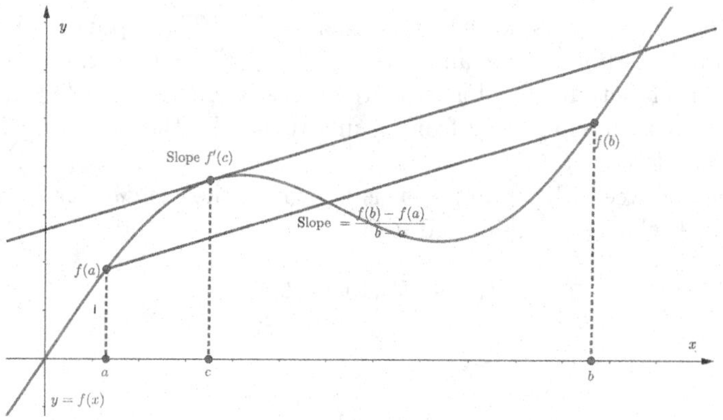

FIGURE 4.5
Geometric interpretation of the mean value theorem of differential calculus.

4.3.2 Taylor polynomials

Smooth functions in $C^\infty(a,b)$ can be represented by Taylor series. Specifically, if $f(x)$ is infinite many times continuously differentiable around a point $x=c$ in the interval (a,b), then the Taylor series of $f(x)$ around $x=c$ is the infinite sum

$$f(x) = f(c) + f'(c)(x-c) + \frac{f''(c)}{2}(x-c)^2 + \cdots ,$$

which can be written in compact form as

$$f(x) = \sum_{k=0}^{\infty} \frac{f^{(k)}(c)}{k!}(x-c)^k ,$$

where $f^{(k)}$ denotes the k-th derivative of f, and $k! = 1 \cdot 2 \cdots k$.

Some commonly used Taylor series (sometimes also referred to as Taylor expansions) for $c=0$ are the following:

$$e^x = 1 + x + \frac{x^2}{2!} + \frac{x^3}{3!} + \cdots = \sum_{n=0}^{\infty} \frac{x^n}{n!} ,$$

$$\sin(x) = x - \frac{x^3}{3!} + \frac{x^5}{5!} - \cdots = \sum_{n=0}^{\infty} \frac{(-1)^n}{(2n+1)!} x^{2n+1} ,$$

$$\cos(x) = 1 - \frac{x^2}{2!} + \frac{x^4}{4!} - \cdots = \sum_{n=0}^{\infty} \frac{(-1)^n}{(2n)!} x^{2n} .$$

Since computers can only compute finite sums, we may wish to approximate smooth functions using polynomials. Polynomials are also easy to integrate and differentiate. One way to find good approximations of functions is by using Taylor's polynomials. Consider a smooth function $f \in C^{n+1}[a,b]$ and let $c \in [a,b]$, then the Taylor polynomial of degree n is the finite sum

$$P_n(x) = f(c) + f'(c)(x-c) + \frac{f''(c)}{2!}(x-c)^2 + \cdots + \frac{f^{(n)}(c)}{n!}(x-c)^n . \tag{4.1}$$

Obviously, this finite sum is not exactly equal to $f(x)$. These polynomials coincide with the function $f(x)$ at the point $x = c$ and it is $P_n(c) = f(c)$. Moreover, the polynomial $P_n(x)$ is very close to the function $f(x)$ for x near c. This means that $P_n(x)$ approximates the function $f(x)$. As x is moving away from c, unfortunately, the polynomial $P_n(x)$ will not necessarily remain close to $f(x)$.

The difference between $P_n(x)$ and $f(x)$ is known as the *residual* or *truncation error* or *remainder* of the Taylor polynomial and it is

$$f(x) - P_n(x) = R_n(x) ,$$

with

$$R_n(x) = \frac{f^{(n+1)}(\xi)}{(n+1)!}(x-c)^{n+1} ,$$

for some $\xi \in [a, b]$.

Equivalently, we can write the function

$$f(x) = P_n(x) + R_n(x) ,$$

where $P_n(x)$ is the truncated Taylor series

$$P_n(x) = \sum_{k=0}^{n} \frac{f^{(k)}(c)}{k!}(x-c)^k ,$$

and $R_n(x)$ the residual of the truncation. The residual is expected to be small for x near c.

Here we present two examples of Taylor polynomials:

- Taylor's approximation for $n = 1$ is $f(x) \approx f(c) + f'(c)(x - c)$.
- Taylor's approximation for $n = 2$ is $f(x) \approx f(c) + f'(c)(x - c) + \frac{f''(c)}{2!}(x - c)$.

In some applications we need to approximate a function f around a point x but not far away from x. For example, assume that we want to compute the polynomial at the point $x + h$ where h is very small ($h \ll 1$). In these cases we consider the Taylor polynomial (4.1) with $x + h$ instead of x and x instead of c so that

$$f(x+h) = f(x) + f'(x)h + \frac{f''(x)}{2!}h^2 + \frac{f'''(x)}{3!}h^3 + \cdots + \frac{f^{(n)}(x)}{n!}h^n + R_n(\xi) ,$$

where

$$R_n(\xi) = \frac{f^{(n+1)}(\xi)}{(n+1)!}h^{n+1} .$$

Since we assumed that h is small, if in addition the $(n + 1)$-th derivative of f is bounded by a constant for all values of x around c, i.e. $|f^{n+1}(x)| \leq M$, then we have

$$|R_n(x)| \leq Ch^{n+1} ,$$

where $C = M/(n + 1)!$.

Note that this formula reveals information about the approximation error. If for example the discretization step is h, then the error of the approximation of f by P_n is of order h^{n+1}. Therefore, if $h \ll 1$ and M small enough for sufficiently large n, then we expect the error to be small.

4.3.3 The big-O notation

The truncation error of the Taylor polynomial of degree n in the previous section was $|R_n(x)| \leq Ch^{n+1}$ where C is a constant that depends on $f(x)$. In order to emphasize the order of this residual, which is $n+1$ and not the fact that it can be less than h^{n+1} we often use the big-O notation by writing $R_n(x) = O(h^{n+1})$. In general we write

$$f(x) = O(g(x)) \quad \text{when} \quad x \in I$$

if and only if

$$|f(x)| \leq M|g(x)| \quad \text{when} \quad x \in I \quad \text{where} \quad M > 0,$$

and we say that f is of order g.

In practice we use Big-O notation to say something about the size of the terms we discard or an error.

4.4 Integration

Assume that we have an object located at a point $x_0 = a$ on the x-axis. This object is moving horizontally due to the action of a force $f(x)$. The final destination of this object is a point $x_N = b$. Between the points x_0 and x_N, there are other points $x_1, \ldots, x_{N-1}$ such that

$$x_0 < x_1 < \cdots < x_{N-1} < x_N \ .$$

We know from classical physics that the work W produced by constant force $f(x)$ from point x_0 to point x_1 is the product of the force times the length of the interval $[x_0, x_1]$. This is $W_0 = f(x)\cdot(x_1 - x_0)$ for $x \in [x_0, x_1]$. Assume that the force changes to a different constant when the object reaches the point x_1 and remains constant until the object reaches x_2. The new work done by the force is now $W_1 = f(x) \cdot (x_2 - x_1)$ for $x \in [x_1, x_2]$. This process is repeated until the object reaches its final destination $x_N = b$. So the force is constant on every interval (x_{i-1}, x_i) for $i = 1, 2, \ldots, N$, and the work produced by the force f in each of these intervals is $W_i = f(x_{i-1}) \cdot \Delta x_i$ with $\Delta x_i = x_i - x_{i-1}$ for $i = 1, 2, \ldots, N$. The total work produced by the force is the sum

$$W = \sum_{i=1}^{N} W_i = \sum_{i=1}^{N} f(x_{i-1})\Delta x_i \ .$$

In Figure 4.6, the y-axis represents the force $f(x)$, while x-axis the displacement. The graph of the function $y = f(x)$ is a piecewise constant line, and the work produced in each interval (x_{i-1}, x_i) is equal to the area of the rectangle formed by the graph of f in the interval (x_{i-1}, x_i) and the interval (x_{i-1}, x_i). Since the force is constant in every interval (x_{i-1}, x_i) we can take any value $z_i \in (x_{i-1}, x_i)$ and write the total work in the form

$$W = \sum_{i=1}^{N} f(z_i)\Delta x_i \ .$$

If the force changes continuously, this means that the lengths of the intervals must tend to 0, and mathematically speaking we need to take $\Delta x_i \to 0$ in the previous sum. This limit is the integral

$$W = \lim_{\Delta x_i \to 0} \sum_{i=1}^{N} f(z_i)\Delta x_i = \int_a^b f(x)\, dx \ .$$

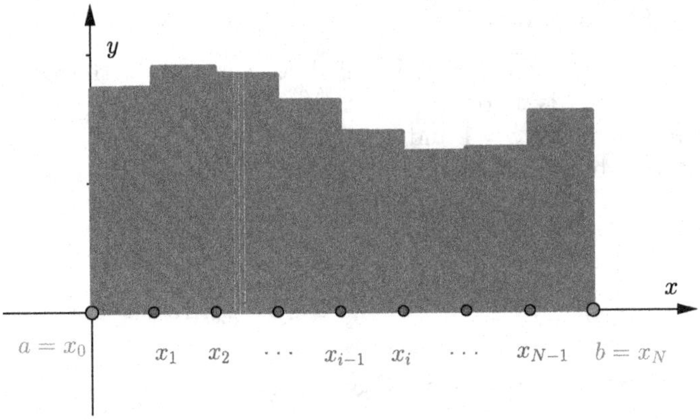

FIGURE 4.6
The work done by a force constant in each interval (x_{i-1}, x_i).

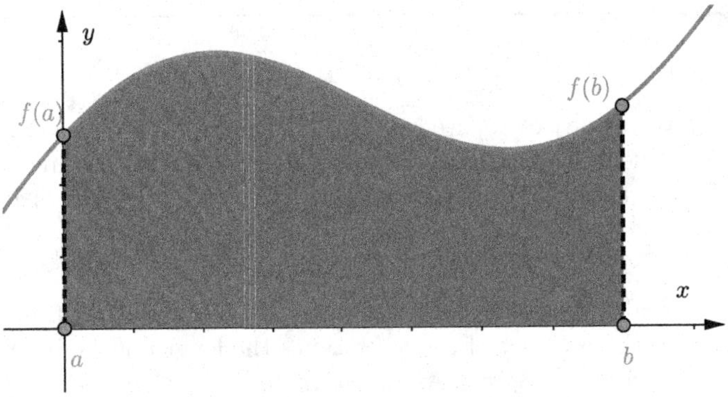

FIGURE 4.7
Integral as the area below a function $f(x)$.

Since each rectangle with sides $f(z_i)$ and Δx_i has area equal to the work W_i, we conclude that the total work done by $f(x)$ is the area described by the function $f(x)$, the x-axis and the lines $x = a$ and $x = b$ as shown in Figure 4.7.

The *Riemann integral* of a function f defined on the interval $[a, b]$, if it exists, is the limit

$$\int_a^b f(x)\,dx = \lim_{\Delta x_i \to 0} \sum_{i=1}^{N} f(z_i)\Delta x_i\ ,$$

where the numbers $x_0, x_1, \ldots, x_N$ satisfy $a = x_0 \leq x_1 \leq \cdots \leq x_N = b$, with $\Delta x_i = x_i - x_{i-1}$ for each $i = 1, 2, \ldots, N$, and z_i is arbitrarily chosen in the interval $[x_{i-1}, x_i]$. A function with a finite integral is called *integrable*. Note that all continuous functions are integrable, which means that their integral is defined and is finite. The function $f(x)$ inside an integral is called *integrant.*

4.4.1 Approximating integrals

Looking at the example with the force in Figure 4.6, we can derive easily a simple way to approximate a definite integral over an interval $[a, b]$. What we need to do first is create a grid $a = x_0 < x_1 < \cdots < x_{N-1} < x_N = b$. As a second step, we approximate the integral of function $f(x)$ in each interval $I_i = (x_{i-1}, x_i)$. In order to approximate an integral in I_i we approximate the function $f(x)$ by a constant line over the interval I_i. This gives a reasonable approximation as long as we take the length of the interval Δx_i very small. For example, we can take $f(x) \approx f(z_i)$ for $x \in (x_{i-1}, x_i)$, where z_i can be taken to be an endpoint of I_i, for example $z_i = x_i$ or even the midpoint $z_i = (x_i + x_{i-1})/2$. The last choice will lead to an approximation formula known as the *midpoint rule*. Because the origin of numerical methods for the approximation of integrals is in the approximation of their area by rectangles, we call these numerical methods *quadrature rules*. Note also that the grid $a = x_0 < x_1 < \cdots < x_{N-1} < x_N = b$ can be uniform in the sense that all the points are *equidistributed* in the interval $[a, b]$ and $\Delta x = x_i - x_{i-1} = (b - a)/N$, for all $i = 1, 2, \ldots, N$.

As an example of the midpoint rule we consider the integral

$$I = \int_0^1 x\, dx = \frac{1}{2}\ .$$

In order to approximate this integral, consider a uniform grid with 5 nodes

$$\{x_0, x_1, x_2, x_3, x_4\} = \{0, 0.25, 0.5, 0.75, 1\}\ ,$$

and step $\Delta x = 0.25$. The midpoints are the points

$$\{z_1, z_2, z_3, z_4\} = \{0.125, 0.375, 0.625, 0.875\}\ .$$

Then the integral can be approximated by

$$\begin{aligned} I &\approx \sum_{i=1}^{4} f(z_i)\Delta x = \Delta x \cdot (f(0.125) + f(0.375) + f(0.625) + f(0.875)) \\ &= 0.25 \cdot (0.125 + 0.375 + 0.625 + 0.875) = 0.5\ . \end{aligned}$$

We observe that although we approximated the integral using such a simple method, the result was equal to the exact value of the integral. This can happen when we approximate integrals of polynomials but this is not true in general. We will study such methods in detail in Chapter 7.

4.4.2 Important theorems of integral calculus

A fundamental property of integrals is the following: We can split an integral over an interval $[a, b]$ into two or more integrals on subintervals of $[a, b]$ and then add the values to calculate the total integral. For example, if we want to compute the integral on the interval $[a, b]$, we take a value $c \in (a, b)$ and we split the interval into two intervals $[a, b] = [a, c] \cup [c, b]$. Then,

$$\int_a^b f(x)\, dx = \int_a^c f(x)\, dx + \int_c^b f(x)\, dx\ .$$

This property will be very helpful in increasing the accuracy of our computations later in Chapter 7.

For theoretical purposes we state here some very important theorems of integral calculus.

Theorem 4.4 (Fundamental theorem of calculus). *If $f'(x)$ is continuous on $[a, b]$, then*

$$\int_a^b f'(x)\, dx = f(b) - f(a)\ .$$

The fundamental theorem of calculus is the main tool we use to compute the value of integrals when we know the derivative of the integrant. A direct consequence of the fundamental theorem of calculus is the formula of integration by parts, which for any $f, g \in C^1(a,b)$ can take the form

$$\int_a^b f(x)g'(x)\,dx = f(b)g(b) - f(a)g(a) - \int_a^b f'(x)g(x)\,dx\ .$$

Integrals with variable limits are differentiable functions. It is easy to compute their derivatives using the following theorem:

Theorem 4.5. *If $f \in C[a,b]$, then the function*

$$g(x) = \int_a^x f(t)\,dt\ ,$$

for $x \in [a,b]$ is a continuous function on $[a,b]$, differentiable on (a,b) and

$$g'(x) = f(x)\ .$$

Finally, let's look at the Mean Value Theorem of integral calculus which again is useful for theoretical calculations.

Theorem 4.6 (Mean value theorem of integral calculus). *Let $f \in C[a,b]$, then there exists $c \in (a,b)$ such that the* average value *of $f(x)$ in the interval $[a,b]$ is*

$$f(c) = \frac{1}{b-a}\int_a^b f(x)\,dx\ .$$

4.4.3 The remainder in Taylor polynomials

We have already seen that for a smooth function $f(x)$ and for x near a value c

$$f(x) = P_n(x) + R_n(x)\ ,$$

where $P_n(x)$ is the Taylor polynomial of degree n

$$P_n(x) = \sum_{i=0}^{n} \frac{f^{(n)}(c)}{n!}(x-c)^n\ ,$$

and $R_n(x)$ the remainder to be discarded

$$R_n(x) = \frac{f^{(n+1)}(\xi)}{(n+1)!}(x-c)^{n+1} \quad \text{for some } \xi \in [a,b]\ .$$

The remainder can be computed alternatively with the help of integrals. For a value x close to c, we can write the functions $f(x)$ as

$$f(x) = f(c) + \int_c^x f'(t)\,dt\ .$$

Integrating by parts yields

$$\begin{aligned} f(x) = f(c) + \int_c^x f'(t)\, dt &= f(c) + [(t-x)f'(t)]_{t=c}^{x} - \int_c^x (t-x)f''(t)\, dt \\ &= f(c) + (x-c)f'(c) - \int_c^x (t-x)f''(t)\, dt\ . \end{aligned}$$

In this way, we obtained the first order Taylor polynomial

$$P_1(x) = f(c) + (x-c)f'(c)\ ,$$

and the remainder

$$R_1(x) = -\int_c^x (t-x)f''(t)\, dt\ .$$

Repeating the same process we can get the second order Taylor polynomial

$$\begin{aligned} f(x) &= f(c) + (x-c)f'(c) - \int_c^x (t-x)f''(t)\, dt \\ &= f(c) - \left[\frac{(t-x)^2}{2}f''(t)\right]_{t=c}^{x} + \int_c^x \frac{(t-x)^2}{2}f'''(t)\, dt \\ &= f(c) + (x-c)f'(c) + \frac{(x-c)^2}{2}f''(c) + \int_c^x \frac{(t-x)^2}{2}f'''(t)\, dt\ . \end{aligned}$$

Continuing this process, we see eventually that

$$f(x) = f(c) + (x-c)f'(c) + \cdots + \frac{(x-c)^n}{n!}f^{(n)}(c) + R_n(x)\ , \tag{4.2}$$

where the remainder $R_n(x)$ is given by the formula

$$R_n(x) = \int_c^x \frac{(x-t)^n}{n!}f^{(n+1)}(t)\, dt\ .$$

This formula will be proven very useful in theoretical computations.

4.5 Further Reading

As this chapter is a refresher of calculus, it does not contain details and examples, even though calculus takes up most of the subjects of this book. For reviewing calculus we suggest textbooks such as [122, 126] or the classical books [3, 4]. For a deeper understanding of calculus and for more formal exploration of analysis, we refer to [133, 134] and the references therein.

Chapter Highlights

- The space of continuous functions in $[a, b]$ is denoted by $C[a, b]$.
- The space of n-times continuously differentiable functions in (a, b) is denoted by $C^n(a, b)$.
- Bolzano's theorem states that if $f(x)$ is continuous in $[a, b]$ and changes sign in $[a, b]$, then there is at least one root of the equation $f(x) = 0$ in (a, b).
- Smooth functions can be approximated around a point $x = c$ using the Taylor polynomial $P_n(x)$ of degree n with

$$P_n(x) = \sum_{i=0}^{n} \frac{f^{(n)}(c)}{n!}(x - c)^n \ .$$

- A Taylor polynomial approximates a function $f(x)$ in the sense that $f(x) - P_n(x) = R_n(x)$ where $R_n(x)$ being the residual (or error). A Taylor polynomial coincides with the function $f(x)$ at the point $x = c$, i.e. $f(c) = P_n(c)$. The error between P_n and f increases as the distance between x and c increases.
- The residual of the n-degree Taylor polynomial is given by the formula

$$R_n(x) = \frac{f^{(n+1)}(\xi)}{(n+1)!}(x - c)^{n+1}$$

  for some $\xi \in [a, b]$.
- An integral of a function f over an interval $[a, b]$ can be thought of as the area included by the function f, the x-axis and the lines $x = a$ and $x = b$.
- An integral can be approximated if we subdivide the interval $[a, b]$ using a grid $a = x_0 < x_1 < \cdots < x_N = b$, approximate the function in each of the subintervals $[x_{i-1}, x_i]$, $i = 1, 2, \ldots, N$ by a constants, and compute the respective integrals.
- An integral on an interval $[a, b] = [a, c] \cup [c, b]$ can be expressed as the sum of two integrals

$$\int_a^b f(x)\ dx = \int_a^c f(x)\ dx + \int_c^b f(x)\ dx \ .$$

- The mean-value of an integral is the value

$$f(c) = \frac{1}{b - a} \int_a^b f(x)\ dx \ ,$$

  for some value $c \in (a, b)$.
- The integral representation for the residual of an n-degree Taylor polynomial is

$$R_n(x) = \int_c^x \frac{(x - t)^n}{n!} f^{(n+1)}(t)\ dt \ .$$

Exercises

1. Prove that the equation $\sin(x) = 2-2x$ has at least one solution in the interval $(\pi/6, \pi/4)$.

2. Prove that the equation $x - e^{-x} = 0$ has a unique solution $x^* \in [0, 1]$.

3. If a function f is continuous in $[a, b]$ and $f(a) \neq f(b)$, then show that there is at least one $x^* \in (a, b)$ such that
$$f(x^*) = \frac{f(a) + f(b)}{2} .$$

4. Consider the function
$$f(x) = \begin{cases} \frac{ax^2+3bx-5}{x-1}, & x \neq 1 \\ 7, & x = 1 \end{cases} .$$
Find the values of $a, b \in \mathbb{R}$ so as f is continuous.

5. Consider the function
$$f(x) = \begin{cases} 3x^2 - a^2x, & x < 1 \\ 5x + a - 4, & x \geq 1 \end{cases} .$$
Find the values of a so as f is differentiable at $x = 1$.

6. Show that if a function f is differentiable at $x = 0$ and
$$\lim_{x\to 0} \frac{f(x)}{x} = 4 ,$$
then $f'(0) = 4$.

7. Find the equation of the tangent line to the graph at $(x_0, f(x_0))$ when $f(x) = 2x^3$ and $x_0 = -1$.

8. Prove that the equation $3x^5 - x^3 + a = 0$ has at most one root in the interval $(-1, 1)$ for any value of a.

9. Show that Rolle's theorem apply to the function $f(x) = x\sin(x)$. Then prove that the equation $\tan(x) = -x$ has at least one root in the interval $(0, \pi)$.

10. Prove that

    (a)
$$2 - \frac{e}{2} < \ln 2 < \frac{2}{e} ,$$

    (b)
$$\cos\frac{5\pi}{18} < \frac{1}{2} + \frac{\pi\sqrt{3}}{36} .$$

11. Prove that

    (a)
$$\frac{x-1}{x} \leq \ln x \leq x - 1 \quad \text{for all } x \in (0, +\infty) ,$$

    (b)
$$\lim_{x\to 1} \frac{\ln x}{x-1} = 1 .$$

12. If two functions f and g satisfy the system of differential equations

$$\begin{cases} f'(x) = g(x), \\ g'(x) = -f(x), \end{cases} \quad \text{for all } x \in \mathbb{R} \ ,$$

then show that the function $f^2(x) + g^2(x)$ is constant.

13. Consider the function $f(x) = \frac{e^x}{x+1}$.

    (a) Find the intervals of monotonicity of $f(x)$.

    (b) Find the minimum and maximum values of $f(x)$.

    (c) Prove that $e^x \geq x + 1$ for all $x \in \mathbb{R}$.

14. Prove that for any two continuous functions in the interval $[a, b]$:

    (a) If $f(x) \geq 0$ for all $x \in [a, b]$ then

$$\int_a^b f(x)\ dx \geq 0 \ .$$

    (b) If $f(x) \leq g(x)$ for all $x \in [a, b]$ then

$$\int_a^b f(x)\ dx \leq \int_a^b g(x)\ dx \ .$$

    (c)

$$\left| \int_a^b f(x)\ dx \right| \leq \int_a^b |f(x)|\ dx \ .$$

15. Prove that

$$1 - \frac{1}{e} \leq \int_1^e \frac{1}{x}\ dx \leq e - 1 \ .$$

16. Prove that for all $x \in \mathbb{R}$

$$\int_0^x |z|\ dz = \frac{1}{2} x|x| \ .$$

17. If g is a continuous function in $\mathbb{R}$, and

$$f(x) = \int_x^{2x+1} g(t)\ dt \ ,$$

    show that $f'(x) = 2g(2x + 1) - g(x)$.

18. The power series of $\sin x$ is

$$\sin x = x - \frac{x^3}{3!} + \frac{x^5}{5!} - \frac{x^7}{7!} + \cdots \ .$$

    The following Python function uses these power series to compute the value of $\sin x$ for some x

```
def powersin(x):
# POWERSIN. Power series for sin(x)
# POWERSIN(x) tries to compute sin(x)
# from a power series
    s = 0
    t = x
    n = 1
    while (s+t != s):
        s = s + t
        t = -x**2/((n+1)*(n+2))*t
        n = n + 2
    return s
```

(a) What causes the `while` loop to terminate?

(b) Answer the following questions for $x = \pi/2$, $11\pi/2$, $21\pi/2$, and $31\pi/2$:

   (i) How accurate is the computed result?

   (ii) How many terms are required?

   (iii) What is the largest term (in magnitude) in the series?

(c) What do you conclude about the use of floating-point arithmetic and power series to evaluate functions?

19. The power series for the function $\arctan(x)$ is

$$\arctan(x) = \sum_{i=1}^{\infty}(-1)^{i+1}\frac{x^{2i-1}}{2i-1} .$$

(a) Show that $\pi = 4\arctan(1)$.

(b) Using the power series of $\arctan(x)$ find value N so as the sum

$$s_N = \sum_{i=1}^{N}(-1)^{i+1}\frac{x^{2i-1}}{2i-1} ,$$

approximates the value of $\pi/4$ with absolute error

$$|s_N - \pi/4| < 10^{-3} .$$

(c) Write a Python function to compute an approximation of π using appropriate power series of $\arctan(x)$.

20. Consider the sequence defined recursively, as follows:

$$x_1 = 2, \quad x_2 = 1, \quad x_n = 2x_{n-1} - \frac{1}{2}x_{n-2} .$$

(a) Write a Python script to compute the value to which the sequence

$$y_n = \frac{x_n}{(1+\sqrt{2}/2)^n} ,$$

converges.

(b) We say that a sequence $\{a_n\}$ converges to a limit ℓ with rate (or order) r if there is a fixed number r such that

$$\lim_{n\to\infty} \frac{|a_{n+1}-\ell|}{|a_n-\ell|^r} = C,$$

where C is constant. If the previous relation holds for any value $r > 0$, then we say that the sequence converges exponentially. If $r = 1$, then we say that the sequence converges linearly, and if $r = 2$ we say that the sequence converges quadratically. Find (numerically) the type of convergence (linear, quadratic, exponential) of the sequence $\{y_n\}$, as accurately as possible.

(c) Find the exact answers to the above questions.

[Hint: Show first that $x_n = c_1(1+\sqrt{2}/2)^n + c_2(1-\sqrt{2}/2)^n$ satisfy the recursive relation for x_n. Then find the values of c_1 and c_2 using the initial conditions. Finally, substitute into y_n and compute the limit.]

21. Consider the sequence $\{y_n\}$ with

$$y_n = 2^n \tan\frac{\pi}{2^n}, \qquad n = 2, 3, \ldots .$$

(a) Prove that y_n is decreasing, and $\lim_{n\to\infty} y_n = \pi$.

(b) Give a geometric explanation of the previous result.

(c) Prove that $\{y_n\}$ can be expressed in terms of the recursive sequence

$$\begin{cases} y_2 = 4\ , \\ y_{n+1} = 2^{2n+1}\frac{\sqrt{1+(2^{-n}y_n)^2}-1}{y_n}, \qquad n = 2, 3, \ldots . \end{cases}$$

(d) Explain why the previous formula is unstable in finite precision arithmetic.

(e) Find an alternative, stable algorithm for the computation of the terms y_n using only addition/subtraction, multiplication/division and square roots.

[Hint: See Section 3.3.4, and use the identity $\tan(2x) = 2\tan x/(1-\tan^2 x)$]

22. Assume that we want to compute members of the sequence

$$y_n = \int_0^1 \frac{x^n}{x+c}\,dx, \quad n = 0, 1, 2, \ldots\ ,$$

where $c > 1$ is a very large constant.

(a) Compute the term y_0.

(b) Prove that the sequence y_n is decreasing, and that $\lim_{n\to\infty} y_n = 0$.

(c) Prove that for all $n \geq 1$

$$y_n = \sum_{k=0}^{n-1} (-1)^k \binom{n}{k} c^k \frac{(1+c)^{n-k} - c^{n-k}}{n-k} + (-c)^n \log\frac{1+c}{c}\ .$$

(d) Will the previous formula lead into a stable method to compute the sequence y_n?

Part II

Introduction to Computational Mathematics

5

Roots of Equations

Scientific problems often lead to algebraic equations of the form

$$f(x) = 0 \ .$$

A solution of such an equation is defined as a value x^* that satisfies $f(x^*) = 0$. Unfortunately, we can solve analytically only linear and certain nonlinear equations, while most of the equations we deal in real-world problems are complicated and cannot be solved analytically. For example, try to solve the simple equation $\cos(x) - x = 0$. In this chapter, we focus on numerical methods for the approximation of solutions of general equations $f(x) = 0$. Specifically, we study the bisection, Newton-Raphson, and secant methods. We study the convergence of these methods, and how to implement and use them in Python.

5.1 Bisection Method

Suppose that $f(x)$ has a root x^* in the interval $[a, b]$. Thus, $x^* \in [a, b]$ is a solution to the equation $f(x) = 0$. The most basic numerical method to approximate this root is the *bisection* method[1]. This method is always convergent, although a bit slow.

5.1.1 Derivation and implementation

The bisection method is based on Bolzano's (intermediate value) theorem (Theorem 4.1) and can be applied only to continuous functions that change sign in the interval $[a, b]$. Because no further smoothness is required, the bisection method is used for its simplicity. Bolzano's theorem states that a continuous function f with

$$f(a) \cdot f(b) < 0 \ ,$$

has a root $x^* \in (a, b)$. The midpoint

$$x_0 = \frac{a+b}{2} \ ,$$

of the interval $[a, b]$ can be close to this root (see Figure 5.1). We will use the midpoint x_0 as a rough first estimation of the root x^*, and nobody stops us from trying to find a better approximation.

Since we introduced a new point $x_0 = \frac{a+b}{2}$, we can split the initial interval $[a, b]$ into two new intervals $[a, x_0]$ and $[x_0, b]$, and repeat the same process in either $[a, x_0]$ or $[x_0, b]$

[1] Bisection method is an example of a *one-dimensional search method*, [23]. Another example of search method is the method of false position (see Algorithm 16) and the golden section search method. More information for the last one can be found in Chapter 11.

DOI: 10.1201/9781003287292-5

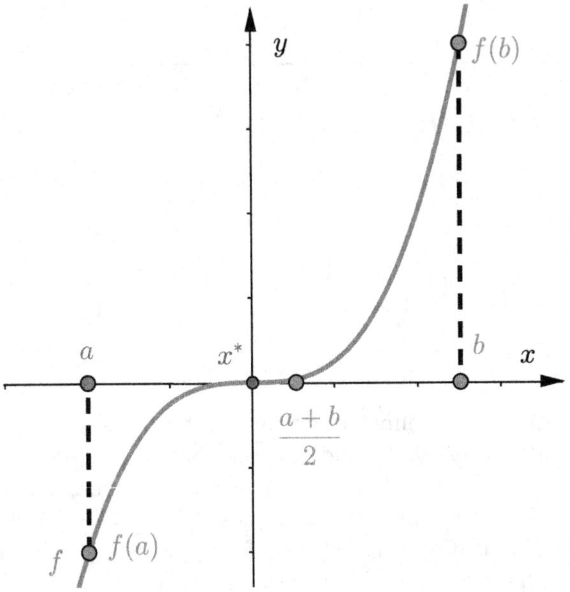

FIGURE 5.1
Approximation of the root x^* by the midpoint of the interval $[a, b]$.

depending on where the root x^* is located. For example, in Figure 5.1 the root x^* is located in the interval $[a, x_0]$. We have that in this interval $f(a) \cdot f(x_0) < 0$ since $f(a) < 0$ and $f(x_0) > 0$. Therefore, we can take an improved approximation x_1 of the root x^* to be the midpoint $x_1 = \frac{a+x_0}{2}$ of the new interval $[a, x_0]$. We can repeat this process until we are satisfied with the accuracy of the approximation.

In order to make this process more suitable for computer implementations and theoretical estimates, we redefine our interval $[a, b]$ at every iteration by calling the working interval $[a, b]$. After the first iteration, the new interval $[a, x_0]$ is denoted again $[a, b]$. Then we set again $x_1 = \frac{a+b}{2}$. In particular, we set $b = x_0$ so as the new interval $[a, b]$ coincides with the interval $[a, x_0]$, and we forget the previous interval. Repeating this process we generate a sequence of approximations $x_k \approx x^*$, each of which is a midpoint of an interval.

Since we repeat this process until certain accuracy has achieved, we need to agree on a tolerance for the accuracy. Usually, we are not aware of the exact solution x^*, so we cannot estimate the error $|x^* - x_k|$. For this reason, we terminate the process when two subsequent approximations are very close to each other. This means that after setting a tolerance TOL for the accuracy we check whether

$$|x_k - x_{k-1}| < TOL ,$$

is satisfied. An alternative stopping criterion could also be the magnitude of the value $|f(x_k)|$. The smallest this value is, the closer to the root we should be. Here, and for the bisection method, we will consider as a measure of convergence the length of the interval $[a, b]$ from which the last midpoint occurred. Usually, we take the tolerance to be a small number of order $O(10^{-5})$ or $O(10^{-10})$ depending on how accurate we need the result. If we take a small tolerance, then the process might require a lot of iterations. The algorithm of this repetitive process can be described with the Algorithm 11. We encourage the interested reader to modify the stopping criteria and test other options.

Algorithm 11 defines a sequence of approximations x_k. For each k we repeat the same iteration process. Such method is characterized as iterative method. To avoid the possibility

Algorithm 11 Bisection method

Set a tolerance TOL for accuracy
Initialize the interval $[a, b]$
Set $k = 0$
Check if $f(a) \cdot f(b) < 0$ (if not then quit)
while $|b - a| > TOL$ **do**
    Set $x_k = (a + b)/2$ (the new approximation)
    Check in which interval $[a, x_k]$ or $[x_k, b]$ the function changes sign
    if $f(x_k) \cdot f(a) < 0$ **then**
        $b = x_k$
    else if $f(x_k) \cdot f(b) < 0$ **then**
        $a = x_k$
    else
        Break the loop because $f(x_k) = 0$
    end if
    $k = k + 1$
end while
Return the approximation of the root x^*

that an iterative method iterates for ever, we count the number of iterations performed and we allow only a certain maximum number of them. This can be done within the `while` loop by adding an additional stopping criterion: The number of iterations cannot exceed a certain number. This may increase the number of inputs in our function but it ensures a smooth iterative procedure. We continue with the implementation of this method in Python.

```
def bisection(f, a, b, tol = 1.e-6):
    iteration = 0 #initialize counter iteration
    if (f(a) * f(b) < 0.0): # check if there is a root
        while ((b-a) > tol): # check if the end-points converge
            iteration = iteration + 1
            x = (a + b)/2
            if (f(a) * f(x) < 0.0):
                b = x
            elif (f(x) * f(b) < 0.0):
                a = x
            else:
                break
            print(iteration, x)
    else:
        print('failure')
    return x
# returns the midpoint of the final interval
```

In this code we record the number of iterations using the variable `iteration`. This counter is increased by `1` every time the `while` loop is being executed. We can define the maximum number of iterations allowed `maxit`, and replace the `while` statement with the statement `while ((b-a) > tol and iterations < maxit):`. This will ensure that the code will not perform too many iterations. The variable `maxit` can be large as `100` or larger. We leave such a modification of the code as an exercise for the reader.

In what follows we test our method and its implementation. Here we choose the nonlinear equation

$$\ln x + x = 0 \ ,$$

in the interval $[0.1, 1]$.

```
1 import numpy as np
2 def f(x):
3     y = np.log(x) + x
4     return y
5 a = 0.1
6 b = 1.0
7 tol = 1.e-4
8 x = bisection(f, a, b, tol)
9 print('The approximate solution x is: ', x)
10 print('And the value f(x) is: ', f(x))
```

```
1 0.55
2 0.775
3 0.6625000000000001
4 0.6062500000000001
5 0.578125
6 0.5640625
7 0.57109375
8 0.567578125
9 0.5658203125000001
10 0.5666992187500001
11 0.567138671875
12 0.5673583984375
13 0.56724853515625
14 0.567193603515625
The approximate solution x is:  0.567193603515625
And the value f(x) is:  0.00013902223881425623
```

This code prints the iteration number and the midpoint at every step. It is noted that in addition to the approximation x_k of the root x^* at the end, we compute the value of the function $f(x_k) \approx f(x^*) = 0$. It was required 14 iterations until the length of the interval $[a, b]$ becomes less than 10^{-4}. Because our tolerance was $TOL = 10^{-4}$, the value $f(x^*)$ we found has only 4 significant digits correct. We can increase the accuracy of the method by choosing smaller value for the tolerance TOL.

☞ In line 1 of the previous code we called NumPy with the acronym **np**. We will not repeat this command, however, it is necessary for all the rest of the chapter.

5.1.2 Proof of convergence

When we design a numerical method we want to know if the generated approximations converge to the correct solution, and how fast. We say that the method converges in the sense that the limit of the error converges to zero as the number of the iterations increases

$$\lim_{k\to\infty} |e_k| = 0 \ .$$

The approximation error $|e_k|$ is defined as

$$|e_k| = |x^* - x_k| ,$$

and is the distance between the exact solution x^* and the approximate solution x_k. For the bisection method we have the following proposition:

Proposition 5.1. *If $f \in C[a,b]$ and $f(a)\cdot f(b) < 0$, then the sequence $\{x_k\}_{k\in\mathbb{N}}$ generated by the bisection method either converges to the root x^* of the equation $f(x) = 0$ as $k \to \infty$ (we write $x_k \to x^*$ as $k \to \infty$), or there is $k = N$ such that $x_k = x^*$. Moreover, the approximation error is estimated to be*

$$|x^* - x_k| \leq \frac{b-a}{2^{k+1}}, \quad k = 0, 1, 2, \ldots .$$

Proof. In order to prove the convergence of the bisection method we modify for convenience our algorithm by setting $a_0 = a$ and $b_0 = b$ the initial endpoints of the interval $[a,b]$ and let $I_i = [a_i, b_i]$, $i = 1, 2, \ldots$ denote the intervals constructed by the bisection method. Observing that the length of the interval $[a_k, b_k]$ is half of the length of $[a_{k-1}, b_{k-1}]$ we have

$$b_k - a_k = \frac{b_{k-1} - a_{k-1}}{2} = \frac{b_{k-2} - a_{k-2}}{4} = \cdots = \frac{b-a}{2^k} ,$$

and

$$x_k = \frac{a_k + b_k}{2} ,$$

while

$$x^* \in [a_k, b_k] .$$

The approximation error is

$$|e_k| = |x^* - x_k| \leq \frac{b_k - a_k}{2} = \frac{b-a}{2^{k+1}} .$$

Taking the limit $k \to \infty$ we get

$$\lim_{k\to\infty} |e_k| \leq \lim_{k\to\infty} \frac{b-a}{2^{k+1}} = 0 ,$$

and therefore we conclude that the sequence x_k produced by the bisection method always converges to the exact root x^*. □

☞ The major advantage of the bisection method is that always converges to a solution and for this reason is often used to provide a good initial approximation for a more efficient procedure. Moreover, it can be used when the function $f(x)$ is just continuous.

5.1.3 Rate of convergence

After verifying that the numerical method converges, it is useful to know how fast the generated sequence of approximations x_k converges to the solution x^*. The speed of convergence is described by the *convergence rate* or *rate of convergence*. The rate of convergence for a sequence of errors $|e_k| = |x^* - x_k|$ is the number $r \geq 1$ for which there is a constant $C > 0$ such that

$$\lim_{k\to\infty} \frac{|e_{k+1}|}{|e_k|^r} = C \ .$$

In order to understand the practical meaning of this limit, think of the error $|e_k| < 1$ to be small. Then, $|e_k|^r$ should be much less than $|e_k|$ for $r > 1$. This can be expressed as $|e_k|^r \ll |e_k|$, for $r > 1$. Thus the error of the next iteration will be much smaller than the error of the previous one and the magnitude will be determined by the exponent r. Specifically, we will have $|e_{k+1}| \approx C|e_k|^r \ll |e_k|$.

The case $r = 1$ requires $C < 1$, otherwise there is no guarantee that the new error will be smaller than the previous one. To see this, assume that we perform k iterations of an iterative method. In practice, we want the error $|e_k| = |x^* - x_k|$ to decrease so that

$$|e_{k+1}| = C|e_k| \ ,$$

which implies

$$|e_{k+1}| = C|e_k| = C^2|e_{k-1}| = \cdots = C^k|e_1| \ .$$

For this reason, we need $C < 1$, otherwise the method will diverge as $k \to \infty$.

When $r = 1$ we say that the method converges linearly, when $r = 2$ quadratically and so on. For the bisection method we proved linear convergence in Proposition 5.1 with $C = 1/2$ and $|e_{k+1}| \leq \frac{1}{2}|e_k|$. Thus the method converges with rate at least 1. We say "at least" because it can happen to converge either instantaneously or in general faster than 1 as the less equal sign suggests. In general, if $|e_{k+1}| \leq C|e_k|^r$, then the rate of convergence is at least of order r meaning that r is the worst case.

When we cannot compute a theoretical estimate for the convergence rate of a numerical method, or if we want to verify our theoretical estimate, we often find approximations of the convergence rate experimentally. In order to compute experimentally the rate of convergence, we compute the errors for three subsequent iterations, let's say e_1, e_2, e_3. Then we assume that

$$|e_2| = C|e_1|^r \ ,$$
$$|e_3| = C|e_2|^r \ .$$

Dividing the previous equations we have

$$\frac{|e_2|}{|e_3|} = \left(\frac{|e_1|}{|e_2|}\right)^r \ .$$

We then solve for r to obtain a formula for approximating the rate of convergence

$$r = \frac{\log \frac{|e_2|}{|e_3|}}{\log \frac{|e_1|}{|e_2|}} \ .$$

We use the natural logarithm but sometimes the use of the logarithm with base 10 can be more informative. Moreover, the experimental estimation of the rate of convergence requires the knowledge of the exact solution. For this reason, we usually choose a problem with known solution for testing purposes. This methodology is sometimes called the method of manufactured solutions.

As an example, we compute the convergence rate for the bisection method numerically by modifying the previous code.

```
def bisection_rates(f, a, b, tol = 1.e-6):
    iteration = 0
    if (f(a) * f(b) < 0.0):
        e1 = abs(b-a) #initialize e1 arbitrarily
        e2 = e1*2 #initialize e2  arbitrarily
        e3 = e1 #initialize e3 arbitrarily
        while ((b-a)>tol):
            e1 = e2
            e2 = e3
            iteration = iteration + 1
            x = (a + b)/2
            if (f(a) * f(x) < 0.0):
                b = x
            elif (f(x) * f(b) < 0.0):
                a = x
            else:
                break
            e3 = np.abs(b-a)
            rate = np.log(e2/e3)/np.log(e1/e2)
            print('iteration = ', iteration, 'rate =', rate)
    else:
        print('failure')

    return x

def f(x):
    y = np.log(x) + x
    return y

a = 0.1
b = 1.0
tol = 1.e-4

x = bisection_rates(f, a, b, tol)
```

```
iteration =  1 rate = 1.0000000000000002
iteration =  2 rate = 0.9999999999999997
iteration =  3 rate = 0.9999999999999993
iteration =  4 rate = 1.0000000000000007
iteration =  5 rate = 1.0000000000000029
iteration =  6 rate = 0.9999999999999971
iteration =  7 rate = 1.0000000000000115
iteration =  8 rate = 0.9999999999999657
iteration =  9 rate = 1.0000000000000682
iteration =  10 rate = 0.9999999999999545
iteration =  11 rate = 0.9999999999998177
iteration =  12 rate = 1.0000000000001823
iteration =  13 rate = 1.0000000000007292
iteration =  14 rate = 0.9999999999992709
```

For this particular example the estimated convergence rate is 1. This is a strong indication that the method converges with rate 1. Note that we do not expect the convergence rate to be correct from the very first iteration since the values `e1`, `e2`, `e3` have arbitrary values.

☞ The convergence of the bisection method can be quite slow and so it may require a lot of iterations k until the error $|x^* - x_k|$ becomes sufficiently small. The bound of the error $|e_k|$ is halved at every step. It is also possible that a good intermediate approximation may be inadvertently discarded.

Estimation of the number of iterations

Due to the simplicity of bisection method, it is easy to estimate the maximum number of iterations required in order to achieve a prescribed accuracy. Assume that we want to find an approximation x_k to the root x^* with accuracy at least ε, in the sense that $|e_k| < \varepsilon$. From Proposition 5.1 we have that $|e_k| \leq (b-a)/2^{k+1}$. To ensure that $|e_k| < \varepsilon$, we set

$$\frac{b-a}{2^{k+1}} < \varepsilon \ .$$

We solve the last inequality in terms of k and we find that we need at least

$$k > \frac{\log(b-a) - \log(2\varepsilon)}{\log(2)} \ ,$$

iterations to achieve the desired accuracy.

For example, consider the initial interval $[0, 1]$ (independent of the equation we solve) and $\varepsilon = 10^{-4}$. The bisection method requires

$$k > \frac{\log(1) - \log(2 \cdot 10^{-4})}{\log(2)} \approx 12.29 \ .$$

We conclude that the bisection method requires $k = 13$, which is to say 14 iterations. This is because we start counting the approximations x_k from $k = 0$.

5.2 Fixed-Point Methods

The bisection method has guaranteed convergence. However it uses almost no information about $f(x)$ beyond its sign. As an alternative to the bisection method we derive other methods, known to as *fixed-point methods* that can converge faster.

The basic idea behind fixed-point methods is the following: Every equation $f(x) = 0$ can be written in the form $x = g(x)$. We can do this in many different ways. For example, we can write the equation $\sin(x) + 1 = 0$ as

$$x = x + \sin(x) + 1 \quad \text{or} \quad x = -x\sin(x) \ .$$

In the first case the function $g(x) = x + \sin(x) + 1$, while in the second case is $g(x) = -x\sin(x)$. Rewriting the original equation $f(x) = 0$ in the form $x = g(x)$ will result in a different numerical method with different convergence properties. This is because we will seek the value x^* such that $x^* = g(x^*)$ instead of seeking the value x^* such that $f(x^*) = 0$.

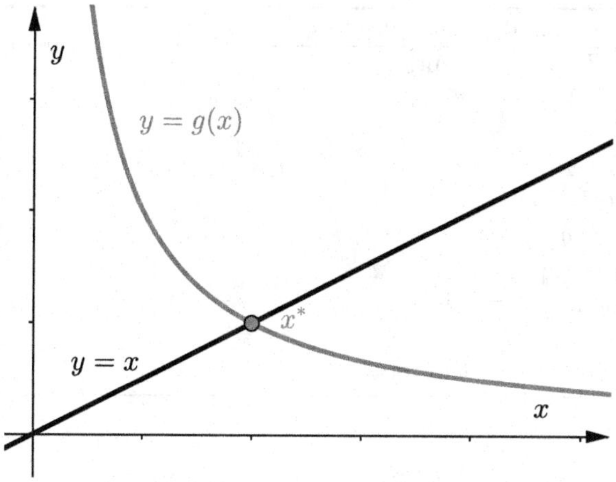

FIGURE 5.2
Geometric interpretation of fixed point $x^* = g(x^*)$.

We saw earlier that the value x^* satisfying $f(x^*) = 0$ is called root of the function $f(x)$. The value x^* that satisfies the relation $x^* = g(x^*)$ is called *fixed point* of g. For this reason, iterative methods we will derive in this book for finding fixed points of functions are called *fixed-point* methods.

☞ A point x^* such that $x^* = g(x^*)$ is called fixed point of g. A fixed-point method is an iterative method for the approximation of fixed points.

5.2.1 Derivation and implementation

The idea behind fixed-point methods is simple. Since a fixed point x^* satisfies $x^* = g(x^*)$, this implies that the point $(x^*, x^*) = (x^*, g(x^*))$ is the point of intersection of the graphs of $y = x$ and $y = g(x)$. This is depicted in Figure 5.2.

Given an equation $f(x) = 0$ we first define an appropriate function $g(x)$, and rewrite our equation as $x = g(x)$. Since there is no unique way to do this, it is a matter of luck as to whether our method is good or not. In the next section we will see a way to derive a fixed-point method that excels in terms of speed of convergence, and thus in accuracy.

Given the fixed-point equation $x = g(x)$, we start with an initial guess x_0 of the fixed point x^*. Yes, an initial guess! We can choose (almost) anything we like. We wrote almost because it is always better to start with a value which is close to the exact solution x^*. If we have no idea how to make an initial, rough estimate, we can use the bisection method for a few iterations to find the initial guess x_0. Then the fixed-point method continues with improvements

$$x_{k+1} = g(x_k) \quad \text{for} \quad k = 0, 1, 2, \ldots \ .$$

The algorithm of a generic fixed-point method is described in Algorithm 12.

In practice, similarly to the bisection method, we usually do not store in memory every approximation in the sequence x_k but only the last approximation. Therefore, the implementation of Algorithm 12 differs slightly from the algorithm itself. In addition to the standard tolerance for the error, as a stopping criterion we also record the total number of iterations.

Algorithm 12 Generic fixed-point method

Set a tolerance TOL for the accuracy
Set $k = 0$
Initialize x_k
Set $Error = TOL + 1$
while $Error > TOL$ **do**
  Compute $x_{k+1} = g(x_k)$
  $Error = |x_{k+1} - x_k|$
  Increase the counter $k = k + 1$
end while
Return the approximation of the root x^*

```
def fixedpoint(g, x0, tol = 1.e-6, maxit = 100):
    # g = the function g(x)
    # x0 = the initial guess of the fixed point x=g(x)
    # tol = tolerance for the absolute error
    #       of two subsequent approximations
    # maxit = maximum number of iterations allowed
    err = 1.0
    iteration = 0
    xk = x0
    while (error > tol and iteration < maxit):
        iteration = iteration + 1
        error = xk
        xk = g(xk)
        error= np.abs(error - xk)
        print ('iteration =', iteration, ', x =', xk)
    return xk
```

To test the code, we consider the equation $f(x) = 0$ with $f(x) = x^2 - x - 1$. This equation has two roots but we are interested in the root $x^* = (1 + \sqrt{5})/2$, which is known to as the *golden ratio*. We can formulate the function $g(x)$ of the fixed-point equation in many ways. One choice could be the following: We first solve for x^2 the original equation $x^2 - x - 1 = 0$ to find $x^2 = x + 1$, and we take the square root to get the fixed-point equation $x = \pm\sqrt{x + 1}$. Then we choose $g(x) = \sqrt{x + 1}$, so the fixed-point iteration will be $x_{k+1} = \sqrt{x_k + 1}$ for $k = 0, 1, 2, \ldots$. As initial guess of the fixed point we take $x_0 = 0$.

```
def f(x):
    y = x**2-x-1.0
    return y
def g(x):
    y = np.sqrt(x+1.0)
    return y
tol = 1.e-4
maxit = 50
x0 = 0.0
x = fixedpoint(g, x0, tol, maxit)
print('The approximate solution x is: ', x)
print('And the value f(x) is: ', f(x))
```

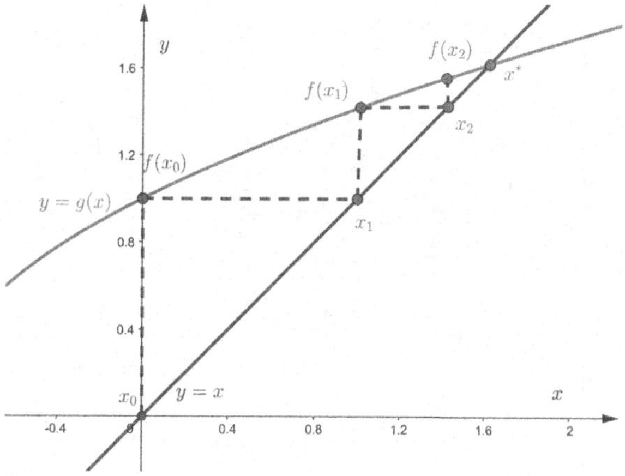

FIGURE 5.3
Geometric representation of the first few iterations $x_{k+1} = g(x_k)$ with $x_0 = 0$ and $g(x) = \sqrt{x+1}$.

```
iteration = 1 , x = 1.0
iteration = 2 , x = 1.4142135623730951
iteration = 3 , x = 1.5537739740300374
iteration = 4 , x = 1.5980531824786175
iteration = 5 , x = 1.6118477541252516
iteration = 6 , x = 1.616121206508117
iteration = 7 , x = 1.6174427985273905
iteration = 8 , x = 1.617851290609675
iteration = 9 , x = 1.6179775309347393
iteration = 10 , x = 1.6180165422314876
The approximate solution x is:  1.6180165422314876
And the value f(x) is:  -3.9011296748103774e-05
```

Although we specified the maximum number of iterations to be `maxit = 50`, the method converged in 10 iterations within the desired tolerance 10^{-4}.

It is worth discussing in some detail how fixed point methods work geometrically. Starting with the initial guess $x_0 = 0$ we compute the value $g(x_0)$, and we set $x_1 = g(x_0)$. Practically, we project the point $y_1 = g(x_0)$ to the straight line $y = x$ by taking $x_1 = y_1$. We continue with the value x_1. We compute the value $g(x_1)$ and we set $y_2 = g(x_1)$. We project again onto the straight line $y = x$ so as $x_2 = y_2$, and so on. This is depicted graphically in Figure 5.3.

Since the equation $x = g(x)$ is not a unique fixed-point representation of the equation $f(x) = 0$, we can proceed with a different choice of $g(x)$. Consider again the equation $x^2 - x - 1 = 0$, but now after solving for x^2, divide by x to obtain the fixed-point equation $x = 1 + \frac{1}{x}$, with $g(x) = 1 + \frac{1}{x}$. For the new fixed-point method we take $x_0 = 1$ to be the initial guess and we obtain the following results:

```
def g(x):
    y = 1.0+1.0/x
    return y
tol = 1.e-4
maxit = 50
```

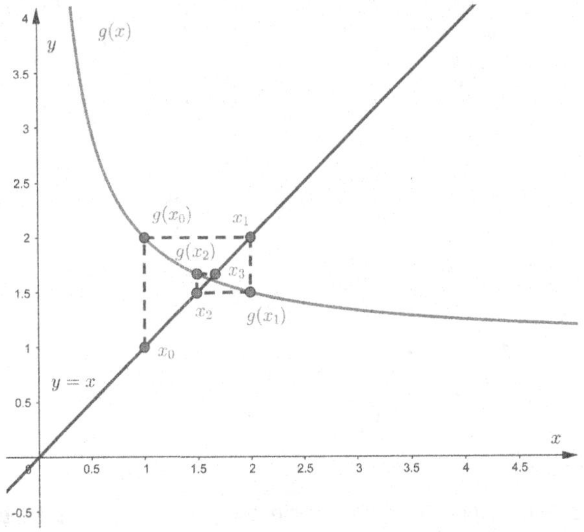

FIGURE 5.4
Geometric representation of the first few iterations $x_{k+1} = g(x_k)$ with $x_0 = 1$ and $g(x) = 1 + \frac{1}{x}$.

```
x0 = 1.0
x = fixedpoint(g, x0, tol, maxit)
print('The approximate solution x is: ', x)
print('And the value f(x) is: ', f(x))
```

```
iteration = 1 , x = 2.0
iteration = 2 , x = 1.5
iteration = 3 , x = 1.6666666666666665
iteration = 4 , x = 1.6
iteration = 5 , x = 1.625
iteration = 6 , x = 1.6153846153846154
iteration = 7 , x = 1.619047619047619
iteration = 8 , x = 1.6176470588235294
iteration = 9 , x = 1.6181818181818182
iteration = 10 , x = 1.6179775280898876
iteration = 11 , x = 1.6180555555555556
The approximate solution x is:  1.6180555555555556
And the value f(x) is:  4.822530864223573e-05
```

We observe that the different function $g(x)$ requires more iterations to converge within the prescribed accuracy. The geometric interpretation of this new iteration can be seen in Figure 5.4. Here the projections form a helix. Without being important for the convergence, it gives an idea of the geometric pattern of the convergence. Specifically, it shows that the convergence depends not only on the initial condition but also on the shape of the graph of the function $g(x)$.

Trying different choices for the function $g(x)$ we can get completely different behavior. For example, taking $g(x) = x^2 - 1$ and starting with $x_0 = 1$ the fixed-point iteration never converges. Specifically, it falls in a square pattern where $x_{2k} = 0$ and $x_{2k+1} = -1$ for $k = 0, 1, 2, \ldots$. By taking $g(x) = x - \frac{x^2-x-1}{2x-1}$ on the other hand the fixed-point method converges in only 5 iterations. The last option is a special case that can be derived following a methodology known as Newton's method. We study Newton's method later in detail.

5.2.2 Theoretical considerations

One question that arose from the previous experiments is the following: What properties of the function $g(x)$ can influence the convergence of a fixed-point method? The answer to this question is given in Proposition 5.2.

Proposition 5.2 (Existence of a fixed point). *Every continuous function* $g(x) : [a,b] \to [a,b]$ *has at least one fixed point in the interval* $[a,b]$.

Proof. The fact that $g(x) : [a,b] \to [a,b]$ implies that the image of g is $g([a,b]) \subset [a,b]$. If $g(a) = a$ or $g(b) = b$, then the existence of a fixed point is obvious. Suppose that $g(a) \neq a$ and $g(b) \neq b$. Then we have only one choice: $g(a) > a$ and $g(b) < b$. We define the function $h(x) = g(x) - x$. The function $h(x)$ is continuous in $[a,b]$, and, moreover, because of our hypothesis we have $h(a) = g(a) - a > 0$ and $h(b) = g(b) - b < 0$. Hence, from Bolzano's theorem, there is at least one point $x^* \in (a,b)$ such that $h(x^*) = 0$ and thus $x^* = g(x^*)$. □

Another interesting question is whether the previous fixed point is unique, or not. This is important to know so that we can be sure that our method converges to the desirable root. In order to prove the uniqueness of the fixed point, we need to assume further regularity for the function $g(x)$. Specifically, we need to assume that $g(x)$ is Lipschitz continuous function, and in particular a contraction. For simplicity we will relax this requirement to $|g'(x)| \leq K < 1$ for all $x \in [a,b]$.

Proposition 5.3 (Uniqueness of a fixed point). *If* $g \in C^1[a,b]$, *i.e.* g *is continuously differentiable function, and* $g(x) \in [a,b]$ *for* $x \in [a,b]$ *and let also*

$$|g'(x)| \leq K < 1, \quad \text{for all } x \in [a,b] \ ,$$

then the function g *has a unique fixed point.*

Proof. The existence of fixed point has been proven in Proposition 5.2. To prove the uniqueness of the fixed point we assume (for contradiction) that there are two fixed points x^* and x^{**} in $[a,b]$ such that $x^* \neq x^{**}$. By the Mean Value Theorem of differential calculus (see Theorem 4.3) there exists $\xi \in (x^*, x^{**})$ such that

$$g'(\xi) = \frac{g(x^*) - g(x^{**})}{x^* - x^{**}} \ .$$

Since $|g'(x)| \leq K < 1$ for all $x \in [a,b]$, we have that

$$|x^* - x^{**}| = |g(x^*) - g(x^{**})| = |g'(\xi)||x^* - x^{**}| < |x^* - x^{**}| \ , \tag{5.1}$$

which is a contradiction. Therefore, the hypothesis of two fixed points is wrong, and thus the fixed point is unique. □

To illustrate the applicability of the previous propositions, consider the function $g(x) = \frac{x^2-1}{3}$ in the interval $[-1,1]$. It is easy to verify that $g(x) \in [-1,1]$ for all values $x \in [-1,1]$. Moreover, $g(x)$ is differentiable and

$$|g'(x)| = \left|\frac{2x}{3}\right| \leq \frac{2}{3} < 1 \ ,$$

for all $x \in (-1, 1)$. We conclude then that the function $g(x)$ has a unique fixed point in the interval $(-1, 1)$.

Note that the requirement for differentiability and for the derivative to be $|g'(x)| < K < 1$ can be replaced by the assumption that the function g is a contraction (Lipschitz continuous with constant $K < 1$). Then for all values of x and y there is a constant $0 < K < 1$ such that $|g(x) - g(y)| < K|x - y|$ and inequality (5.1) can be replaced by the inequality

$$|x^* - x^{**}| = |g(x^*) - g(x^{**})| < K|x^* - x^{**}| < |x^* - x^{**}| .$$

The rest of the proof need no modifications. To review Lipschitz continuous functions see Section 4.2.

5.2.3 Convergence of fixed-point methods

A fixed-point method $x_{k+1} = g(x_k)$ for $k = 0, 1, 2, \ldots$ with an initial guess x_0 and appropriate function $g(x)$ can generate a sequence of approximations to the solution of the fixed-point equation $x^* = g(x^*)$. We saw that the condition $|g'(x)| \leq K < 1$ (or if g is a contraction) guarantees the uniqueness of the fixed point. Under the same condition the fixed-point method generates a convergent sequence to the fixed point x^*. The following proposition gives a sufficient condition for the convergence of the generic fixed-point method.

Proposition 5.4. *Let $g(x)$ has a fixed point x^* and is such that*

$$|g'(x)| \leq K < 1 .$$

Then the sequence $\{x_k\}_{k\in\mathbb{N}}$ generated by the fixed-point method $x_{k+1} = g(x_k)$ converges to x^* for any given initial guess x_0 in the domain of the function g.

Proof. In order to prove convergence we consider the error

$$e_k = x_k - x^* , \tag{5.2}$$

and we show that the sequence e_k tends to zero as $k \to \infty$.

From (5.2) we have that $x_k = x^* + e_k$ and $x_{k+1} = x^* + e_{k+1}$. Since $x_{k+1} = g(x_k)$ for $k = 0, 1, 2, \ldots$, we have that

$$x^* + e_{k+1} = g(x^* + e_k) .$$

With the help of the Mean Value Theorem of differential calculus and the hypothesis $|g'(x)| \leq K < 1$ we get

$$\begin{aligned} |e_{k+1}| &= |x_{k+1} - x^*| \\ &= |g(x_k) - g(x^*)| \\ &= |g'(c_k)|\ |x_k - x^*| \quad [\text{ for some } c_k \text{ between } x_k \text{ and } x^*] \\ &\leq K|x_k - x^*| \\ &= K|e_k| . \end{aligned}$$

Thus,

$$|e_{k+1}| \leq K|e_k| \leq K^2|e_{k-1}| \leq \cdots \leq K^k|e_0| .$$

Taking the limits $k \to \infty$ in the above inequality leads to

$$\lim_{k\to\infty} |e_{k+1}| \leq \lim_{k\to\infty} K^k|e_0| = 0 ,$$

since $K < 1$. This means that the error converges to zero and thus the iteration converges to x^*. □

The condition $|g'(x)| < K < 1$ was used only after the Mean Value Theorem to show that

$$|g(x_k) - g(x^*)| = |g'(c_k)|\ |x_k - x^*| \leq K|x_k - x^*|\ .$$

The middle step can be omitted without modifying the proof essentially if we assume that g is a contraction. Specifically, if $|g(x) - g(y)| \leq K|x - y|$ for $0 < K < 1$ and for all x, y, then we have the same result without assuming that g is differentiable. Notice also that in the proof of the Proposition 5.4 the constant c_k is between x_k and x^*, and since $x_k \to x^*$ then $c_k \to x^*$. Thus, if g' is continuous, then $g'(c_k) \to g'(x^*)$. This yields

$$\lim_{k\to\infty} \frac{|e_{k+1}|}{|e_k|} = |g'(x^*)|\ .$$

We observe that if $g'(x^*) = 0$, then it is expected that the method converges with $r > 1$.

☞ It is impossible to describe the convergence rate of a generic fixed-point method since it depends on the choice of $g(x)$. To estimate experimentally the convergence rate of a generic fixed-point method, we can use the numerical procedure described in Section 5.1.3. In the next section we describe a fixed-point method with quadratic convergence rate.

5.3 Newton's Method

Newton's method (also known as the Newton-Raphson method named after Sir Isaac Newton (1662–1726) and Joseph Raphson (1668–1712)) is a particular fixed-point method with quadratic convergence rate. Newton's method is considered one of the most efficient iterative methods for the approximation of roots of functions. The success of this method is based on the fact that the tangent line of f at some point $(x_k, f(x_k))$ is a good approximation of $f(x)$ around that point[2]. So instead of computing the exact zero of $f(x)$ we compute the root of its tangent line. This idea is depicted in Figure 5.5.

5.3.1 Derivation and implementation

Newton's method is a fixed-point method. For this reason, we start again with an initial guess x_0 close to the actual root x^*. Then we set up a sequence of approximations $\{x_k\}_{k\in\mathbb{N}}$ to the root x^* by approximating the function f using its tangent line. Assume that we have computed the approximation x_k for some $k \in \mathbb{N}$, and that the function f is differentiable. The tangent line to the function $f(x)$ at the point $(x_k, f(x_k))$ is described by the equation

$$y = f'(x_k)x + b\ , \tag{5.3}$$

where b has to be found. We can find the constant b using the fact that the tangent line passes through the point $(x_k, f(x_k))$. Specifically, substituting $x = x_k$ and $y = f(x_k)$ in (5.3) yields $f(x_k) = f'(x_k)x_k + b$. Solving the last equation for b we get

$$b = f(x_k) - f'(x_k)x_k\ .$$

[2]The approximation of $f(x)$ by its tangent is also called linearization.

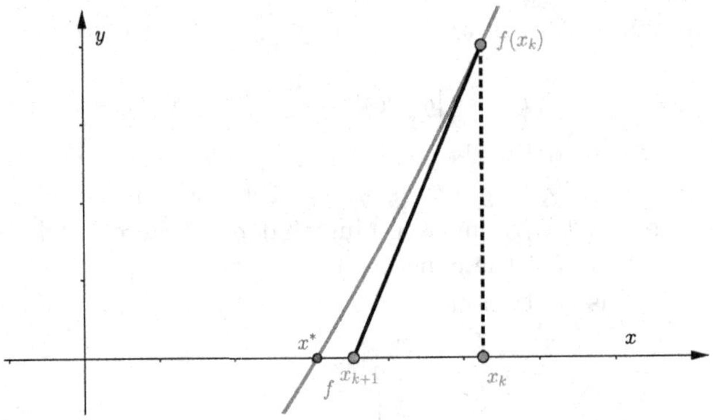

FIGURE 5.5
Geometric interpretation of Newton's method. The function f and its tangent at $(x_k, f(x_k))$.

Substituting b into (5.3) we obtain the equation for the tangent line $y = f'(x_k)x + f(x_k) - f'(x_k)x_k$. Rearranging the terms in the last formula we have

$$y = f'(x_k)(x - x_k) + f(x_k) \ . \tag{5.4}$$

Now we can approximate the root of the function f with the root of its tangent line. Denote the root of the tangent line by x_{k+1} such as

$$0 = f'(x_k)(x_{k+1} - x_k) + f(x_k) \ .$$

The root of the tangent line x_{k+1} is our new approximation to x^*. Solving the last equation for the new approximation x_{k+1} we obtain the iteration

$$x_{k+1} = x_k - \frac{f(x_k)}{f'(x_k)} \ , \tag{5.5}$$

known to as *Newton's iteration.* Newton's iteration is a fixed-point iteration $x_{k+1} = g(x_k)$ with

$$g(x) = x - \frac{f(x)}{f'(x)} \ .$$

The form of this function leads also to the idea of a more general class of fixed-point methods (*relaxation* methods) based on functions g of the form $g(x) = x - M(x) \cdot f(x)$ where $M(x)$ can be any appropriate function leading to convergence of the iterative method, [132].

Alternative derivation

The previous derivation is based on our geometric intuition. Here we derive the same method using a more formal derivation which also gives us an idea of the error in the approximation. This alternative derivation is based on Taylor's polynomials.

Let x_k be an approximation of the root x^*. The Taylor polynomial of degree one about the current Newton iteration x_k is

$$f(x) = f(x_k) + f'(x_k)(x - x_k) + O((x - x_k)^2) \ ,$$

where the $O((x - x_k)^2)$ denotes terms of the specific order. Since x is near x_k these terms will not be important. Evaluating the first order Taylor polynomial at $x = x^*$ we get

$$f(x^*) = f(x_k) + f'(x_k)(x^* - x_k) + O(e_k^2) \ ,$$

where $e_k = x^* - x_k$. Since x^* is the exact root of the function $f(x)$ we have $f(x^*) = 0$, and thus

$$0 = f(x_k) + f'(x_k)(x^* - x_k) + O(e_k^2)\ . \tag{5.6}$$

From this formula we conclude that the anticipated error is expected to be of $O(e_k^2)$, which is of second order. Discarding the error term and solving equation (5.6) in terms of x^* we get

$$x^* \approx x_k - \frac{f(x_k)}{f'(x_k)}\ .$$

Setting the right-hand side of the previous formula equal to x_{k+1} yields again the Newton iteration

$$x_{k+1} = x_k - \frac{f(x_k)}{f'(x_k)}\ .$$

A complete algorithm of Newton's method is presented in Algorithm 13.

Algorithm 13 Newton's method

Set a tolerance TOL for the accuracy
Set the maximum number of iterations $MAXIT$
Set $k = 0$
Initialize x_k and $Error = TOL + 1$
while $Error > TOL$ and $k < MAXIT$ **do**
  if $f'(x_k) \neq 0$ **then**
    Compute $x_{k+1} = x_k - \frac{f(x_k)}{f'(x_k)}$
    $Error = |x_{k+1} - x_k|$
    Increase the counter $k = k + 1$
  end if
end while
Return the approximation of the root x^*

Newton's method requires the knowledge and the computation of the derivative f' at each step, and this can cause delays. It also requires division by $f'(x_k)$. For this reason, every time we perform the division we need to check if the derivative is zero or not. Because of floating point errors, instead of checking whether $f'(x_k) \neq 0$ it is better practice to check whether $f'(x_k)$ is greater than a tolerance. For example, we usually check if $|f'(x_k)| > \varepsilon$ for some very small value ε. The parameter ε can be chosen quite small such as 10^{-10}. Alternatively, we can check if $|f'(x_k)|$ is close to zero using the NumPy function `isclose` where again we need to provide a tolerance. More information about this function can be found online.

To keep the algorithm simple we do not verify if the derivative becomes zero. This task is left as an exercise for the reader. An implementation of the previous algorithm in Python can be the following

```
def newton(f, df, x0, tol = 1.e-6, maxit = 100):
    # f = the function f(x)
    # df = the derivative of f(x)
    # x0 = the initial guess of the solution
    # tol = tolerance for the absolute error
    # maxit = maximum number of iterations
    err = 1.0
    iteration = 0
```

```
    xk = x0
    while (err > tol and iteration < maxit):
        iteration = iteration + 1
        err = xk # store previous approximation to err
        xk = xk - f(xk)/df(xk) # Newton's iteration
        err = np.abs(err - xk) # compute the new error
        print(iteration, xk)
    return xk
```

We test our new method for the equation

$$\ln x + x = 0 \ .$$

This is the same equation we used to test the bisection method before. The difference here is that we do not need to specify any initial interval but an initial guess of x^*. We take the value $x_0 = 1$ as an initial guess. In this case, $f(x) = \ln x + x$ and $f'(x) = \frac{1}{x} + 1$.

```
def f(x):
    y = np.log(x) + x
    return y
def df(x):
    y = 1.0 / x + 1.0
    return y
tol = 1.e-4
maxit = 50
x0 = 1.0
x = newton(f, df, x0, tol, maxit)
print('The aproximate solution is: ', x)
print('And the error is: ', f(x))
```

```
1 0.5
2 0.5643823935199818
3 0.5671389877150601
4 0.5671432903993691
The approximate solution x is:  0.5671432903993691
And the value f(x) is:  -2.877842408821607e-11
```

Observe that even if the tolerance `tol=1.e-4` was the same as in the bisection method, Newton's method converged to a better approximation with only 4 iterations! This is an indication (in accordance with the derivation of the method) that the method is quadratically convergent.

We have already noted that Newton's method is a fixed-point method with

$$g(x) = x - \frac{f(x)}{f'(x)} \ .$$

If f is twice-differentiable in a neighborhood of the root x^*, then we have

$$g'(x) = 1 - \frac{(f'(x))^2 - f(x)\, f''(x)}{(f'(x))^2} = \frac{f(x)\, f''(x)}{(f'(x))^2} \ .$$

Thus, if $f'(x^*) \neq 0$, then $g'(x^*) = 0$, and according to the remarks in Section 5.2.3 we expect the convergence rate to be $r > 1$. We prove that the convergence rate is indeed $r = 2$.

> ☞ Newton's method is a particular fixed-point method that is expected to converge with rate of order 2. Downsides of Newton's method include the requirement of the derivative $f'(x_k)$ at every iteration. If f is not differentiable, Newton's method cannot be applied.

5.3.2 Rate of convergence

In this section we prove that Newton's method converges with quadratic rate. We saw that Newton's method is a fixed-point method of the form

$$x_{k+1} = g(x_k) \quad \text{with} \quad g(x) = x - \frac{f(x)}{f'(x)} .$$

Denoting the errors of two successive iterations by

$$e_{k+1} = x_{k+1} - x^* \quad \text{and} \quad e_k = x_k - x^* ,$$

we have that

$$x_{k+1} = x^* + e_{k+1} \quad \text{and} \quad x_k = x^* + e_k .$$

Substitution into $x_{k+1} = g(x_k)$ and using Taylor's expansion yields

$$x^* + e_{k+1} = g(x^* + e_k) = g(x^*) + g'(x^*)e_k + \frac{g''(\xi)}{2!}e_k^2 ,$$

where ξ is between x_k and x_{k+1}. Because x^* and $g(x^*)$ are canceled out we obtain

$$e_{k+1} = g'(x^*)e_k + \frac{g''(\xi)}{2!}e_k^2 . \tag{5.7}$$

We saw that

$$g'(x) = 1 - \frac{(f'(x))^2 - f(x)f''(x)}{(f'(x))^2} = \frac{f(x)f''(x)}{(f'(x))^2} ,$$

which evaluated at $x = x^*$ becomes

$$g'(x^*) = \frac{f(x^*)f''(x^*)}{f'(x^*)^2} = 0 ,$$

since $f(x^*) = 0$ by definition (assuming $f''(x^*)$ and $f'(x^*)$ are behaved appropriately). The expansion (5.7) simplifies to

$$e_{k+1} = \frac{g''(\xi)}{2!}e_k^2, \qquad \text{and thus} \qquad |e_{k+1}| = \left|\frac{g''(\xi)}{2!}\right| |e_k|^2.$$

Newton's method is therefore quadratically convergent where the constant $\frac{g''(\xi)}{2}$ is controlled by the second derivative of g. The requirement for the function g to be twice differentiable is not necessary. It suffices that f is twice continuously differentiable to establish convergence of Newton's method with appropriate initial guess. This can be seen in the following theorem.

Theorem 5.5. *Let f be twice continuously differentiable function and x^* a simple root of f. That is $f(x^*) = 0$ and $f'(x^*) \neq 0$. Then, there is a closed interval $[a, b]$ such that the sequence $\{x_k\}_{k\in\mathbb{N}}$ of Newton's iterations given by the recursive formula*

$$x_{k+1} = x_k - \frac{f(x_k)}{f'(x_k)} ,$$

converges to the root x^* for any initial guess $x_0 \in [a, b]$. Moreover, the convergence is quadratic with

$$\lim_{k\to\infty} \frac{e_{k+1}}{e_k^2} = \frac{f''(x^*)}{2f'(x^*)} ,$$

where $e_k = x_k - x^*$ and $e_{k+1} = x_{k+1} - x^*$.

Proof. First we prove the convergence with the relaxed assumptions required by the theorem. Since f is twice continuously differentiable, then the function $g(x) = x - f(x)/f'(x)$ is just continuously differentiable with $g'(x^*) = 0$. Since $g'(x^*) = 0$ we can find a closed interval $[a, b]$ that contains $x^* \in [a, b]$ by choosing a and b to be very close to x^* if necessary such that

$$\max_{x\in[a,b]} |g'(x)| \leq K < 1 .$$

From Proposition 5.4 we have that the sequence $x_{k+1} = g(x_k)$ converges to x^*, with error $e_k = x_k - x^* \to 0$ as $k \to \infty$.

Now we are ready to estimate the convergence rate of Newton's iterations. Since f is twice continuously differentiable we have the following Taylor expansions for the function f and its derivative

$$\begin{aligned} f(x_k) &= f(x^*) + (x_k - x^*)f'(x^*) + \frac{(x_k - x^*)^2}{2} f''(c_k) , \\ f'(x_k) &= f'(x^*) + (x_k - x^*)f''(d_k) , \end{aligned}$$

where c_k and d_k between x_k and x^*. Substituting these expansions into the Newton iteration $x_{k+1} = x_k - f(x_k)/f'(x_k)$ we get

$$x_{k+1} = x_k - \frac{e_k f'(x^*) + \frac{1}{2}e_k^2 f''(c_k)}{f'(x^*) + e_k f''(d_k)} ,$$

which can be written as

$$x_{k+1} - x^* = x_k - x^* - \frac{e_k f'(x^*) + \frac{1}{2}e_k^2 f''(c_k)}{f'(x^*) + e_k f''(d_k)} .$$

Using the notation $e_k = x_k - x^*$, we write the previous relationship in the form

$$e_{k+1} = e_k - \frac{e_k f'(x^*) + \frac{1}{2}e_k^2 f''(c_k)}{f'(x^*) + e_k f''(d_k)} = \frac{e_k^2 f''(d_k) + \frac{1}{2}e_k^2 f''(c_k)}{f'(x^*) + e_k f''(d_k)} .$$

Dividing both sides by e_k^2 we obtain

$$\frac{e_{k+1}}{e_k^2} = \frac{f''(d_k) - \frac{1}{2}f''(c_k)}{f'(x^*) + e_k f''(d_k)} .$$

Taking the limit as $k \to \infty$ we have

$$\lim_{k\to\infty} \frac{e_{k+1}}{e_k^2} = \frac{f''(x^*)}{2f'(x^*)} ,$$

because $c_k \to x^*$ and $d_k \to x^*$ since c_k and d_k are between x_k and x^*. □

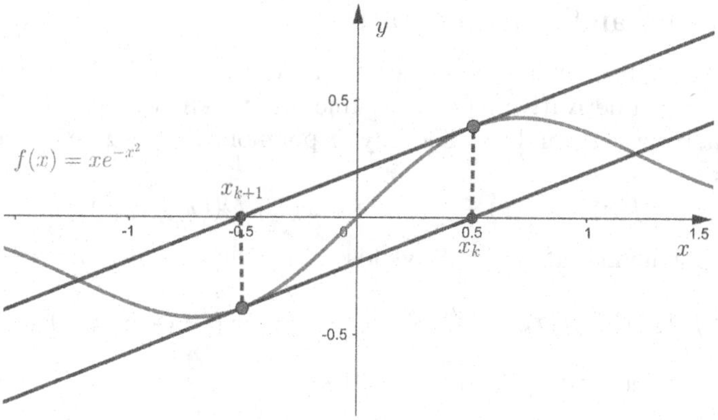

FIGURE 5.6
An example of Newton's method failure.

The existence of a particular interval $[a, b]$ from which we can choose initial guess according to Theorem 5.5 is very important. For example, let's consider the function $f(x) = xe^{-x^2}$, which has the root $x^* = 0$. Starting with the initial guess $x_0 = 1/2$, Newton's method fails to converge as it leads to an infinite loop generating a sequence of repeating approximations $x_{2k} = 1/2$ and $x_{2k+1} = -1/2$ for all $k = 0, 1, 2, \ldots$. Starting with any value x_0 larger than $1/2$ (or smaller than $-1/2$) results in a divergent sequence because the initial guess is far away from the exact root. On the contrary, starting with any value x_0 with $|x_0| < 1/2$ the method converges rapidly.

This particular case of convergence failure is depicted in Figure 5.6. It is observed that starting from $x_0 = \pm 0.5$, the tangent lines are parallel and they return to the same point at every iteration. If we don't implement a stopping criterion for the maximum number of iterations, the code would never stop. Another case where Newton's method can fail is in the same example when we choose $x_0 = \pm 1/\sqrt{2}$. In this case, we have that $f'(x_0) = 0$, and thus the tangent line is parallel to the x-axis.

In all these unfortunate situations we need to choose a different value x_0 closer to the value x^*. Initial guesses, lying closer to x^*, can be obtained with a few iterations of the bisection method.

☞ To ensure Newton's method convergence choose an initial guess close to the exact root x^* where $f'(x_0) \neq 0$. Such initial guess can be obtained using the bisection method.

5.4 Secant Method

One of the main disadvantages of Newton's method is the involvement of the derivative $f'(x)$ in its recursive formula. This can be a significant drawback especially when we solve large systems of nonlinear equations. In order to overcome this requirement we can substitute the first derivative with an appropriate approximation of it, leading to a different fixed-point method. One such method is the *secant method.*

5.4.1 Derivation and implementation

The secant method relies on the modification of Newton's method using a particular approximation of the first derivative $f'(x_k)$. Assume that we know at least two approximations of the root x^*, namely x_{k-1} and x_k. The Taylor polynomial of degree one about x_k for the function $f(x)$ is

$$f(x) = f(x_k) + f'(x_k)(x - x_k) + O((x - x_k)^2) \ .$$

Evaluating this polynomial at $x = x_{k-1}$ yields

$$f(x_{k-1}) = f(x_k) + f'(x_k)(x_{k-1} - x_k) + O((x_k - x_{k-1})^2) \ .$$

Solving for the first derivative $f'(x_k)$ we get that

$$f'(x_k) = \frac{f(x_k) - f(x_{k-1})}{x_k - x_{k-1}} + O(x_k - x_{k-1}) \ .$$

Discarding the residual in the last equation, we obtain the approximation of the first derivative

$$f'(x_k) \approx \frac{f(x_k) - f(x_{k-1})}{x_k - x_{k-1}} \ . \tag{5.8}$$

Such a formula is called *finite difference approximation* of the first derivative. The specific approximation can be quite accurate when the values x_{k-1} and x_k are very close. This can be seen from the residual, which is $O(x_k - x_{k-1})$. Alternatively, the same approximation can be obtained by the definition of the first derivative

$$f'(x_k) = \lim_{x \to x_k} \frac{f(x) - f(x_k)}{x - x_k} \ .$$

From this definition, if we choose $x = x_{k-1}$ very close to x_k, we get the same approximation as before.

Substituting the approximation of the derivative (5.8) into Newton's iteration (5.5) $x_{k+1} = x_k - \frac{f(x_k)}{f'(x_k)}$, we obtain

$$x_{k+1} = x_k - \frac{f(x_k)}{\frac{f(x_k)-f(x_{k-1})}{x_k - x_{k-1}}} \ ,$$

or in a more compact form

$$x_{k+1} = x_k - \frac{x_k - x_{k-1}}{f(x_k) - f(x_{k-1})} f(x_k) \ . \tag{5.9}$$

This can be written also as

$$x_{k+1} = \frac{f(x_k)x_{k-1} - f(x_{k-1})x_k}{f(x_k) - f(x_{k-1})} \ ,$$

for $k = 0, 1, 2, \ldots$.

There is one small issue: In order to compute the first improved approximation x_1 of the root x^* we need two initial guesses of the root x^*, namely x_{-1} and x_0 with $x_{-1} \neq x_0$. If we have only one guess for the value x_0, then we can take the second guess to be $x_{-1} = x_0 \pm \Delta x$ for a small value Δx.

The secant method is a relaxation-type method of the form $x_{k+1} = x_k - M(x_k, x_{k-1}) f(x_k)$. The main difference between the secant method and Newton's method

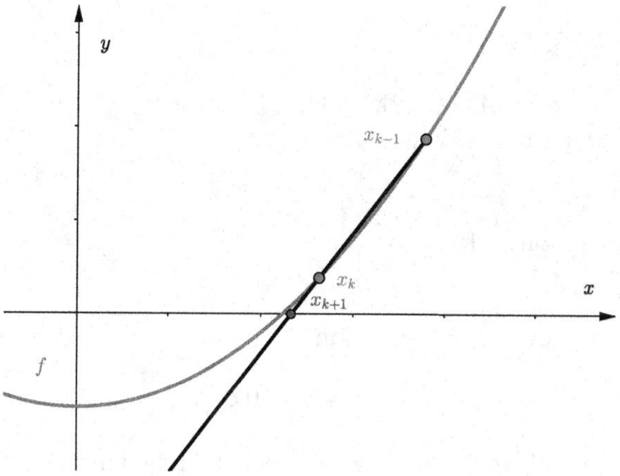

FIGURE 5.7
Geometric interpretation of the secant method.

is that the secant method approximates the root with the point of intersection between the x-axis and a secant line with slope $\frac{f(x_k)-f(x_{k-1})}{x_k-x_{k-1}}$. This is depicted in Figure 5.7.

The algorithm for the secant method is summarized in Algorithm 14.

Algorithm 14 Secant method

Set a tolerance TOL for the accuracy
Set the maximum number of iterations $MAXIT$
Initialize x_{k-1} and x_k and set $k = 0$
Set $Error = |x_k - x_{k-1}|$
while $Error > TOL$ and $k < MAXIT$ **do**
  Compute new $x_{k+1} = x_k - \frac{x_k-x_{k-1}}{f(x_k)-f(x_{k-1})} f(x_k)$
  Set $Error = |x_{k+1} - x_k|$
  Increase the counter $k = k + 1$
end while
Return the approximation x_k of the root x^*

In this algorithm we assume that the values $x_{k-1} \neq x_k$. In some examples, the initial guesses are computed using bisection or other methods and so we should not worry about their choice. To keep things simple here, we consider the initial guesses manually.

The algorithm for the secant method can be implemented in Python as follows

```
def secant(f, x1, x2, tol = 1.e-6, maxit = 100):
    # f = the function f(x)
    # x1 = an initial guess of the solution
    # x2 = another initial guess of the solution
    # tol = tolerance for the absolute error
    # maxit = maximum number of iterations
    err = 1.0
    iteration = 0
    while (err > tol and iteration < maxit):
        xk = x1
        xk1 = x2
```

```
        iteration = iteration + 1
        err = xk1
        xk1 = xk - (xk-xk1)/(f(xk)-f(xk1))*f(xk)
        err = np.abs(err - xk1)
        x1 = x2
        x2 = xk1
        print(iteration, xk1)
    return xk1
```

Testing this new method for the equation

$$\ln x + x = 0 ,$$

with initial guesses $x_{-1} = 1$ and $x_0 = 2$, we obtain the following results:

```
def f(x):
    y = np.log(x) + x
    return y
tol = 1.e-4
maxit = 50
x1 = 1.0
x2 = 2.0
x = secant(f, x1, x2, tol, maxit)
print('The approximate solution is: ', x)
print('And the error is: ', f(x))
```

```
1 0.40938389085035876
2 0.651575386390747
3 0.5751035382227284
4 0.5667851889083253
5 0.5671448866112347
6 0.5671432907314143
The approximate solution x is:  0.5671432907314143
And the value f(x) is:  8.887366398369068e-10
```

Observe that the secant method required 6 iterations to satisfy the same stopping criteria with Newton's method. This indicates that the secant method is slower than Newton's method. This is indeed the price we pay for approximating the derivative $f'(x_k)$ using a finite difference quotient. The secant method converges with convergence rate $r \approx 1.62$, which is smaller than Newton's method convergence rate but still greater than 1. This fact makes the secant method appealing.

> ☞ The secant method is based on Newton's iteration with the difference that the first derivative is approximated by a finite difference quotient. The price we pay for using the finite difference approximation to the derivative is a few extra iterations for convergence. The convergence rate of the secant method is equal to the golden ratio $r = 1.6180\cdots$.

5.4.2 Rate of convergence

Proving convergence for the secant method is a difficult task. For this reason, first we estimate the convergence rate and then we proceed with a formal proof of convergence. Because the formal proof is too technical you can skip it and proceed with an experimental verification of the convergence rate as described in Section 5.1.3.

In order to estimate the rate of convergence of the secant method we define the errors $e_{k-1} = x_{k-1} - x^*$, $e_k = x_k - x^*$ and $e_{k+1} = x_{k+1} - x^*$. We will show that

$$\lim_{k\to\infty} \frac{|e_{k+1}|}{|e_k|^r} = C ,$$

where $r = (1+\sqrt{5})/2$ and $C > 0$ constant.

Starting with the error e_{k+1} we have that

$$\begin{aligned}
e_{k+1} &= x_{k+1} - x^* \\
&= x_k - \frac{x_k - x_{k-1}}{f(x_k) - f(x_{k-1})} f(x_k) - x^* \\
&= e_k - \frac{x_k - x_{k-1}}{f(x_k) - f(x_{k-1})} f(x_k) \\
&= \frac{x_k - x_{k-1}}{f(x_k) - f(x_{k-1})} \left[\frac{f(x_k) - f(x_{k-1})}{x_k - x_{k-1}} e_k - f(x_k) \right] \\
&= \frac{x_k - x_{k-1}}{f(x_k) - f(x_{k-1})} \left[\frac{f(x_k) - f(x_{k-1})}{x_k - x_{k-1}} - \frac{f(x_k)}{e_k} \right] e_k \\
&= \frac{x_k - x_{k-1}}{f(x_k) - f(x_{k-1})} \left[\frac{f(x_k) - f(x_{k-1}) - \frac{f(x_k)}{e_k}(x_k - x_{k-1})}{x_k - x_{k-1}} \right] e_k \\
&= \frac{x_k - x_{k-1}}{f(x_k) - f(x_{k-1})} \left[\frac{f(x_k) - f(x_{k-1}) - \frac{f(x_k)}{e_k}(x_k - x^* + x^* - x_{k-1})}{x_k - x_{k-1}} \right] e_k \\
&= \frac{x_k - x_{k-1}}{f(x_k) - f(x_{k-1})} \left[\frac{f(x_k) - f(x_{k-1}) - \frac{f(x_k)}{e_k}(e_k - e_{k-1})}{x_k - x_{k-1}} \right] e_k \\
&= \frac{x_k - x_{k-1}}{f(x_k) - f(x_{k-1})} \left[\frac{-f(x_{k-1}) + \frac{f(x_k)}{e_k} e_{k-1}}{x_k - x_{k-1}} \right] e_k ,
\end{aligned}$$

or equivalently,

$$e_{k+1} = \frac{x_k - x_{k-1}}{f(x_k) - f(x_{k-1})} \left[\frac{\frac{f(x_k)}{e_k} - \frac{f(x_{k-1})}{e_{k-1}}}{x_k - x_{k-1}} \right] e_k e_{k-1} . \tag{5.10}$$

Now we simplify the fractions in the previous relation. The second order Taylor expansion of $f(x_k)$ about x^* is

$$f(x_k) = f(e_k + x^*) = f(x^*) + e_k f'(x^*) + \frac{1}{2} e_k^2 f''(x^*) + O(e_k^3) .$$

Since $f(x^*) = 0$ we have that $f(x_k) = e_k f'(x^*) + \frac{1}{2} e_k^2 f''(x^*) + O(e_k^3)$. After discarding the residual and dividing by e_k we get

$$\frac{f(x_k)}{e_k} \approx f'(x^*) + \frac{1}{2} e_k f''(x^*) . \tag{5.11}$$

Similarly we have that

$$\frac{f(x_{k-1})}{e_{k-1}} \approx f'(x^*) + \frac{1}{2}e_{k-1}f''(x^*) \ . \tag{5.12}$$

Eliminating the term $f'(x^*)$ from the relations (5.11) and (5.12) yields

$$\frac{f(x_k)}{e_k} - \frac{f(x_{k-1})}{e_{k-1}} \approx \frac{1}{2}(e_k - e_{k-1})f''(x^*) \ .$$

Since $e_k - e_{k-1} = x_k - x^* - x_{k-1} + x^* = x_k - x_{k-1}$ we get

$$\frac{\frac{f(x_k)}{e_k} - \frac{f(x_{k-1})}{e_{k-1}}}{x_k - x_{k-1}} \approx \frac{1}{2}f''(x^*) \ . \tag{5.13}$$

For the first ratio in (5.10) we have that

$$\frac{x_k - x_{k-1}}{f(x_k) - f(x_{k-1})} \approx \frac{1}{f'(x^*)} \ . \tag{5.14}$$

Substituting (5.13) and (5.14) into (5.10) we get

$$e_{k+1} \approx \frac{1}{2}\frac{f''(x^*)}{f'(x^*)}e_k e_{k-1} \ , \tag{5.15}$$

and thus

$$|e_{k+1}| \approx C|e_k||e_{k-1}| \ , \tag{5.16}$$

with

$$C = \frac{1}{2}\frac{|f''(x^*)|}{|f'(x^*)|} \ .$$

If $f''(x^*) \neq 0$ and $f'(x^*) \neq 0$, then $C > 0$. If the rate of convergence is $r > 1$, then dividing both sides of (5.16) by $|e_k|^r$ gives

$$\frac{|e_{k+1}|}{|e_k|^r} \approx C|e_k|^{1-r}|e_{k-1}| = C\left(\frac{|e_k|}{|e_{k-1}|^r}\right)^{\alpha} \ , \tag{5.17}$$

where the last equality holds for $\alpha = 1 - r$ and $\alpha = -1/r$. Thus, $1 - r = -1/r$ and so $r^2 - r - 1 = 0$, which has root $r > 1$,

$$r = \frac{1+\sqrt{5}}{2} \approx 1.62 \ .$$

Therefore, if we knew that the sequence $y_k = \frac{|e_{k+1}|}{|e_k|^r}$ converges to some constant $C > 0$ then we get the desired convergence

$$\lim_{k\to\infty} \frac{|e_{k+1}|}{|e_k|^r} = C \ ,$$

with rate is $r = (1+\sqrt{5})/2$.

The number $r = (1+\sqrt{5})/2$ is called *the golden ratio* and is known since the era of Euclid. A ratio r between two positive numbers $a > b > 0$ is called golden ratio if it is

$$r = \frac{a}{b} = \frac{a+b}{a} \ .$$

In other words, if a and b are lengths in which we split a straight line with $a > b > 0$, then the ratio $r = a/b$ is called golden if it is equal to the ratio between the length of the whole line $a + b$ and the largest length a.

Proof of convergence

While the previous estimation is very enlightening it does not suggest a formal proof of convergence. On the other hand, it is crucial to know that our numerical method converges to the correct solution. In order to have a complete theory for the secant method we proceed with its rigorous proof of convergence. First we present a lemma which makes the proof a little easier. This lemma estimates the difference between a function f and its secant passing through two points of f.

Lemma 5.6. *If $f \in C^2[a,b]$, $x_{k-1}, x_k \in [a,b]$, $x_{k-1} \neq x_k$ and there is a polynomial $p \in \mathbb{P}_1$ of degree at most one such that $p(x_{k-1}) = f(x_{k-1})$ and $p(x_k) = f(x_k)$, then for all $x \in [a,b]$ there is a $\xi \in J := (\min\{x, x_{k-1}, x_k\}, \max\{x, x_{k-1}, x_k\})$ such that*

$$f(x) - p(x) = \frac{(x_{k-1} - x)(x_k - x)}{2} f''(\xi) .$$

Proof. For $x = x_{k-1}$ and $x = x_k$ the lemma is obviously true since for such values of x we have $f(x) - p(x) = 0$. For $x \in [a,b]$ with $x \neq x_{k-1}, x_k$, we define the function $h : [a,b] \to \mathbb{R}$ such as

$$h(t) = f(t) - p(t) - \frac{f(x) - p(x)}{(x_{k-1} - x)(x_k - x)} (x_{k-1} - t)(x_k - t) .$$

We observe that $h \in C^2[a,b]$ and $h(x_{k-1}) = h(x_k) = h(x) = 0$. According to Rolle's Theorem (Theorem 4.2) the second derivative $h''(t)$ has at least one root ξ in the interval J. Since p is a linear polynomial, $p''(t) = 0$ and thus

$$h''(t) = f''(t) - 2\frac{f(x) - p(x)}{(x_{k-1} - x)(x_k - x)} ,$$

which implies

$$0 = h''(\xi) = f''(\xi) - 2\frac{f(x) - p(x)}{(x_{k-1} - x)(x_k - x)} .$$

Solving for $f(x) - p(x)$ we get the relationship

$$f(x) - p(x) = \frac{(x_{k-1} - x)(x_k - x)}{2} f''(\xi) ,$$

and this completes the proof. □

Now we state the theorem of convergence of the secant method.

Theorem 5.7. *If $f \in C^2(a,b)$ with a simple root x^* such that $f'(x^*) \neq 0$ and $f''(x^*) \neq 0$, then there is an interval I, which contains the root x^*, such that for $x_{-1}, x_0 \in I$, with $x_{-1} \neq x_0$, the sequence $\{x_k\}_{k \in \mathbb{N}}$ generated by the secant method*

$$x_{k+1} = x_k - \frac{x_k - x_{k-1}}{f(x_k) - f(x_{k-1})} f(x_k) , \tag{5.18}$$

for $k = 0, 1, \ldots$, converges to x^* as $k \to \infty$ with rate the golden ratio $r = (1 + \sqrt{5})/2$.

Proof. We first define the polynomial

$$p(x) := f(x_k) - \frac{f(x_k) - f(x_{k-1})}{x_k - x_{k-1}} (x_k - x) . \tag{5.19}$$

We observe that $p(x_k) = f(x_k)$ and also $p(x_{k-1}) = f(x_{k-1})$. Taking $x = x_{k+1}$ in (5.19) and using (5.18) we can verify that $p(x_{k+1}) = 0$. From Lemma 5.6 there is a $\xi_k \in (\min\{x, x_{k-1}, x_k\}, \max\{x, x_{k-1}, x_k\})$ for any $k = 0, 1, 2, \ldots$ such that

$$f(x^*) - p(x^*) = \frac{(x_{k-1} - x^*)(x_k - x^*)}{2} f''(\xi_k) \ .$$

Since $f(x^*) = 0$ and $p(x_{k+1}) = 0$ we rewrite the previous relations as

$$p(x^*) - p(x_{k+1}) = -\frac{(x_{k-1} - x^*)(x_k - x^*)}{2} f''(\xi_k) \ . \tag{5.20}$$

We subtracted $p(x_{k+1})$ because

$$\begin{aligned} p(x^*) - p(x_{k+1}) &= f(x_k) - \frac{f(x_k) - f(x_{k-1})}{x_k - x_{k-1}}(x_k - x^*) - \\ &\quad - \left[f(x_k) - \frac{f(x_k) - f(x_{k-1})}{x_k - x_{k-1}}(x_k - x_{k+1}) \right] \\ &= -\frac{f(x_k) - f(x_{k-1})}{x_k - x_{k-1}} \left[(x_k - x^*) - (x_k - x_{k+1}) \right] \\ &= \frac{f(x_k) - f(x_{k-1})}{x_k - x_{k-1}} (x_{k+1} - x^*) \ , \end{aligned}$$

and thus (5.20) becomes

$$(x_{k+1} - x^*) \frac{f(x_k) - f(x_{k-1})}{x_k - x_{k-1}} = -\frac{(x_{k-1} - x^*)(x_k - x^*)}{2} f''(\xi_k) \ ,$$

or in other words

$$e_{k+1} = -\frac{1}{2} \frac{f''(\xi_k)}{\frac{x_k - x_{k-1}}{f(x_k) - f(x_{k-1})}} e_k e_{k-1} \ .$$

From the Mean Value Theorem of differential calculus (see Theorem 4.3), we deduce that for all $k = 0, 1, \ldots$ there is a $\zeta_k \in (\min\{x_{k-1}, x_k\}, \max\{x_{k-1}, x_k\})$ such that

$$\frac{f(x_k) - f(x_{k-1})}{x_k - x_{k-1}} = f'(\zeta_k) \ .$$

Since $f'(x^*) \neq 0$, $f' \in C[a, b]$, and x_{k-1}, x_k are close to x^*, then $f'(\zeta_k) \neq 0$. Therefore, we have that

$$e_{k+1} = -\frac{1}{2} \frac{f''(\xi_k)}{f'(\zeta_k)} e_k e_{k-1} \ .$$

Since we want to estimate the rate of the absolute error we get

$$|e_{k+1}| = \frac{1}{2} \frac{|f''(\xi_k)|}{|f'(\zeta_k)|} |e_k| |e_{k-1}| \ . \tag{5.21}$$

Observe that there is $0 < \delta < 1$ such that

$$\frac{1}{2} \max_{x, y \in I} \left| \frac{f''(x)}{f'(y)} \right| \leq \frac{c}{\delta} \ ,$$

for all $x, y \in I = [x^* - \delta, x^* + \delta]$ with $c < 1$. Then, for $x_{-1}, x_0 \in I$, (5.21) implies that $|e_1| = |x_1 - x^*| < \delta$ and thus $x_2 \in I$. Similarly we conclude that $x_k \in I$ for all $k = 1, 2, \ldots$. Moreover, from (5.21) we have that

$$|e_{k+1}| \le c|e_k| \le c^k |e_0| ,$$

and since $c < 1$ we get that $x_k \to x^*$ as $k \to \infty$. Since ξ_k and ζ_k are in between x_{k-1} and x_k it holds also that $\xi_k \to x^*$, $\zeta_k \to x^*$ and therefore the sequence

$$C_k = \frac{1}{2}\left|\frac{f''(\xi_k)}{f'(\zeta_k)}\right| \quad \text{converges to} \quad C = \frac{1}{2}\left|\frac{f''(x^*)}{f'(x^*)}\right| .$$

Moreover, we can write (5.21) in the form $|e_{k+1}| = C_k|e_k||e_{k-1}|$, with $\lim_{k\to\infty} C_k = C$.

As in (5.17), if r is the rate of convergence, then

$$\frac{|e_{k+1}|}{|e_k|^r} = C_k \left(\frac{|e_k|}{|e_{k-1}|^r}\right)^{-1/r} , \tag{5.22}$$

with $r = (1+\sqrt{5})/2$. Denoting the sequence $y_k = \frac{|e_k|}{|e_{k-1}|^r}$, we rewrite (5.22) $y_{k+1} = C_k y_k^q$, with $q = -1/r$. We prove that the sequence $\{y_k\}_{k\in\mathbb{N}}$ converges with $\lim_{k\to\infty} y_k = \ell \neq 0$. In fact we have

$$y_{k+1} = C_k C_{k-1}^q C_{k-2}^{q^2} \cdots C_0^{q^k} y_0^{q^{k+1}} ,$$

or after setting $c_k = C_k/C$,

$$y_{k+1} = C^{1+q+\cdots q^k} c_k c_{k-1}^q c_{k-2}^{q^2} \cdots c_0^{q^k} y_0^{q^{k+1}} .$$

Because $|q| = 1/r < 1$ and $1 - r = q = -1/r$, we have

$$C^{1+q+\cdots q^k} c_k c_{k-1}^q c_{k-2}^{q^2} \cdots c_0^{q^k} = C^{\frac{1-q^{k+1}}{1-q}} \to C^{\frac{1}{1-q}} = C^{\frac{1}{r}}, \text{ as } k \to \infty ,$$

and

$$\lim_{k\to\infty} y_0^{q^{k+1}} = y_0^0 = 1 .$$

Denoting $s = |q| < 1$ then we have

$$\begin{aligned} \left|\log\left(c_k c_{k-1}^q c_{k-2}^{q^2} \cdots c_0^{q^k}\right)\right| &= |\log c_k + q \log c_{k-1} + \cdots + q^k \log c_0| \\ &\le |\log c_k| + s|\log c_{k-1}| + \cdots + s^k |\log c_0| . \end{aligned}$$

For $\varepsilon > 0$, let $K > 0$ such that $|\log c_k| < \varepsilon$ for $k \ge K$. Then

$$\left|\log\left(c_k c_{k-1}^q c_{k-2}^{q^2} \cdots c_0^{q^k}\right)\right| \le \varepsilon(1 + s + s^2 + \cdots s^{k-K}) + s^{k-K+1} \max_i \log |c_i| ,$$

from which we conclude that

$$\lim_{k\to\infty} \left(c_k c_{k-1}^r \cdots c_0^{r^k}\right) = 1 .$$

Thus

$$\lim_{k\to\infty} y_k = C^{\frac{1}{r}} = \ell ,$$

and this completes the proof. □

5.5 Other Methods and Generalizations

So far we have discussed three classical numerical methods for the approximation of simple roots of functions. However, there are problems where we need to compute roots of multiplicity $m > 1$ or solve systems of nonlinear equations. These issues among many others require further investigation. Here we present briefly extensions to some of the previous methods and how to deal with more general problems.

5.5.1 Stopping criteria

All iterative methods we studied in this chapter generate sequences of approximations x_k. In practice we need to terminate the iterations when the value of x_k is close to the root x^*. Because we do not usually know the exact value of x^* we are not able to measure the error $|x_k - x^*|$. On the other hand, if the sequence $\{x_k\}_{k\in\mathbb{N}}$ converges to x^*, then for values of $k > K$ for some $K \in \mathbb{N}$ the error $|x_k - x^*| < \varepsilon$ becomes small. This inequality yields

$$|x_{k+1} - x_k| = |x_{k+1} - x^* + x^* - x_k| \le |x_{k+1} - x^*| + |x_k - x^*| < 2\varepsilon \ ,$$

for $k > K$. If the difference $|x_k - x^*|$ is small, then the difference $|x_{k+1} - x_k|$ is small too. Thus, instead of checking whether the error $|x_k - x^*|$ is small, we check whether the difference between subsequent approximations $|x_{k+1} - x_k|$ is small. In some cases, it is more reliable to estimate the relative error $|x_{k+1} - x_k|/|x_k|$ instead of the absolute error.

Because of all these, we consider tolerance $TOL > 0$, and at every iteration step we check whether $|x_{k+1} - x_k| < TOL$. Depending on the required accuracy, we give appropriate values to the tolerance TOL, for example, something between 10^{-5} to 10^{-15}.

As we discussed earlier, there is a possibility of a failure in convergence due to an unfortunate initial guess x_0. It is then useful to include an additional stopping criterion that restricts the total number of iterations. In the unfortunate event where the maximum number of iterations has been reached, we can terminate the program with an error message. The error message can indicate failure in achieving the required accuracy described by the tolerance TOL within the prescribed number of iterations.

5.5.2 Steffensen's method

Steffensen's method is a modification of Newton's method that bypasses the use of the first derivative f'. The advantage of Steffensen's method compared to the secant method is that the convergence is quadratic. The Steffensen iteration is

$$x_{k+1} = x_k - \frac{f(x_k)}{h(x_k)} \ ,$$

with

$$h(x) = \frac{f(x + f(x))}{f(x)} - 1 \ .$$

In order to see the equivalence with Newton's method we take the Taylor expansion of the term $f(x_k + f(x_k))$, which is

$$f(x_k + f(x_k)) = f(x_k) + f'(x_k)f(x_k) + O\left(f^2(x_k)\right) \ ,$$

and thus

$$h(x_k) = \frac{f(x_k + f(x_k))}{f(x_k)} - 1 \approx f'(x_k) \ .$$

The function h is practically the average value for the slope of f in the interval $[x_k, x_k + f(x_k)]$. Taking into account that as x_k approximates the root x^*, then the value $f(x_k)$ becomes very small, and this approximation is very close to the actual slope of the tangent $f'(x_k)$. The requirement for the Steffensen iteration to converge is that $-1 < f'(x^*) < 0$, which is also the major disadvantage of this method. For other methods that overcome the barrier of the first derivative please see the exercises at the end of this chapter.

5.5.3 Aitken's delta-squared process

Aitken's delta-squared process, also known as the Aitken extrapolation, is a technique to accelerate the convergence of an iterative method. This technique requires the computation of three approximations x_k, x_{k+1} and x_{k+2} of the exact solution x^*. Then, assuming that

$$\frac{x_{k+1} - x^*}{x_k - x^*} \approx \frac{x_{k+2} - x^*}{x_{k+1} - x^*} ,$$

we have that

$$(x_{k+1} - x^*)^2 \approx (x_{k+2} - x^*)(x_k - x^*) ,$$

and solving for x^* we get

$$x^* \approx \frac{x_{k+2}x_k - (x_{k+1})^2}{x_{k+2} - 2x_{k+1} + x_k} \quad \text{or equivalently} \quad x^* \approx x_k - \frac{(x_{k+1} - x_k)^2}{x_{k+2} - 2x_{k+1} + x_k} .$$

Setting

$$x_{k+3} = x_k - \frac{(x_{k+1} - x_k)^2}{x_{k+2} - 2x_{k+1} + x_k} ,$$

and without using the actual iterative method, we derived an improved approximation. For this reason, Aitken's delta-squared process can be used with any of the previously mentioned numerical methods.

It is worth mentioning that Aitken's process can lead to Steffensen's method. Suppose we know an approximate solution x_k of the equation $f(x) = 0$. Also, assume we write the last equation as $x = g(x)$. Then, we define $x_{k+1} = g(x_k)$, $x_{k+2} = g(x_{k+1})$ as usual, and we redefine the iteration x_{k+1} using Aitken's delta squared method as

$$x_{k+1} = x_k - \frac{(x_{k+1} - x_k)^2}{x_{k+2} - 2x_{k+1} + x_k} = x_k - \frac{(g(x_k) - x_k)^2}{g(g(x_k)) - 2g(x_k) + x_k} ,$$

for $k = 0, 1, 2, \ldots$. It remains to choose appropriate function g. By taking the naive fixed-point function $g(x) = x + f(x)$ in the previous iteration and after simplifying same terms we obtain

$$x_{k+1} = x_k - \frac{[f(x_k)]^2}{f(x_k + f(x_k)) - f(x_k)} ,$$

which is exactly the Steffensen method. Therefore, the Steffensen method can be considered a generic fixed-point method that is similar to Aitken's delta-squared process.

5.5.4 Multiple roots

So far we assumed that x^* is a simple root of the equation $f(x) = 0$. This means that we required $f(x^*) = 0$ and $f'(x^*) \neq 0$. We say that the root x^* has multiplicity m if $f(x^*) = f'(x^*) = \cdots = f^{(m-1)}(x^*) = 0$ and $f^{(m)}(x^*) \neq 0$. In such a case, the convergence of Newton's method is not quadratic. When we search for roots with multiplicity $m > 1$ we need to modify the formula (5.5) appropriately in order to achieve quadratic convergence rate.

The modification of Newton's method is based on the following observation: If x^* is a root of $f(x)$ with multiplicity m, then $f(x) = (x - x^*)^m h(x)$, where $h(x)$ is a function with no root. The function $F(x) = \sqrt[m]{f(x)}$ has the simple root x^* and thus, according to Theorem 5.5, we can apply Newton's method with quadratic convergence rate. Applying Newton's method to the function $F(x)$ we get

$$x_{k+1} = x_k - \frac{F(x_k)}{F'(x_k)} = x_k - \frac{\sqrt[m]{f(x_k)}}{\frac{\sqrt[m]{f(x_k)}}{mf(x_k)} f'(x_k)} ,$$

which can be simplified to the following modification of Newton's method

$$x_{k+1} = x_k - m\frac{f(x_k)}{f'(x_k)} .$$

The modified Newton's method has similar convergence properties with the classical Newton's method and of course quadratic convergence rate.

5.5.5 High-order methods

It is possible to construct fixed-point methods that converge with cubic rate. One such example is Halley's method. The derivation of Halley's method is based on the observation that the functions $f(x)$ and $F(x) = f(x)/\sqrt{|f'(x)|}$ have the same simple roots. Instead of using Newton's method applied to the function $f(x)$ we can use Newton's method for the function $F(x)$. This can be written as

$$x_{k+1} = x_k - \frac{F(x_k)}{F'(x_k)} ,$$

which can be simplified into the iteration

$$x_{k+1} = x_k - \frac{2f(x_k)f'(x_k)}{2[f'(x_k)]^2 - f(x_k)f''(x_k)} .$$

This method converges faster than the classical Newton iteration. However, it requires in addition to the first derivative, the second derivative of f. Newton's method usually converges very fast even with quadratic convergence, and taking into account the increased complexity of Halley's method, it is nevertheless preferable to use the simple Newton's method.

5.5.6 Systems of equations

Suppose that we want to find the solution (x^*, y^*) of the system of simultaneous equations

$$\begin{cases} x - y^3 = 1 \\ \sin(x) + y = 1 \end{cases} . \tag{5.23}$$

Each equation corresponds to a curve on the xy−plane. The solution (x^*, y^*) is the point of intersection of these curves.

For simplicity, we consider the vector $\boldsymbol{x} = (x, y)^T$ and we write our system in the form $\boldsymbol{F}(\boldsymbol{x}) = 0$, where $\boldsymbol{F} : \mathbb{R}^2 \to \mathbb{R}^2$ is

$$\boldsymbol{F}(\boldsymbol{x}) = \begin{pmatrix} f_1(x, y) \\ f_2(x, y) \end{pmatrix} .$$

In our case, with only two unknowns, we can write (5.23) in the form $\boldsymbol{F}(\boldsymbol{x}) = 0$, with

$$\boldsymbol{F}(\boldsymbol{x}) = \begin{pmatrix} x - y^3 - 1 \\ \sin(x) + y - 1 \end{pmatrix} .$$

Most of the iterative methods discussed in this chapter can be also applied to systems. The easiest way to solve such a system is by using the generic fixed-point method. We rewrite the system as a fixed-point equation $\boldsymbol{x} = \boldsymbol{G}(\boldsymbol{x})$, where $\boldsymbol{x}$ is now a vector and $\boldsymbol{G}$ a vector function. The corresponding fixed-point method is $\boldsymbol{x}^{(k+1)} = \boldsymbol{G}(\boldsymbol{x}^{(k)})$ for $k = 0, 1, 2, \dots$ and a given initial guess $\boldsymbol{x}^{(0)}$.

In our example a straightforward fixed-point system is

$$\begin{cases} x = y^3 + 1 \\ y = -\sin(x) + 1 \end{cases} .$$

The fixed-point method can be written as

$$\begin{cases} x_{k+1} = (y_k)^3 + 1 \\ y_{k+1} = -\sin(x_k) + 1 \end{cases} ,$$

for let's say $(x_0, y_0) = (0, 0)$. Note that the iteration index is now appeared as a superscript to avoid confusion with vector coordinates.

Newton's method is a little bit more complicated, but is again based on Taylor polynomial for multivariable functions. The Taylor polynomial of degree one for a function $\boldsymbol{F}(\boldsymbol{x})$ is the vector function $\boldsymbol{P}_1(\boldsymbol{x})$ such that

$$\boldsymbol{F}(\boldsymbol{x}) \approx \boldsymbol{P}_1(\boldsymbol{x}) = \boldsymbol{F}(\boldsymbol{x}^{(k)}) + \boldsymbol{DF}(\boldsymbol{x}^{(k)})(\boldsymbol{x} - \boldsymbol{x}^{(k)}) ,$$

where

$$\boldsymbol{DF}(\boldsymbol{x}) = \begin{pmatrix} \frac{\partial f_1}{\partial x} & \frac{\partial f_1}{\partial y} \\ \frac{\partial f_2}{\partial x} & \frac{\partial f_2}{\partial y} \end{pmatrix} ,$$

is the Jacobian matrix. Here, $\boldsymbol{x} = (x, y)^T$ and $\boldsymbol{x}^{(k)} = (x_k, y_k)^T$. The Jacobian matrix represents the derivative of a vector function. Taking $\boldsymbol{x} = \boldsymbol{x}^* = (x^*, y^*)^T$ in Taylor's polynomial we get

$$0 \approx \boldsymbol{F}(\boldsymbol{x}^{(k)}) + \boldsymbol{DF}(\boldsymbol{x}^{(k)})(\boldsymbol{x}^* - \boldsymbol{x}^{(k)}) .$$

Solving the last equation for $\boldsymbol{x}^*$ we get the linear system

$$\boldsymbol{DF}(\boldsymbol{x}^{(k)})\boldsymbol{x}^* \approx \boldsymbol{DF}(\boldsymbol{x}^{(k)})\boldsymbol{x}^{(k)} - \boldsymbol{F}(\boldsymbol{x}^{(k)}) .$$

We solve this linear system by multiplying both sides with the inverse of the Jacobian matrix to obtain

$$\boldsymbol{x}^* \approx \boldsymbol{x}^{(k)} - [\boldsymbol{DF}(\boldsymbol{x}^{(k)})]^{-1}\boldsymbol{F}(\boldsymbol{x}^{(k)}) .$$

The new approximation of the solution $\boldsymbol{x}^* = (x^*, y^*)^T$ is then defined

$$\boldsymbol{x}^{(k+1)} = \boldsymbol{x}^{(k)} - [\boldsymbol{DF}(\boldsymbol{x}^{(k)})]^{-1}\boldsymbol{F}(\boldsymbol{x}^{(k)}) ,$$

for $k = 0, 1, 2, \dots$ with a given initial guess $\boldsymbol{x}^{(0)}$. Newton's iteration can be expressed in a more analytical form as

$$\begin{pmatrix} x_{k+1} \\ y_{k+1} \end{pmatrix} = \begin{pmatrix} x_k \\ y_k \end{pmatrix} - [\boldsymbol{DF}(x_k, y_k)]^{-1}\boldsymbol{F}(x_k, y_k) .$$

It is noted that Newton's method applied to a system of equations requires the computation of the inverse of the Jacobian matrix or equivalently the solution of a linear system with matrix $\boldsymbol{DF}(\boldsymbol{x}^{(k)})$.

In our example

$$\boldsymbol{F}(\boldsymbol{x}) = \begin{pmatrix} x - y^3 - 1 \\ \sin(x) + y - 1 \end{pmatrix} \text{ with Jacobian matrix } \boldsymbol{DF}(\boldsymbol{x}) = \begin{pmatrix} 1 & -3y^2 \\ \cos(x) & 1 \end{pmatrix} .$$

The inverse of the Jacobian matrix is

$$[\boldsymbol{DF}(\boldsymbol{x})]^{-1} = \frac{1}{1 + 3y^2\cos(x)} \begin{pmatrix} 1 & 3y^2 \\ -\cos(x) & 1 \end{pmatrix} .$$

Then Newton's method can be formulated as

$$\begin{pmatrix} x_{k+1} \\ y_{k+1} \end{pmatrix} = \begin{pmatrix} x_k \\ y_k \end{pmatrix} - \frac{1}{1 + 3y_k^2\cos(x_k)} \begin{pmatrix} 1 & 3y_k^2 \\ -\cos(x_k) & 1 \end{pmatrix} \begin{pmatrix} x_k - y_k^3 - 1 \\ \sin(x_k) + y_k - 1 \end{pmatrix} ,$$

for let's say $(x_0, y_0) = (0, 0)$. After performing the matrix-vector multiplication and simplifying same terms, we can write the Newton iteration in the form

$$x_{k+1} = x_k - \frac{x_k - 1 + y_k^2(2y_k - 3) + 3y_k^2\sin(x_k)}{1 + 3y_k^2\cos(x_k)} ,$$

$$y_{k+1} = y_k - \frac{y_k - 1 + (1 - x_k - y_k^3)\cos(x_k) + \sin(x_k)}{1 + 3y_k^2\cos(x_k)} ,$$

for $k = 0, 1, 2, \ldots$ with $(x_0, y_0) = (0, 0)$. The solution to this system is $\boldsymbol{x}^* \approx (1.00382207, 0.160754)^T$.

5.6 The Module `scipy.optimize`

In this chapter, we studied to some extent four numerical methods for solving approximately equations of the form $f(x) = 0$. In particular, we studied the bisection, fixed-point, Newton's and secant methods. Python provides implementations of all these methods in the module `scipy.optimize` of SciPy. The module is called `optimize` because the field of optimization in computational mathematics includes the approximation of roots of functions. For example, in optimization we can compute the minimum of a function $f(x)$ by solving the equation of $f'(x) = 0$. The bisection method is implemented in the function `bisect`, the fixed-point method in the function `fixed_point` and the Newton and secant methods are implemented in the function `newton`. (The names of the functions and their variables that we present here are in accordance to the original names of the SciPy functions.)

5.6.1 Bisection method

The function `scipy.optimize.bisect` implements the bisection method. A basic call of this function is

```
bisect(f, a, b, args, xtol, rtol, maxiter, full_output, disp)
```

where

`f`: The name of the function $f(x)$

`a` and `b`: The end-points of the interval $[a, b]$ where the root of f is located

`args`: A tuple with additional arguments and parameters for the functions f. (Optional with default value `args=()`)

`xtol` and `rtol`: The absolute and relative error tolerances for the termination of the method. (Optional with default values `xtol` = 2.e − 12 and `rtol` = 8.88e − 16)

`maxiter`: An integer with the maximum number of iterations allowed. (Optional with default value `maxiter=100`)

`full_output`: Is a boolean variable determining the output. If `full_output=False`, then the function returns only the approximation of the root. If `full_output=True`, then the function returns as output additional information about the convergence of the method. (Optional with default value is `full_output=False`)

`disp`: If it is `True`, then the function displays messages in case of possible error. (Optional with default value `disp=True`)

The important input arguments are the function `f`, the end-points `a` and `b` of the initial interval $[a, b]$, the parameters `xtol` and `rtol` for the absolute and relative error tolerances and the maximum number of iterations `maxiter`. The argument `args` takes extra arguments for the function `f` if there are any. This allows to apply the bisection method on a specific argument of a multivariable function.

On the output of the function `bisect` we receive the approximation of the solution `x0` and the object (class) `r` containing information about the convergence. For example `r.converged` is `True` if the method converged.

We demonstrate the usage of this function for the same example we used before, which is the equation $\ln x + x = 0$ in the interval $[0.1, 1]$

```python
import scipy.optimize as spo
def f(x):
    y = np.log(x)+x
    return y
a = 0.1
b = 1.0
tol = 1.e-4
x = spo.bisect(f, a, b, () , tol)
print('The approximate solution x is: ', x)
print('And the value f(x) is: ', f(x))
```

```
The approximate solution x is:  0.567193603515625
And the value f(x) is:  0.0001390223881425623
```

We observe that the SciPy function and our implementation give the same results. Of course, the SciPy implementation has more capabilities and is faster as it is precompiled.

5.6.2 Fixed-point methods

The function `scipy.optimize.fixed_point` implements the fixed-point iteration for the numerical solution of the equation $x = g(x)$. A basic call of the function is

```
fixed_point(func, x0, args, xtol, maxiter, method)
```

with arguments

`func`: The name of the function $g(x)$

`x0`: The initial guess x_0 of the root x^*

`args`: A tuple with additional arguments and parameters for the functions f. (Optional with default value `args=()`)

`xtol`: The absolute error tolerance. (Optional with default value is `xtol` $=$ `1.e` $-$ `08`)

`maxiter`: The maximum number of iterations. (Optional with default values is `maxiter=500`)

`method`: Is a string determining the method. If `method='iteration'` then the function uses the general fixed-point iteration $x_{k+1} = g(x_k)$, while if `method='del2'` then the function uses the Steffensen's method with Aitken's delta-squared process. For more information see Sections 5.5.2 and 5.5.3. (Optional with default value `method='del2'`)

The method returns only the approximation of the root x^* in the output. In order to demonstrate the use of the `fixed_point` function we consider the equation $x^2 - x - 1 = 0$ with the initial guess $x_0 = 0$ and $g(x) = \sqrt{x+1}$.

```
import scipy.optimize as spo
def f(x):
    y = x**2-x-1.0
    return y
def g(x):
    y = np.sqrt(x+1.0)
    return y
x0 = 1.0
tol = 1.e-4
maxit = 50
x = spo.fixed_point(g, x0, (), tol, maxit)
print('The approximate solution x is: ', x)
print('And the value f(x) is: ', f(x))
```

```
The approximate solution x is:  1.6180339887498991
And the value f(x) is:  9.547918011776346e-15
```

We observe that the accuracy is better with Steffensen's method. Unfortunately, the specific function does not return the number of iterations required for the method to satisfy the prescribed stopping criterion. On the other hand, the use of Aitken's extrapolation method allows expectations for fast convergence.

5.6.3 Newton and secant methods

Newton's and secant methods are implemented in the function `scipy.optimize.newton`. A basic call of this function is

```
newton(func, x0, fprime, args, tol, maxiter, fprime2,
    x1, rtol, full_output, disp)
```

where here

`func`: The name of the function $f(x)$

`x0`: The initial guess x_0 of the root x^*

`fprime`: The name of the first derivative $f'(x)$. If we provide `fprime = None`, then the secant method is used instead of Newton's method. (Optional with default value `fprime=None`)

`args`: A tuple with additional arguments and parameters for the function f. (Optional with default value `args=()`)

`tol`: The absolute error tolerance. (Optional with default value `xtol` = 1.48e − 08)

`maxiter`: The maximum number of iterations. (Optional with default values `maxiter=50`)

`fprime2`: Is a callable function for the second order derivative of f. When `fprime2` is provided, then the function uses Halley's method of Section 5.5.5. (Optional with default value `fprime2=None` for the secant method or Newton's method)

`x1`: A second guess of the root for use with secant method. (Optional with default value `x1=None`)

`rtol`: The relative error tolerance. (Optional with default value `rtol` = 0.0)

`full_output`: Boolean input. If `False` then the function returns the root. If `True` then the function returns the tuple (x,r) where `x` is the root and `r` a class with further information. (Optional with default value `False`)

`disp`: Boolean input that specifies the level of communication. (Optional with default value `xtol` = 1.48e − 08)

In order to demonstrate the usage of this function we consider the example with $f(x) = \ln(x) + x$ and initial guess $x_0 = 1$.

```
import scipy.optimize as spo
def f(x):
    y = np.log(x)+x
    return y
def df(x):
    y = 1.0/x+1.0
    return y
x0 = 1.0
x = spo.newton(f, x0, df, tol=1.e-4, maxiter=50)
print('The approximate solution x is: ', x)
print('And the value f(x) is: ', f(x))
```

```
The approximate solution x is:  0.5671432903993691
And the value f(x) is:  -2.877842408821607e-11
```

Newton's method is faster and more accurate compared to the one obtained by the bisection method. It is noted that when the solution is printed on the screen it does not contain all the significant digits computed by the method, and the full answer can be displayed by using a formatted print command.

5.6.4 A hybrid method

There is a general purpose Python function that combines various methods for fast and guaranteed convergence, independent of the initial guess x_0 or the properties of the function f. This function is the `scipy.optimize.fsolve`, which is a wrapper of specific algorithms implemented in Fortran's library minpack, [91]. This function can be used also to solve systems of equations. If the derivative of f (or Jacobian matrix for systems) is not specified in the input arguments, the method computes good approximations by using similar finite-difference approximations like those we used in the derivation of the secant method.

The basic call of this function is:

```
fsolve(func, x0, args, fprime, full_output, col_deriv,
    xtol, maxfev, band, epsfcn, factor, diag)
```

The necessary input arguments are the following

`func`: The function f

`x0`: The initial guess x_0

The rest of the input arguments are optional and similar to the function `newton`.

`fprime`: A function to compute the Jacobian of `func` with derivatives across rows. (Optional with default value `fprime=None`)

`xtol`: Floating point variable determining the tolerance for terminating the algorithm. (Optional with default value `xtol=1.49012e-08`)

`full_output`: Specifies if we want the method to display detailed information about the convergence of the method. (Optional with default value `full_output=FALSE`)

`col_deriv`: Specifies whether the Jacobian matrix/function computes derivatives using columns. This results in faster implementation as there is no need to use of the transpose of matrices. (Optional with default value `col_deriv=FALSE`)

`maxfev`: Is the maximum number of calls of f. (Optional with default value is `maxfev=0`. This means that `100*(N+1)` is the maximum number, where `N` is the number of the entries in `x0`)

`band`: Specifies whether the Jacobian matrix is a band matrix by providing the number of sub- and super-diagonals within the band in a tuple. This argument can be used only if we do not provide the derivative (Jacobian) of the function (`fprime=None`). (Optional with default value is `band=None`)

`epsfcn`: Defines a suitable length for the forward-difference approximation of the derivative (or Jacobian matrix). Again we can use this argument if we do not provide the exact derivative of the function f. (Optional with default value `epsfcn=None`)

`factor`: Is a number in the interval $(0.1, 100)$ and determines the initial step bound. (Optional with default value `factor=100`)

`diag`: Is a sequence of N positive entries that serve as scale factors for the variables. (Optional with default value `diag=None`)

The function `fsolve` returns as output the approximate solution `x` and optionally a dictionary with the following keys `nfev` and `njeb`: The number of function and Jacobian calls respectively

`fvec`: The value `func(x)`

`fjac` and `r`: Are matrices where their product is equal to the approximation of the Jacobian matrix

`qtf`: Is the product `transpose(fjac)*fvec`

`ier`: Is 1 if the method converged to the solution, otherwise the last key:

`mesg`: Returns a string with the cause of the failure

We demonstrate here the use of the function `fsolve` in our standard example:

```
import scipy.optimize as spo
def f(x):
    y = np.log(x)+x
    return y
def df(x):
    y = 1.0/x+1.0
    return y
x0 = 1.0
x = spo.fsolve(f, x0, fprime=df, xtol=1.e-4)
print('The approximate solution x is: ', x)
print('And the value f(x) is: ', f(x))
```

```
The approximate solution x is:  [0.56714329]
And the value f(x) is:  [3.4803842e-09]
```

The results are similar to the results obtained with the previous methods. The advantage of `fzero` is its convergence and applicability to systems of nonlinear equations. For an algorithm of a hybrid method that combines the bisection method with Newton's method, we refer to Algorithm 15 in the exercises.

5.6.5 Application in astrophysics

Nonlinear equations occur naturally in many physical applications. An example of a well-known nonlinear equation in astrophysics is Kepler's equation. The famous German astronomer and mathematician Johannes Kepler (1571–1630) described planetary orbits around the Sun with great accuracy. Kepler also stated that planetary orbits should be ellipses with the Sun located at one of the two foci. Figure 5.8 presents an idealized orbit, where the ellipse has eccentricity $e = \sqrt{1 - b^2/a^2}$, semi-major axis a and semi-minor axis b. The angle θ as shown in Figure 5.8 is the *eccentric anomaly.* (The angle ϕ is called the *true anomaly*). If angle ψ is the *mean anomaly* defined as $\psi = \nu\, t$, where $\nu = 2\pi/T$ is the *mean angular velocity*, and t is the time since perihelion (point A), then Kepler's equation is the (fixed point) equation

$$\theta = \psi + e \sin\theta\ .$$

Because ψ is a function of time t, Kepler's equation determines the eccentric anomaly and therefore we can compute the position of a planet as a function of time. If we compute the eccentric anomaly θ for a given time t, then we can compute the true anomaly ϕ by solving the equation

$$(1 - e)\tan^2\frac{\phi}{2} = (1 + e)\tan^2\frac{\theta}{2}\ .$$

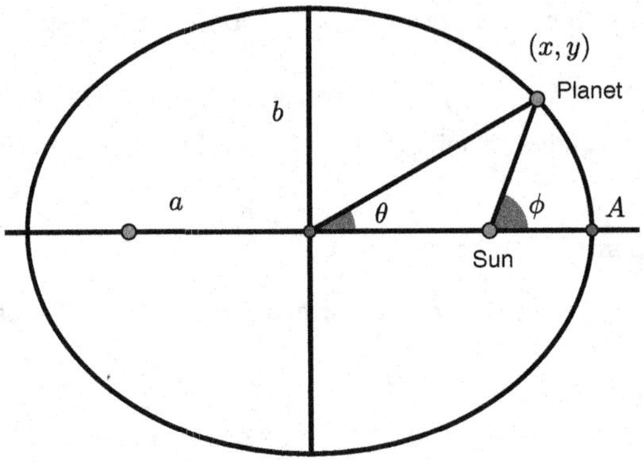

FIGURE 5.8
Planetary orbit and its anomalies.

Moreover, the heliocentric distance r between the planet and the Sun can be estimated using the formula

$$r = a(1 - e\cos\theta)\ .$$

Given the parameters e and ψ, we compute the eccentric anomaly θ by solving directly the fixed point equation $\theta = g(\theta)$ where $g(\theta) = \psi + e\sin\theta$. For example, for a planetary orbit of eccentricity $e = 10^{-6}$ (almost circular orbit) and mean anomaly $\psi = \pi/6$, we take as initial guess of the eccentric anomaly $\theta_0 = \psi$, which is not very far from the true value. As we can see in the following code, using the function `fixed_point` of module `scipy.optimize` we obtain the solution `0.5235992755987319`.

```
import numpy as np
import scipy.optimize as spo
def g(theta):
    e = 1.e-6
    psi = np.pi/6.0
    return psi+e*np.sin(theta)
theta0 = np.pi/6.0
theta = spo.fixed_point(g, theta0)
print('eccentric anomaly=', theta)
```

```
eccentric anomaly = 0.5235992755987319
```

We obtain similar results when we use the other functions of `scipy.optimize` for solving the equation $f(\theta) = 0$ where $f(\theta) = \theta - g(\theta)$. Note that Newton's iteration can be written explicitly in this case as

$$\theta_{k+1} = \theta_k + \frac{\psi + e\sin\theta_k - \theta_k}{1 - e\cos\theta_k}, \quad k = 0, 1, 2, \dots\ .$$

We leave the implementation of this method as well as the use of the other Python functions of the module `scipy.optimize` as an exercise for the reader.

5.7 Further Reading

The numerical solution of nonlinear equations is discussed in all classic numerical analysis textbooks [18, 22, 76, 108, 109, 120, 26, 27, 25, 70, 132, 7, 89, 90, 57, 66, 116]. More detailed analysis of the methods can be found in advanced textbooks such as [127, 1, 34, 94, 125] including proofs of convergence and error estimates. Books specialized on iterative methods and their convergence include [5, 6, 96]. Because of the influence of this topic in unconstrained optimization and other fields of sciences, there are several specialized books focused on the numerical solution of nonlinear equations. One such book is the classical book [95].

Chapter Highlights

- In this chapter, we studied numerical methods for the approximation of roots of functions by solving the equation $f(x) = 0$.
- The simplest method for approximating roots of functions is the bisection method. The bisection method is based on Bolzano's theorem and requires only the function f to be continuous and to change sign around the root.
- The bisection method always converges with a convergence rate at least 1.
- It is always preferable to have methods that converge with a rate grater than 1.
- Fixed point methods are based on the transformation of the equation $f(x) = 0$ into a fixed point equation $x = g(x)$. This can be done in many different ways. The generic fixed point method defines an iteration $x_{k+1} = g(x_k)$ and requires an initial guess x_0 of the root x^*.
- The convergence rate of fixed point methods depends on the choice of function g.
- Newton's method is a particular case of fixed point method with

$$g(x) = x - \frac{f(x)}{f'(x)} \ .$$

- Newton's method converges quadratically but requires the knowledge of the derivative $f'(x)$ and a good initial guess x_0 of the root x^*.
- Secant method is based on Newton's method but uses a finite difference approximation of the derivative.
- The convergence rate of secant method is equal to the golden ratio $r = (1 + \sqrt{5})/2$.
- To avoid infinite iterations in the case of convergence failure we usually use stopping criteria such as $|x_{k+1} - x_k| < TOL$ and k to be less than a maximum number of iterations.
- Steffensen's method is a modified Newton's method that converges quadratically without the need of the first derivative.
- Steffensen's method converges only when $-1 < f'(x^*) < 0$.
- Aitken's method can be used instead to improve the accuracy of an iterative method.
- Fixed point methods such as Newton's method can be generalized for systems of equations.
- The module `scipy.optimize` contains implementations of most of the numerical methods discussed in this chapter, including the bisection, fixed point, Newton and secant methods.

Exercises

1. Consider the function
$$f(x) = x^3 - x - 1 \, .$$
   (a) Prove that the equation $f(x) = 0$ has a unique solution in the interval $[1, 2]$.
   (b) Estimate the root of this equation using the bisection method with 3 iterations.
   (c) How many iterations of the bisection method do you need in order to approximate the exact solution with error of order 10^{-6}?

2. In this problem consider the bisection method.
   (a) Write a Python function that computes an approximation of the solution x^* of the equation $f(x) = 0$ in the interval $[a, b]$ using the bisection method. For a stopping criterion use the following: If $|x_{k+1} - x_k| \leq \texttt{TOL}$ for the first time, then return x_{k+1} as approximation of the root x^*. Allow the code to do only `KMAX` iterations.
   (b) Test your code by finding an approximate solution to the equation $\ln(x) + x = 0$ in the interval $[0.1, 1]$.
   (c) Compare your results with those obtained using the function `bisect` of `scipy.optimize`.

3. Consider the function
$$f(x) = x^3 + x - 1 \, .$$
   (a) Prove that the equation $f(x) = 0$ has a unique solution x^* in the interval $(0, 1)$.
   (b) Estimate the root of this equation using the bisection method with 3 iterations.
   (c) How many iterations of the bisection method do you need in order to approximate the exact solution with error of the order 10^{-6}?
   (d) Write Newton's method for the approximation of the solution x^*.
   (e) Perform using pen and paper 3 iterations of Newton's method and compare with the result of bisection method.
   (f) Prove that the sequence x_k generated by Newton's method converges to x^* for any $x_0 \in \mathbb{R}$ and that
$$\lim_{k\to\infty} \frac{x_{k+1} - x^*}{(x_k - x^*)^2} = \frac{3x^*}{3(x^*)^2 + 1} \, .$$
   (g) Interpret the last question in terms of discretization errors and convergence rates.

4. Consider the function
$$f(x) = x^2 - 2 \, .$$
   (a) Prove that if $x_0 > \sqrt{2}$, then the sequence
$$x_{k+1} = \frac{1}{2}x_k + \frac{1}{x_k}, \quad k = 0, 1, \ldots \, ,$$
   converges to $\sqrt{2}$.

(b) Observe that if $0 < x_0 < \sqrt{2}$, then $x_1 > \sqrt{2}$. What happens to the convergence of the sequence x_k if $x_0 \in (0, \sqrt{2}]$?

(c) What if $x_0 \leq 0$?

5. Consider the function

$$f(x) = xe^{x+2} + 2 \ .$$

(a) Prove that the equation $f(x) = 0$ has exactly two roots $x^* \in (-1, 0)$ and $x^{**} = -2$.

(b) Prove that the sequence x_k generated by Newton's method for the equation $f(x) = 0$ converges to the root x^* only if $x_0 > -1$.

(c) What happens when $x_0 \leq 1$?

6. Let $g : \mathbb{R} \to \mathbb{R}$ a function such that $g(x^*) = x^*$ for some $x^* \in \mathbb{R}$. In addition assume that

$$g'(x^*) = g''(x^*) = \cdots = g^{(p-1)}(x^*) = 0 \quad \text{and} \quad g^{(p)}(x^*) \neq 0 \ .$$

(a) Prove that the fixed point iteration $x_{k+1} = g(x_k)$ converges to the fixed point x^*.

(b) Show that the fixed point iteration converges with order p, in the sense that

$$\lim_{k\to\infty} \frac{x_{k+1} - x^*}{(x_k - x^*)^p} = \frac{1}{p!} g^{(p)}(x^*) \ .$$

7. In this problem consider Newton's method.

(a) Write a Python function that computes an approximation of the root x^* of the equation $f(x) = 0$ using Newton's method. Specifically the code shall compute the terms x_k of the sequence

$$x_{k+1} = x_k - \frac{f(x_k)}{f'(x_k)}, \qquad k = 0, 1, 2, \ldots \ ,$$

where x_0 is a given initial guess (or approximation) of the root x^*. For a stopping criterion use the following: If $|x_{k+1} - x_k| \leq$ `TOL` for the first time, then return x_{k+1} as approximation of the root x^*. Allow the code to do only `KMAX` iterations.

At every step k, the code should print on screen the values k, x_k, $f(x_k)$ and $|x_{k+1} - x_k|$. If the number of performed iterations reaches the maximum allowed number without satisfying the stopping criterion, then the code shall print on screen an error message, for example: "Failure: Algorithm failed to converge using only `KMAX` iterations".

(b) Test your code by finding an approximate solution to the equation $\ln(x) + x = 0$ with an initial guess of your choice.

8. Consider the function

$$f(x) = \arctan(2(x - 1)) - \ln|x| \ .$$

(a) Plot (using Python) the graph of $f(x)$ and describe the intervals of monotonicity. What conclusions can be derived about the solutions of the equation $f(x) = 0$.

(b) Prove by analytical means (using calculus), that the equation $f(x) = 0$ has exactly four real roots $\rho_1 < \rho_2 < \rho_3 < \rho_4$. [Hint: Use Bonzano's theorem in appropriate intervals selected. Then use the monotonicity of the function f in each of these intervals to show uniqueness of roots.]

(c) Find approximations of the roots using the codes of Problems 2 and 7, with input parameters x_0, TOL and KMAX of your choice (try x_0 values close to the real roots).

(d) Compare the results and the CPU times elapsed with the results obtained by using the Python functions bisect, newton and fsolve of scipy.optimize module. What do you observe?

(e) Finally, take TOL=1.e-6, KMAX=50 and execute your implementation of Newton's method with $x_0 = -1, 0.65, 0.7, 1.7, 1.8, 1.9, 5$ and 10. Comment on the results.

9. The following algorithm for approximating roots combines the bisection method with Newton's method to ensure the convergence even when Newton's method fails to converge.

Algorithm 15 A hybrid Newton-bisection method

Choose an interval $[a, b]$ such that $f(a)f(b) \leq 0$
Compute initial approximation x_0 of the root using bisection method.
for $i = 1, 2, \ldots$ **do**
  Compute x_i using Newton's method and x_{i-1}
  If $x_i \not\in [a, b]$, set $x_i = (a + b)/2$ (from bisection)
  Check for convergence
  If $f(a)f(x_i) \leq 0$ set $b = x_i$, else set $a = x_i$
end for

(a) Implement this algorithm in a Python function. Call your new function hybridNewton and use the following specifications:

```
def  hybridNewton(a,b,tol,maxit,f,df):
# Input:
# a, b = The endpoints of the interval
# tol = The required tolerance
# maxit = Maximum number of iterations
# f, df = The function and its derivative
# Output:
# xstar = approximation of root
# niter = number of iterations for convergence
# ierr =
#          0, the method converged
#          1, df returned zero value
#          2, maximum number of iterations has been reached

    return xstar,niter,ierr
```

(b) Use this function with input parameters of your choice to find approximations of the roots of Problem 8.

10. Consider the secant method.

(a) Write a Python function using the same criteria as in Problem 7 that will compute an approximation of a solution x^* of an equation $f(x) = 0$ using the secant method.

(b) Estimate the rate of convergence using the equation $x(e^{x/2} + 1) = 0$, $x_0 = 2.5$ and the exact root $x^* = 0$. Confirm that the convergence rate is equal to the golden ratio $r = (1 + \sqrt{5})/2 \approx 1.62$.

11. Consider the Halley iteration defined as

$$x_{k+1} = x_k - \frac{2f(x_k)f'(x_k)}{2[f'(x_k)]^2 - f(x_k)f''(x_k)}, \quad k = 0, 1, \ldots ,$$

and the method defined by the iteration

$$x_{k+1} = x_k - \frac{2f(x_k)[f'(x_k)]^2 + [f(x_k)]^2 f''(x_k)}{2[f'(x_k)]^3}, \quad k = 0, 1, \ldots ,$$

for the numerical solution of a equation $f(x) = 0$. Consider also appropriate initial guess x_0 for the exact root x^*.

(a) Write Python functions implementing the two methods.

(b) Study the convergence of both methods by finding experimentally the convergence rates.

(c) For each method describe necessary conditions that guarantee cubic order convergence rates.

12. Consider the function

$$f(x) = \frac{3x}{x^2 - 2x + 4} .$$

(a) Using the module `matplotlib` plot the graph of $f(x)$ in the interval $[-10, 10]$.

(b) Compute the maximum and minimum values of $f(x)$ using functions from `scipy.optimize`.

(c) Study the monotonicity intervals of $f(x)$ with the help of calculus.

(d) Verify the numerical results by solving theoretically the equation $f'(x) = 0$.

13. Galileo Galilei performed several experiments from the leaning tower of Pisa to understand the free fall of objects and the effects of gravity. Experiments showed that the air resistance $F_d(t)$ is proportional to the speed $v(t)$ of a falling object, where t denotes time in seconds. Specifically, we assume that

$$F_d(t) = b \cdot v(t) ,$$

where b is the *air resistance coefficient.* Since the air resistance increases as the speed of the falling object increases, eventually, the gravitational force becomes equal to the air resistance and the object falls at a constant speed V. The speed V is known to as the *terminal speed.* Applying Newton's second law of motion to the falling object with terminal speed we have that $F_d = mg$, and thus we compute the terminal speed

$$V = \frac{mg}{b} ,$$

where $g \approx 9.81\ m/s^2$ is the acceleration due to gravity. We can also estimate the velocity of the falling object as a function of time given by

$$v(t) = V \cdot \left(1 - e^{-\frac{b}{m}t}\right) .$$

Consider the free fall of three objects with masses $m_1 = 0.1\ kg$, $m_2 = 1\ kg$ and $m_3 = 3\ kg$ from the top of the tower of Pisa. Assume that the coefficients of air resistance for the three objects are $b_1 = 1\ kg/s$, $b_2 = 0.8\ kg/s$ and $b_3 = 0.5\ kg/s$, respectively.

(a) Given that the height of Pisa's tower is $h_0 = 55.86\ m$, show by integrating the formula for the velocity that the height of the falling object measured from the ground is given by the formula

$$h(t) = h_0 - Vt + V\frac{m}{b}\left(1 - e^{-\frac{b}{m}t}\right) .$$

(b) Use the module `scipy.optimize` to estimate the time it takes for the objects to fall to the ground.

(c) Compute the falling time in the absence of air resistance, and comment on the effect of air resistance to the free fall of these objects.

14. Consider the following generalization of the Bisection method called *the method of false position* or sometimes *Regula Falsi* as described in Algorithm 16.

Algorithm 16 Method of false position

Set a tolerance TOL for accuracy
Initialize the interval $[a, b]$
Set $k = 0$
Check if $f(a) \cdot f(b) < 0$ (if not then quit)
while $|b - a| > TOL$ **do**
    Set $x_k = \frac{bf(a)-af(b)}{f(a)-f(b)}$ (the new approximation)
    Check in which interval $[a, x_k]$ or $[x_k, b]$ the function changes sign
    if $f(x_k) \cdot f(a) < 0$ **then**
        $b = x_k$
    else if $f(x_k) \cdot f(b) < 0$ **then**
        $a = x_k$
    else
        Break the loop because $f(x_k) = 0$
    end if
    $k = k + 1$
end while
Return the approximation of the root x^*

(a) Observe that this algorithm combines the bisection method and another method. Describe what other method is combined and why this is promising.

(b) Implement the method of false position in Python.

(c) Describe your expectations for the convergence of this method.

(d) Use your computer code to approximate the root $x^* = 0$ of the function $f(x) = 2x^3 - 4x^2 + 3x$ in $[-1, 1]$. Describe the convergence of the method of false position in this case and compare with the bisection method.

15. In this problem you will need to use complex number arithmetic. Assume that the equation $f(x) = 0$ has a unique solution in $[a, b]$, and that $f \in C^3[a, b]$.

(a) Let $0 < h \ll 1$ be a small number. Using appropriate Taylor expansion show that for any x in the domain of f we have

$$f'(x) = \mathrm{Im}(f(x + \mathrm{i}h))/h + O(h^2) ,$$

where i is the imaginary unit such that $\mathrm{i}^2 = -1$.

(b) Since $f'(x) \approx \mathrm{Im}(f(x+ih))/h$, consider the following modification of Newton's method with complex step

$$x_{k+1} = x_k - h\frac{f(x_k)}{\mathrm{Im}(f(x_k + ih))}, \quad k = 0, 1, 2, \ldots ,$$

with given x_0. Implement this complex-step Newton's method.

(c) Assume that

$$\lim_{k\to\infty} \frac{|x_{k+1} - x^*|}{|x_k - x^*|^r} = C ,$$

where x^* is such that $f(x^*) = 0$. This means that the convergence rate of the new method is r. Estimate the convergence rate r using the equation $x(e^{x/2} + 1) = 0$ and $x_0 = 2.5$. For implementation purposes you can use the default value $h = 10^{-10}$ but you can try even smaller such as $h = 10^{-20}$ or even smaller.

(d) Study the influence of the parameter h in the convergence rate.

[The approximation of the derivative in this problem is known to as the *complex-step* derivative approximation and works for even tiny values of $h > 0$. For more information, we refer to the publications [123, 85] and the problems of Chapter 8]

6

Interpolation and Approximation

Suppose that we are in possession of data points $(x_0, y_0), (x_1, y_1), \dots, (x_N, y_N)$ representing measurements of a physical quantity y at location x, but we need an estimate of the quantity y_p at some point x_p other than $x_0, x_1, \dots, x_N$. To formulate a mathematical problem we assume that the given data obey to a function $y = f(x)$, in the sense that $y_i = f(x_i)$ for $i = 0, 1, \dots, N$. Of course, we do not know the formula $f(x)$. The question is whether we can compute an approximation of $f(x)$ at some $x \neq x_i$, $i = 0, 1, \dots, N$. To tackle such problems we usually use interpolation. Interpolation is a method of constructing a function whose graph passes through a specified set of points. When the data are too noisy to interpolate with smooth functions, we proceed with approximating the dataset. Specifically, we generate a function whose graph is close to the data points without necessarily passing through them. Such a method is the least squares approximation (also known as regression). This chapter is an introduction to interpolation and approximation.

6.1 Introduction

In some applications there is a finite set of data points $(x_i, y_i) = (x_i, f(x_i))$, $i = 0, 1, \dots, N$ of an unknown function $f(x)$, and we require an estimate of this function at a different point. For example, assume that we have the population of a city for the years $x_0 = 2000$, $x_1 = 2002$ and $x_2 = 2004$, which is $y_0 = 12,900$, $y_1 = 13,400$ and $y_2 = 13,800$ respectively, and we need to estimate the population of the city for the year 2001. In this problem the unknown function $f(x)$ describes the population of a city at a year x. To find an approximate value for $f(2001)$, we compute appropriate polynomial $P(x)$ that satisfies $P(x_i) = y_i$. If this polynomial approximates the function $f(x)$, then we can estimate the unknown quantity by computing the value $P(2001) \approx f(2001)$. Sometimes, it is beneficial to use sophisticated interpolants $P(x)$, such as piecewise polynomial functions, but for the moment we consider simple polynomials.

When the graph of the polynomial $P(x)$ passes through the given (x_i, y_i) for all values of i, we say that $P(x)$ *interpolates* the points (x_i, y_i). The polynomial $P(x)$ is called *interpolant.* The procedure to estimate the values (x, y) other than (x_i, y_i) using $P(x)$ is called *interpolation.* The points x_i are called *nodes* (also known as *sites* or *knots*). A set of nodes $\{x_0, x_1, \dots, x_N\}$ is called *grid* or *mesh.* In principle, interpolation is when $\min x_i \leq x \leq \max x_i$ otherwise we call it *extrapolation* even if we use the same methods. If the polynomial does not necessarily interpolate the points (x_i, y_i), then we speak about approximation in general. Figure 6.1 shows the idea of interpolation. Observe that the interpolating polynomial $P(x)$ approximates the function $f(x)$ quite well for $x \in [x_0, x_3]$ and diverges for values of x outside the interval $[x_0, x_3]$. This phenomenon can be a problem especially if the dataset consists of only a few points. The situation can be different when there are more data points to interpolate or in cases of sophisticated interpolation methods.

DOI: 10.1201/9781003287292-6

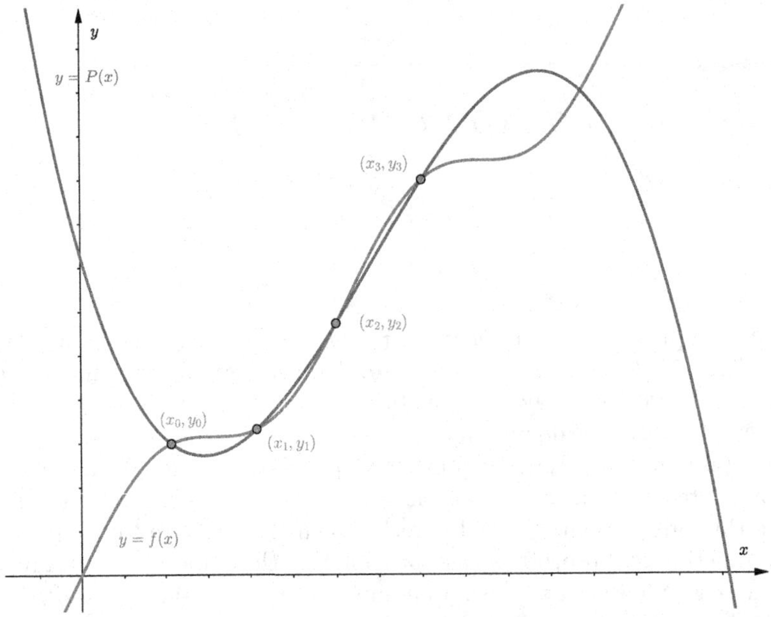

FIGURE 6.1
The notion of polynomial interpolation for the function $f(x) = x + \sin(x)$ with four nodes (N=3).

The reason for using polynomials to approximate functions is because of the *Weierstrass Approximation Theorem.* This theorem guarantees that every continuous function can be approximated arbitrarily well by a polynomial. Mathematically speaking this can be stated as:

Theorem 6.1 (Weierstrass Approximation Theorem). *For any function $f \in C[a,b]$, and for each $\varepsilon > 0$, there is a polynomial $P(x)$, such that $|f(x) - P(x)| < \varepsilon$ for all $x \in [a,b]$.* [1]

Any linear combination of two polynomials is a polynomial again of the same degree at most. For this reason, we say that the polynomials of degree N form a linear space denoted by $\mathbb{P}_N$.

Interpolation and approximation can be useful in problems such as data fitting and function approximation, as well as in other numerical methods, including root finding, optimization, numerical integration and differentiation. In this chapter, we discuss interpolation methods such as the Lagrange, Hermite and Spline interpolation. We will close this chapter by discussing the method of least squares for approximating datasets.

> ☞ Interpolation is the procedure of finding a function (usually a polynomial or a piecewise polynomial function) with graph passing through some data points. When the data are very noisy, we usually approximate them using different procedures, such as the method of least squares.

[1] A proof of Weierstrass Approximation Theorem requires advanced tools of mathematical analysis and can be found, for example, in [81].

6.2 Lagrange Interpolation

Suppose that we have $N+1$ values $y_0, y_1, \ldots, y_N$ and $x_0, x_1, \ldots, x_N$ such that $y_i = f(x_i)$ for $i = 0, 1, \ldots, N$, and that f is a smooth function. Recall that smooth functions are differentiable functions. The procedure of finding a unique polynomial $P_N(x)$ of degree N such that $p(x_i) = y_i$, for $i = 0, 1, \ldots, N$ is called *Lagrange interpolation.* The polynomial we obtain with Lagrange interpolation is called *Lagrange interpolating polynomial* or in short *Lagrange interpolant* named after the French mathematician Joseph-Louis Lagrange (1736–1813). Such an interpolant is depicted in Figure 6.1 where $f(x) = x + \sin(x)$ and $N = 3$. We first present a naive construction of Lagrange interpolating polynomials.

6.2.1 Naive construction of the interpolant

Consider first the simplest case with only two points (x_0, y_0) and (x_1, y_1). In this case, there is a unique straight line $P_1(x)$ passing through these points. Define

$$P_1(x) = a_0 + a_1 x \; , \tag{6.1}$$

the interpolation polynomial with unknown coefficients a_0 and a_1. To find the coefficients a_0 and a_1 we use the fact that $P_1(x)$ interpolates the points (x_i, y_i), $i = 0, 1$, which is the same to say that

$$y_0 = P_1(x_0) = a_0 + a_1 x_0 \quad \text{and} \quad y_1 = P(x_1) = a_0 + a_1 x_1 \; .$$

Both equations form the system

$$\begin{cases} y_0 = a_0 + a_1 x_0 \\ y_1 = a_0 + a_1 x_1 \end{cases} .$$

Solving this system (simultaneous equations) we obtain the coefficients

$$a_1 = \frac{y_1 - y_0}{x_1 - x_0} \quad \text{and} \quad a_0 = y_0 - \frac{y_1 - y_0}{x_1 - x_0} x_0 \; .$$

Substituting a_0 and a_1 into (6.1) we obtain the formula of $P_1(x)$ as

$$P_1(x) = \frac{y_1 - y_0}{x_1 - x_0} x + y_0 - \frac{y_1 - y_0}{x_1 - x_0} x_0 = \frac{y_1 - y_0}{x_1 - x_0}(x - x_0) + y_0 \; .$$

This is the linear Lagrange interpolant, and its construction requires two points $(N = 1)$.

Moving one step further, we consider three points (x_0, y_0), (x_1, y_1) and (x_2, y_2). Since we have three points, the interpolating polynomial should be a quadratic polynomial of the form $P_2(x) = a_0 + a_1 x + a_2 x^2$ with $y_0 = P_2(x_0)$, $y_1 = P_2(x_1)$ and $y_2 = P_2(x_2)$. These equations form the following system of simultaneous equations

$$\begin{cases} y_0 = a_0 + a_1 x_0 + a_2 x_0^2 \\ y_1 = a_0 + a_1 x_1 + a_2 x_1^2 \\ y_2 = a_0 + a_1 x_2 + a_2 x_2^2 \end{cases} . \tag{6.2}$$

To solve this system for a_0, a_1 and a_2 it is not as easy unless we use techniques from linear algebra. For example, we can use the NumPy module and its functions. System (6.2) can be written in matrix form as

$$\begin{pmatrix} 1 & x_0 & x_0^2 \\ 1 & x_1 & x_1^2 \\ 1 & x_2 & x_2^2 \end{pmatrix} \begin{pmatrix} a_0 \\ a_1 \\ a_2 \end{pmatrix} = \begin{pmatrix} y_0 \\ y_1 \\ y_2 \end{pmatrix} .$$

The matrix

$$\boldsymbol{V} = \begin{pmatrix} 1 & x_0 & x_0^2 \\ 1 & x_1 & x_1^2 \\ 1 & x_2 & x_2^2 \end{pmatrix} ,$$

is a 3×3 *Vandermonde* matrix. We write the previous system in matrix form as $\boldsymbol{V}\boldsymbol{a} = \boldsymbol{y}$, with

$$\boldsymbol{y} = \begin{pmatrix} y_0 \\ y_1 \\ y_2 \end{pmatrix} , \quad \boldsymbol{a} = \begin{pmatrix} a_0 \\ a_1 \\ a_2 \end{pmatrix} .$$

This procedure can be generalized for $N+1$ distinct points (x_i, y_i), $i = 0, 1, 2, \ldots, N$ with $x_i \neq x_j$ if $i \neq j$. Then the coefficients of the polynomial of degree N

$$P_N(x) = a_0 + a_1 x + a_2 x^2 + \cdots + a_N x^N ,$$

can be found by solving the analogous $(N+1) \times (N+1)$ system of linear equations $\boldsymbol{V}\boldsymbol{a} = \boldsymbol{y}$ with $\boldsymbol{V}$ the Vandermonde matrix

$$\boldsymbol{V} = \begin{pmatrix} 1 & x_0 & x_0^2 & \cdots & x_0^N \\ 1 & x_1 & x_1^2 & \cdots & x_1^N \\ \vdots & \vdots & \vdots & & \vdots \\ 1 & x_N & x_N^2 & \cdots & x_N^N \end{pmatrix} .$$

The determinant of the Vandermonde matrix is

$$\det(\boldsymbol{V}) = \prod_{0 \leq i < j \leq N} (x_j - x_i) ,$$

which is $\det(\boldsymbol{V}) \neq 0$ since the points x_i are distinct, i.e. $x_i \neq x_j$ for $i \neq j$. Thus, $\boldsymbol{V}$ is invertible, and the linear system $\boldsymbol{V}\boldsymbol{a} = \boldsymbol{y}$ has a unique solution $\boldsymbol{a} = \boldsymbol{V}^{-1}\boldsymbol{y} = (a_0, a_1, \ldots, a_N)^T$.

> ☞ For $N+1$ points (x_i, y_i) there is always a unique interpolating polynomial P_N of degree N such that $y_i = P_N(x_i)$.

There are two main disadvantages in solving linear systems with Vandermonde matrix. The first is that Vandermonde matrices are dense. Thus, in the case of large amount of data solving the corresponding dense linear system will be a very slow procedure. The other drawback is that computations with Vandermonde matrices are very sensitive to floating point errors. Matrices that are sensitive to floating point errors are called ill-conditioned, and we should be very careful when we use them. More information about the condition of matrices can be found in Chapter 9.

In Python we can construct Vandermonde matrices in any dimension using the NumPy function

```
numpy.vander(x, N, increasing)
```

where N is the dimension of the vector x, and the argument `increasing` specifies the form of the Vandermonde matrix. When `increasing=True`, then the entries of the corresponding matrix start from 1, otherwise they start from x_i^N. In our case we need `N=None` and `increasing=True`. The linear system $\boldsymbol{Va} = \boldsymbol{y}$ can be solved using the function `numpy.linalg.solve`. This can be quite safe since the function `solve` takes into account the condition of matrices and performs appropriate techniques to minimize floating point errors.

As an example we consider the points $(0, 1)$, $(0.5, -1)$, $(1, 2)$ and $(1.5, 1.5)$. Since we have four points we can construct the unique cubic interpolating polynomial ($N = 3$). We store the values x_i in a vector x and the values y_i in a vector y. We construct the Vandermonde matrix and solve the corresponding linear system using the function `solve`.

```
1  import numpy as np
2  import numpy.linalg as npl
3  import matplotlib.pyplot as plt
4
5  # interpolation data (x,y)
6  x = np.array([0.0, 0.5, 1.0, 1.5])
7  y = np.array([1.0, -1.0, 2.0, 1.5])
8  # assembly of Vandermonde matrix V
9  V = np.vander(x, increasing=True)
10 # solution of linear system Va=y
11 a = npl.solve(V,y)
12 # define more points for plotting
13 xx = np.linspace(0.0,1.5,100)
14 # evaluate the interpolating polynomial at xx
15 p = a[0]+a[1]*xx+a[2]*xx**2+a[3]*xx**3
16 plt.plot(xx,p)
17 plt.plot(x,y,'o')
18 plt.xlabel('x'); plt.ylabel('y'); plt.show()
```

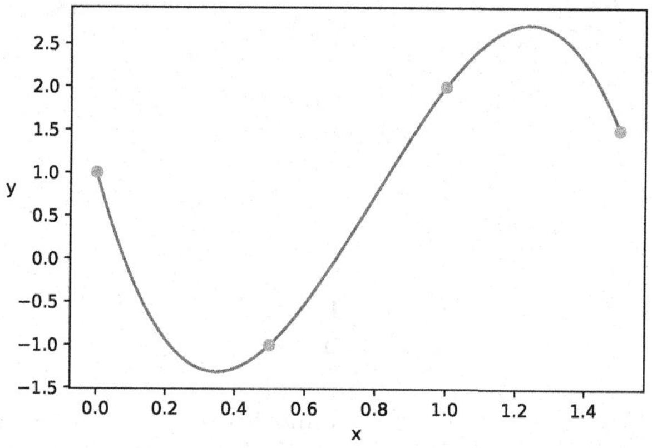

The 6th and 7th lines of this code define the vectors $\boldsymbol{x}$ and $\boldsymbol{y}$ with the data (x_i, y_i). The 9th line performs the assembly of the Vandermonde matrix $\boldsymbol{V}$ and the 11th line solves the corresponding system $\boldsymbol{Va} = \boldsymbol{y}$. In order to plot the resulting polynomial we define a fine mesh of 100 points (line 13) and we store these points in the variable `xx`. We evaluate the cubic polynomial `p` at the points `xx` (line 15). Finally, we plot the polynomial `p` over the points `xx` (line 16) and we draw the data points (x_i, y_i) using circles (line 17).

We observe that the cubic polynomial interpolates all the four points. Also if we want to evaluate the polynomial $P_3(x)$ at any value x we can do it easily since we have already computed the coefficients a_0, a_1, a_2 and a_3.

Imagine we have the experimental data (x_i, y_i) for $i = 0, 1, \ldots 9, 999$. The Vandermonde matrix will be $10,000 \times 10,000$ dense with $100,000,000$ entries. This dense matrix is very large and thus the solution of the system $\boldsymbol{Va} = \boldsymbol{y}$ will be a slow process. The same is true for the evaluation of the interpolating polynomial since it involves a very large and perhaps unstable sum. In order to construct the interpolating polynomial efficiently, we make use of some alternative basis functions, in place of the naive monomial basis $\{1, x, x^2, \ldots, x^N\}$.

6.2.2 Lagrange polynomials

Given $N+1$ points $(x_0, y_0), (x_1, y_1), \ldots, (x_N, y_N)$ with $x_i \neq x_j$ for $i \neq j$, the interpolating polynomial $P_N(x)$ can be written as

$$P_N(x) = \sum_{i=0}^{N} a_i x^i \ . \tag{6.3}$$

This is a linear combination of the monomial basis functions $1, x, x^2, \ldots, x^N$ with unknown coefficients a_i. An alternative way of expressing the same polynomial using different basis functions is by rewriting the polynomial in the form

$$P_N(x) = \sum_{i=0}^{N} y_i \ell_i(x) \ , \tag{6.4}$$

where the coefficients are the given interpolating data. Of course, we need to find the basis function $\ell_i(x)$. This is the Lagrange representation of the interpolating polynomial.

Since $P_N(x_i) = y_i$ for all values of i we have that

$$y_i = P_N(x_i) = \sum_{i=0}^{N} y_i \ell_i(x_i) = y_0 \ell_0(x_i) + \cdots + y_i \ell_i(x_i) + \cdots + y_N \ell_N(x_i) \ .$$

This implies that $\ell_i(x_j) = \delta_{ij}$, where

$$\delta_{ij} = \begin{cases} 0, & \text{if } i \neq j \\ 1, & \text{if } i = j \end{cases} ,$$

is the Kronecker delta.

Because of that, the polynomials $\ell_i(x)$ are of the form

$$\ell_i(x) = C \prod_{\substack{j=0 \\ j \neq i}}^{N} (x - x_j) \ ,$$

for some constant C. The constant C can be found from the relation $\ell_i(x_i) = 1$. We have

$$\ell_i(x_i) = C \prod_{\substack{j=0 \\ j \neq i}}^{N} (x_i - x_j) = 1 \ ,$$

which implies

$$C = \prod_{\substack{j=0 \\ j\neq i}}^{N} \frac{1}{x_i - x_j} = \frac{1}{(x_i - x_0)(x_i - x_1)\cdots(x_i - x_{i-1})(x_i - x_{i+1})\cdots(x_i - x_N)} .$$

Subsequently we have that

$$\ell_i(x) = \frac{x - x_0}{x_i - x_0}\frac{x - x_1}{x_i - x_1}\cdots\frac{x - x_{i-1}}{x_i - x_{i-1}}\frac{x - x_{i+1}}{x_i - x_{i+1}}\cdots\frac{x - x_N}{x_i - x_N} = \prod_{\substack{j=0 \\ j\neq i}}^{N} \frac{x - x_j}{x_i - x_j} .$$

The polynomials $\ell_i(x)$ are called Lagrange polynomials.

If we write

$$Q_{N+1}(x) = \prod_{j=0}^{N}(x - x_j) \quad \text{then} \quad Q'_{N+1}(x_i) = \prod_{\substack{j=0 \\ j\neq i}}^{N}(x_i - x_j) ,$$

then we have the alternative formulas for the Lagrange polynomials and the corresponding interpolating polynomial

$$\ell_i(x) = \frac{Q_{N+1}(x)}{(x - x_i)Q'_{N+1}(x_i)} \quad \text{and} \quad P_N(x) = \sum_{i=0}^{N} y_i \ell_i(x) . \tag{6.5}$$

☞ An advantage of the interpolating polynomial written in the Lagrange form is that there is no need for the computation of the coefficients a_i as long as we have the Lagrange basis functions evaluated at points x.

For example, if $N = 1$ and given the two distinct points (x_0, y_0) and (x_1, y_1), the Lagrange polynomials are

$$\ell_0(x) = \frac{x - x_1}{x_0 - x_1} \quad \text{and} \quad \ell_1(x) = \frac{x - x_0}{x_1 - x_0} .$$

The polynomial $P_1(x)$ is given by the formula

$$P_1(x) = \ell_0(x)\cdot y_0 + \ell_1(x)\cdot y_1 = \frac{x - x_1}{x_0 - x_1}\cdot y_0 + \frac{x - x_0}{x_1 - x_0}\cdot y_1 .$$

We could work similarly to compute the quadratic polynomial $P_2(x)$ with three distinct points. However, it is easier to write a Python program to do it for us. Let us write a Python function first to compute the $N+1$ Lagrange polynomials. Since each polynomial is a product, the implementation requires a nested loop. The procedure for the computation of the Lagrange polynomials is presented in Algorithm 17.

Algorithm 17 Evaluation of Lagrange polynomials

Given a point z and the nodes $x_0, x_1, \ldots, x_N$
for $i = 0 : N$ **do**
  for $j = 0 : N$ **do**
    if $i \neq j$ **then**
      $\ell_i = \ell_i \cdot \frac{z - x_j}{x_i - x_j}$
    end if
  end for
end for
Return the values of the polynomials ℓ_i at the point z

The previous algorithm works even if z is a vector. We implement Algorithm 17 in the function `lagrange_basis`. Suppose that we want to compute the values $P_N(z_k)$ of the Lagrange interpolant at m points z_k for $k = 0, 1, 2, \ldots, m-1$. We store all these points z_k in a one-dimensional array `z`. We also store the data points x_i in an array called `x`. Then, for each $i = 0, 1, 2, \ldots, N$ we compute the i-th Lagrange polynomial as a product of the form

$$\ell_i(x) = \prod_{\substack{j=0 \\ j \neq i}}^{N} \frac{x - x_j}{x_i - x_j} . \tag{6.6}$$

The amount of points x_i is equal to the number of Lagrange basis functions ℓ_i and we store this number in the variable `n`, which is practically equal to $N + 1$. After computing the m values of the Lagrange polynomial $\ell_i(z_k)$ for $k = 0, 1, \ldots, m-1$, we store them in the i-th row of an array called `basis`. In the end, each row i of the array `basis` will contain the m values of the polynomial $\ell_i(z_k)$.

```
1  def lagrange_basis(z, x):
2  # Compute the Lagrange basis l_i(z)
3  # given the nodes x_i stored in vector x
4      n = len(x)
5      m = len(z)
6      basis = np.ones((n, m))
7      for i in range(n):
8          for j in range(n):
9              if i != j:
10                 basis[i,:] *= (z-x[j])/(x[i]-x[j])
11     return basis
```

The 4th line of this code defines the number of Lagrange polynomials required for the interpolation. The 5th line defines the amount m of the points on which we want to evaluate the interpolating polynomial. In the 6th line we reserve the memory for the `n` $\times$ `m` array `basis`. Finally, using the loop of line 7 we compute the polynomials ℓ_i at the m points. Using the loop of line 8 we compute the product over j. The function `lagrange_basis` returns an array with the values of the required basis functions ℓ_i at all points `z`.

Since the interpolating polynomial of degree N is given by the formula

$$P_N(x) = \sum_{i=0}^{N} y_i \ell_i(x) ,$$

we write one more function called `lagrange_interpolant` that evaluates the interpolating polynomial at the given m points stored in the array `z`. In this function we need the values of y_i in addition to the values x_i. It is noted that the values z_i are only needed for the computation of the Lagrange polynomials, while the values y_i are only needed for the computation of the value of the interpolant. The values y_i are stored in the one-dimensional array `y`. Again, since we want to compute the interpolant at m different values z_k, we store the values $P_N(z_k)$ for all k in a one-dimensions array `P`.

```
1 def lagrange_interpolant(z, x, y):
2 #Compute the interpolant using Lagrange polynomials
3     n = len(x)
4     m = len(z)
5     P = np.zeros(m)
6     basis = lagrange_basis(z, x)
7     for i in range(n):
8         P += basis[i,:] * y[i]
9     return P
```

The actual computation of the values of the polynomial P_N is implemented in line 8, where y_i is the `y[i]` and the values $\ell_i(z_k)$ are the entries of the vector `basis[i,:]`.

As an example, consider the function $f(x) = \frac{1}{x}$. We will compute the Lagrange interpolating polynomial using the points $x_0 = 1$, $x_1 = 1.5$ and $x_2 = 4$. Note that for Lagrange interpolation there is no requirement for uniformly distributed points x_i. Here, $N = 2$ and therefore the unique polynomial will be quadratic. We first compute the Lagrange polynomials

$$\ell_0(x) = \frac{(x-x_1)(x-x_2)}{(x_0-x_1)(x_0-x_2)} = \frac{(x-1.5)(x-4)}{(1-1.5)(1-4)} = \frac{2}{3}x^2 - \frac{11}{3}x + 4\ ,$$
$$\ell_1(x) = \frac{(x-x_0)(x-x_2)}{(x_1-x_0)(x_1-x_2)} = \frac{(x-1)(x-4)}{(1.5-1)(1.5-4)} = -\frac{4}{5}x^2 + 4x - \frac{16}{5}\ ,$$
$$\ell_2(x) = \frac{(x-x_0)(x-x_1)}{(x_2-x_0)(x_2-x_1)} = \frac{(x-1)(x-1.5)}{(4-1)(4-1.5)} = \frac{2}{15}x^2 - \frac{1}{3}x + \frac{1}{5}\ .$$

Moreover, $y_0 = f(x_0) = 1$, $y_1 = f(x_1) = \frac{2}{3}$ and $y_2 = f(x_2) = \frac{1}{4}$. The interpolating polynomial is

$$P_2(x) = \sum_{i=0}^{N} y_i\ell_i(x) = 1\cdot\ell_0(x) + \frac{2}{3}\ell_1(x) + \frac{1}{4}\ell_2(x) = \frac{1}{6}x^2 - \frac{13}{12}x + \frac{23}{12}\ .$$

The same interpolating polynomial can be computed using the naive interpolation method but this requires the solution of a linear system with Vandermonde matrix, which we avoid using the Lagrange polynomials.

Now we compute the interpolating polynomial using our previous functions. We compute the Lagrange polynomials first and then the interpolating polynomial in the interval $[1,4]$, using 100 points z_i. These points are generated by using the function `linspace(1, 4, 100)`. Then we define the arrays `x` and `y` to interpolate, and we generate the values of the interpolating polynomial using the function `lagrange_interpolant`. The results are plotted with the help of `matplotlib`. We also add a legend for the first time in order to distinguish the interpolating polynomial, the data points and the actual function.

```
def f(x):
    return 1.0/x
# define points for creating plot
z = np.linspace(1, 4, 100)
# define data points and store them in x and y
x = np.array([1, 1.5, 4])
y = f(x)
# compute Lagrange interpolant and store it to yz
yz = lagrange_interpolant(z, x, y)
# plot the results
plt.plot(z,yz,'-')
plt.plot(x,y,'o')
plt.plot(z,f(z),'-.')
plt.legend(["$P_2(x)$","data","$f(x)$"])
plt.xlabel('x'); plt.ylabel('y'); plt.show()
```

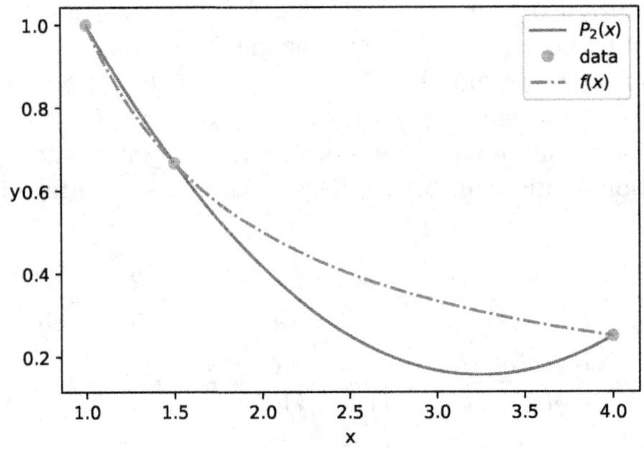

The interpolating polynomial coincides with $f(x)$ at the points (x_i, y_i). However, there is a notable difference between the graphs of the function and its interpolant. We can achieve better approximation using interpolating polynomials of higher degree N. For example, if we consider the interpolating polynomial P_4 by taking 5 points (x_i, y_i) the interpolating polynomial is almost identical to the function $f(x)$, as shown in the following figure.

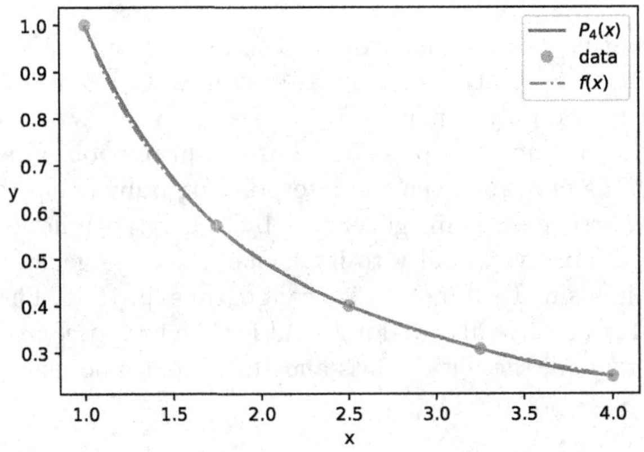

Taking higher degree interpolating polynomials the situation will not be always better. A counter example is the well-known Runge's example.

6.2.3 Failure of Lagrange interpolation

The function

$$f(x) = \frac{1}{1 + 25x^2} ,$$

is known as the Runge function. Surprisingly, the Lagrange interpolation fails for this function, in the sense that the interpolant cannot be as close to $f(x)$ as we would like to be, [111]. The shape of Runge's function is like a Gaussian bell but it decays with a polynomial rate. We will interpolate this function in the interval $[-1, 1]$ using 10 and 15 points. Using $N = 9$ we observe that the interpolating polynomial oscillates around Runge's function. With the hope that the approximation will be improved we take $N = 15$. To our surprise, we observe that the polynomial P_{14} is far away from f, especially for x closer to the endpoints ± 1.

The explanation of this phenomenon lies in the error of polynomial interpolation. As in the case of Taylor polynomials, the error of the Lagrange interpolation depends on the magnitude of the $N + 1$-th derivative of f. If this derivative is large, then the interpolation error will also be large. And in the case of Runge's function, the magnitude of the derivatives of f is increasing with N. This means that the higher the degree of the interpolating polynomial, the larger the interpolation error.

A remedy to this problem is to use a non-uniform grid of nodes. A standard example of such nodes are the Chebyshev nodes

$$x_k = \cos\left(\frac{2k+1}{2(N+1)}\pi\right), \quad k = 0, 1, \ldots, N .$$

☞ Runge's phenomenon demonstrates the fact that high degree polynomials are in general not suitable for interpolation problems.

The reader who is interested in the error analysis of polynomial interpolation can proceed with the next section. Otherwise, we consider efficient and more stable ways of computing the interpolating polynomial in Section 6.2.5. We also discuss the interpolation with piecewise polynomial functions, which solves Runge's phenomenon in Section 6.3.

6.2.4 Error analysis

Thus far, we have proved the existence of a unique interpolating polynomial as a consequence of Vandermonde matrix properties. We have also seen that there are cases where the Lagrange interpolation can fail. Here we estimate the error between the interpolating polynomial and function f, and explain why Lagrange interpolation can fail.

We easily understand the distance between two numbers to be the absolute value of their difference. The distance between two functions (polynomials are functions too) is harder to conceptualize. We measure the distance between functions using appropriate norms. A norm generalizes the notion of absolute value and is denoted by $\|\cdot\|$ instead of $|\cdot|$ which we reserve for absolute values. So the distance between two functions f and g will be the norm $\|f - g\|$ and is a positive number (or 0 if the two functions are identically the same).

In particular, a norm is a positive function $\|\cdot\| : S \to [0, \infty)$, where S is a space of functions like $C(a, b)$. A norm should also satisfy the following properties: For all $f, g \in S$ and all $a \in \mathbb{R}$,

- $\|f\| \geq 0$
- If $\|f\| = 0$ then $f(x) = 0$ for all x in the domain of f
- $\|f + g\| \leq \|f\| + \|g\|$
- $\|af\| = |a|\|f\|$

Here we consider the $\|f\|_\infty = \max_{a\leq x\leq b} |f(x)|$ which satisfies all the previous properties and is known as the L^∞-norm. If we compute the L^∞-norm of the difference of two functions

$$\|f - g\|_\infty = \max_{a\leq x\leq b} |f(x) - g(x)| ,$$

we practically compute the maximum vertical distance between the two functions. This norm is also called the uniform or maximum norm.

Among other norms for functions we use very often the L^2-norm

$$\|f\|_2 = \left(\int_a^b |f(x)|^2\, dx\right)^{1/2} .$$

The last one again is a positive number and measures the square root of the area of the function $|f(x)|^2$. Taking again the difference of two functions $\|f - g\|_2$ we compute the square root of the area of the function $|f(x) - g(x)|^2$. If the two functions are close together, then the area of the difference squared will be small[2].

We proceed now with the estimation of the interpolation error. This is practically equivalent with estimating the remainder of the interpolating polynomial.

Theorem 6.2. *Let $N \in \mathbb{N}$, $f \in C^{N+1}[a,b]$ and the nodes $x_0, \dots, x_N \in [a,b]$. Then, for the Lagrange interpolating polynomial $P_N(x)$ of the function $f(x)$ there is $\xi \in (a,b)$ such that*

$$f(x) - P_N(x) = \frac{f^{N+1}(\xi)}{(N+1)!}\prod_{i=0}^{N}(x - x_i) \quad \textit{for all } x \in [a,b] ,$$

and

$$\|f - P_N\|_\infty \leq \max_{a\leq x\leq b}\left|\prod_{i=0}^{N}(x - x_i)\right| \frac{\|f^{(N+1)}\|_\infty}{(N+1)!} ,$$

where

$$\|g\|_\infty = \max_{a\leq x\leq b} |g(x)| .$$

Proof. For $x \in \{x_0, \dots, x_N\}$ the result is obvious since $f(x_i) - p(x_i) = 0$. Let $x \neq x_i$ and let

$$Q_{N+1}(x) = \prod_{i=0}^{N}(x - x_i) .$$

We define the function

$$h(t) := f(t) - P_N(t) - \frac{f(x) - P_N(x)}{Q_{N+1}(x)} Q_{N+1}(t) ,$$

with $t \in [a,b]$. Obviously, $h \in C^{N+1}[a,b]$. For $t = x_i$ we have

$$h(x_i) = f(x_i) - P_N(x_i) = 0, \quad i = 0, \dots, N ,$$

[2]The absolute value in the definition of the L^2-norm is required only if f is a complex function.

and for $t = x$ we have

$$h(x) = f(x) - P_N(x) - \frac{f(x) - P_N(x)}{Q_{N+1}(x)} Q_{N+1}(x) = 0 \ .$$

This means that $h(t)$ has at least $N+2$ distinct roots in $[a, b]$, namely, x_i, for $i = 0, \ldots, N$ and x. According to Rolle's Theorem 4.2, h' has at least $N+1$ roots in (a, b), and so h'' has at least n roots, e.t.c. and $h^{(N+1)}$ has at least one root $\xi \in (a, b)$. Then, for $t \in [a, b]$

$$h^{(N+1)}(t) = f^{(N+1)}(t) - \frac{f(x) - P_N(x)}{Q_{N+1}(x)}(N+1)! \ ,$$

and so

$$0 = f^{(N+1)}(\xi) - \frac{f(x) - P_N(x)}{Q_{N+1}(x)}(N+1)! \ .$$

Solving for $f(x) - P_N(x)$ yields

$$f(x) - P_N(x) = \frac{f^{(N+1)}(\xi)}{(N+1)!} \prod_{i=0}^{N} (x - x_i) \ , \tag{6.7}$$

for all $x \in [a, b]$.

Taking the maximum of $|f(x) - P_N(x)|$ in (6.7) we see that

$$\|f - P_N\|_\infty \leq \max_{a \leq x \leq b} \left| \prod_{i=0}^{N} (x - x_i) \right| \frac{\|f^{(N+1)}\|_\infty}{(N+1)!} \ ,$$

which gives an upper bound for the interpolation error in the L^∞-norm. □

What we have practically showed is the following: Let $f(x) \in C^{N+1}[a, b]$, then

$$f(x) = P_N(x) + R_N(x) \ ,$$

where $P_N(x)$ is the interpolating polynomial and

$$R_N(x) = \frac{f^{(N+1)}(\xi)}{(N+1)!} Q_{N+1}(x) \quad \text{with} \quad \xi \in (a, b) \ ,$$

with

$$Q_{N+1}(x) = \prod_{i=0}^{N} (x - x_i) = (x - x_0)(x - x_1) \cdots (x - x_N) \ ,$$

is the remainder. It is worth mentioning the following observations:

- For Taylor polynomials $Q_{N+1}(x) = (x - x_0)^{N+1}$ and the error vanishes only at the point x_0.
- For Lagrange interpolating polynomials the error vanishes at all nodes x_i, $i = 0, 1, \ldots, N$.
- To minimize $R_N(x)$ requires minimizing $|Q(x)|$ for $x \in [a, b]$.

The reason for possible bad performance of the Lagrange interpolation is the high-order derivatives in the formula of the error $R_N(x)$. In Runge's example the high-order derivatives are large near the boundaries of the interval $[-1, 1]$, which results to large errors in the approximation.

6.2.5 Newton's representation with divided differences

The representation of the interpolating polynomial using Lagrange polynomials is very convenient especially for theoretical considerations. On the other hand, the computation of Lagrange polynomials can be proved inefficient in practice. For this reason, we will present an alternative representation of the same interpolating polynomial using Newton's form.[3]

Suppose that $P_N(x)$ is the Lagrange polynomial of degree N interpolating $f(x)$ at the nodes x_0, $x_1, \ldots, x_N$. Newton's form for writting the polynomial P_N is

$$P_N(x) = a_0 + a_1(x - x_0) + a_2(x - x_0)(x - x_1) + \cdots + a_N(x - x_0)\cdots(x - x_{N-1}) \ ,$$

where $a_0, a_1, \ldots, a_N$ have to be computed. Note that these coefficients are different from the coefficients of the polynomial in the naive interpolation.

The main idea is to write

$$P_N(x) = P_{N-1}(x) + q_N(x) \ ,$$

where $P_{N-1}(x)$ is the Lagrange interpolating polynomial of degree $N-1$, and $q_N(x)$ a polynomial of degree N. Apparently, $q_N(x) = a_N(x - x_0)\cdots(x - x_{N-1})$. This form has the advantage that we can add nodes without the need of computing the whole interpolating polynomial from the beginning. Newton's form of polynomials leads to the efficient evaluation of polynomials. To see how we can evaluate these polynomials efficiently in Newton's form consider a polynomial of degree 3 written as

$$\begin{aligned} P_3(x) &= a_0 + a_1(x - x_0) + a_2(x - x_0)(x - x_1) + a_3(x - x_0)(x - x_1)(x - x_2) \\ &= a_0 + (x - x_0)\{a_1 + (x - x_1)[a_2 + a_3(x - x_2)]\} \ . \end{aligned}$$

This can be evaluated with the following recursive relations

$$\begin{aligned} p_0(x) &= a_3 \ , \\ p_1(x) &= a_2 + (x - x_2)p_0(x) \ , \\ p_2(x) &= a_1 + (x - x_1)p_1(x) \ , \\ p_3(x) &= a_0 + (x - x_0)p_2(x) \ , \end{aligned}$$

and finally $P_3(x) = p_3(x)$.

The recursive relation for any value of N is

$$\begin{aligned} p_0(x) &= a_N \ , \\ p_k(x) &= a_{N-k} + (x - x_{N-k})p_{k-1}(x), \ k = 1, 2, \ldots, N \ . \end{aligned}$$

Given the coefficients $a_0, a_1, \ldots, a_N$ the evaluation of a polynomial in its Newton's form at the point x can be summarized in Algorithm 18.

Algorithm 18 Evaluation of a polynomial in its Newton's form

Given the point z and the nodes $x_0, x_1, \ldots, x_N$
Initialize $p = a_N$
for $k = 1 : N$ **do**
  $p = a_{N-k} + (z - x_{N-k})p$
end for
On return the variable p contains the value $P_N(z)$

[3]This should not be confused with Newton's method as it is totally irrelevant. The name comes from the way we write the polynomial and which is due to Newton again!

The implementation of Algorithm 18 in Python can be carried out in the following way:

```
def poly_evaluation(a,x,z):
    #Evaluation of the polynomial with coefficients a at the points z
    N = len(x) - 1 #Degree of polynomial
    p = a[N]
    for k in range(1,N+1):
        p = a[N-k] + (z - x[N-k])*p
    return p
```

Efficient evaluation of the coefficients

The remaining issue in Newton's form of interpolation is to determine the coefficients a_i for $i = 0, 1, \ldots, N$. Since the graph of the interpolation polynomial P_N passes through the data (x_i, y_i) we have

$$y_i = P_N(x_i),\ i = 0, 1, \ldots, N\ .$$

This implies that

$$\begin{aligned}
y_0 &= a_0\ ,\\
y_1 &= a_0 + a_1(x_1 - x_0)\ ,\\
y_2 &= a_0 + a_1(x_2 - x_0) + a_2(x_2 - x_0)(x_2 - x_1)\ ,\\
&\ \vdots\\
y_N &= a_0 + a_1(x_N - x_0) + \cdots + a_N(x_N - x_0)(x_N - x_1)\cdots(x_N - x_{N-1})\ ,
\end{aligned}$$

and in general

$$y_k = \sum_{i=0}^{k} a_i \prod_{j=0}^{i-1}(x_k - x_j), \qquad \text{for } k = 0, 1, \ldots, N\ .$$

Note that

$$y_k = P_{k-1}(x_k) + a_k\, Q_k(x_k) \quad \text{where} \quad Q_k(x) = \prod_{j=0}^{k-1}(x - x_j)\ .$$

Solving for a_k we have

$$a_k = \frac{y_k - P_{k-1}(x_k)}{Q_k(x_k)}, \qquad k = 0, 1, \ldots, N\ .$$

Because $y_k = P_k(x_k) = f(x_k)$ we have

$$a_k = \frac{f(x_k) - P_{k-1}(x_k)}{Q_k(x_k)}, \qquad k = 0, 1, \ldots, N\ .$$

More analytically, we have

$$a_0 = y_0 = f(x_0)\ ,$$

and

$$a_1 = \frac{y_1 - a_0}{x_1 - x_0} = \frac{y_1 - y_0}{x_1 - x_0} = \frac{f(x_1) - f(x_0)}{x_1 - x_0}\ ,$$

TABLE 6.1
Divided differences tableau for the computation of the coefficients of the Newton's polynomial

x_i	$f(x_i)$	**1-st order**	**2-nd order**	**3-rd order**	$\cdots$	***N*-th order**
x_0	$\boxed{y_0}$					
x_1	y_1	$\boxed{\Delta^1 y_1}$				
x_2	y_2	$\Delta^1 y_2$	$\boxed{\Delta^2 y_2}$			
x_3	y_3	$\Delta^1 y_3$	$\Delta^2 y_3$	$\boxed{\Delta^3 y_3}$		
$\vdots$	$\vdots$	$\vdots$	$\vdots$	$\vdots$	$\ddots$	
x_N	y_N	$\Delta^1 y_N$	$\Delta^2 y_N$	$\Delta^3 y_N$	$\cdots$	$\boxed{\Delta^N y_N}$

$$a_2 = \frac{y_2 - a_0 - a_1(x_2 - x_0)}{(x_2 - x_0)(x_2 - x_1)} = \frac{\frac{f(x_2)-f(x_0)}{x_2-x_0} - \frac{f(x_1)-f(x_0)}{x_1-x_0}}{x_2 - x_1} ,$$

$$\vdots$$

The fractions of the form $\frac{f(x_i)-f(x_j)}{x_i-x_j}$, appearing in the right-hand sides of these formulas are called *divided differences.* Divided differences approximate derivatives of a function f. In particular, a_1 seems to approximate the first derivative of f at some point and the coefficient a_2 the second derivative of f at some other point. It is very common also to denote $a_k = f[x_0, x_1, \ldots, x_k]$ to highlight the dependence of the coefficient on the nodes x_i. At this stage, in order to simplify notation, we will denote the divided difference approximations of the first order derivatives by $\Delta^1 y_i$, the divided difference approximation of the second order derivatives by $\Delta^2 y_i$ and that of the N-th order by $\Delta^N y_i$ and are given by the formulas

$$\Delta^1 y_i = \frac{y_i - y_0}{x_i - x_0}, \quad i = 1, 2, \ldots, N ,$$
$$\Delta^2 y_i = \frac{\Delta y_i - \Delta y_1}{x_i - x_1}, \quad i = 2, \ldots, N ,$$
$$\Delta^3 y_i = \frac{\Delta^2 y_i - \Delta^2 y_2}{x_i - x_2}, \quad i = 3, \ldots, N ,$$
$$\vdots$$
$$\Delta^N y_i = \frac{\Delta^{N-1} y_i - \Delta^{N-1} y_{i-1}}{x_i - x_{i-1}}, \quad i = N .$$

It is convenient to arrange all the divided differences in the columns of a tableau like the one in Table 6.1. In such a tableau the coefficients a_i are the diagonal entries $\Delta^i y_i$ shown in boxes.

For example, consider the data

x_i	1	2	4	5
y_i	1	4	16	25

In this case, the interpolating polynomial is of degree at most 3 (since $N = 3$) and it can be written in Newton's form as

$$P_3(x) = a_0 + a_1(x - x_0) + a_2(x - x_0)(x - x_1) + a_3(x - x_0)(x - x_1)(x - x_2) .$$

TABLE 6.2
Example of divided differences tableau

x_i	$f(x_i)$	**1-st order**	**2-nd order**	**3-rd order**
1	$\boxed{1}$			
2	4	$\boxed{3}$		
4	16	5	$\boxed{1}$	
5	25	6	1	$\boxed{0}$

To find the coefficients a_0, a_1, a_2, a_3 we fill in the divided differences Table 6.2 and we conclude that the interpolating polynomial is actually the quadratic polynomial

$$P_2(x) = 1 + 3(x-1) + 1(x-1)(x-2) + 0(x-1)(x-2)(x-4) = x^2 \ .$$

Algorithm 19 Interpolating polynomial coefficients using divided differences

Given the data (x_i, y_i), $i = 0, 1, \ldots, N$
for $i = 0 : N$ **do**
    $a_i = y_i = f(x_i)$
end for
for $k = 1 : N$ **do**
    for $j = k : N$ **do**
        $a_j = (a_j - a_{k-1})/(x_j - x_{k-1})$
    end for
end for

The algorithm for the computation of the coefficients a_i using divided differences is summarized in Algorithm 19. In this algorithm we store only the diagonal entries of Table 6.3. Initially the coefficient a_0 is the correct polynomial coefficient. In the next step we compute the first order divided differences, and we store them in the entries $a_1, a_2, \ldots, a_n$ where a_1 is the correct coefficient for the interpolating polynomial. We continue in the same way modifying in each step only the coefficients that they are not correct. Assuming that the interpolating data (x_i, y_i) are stored in the arrays (vectors) `x` and `y`, the implementation of this algorithm in Python can be straightforward and is presented in the function `poly_coeffs`.

```
def poly_coeffs(x, y):
    n = len(x)
    a = y.copy()
    for k in range(1,n):
        a[k:n] = (a[k:n] - a[k-1])/(x[k:n] - x[k-1])
    return a
```

The interpolating polynomial in Newton's form can be written as

$$P_N(x) = \sum_{i=0}^{N} a_i \prod_{j=0}^{i-1} (x - x_j) \ .$$

This form has been implemented in the function `poly_evaluation`. In order to test our code we can try to interpolate Runge's function to obtain the same results as before.

```
def f(x):
    return 1.0 / (1.0 + 25.0 * x**2)
# z are the points we use to plot the interpolating polynomial
z = np.linspace(-1, 1, 100)
x = np.linspace(-1, 1, 10)
y = f(x)
a = poly_coeffs(x, y)
# yz are the values of the interpolating polynomial at z
yz = poly_evaluation(a,x,z)
plt.plot(z,yz,'-')
plt.plot(x,y,'o')
plt.plot(z,f(z),'-.')
plt.legend(["$P_9(x)$","data","$f(x)$"])
plt.xlabel('x'); plt.ylabel('y'); plt.show()
```

An advantage of Newton's form for the interpolating polynomial is that we can add an extra point in the interpolation data without evaluating the polynomial from the beginning. We simply add an extra term in the polynomial formula. Moreover, the evaluation of the polynomial is faster, which is important especially in high-order interpolation. Although there are big advantages in the use of Newton's form, there is no way to overcome Runge's problem due to the interpolation error.

6.2.6 More on divided differences

In this section we present an alternative divided differences algorithm for the computation of the coefficients of the interpolating polynomial in Newton's form. We saw that the interpolating polynomial $P_N(x)$ written in Newton's form is

$$\begin{aligned} P_N(x) &= a_0 + a_1(x-x_0) + a_2(x-x_0)(x-x_1) + \cdots + a_N(x-x_0)\cdots(x-x_{N-1}) \\ &= P_{N-1}(x) + a_N\, Q_N(x) , \end{aligned}$$

where

$$Q_N(x) = (x-x_0)\cdots(x-x_{N-1}) .$$

As a consequence of (6.5), the polynomial coefficients a_k can be computed using finite differences via the formula

$$a_k = \sum_{i=0}^{k} \frac{y_i}{Q'_{N+1}(x_i)} .$$

These coefficients can be written as

$$a_0 = y_0 = f(x_0) ,$$

and

$$\begin{aligned} a_1 &= \frac{y_1 - a_0}{x_1 - x_0} = \frac{y_1 - y_0}{x_1 - x_0} = \frac{f(x_1) - f(x_0)}{x_1 - x_0} , \\ a_2 &= \frac{y_2 - a_0 - a_1(x_2 - x_0)}{(x_2 - x_0)(x_2 - x_1)} = \frac{\frac{f(x_2)-f(x_0)}{x_2-x_0} - \frac{f(x_1)-f(x_0)}{x_1-x_0}}{x_2 - x_1} , \\ &\vdots \end{aligned}$$

Rearranging the fractions appeared in a_2 we can write it as

$$\begin{aligned} a_2 &= \frac{\frac{f(x_2)-f(x_0)}{x_2-x_0} - \frac{f(x_1)-f(x_0)}{x_1-x_0}}{x_2 - x_1} \\ &= \frac{\frac{f(x_2)-f(x_1)}{x_2-x_0} + \frac{f(x_1)-f(x_0)}{x_2-x_0} - \frac{f(x_1)-f(x_0)}{x_1-x_0}}{x_2 - x_1} \\ &= \frac{\frac{f(x_2)-f(x_1)}{x_2-x_0}}{x_2 - x_1} + \frac{[f(x_1) - f(x_0)](x_2 - x_1)}{(x_2 - x_0)(x_2 - x_1)(x_1 - x_0)} , \end{aligned}$$

thus

$$a_2 = \frac{\frac{f(x_2)-f(x_1)}{x_2-x_1} - \frac{f(x_1)-f(x_0)}{x_1-x_0}}{x_2 - x_0} .$$

Similarly, we can modify the rest of the formulas to obtain a new formulation for our interpolation algorithm. Specifically, the coefficients a_i can be defined as

$$a_i = f[x_0, x_1, \ldots, x_i] ,$$

where

$$f[x_0, x_1, \ldots, x_i] = \frac{f[x_1, x_2, \ldots, x_i] - f[x_0, x_1, \ldots, x_{i-1}]}{x_i - x_0}, \qquad x_i \neq x_0 ,$$

and

$$f[x_i, x_{i+1}, \ldots, x_j] = \frac{f[x_{i+1}, x_{i+2}, \ldots, x_j] - f[x_i, x_{i+1}, \ldots, x_{j-1}]}{x_j - x_i}, \qquad x_i \neq x_j . \tag{6.8}$$

In this notation we have that

$$\begin{aligned} f[x_i] &= f(x_i) , \\ f[x_i, x_{i+1}] &= \frac{f[x_{i+1}] - f[x_i]}{x_{i+1} - x_i} , \\ f[x_i, x_{i+1}, x_{i+2}] &= \frac{f[x_{i+1}, x_{i+2}] - f[x_i, x_{i+1}]}{x_{i+2} - x_i} , \\ &\cdots \end{aligned}$$

while $f[x_0, x_1, \ldots, x_i]$ is the coefficient a_i of the interpolating polynomial P_i. Furthermore, $f[x_1, x_2, \ldots, x_i]$ is the coefficient of the term x^{i-1} of the polynomial of degree less than or equal to $i - 1$ that interpolates f at $x_1, x_2, \ldots, x_i$.

In this book we define the divided differences with one of their properties:

Definition 6.3. *For $x_0 \leq x_1 \leq \cdots \leq x_N$, the divided differences of a smooth function f are defined as*

$$f[x_i, x_{i+1}, \ldots, x_j] = \begin{cases} \frac{f[x_{i+1},\ldots,x_j]-f[x_i,\ldots,x_{j-1}]}{x_j-x_i}, & x_i \neq x_j \\ \frac{f^{(j-i)}(x_i)}{(j-i)!}, & x_i = x_j \end{cases} .$$

In this definition we have included the case $x_i = x_j$, which will be useful later for the Hermite interpolation. There are other more general definitions but here, and at the particular stage of this book, we focus on the computational part of this technique. Needless to say, Newton's form of polynomials is in general very useful for theoretical purposes as well.

TABLE 6.3
Divided differences tableau for the computation of the coefficients of Newton's polynomial

x_i	$f(x_i)$	1st order	2nd order	3rd order	4th order
x_0	$\boxed{y_0 = f[x_0]}$				
		$\boxed{f[x_0, x_1]}$			
x_1	$y_1 = f[x_1]$		$\boxed{f[x_0, x_1, x_2]}$		
		$f[x_1, x_2]$		$\boxed{f[x_0, x_1, x_2, x_3]}$	
x_2	$y_2 = f[x_2]$		$f[x_1, x_2, x_3]$		$\boxed{f[x_0, x_1, x_2, x_3, x_4]}$
		$f[x_2, x_3]$		$f[x_1, x_2, x_3, x_4]$	
x_3	$y_2 = f[x_3]$		$f[x_2, x_3, x_4]$		
		$f[x_3, x_4]$			
x_4	$y_3 = f[x_4]$				

It is not hard to see that the first three divided differences a_0, a_1 and a_2 are given by the formulas

$$f[x_0] = f(x_0) \ ,$$
$$f[x_0, x_1] = \frac{f(x_1) - f[x_0]}{x_1 - x_0} \ ,$$
$$f[x_0, x_1, x_2] = \frac{f(x_2) - f[x_0] - f[x_0, x_1](x_2 - x_0)}{(x_2 - x_0)(x_2 - x_1)} \ .$$

Using this traditional notation, Newton's form of the interpolating polynomial is written in the form

$$P_N(x) = \sum_{i=0}^{N} \left[f[x_0, x_1, \ldots, x_i] \prod_{j=0}^{i-1} (x - x_j) \right] \ .$$

For example, if we consider the case with only 2 nodes ($N = 1$), the linear polynomial interpolating the points $(x_0, f(x_0))$ and $(x_1, f(x_1))$ is

$$P_1(x) = f[x_0] + f[x_0, x_1](x - x_0) \ .$$

This is the secant of f passing through the two interpolation points

$$P_1(x) = f(x_0) + \frac{f(x_1) - f(x_0)}{x_1 - x_0}(x - x_0) \ .$$

It is also noted that if $f(x)$ is continuously differentiable function, then

$$\lim_{x_{i+1} \to x_i} f[x_i, x_{i+1}] = \lim_{x_i \to x_{i+1}} \frac{f(x_{i+1}) - f(x_i)}{x_{i+1} - x_i} = f'(x_i) \ .$$

If $f(x)$ is polynomial of degree at most k, then $f[x_i, \ldots, x_{i+k}] = 0$.

Because of this new formulation it is convenient to rearrange Table 6.1 to a new Table 6.3, where the coefficients of the interpolating polynomial are the same as before but there are differences in the other entries.

For example, if we consider the data of Table 6.2, the new divided differences tableau is presented in Table 6.4. This is different from Table 6.1 but the interpolating polynomial remains the same as the diagonal entries are identically the same.

TABLE 6.4
Example of divided difference method

x_i	$f(x_i)$	**1st order**	**2nd order**	**3rd order**
1	$\boxed{1}$			
		$\boxed{3}$		
2	4		$\boxed{1}$	
		6		$\boxed{0}$
4	16		1	
		9		
5	25			

The algorithm for the computation of the coefficients of the divided difference table in the last form can be written as follows: Since we are interested only in the computation of the diagonal entries a_i, we apply formula (6.8) directly to obtain the Algorithm 20.

Algorithm 20 Divided differences coefficients

Given the data (x_i, y_i), $i = 0, 1, \dots, N$
for $i = 0 : N$ **do**
  Set $a_i = y_i = f(x_i)$
end for
for $j = 1 : N$ **do**
  for $i = n : j : -1$ **do**
    $a_i = (a_i - a_{i-1})/(x_{i+j} - x_i)$
  end for
end for

The notation in the loop conditions $i = a : b : c$ of the previous pseudo-code means that a is the starting value, b is the stop value and c is the step. The implementation of Algorithm (20) is straightforward and is left as an exercise for the reader.

In this notation, the polynomial $P_N(x)$ is then

$$P_N(x) = \sum_{i=0}^{N} a_i \prod_{j=0}^{i-1} (x - x_j) \ .$$

Having developed most of the ideas of polynomial interpolation, we move on with another interpolation method that takes into account the smoothness of f.

6.3 Hermite Interpolation

Assume that we have at our disposal in addition to data (x_i, y_i), for $i = 0, 1, \dots, N$ with $y_i = f(x_i)$, all the values of the derivative $f'(x_i)$ at the nodes x_i. This means that in addition to (x_i, y_i) we have at our disposal the data (x_i, y_i') where $y_i' = f'(x_i)$, for $i = 0, 1, \dots, N$. Particularly, in this case, we have $2N + 2$ data that we can use to find an interpolating polynomial of degree $2N + 1$ with $2N + 2$ coefficients. The new polynomial will interpolate the values $P_{2N+1}(x_i) = f(x_i)$ and $P'_{2N+1}(x_i) = f'(x_i)$ for $i = 0, 1, \dots, N$. This extension of the Lagrange interpolation that respects the derivative of the function at the nodes is called *Hermite interpolation* and was introduced by Charles Hermite (1822–1901). In the sequel we denote the Hermite interpolation polynomial by H_{2N+1} instead of P_{2N+1} to distinguish it from the Lagrange interpolant.

6.3.1 Computation of the Hermite interpolant

In Hermite interpolation is rather convenient to write the interpolating polynomial in the form

$$H_{2N+1}(x) = \sum_{i=0}^{N} [\alpha_i(x) f(x_i) + \beta_i(x) f'(x_i)] \ . \tag{6.9}$$

In order for H_{2N+1} to interpolate the points (x_i, y_i) we need

$$\alpha_i(x_j) = \delta_{ij}, \quad \text{and} \quad \beta_i(x_j) = 0 \ . \tag{6.10}$$

Furthermore, in order for the derivative of the polynomial to interpolate the data (x_i, y_i') we need

$$\alpha_i'(x_j) = 0, \quad \text{and} \quad \beta_i'(x_j) = \delta_{ij} \ . \tag{6.11}$$

Thus, we expect that the unique Hermite polynomial $H_{2N+1}(x)$ will satisfy the constraints (6.10)–(6.11). In order to construct the coefficients α_i and β_i we make use of the usual Lagrange polynomials

$$\ell_i(x) = \prod_{\substack{j=0 \\ j\neq i}}^{N} \frac{x - x_j}{x_i - x_j} \ ,$$

that satisfy $\ell_i(x_j) = \ell_i^2(x_j) = \delta_{ij}$. The degree of the polynomial $\ell_i(x)$ is N, implying that the degree of $\ell_i^2(x)$ is $2N$. Thus, in order for the degree of $\alpha_i(x)$ and $\beta_i(x)$ to be $2N+1$, we write

$$\alpha_i(x) = r_i(x)\ell_i^2(x), \qquad \beta_i(x) = s_i(x)\ell_i^2(x) \ ,$$

where $r_i(x)$ and $s_i(x)$ should be both linear polynomials. Using (6.10) we have

$$\delta_{ij} = \alpha_i(x_j) = r_i(x_j)\ell_i^2(x_j) = r_i(x_j)\delta_{ij} \ ,$$

and thus

$$r_i(x_i) = 1 \ . \tag{6.12}$$

Similarly, using (6.11) we have

$$0 = \alpha_i'(x_j) = r_i'(x_j)\ell_i^2(x_j) + 2r_i(x_j)\ell_i(x_j)\ell_i'(x_j) = r_i'(x_j)\delta_{ij} + 2r_i(x_j)\delta_{ij}\ell_i'(x_j) \ ,$$

which implies

$$r_i'(x_i) = -2\ell_i'(x_i) \ . \tag{6.13}$$

Writing $r_i(x) = ax + b$ (since it is a linear polynomial) we compute the coefficients a and b through (6.12) and (6.13) as

$$a = -2\ell_i'(x_i), \quad b = 1 + 2\ell_i'(x_i)x_i \ .$$

Therefore, we obtain

$$\alpha_i(x) = [1 - 2\ell_i'(x_i)(x - x_i)]\ell_i^2(x) \ . \tag{6.14}$$

Similarly, from (6.10) we obtain $s_i(x_i) = 0$ and from (6.11) that $s_i'(x_j) = 1$, which implies immediately that

$$s_i(x) = x - x_i \ ,$$

and thus

$$\beta_i(x) = (x - x_i)\ell_i^2(x) \ . \tag{6.15}$$

☞ Given $N+1$ nodes $x_0 < x_1 < \cdots < x_N$ and the values $f(x_i)$ and $f'(x_i)$ for $i = 0, 1, \ldots, N$, the Hermite interpolating polynomial is the polynomial

$$H_{2N+1}(x) = \sum_{i=0}^{N} [\alpha_i(x) f(x_i) + \beta_i(x) f'(x_i)] ,$$

where α_i and β_i are given in terms of the Lagrange polynomials as

$$\alpha_i(x) = [1 - 2\ell_i'(x_i)(x - x_i)]\ell_i^2(x) \quad \text{and} \quad \beta_i(x) = (x - x_i)\ell_i^2(x) .$$

Here we present an implementation of the previous formulas. In order to compute the polynomials $\ell_i(x)$ we use the NumPy function `numpy.poly1d`, which given the polynomial coefficients in decreasing powers returns a polynomial function. We then compute the first derivative of this function with the function `numpy.polyder`.

```
def Hermite(x,y,z):
# x, y: are the interpolating data
# z: values to the interpolant
    N = len(x)
    p = np.poly1d([0]) # initialize p
    for i in range(N):
        #derive L_k
        L = np.poly1d([1])
        for j in range(N):
            if j!=i:
                L = L*np.poly1d([1.0/(x[i]-x[j]), - x[j]/(x[i]-x[j])])
        #Derivative of the Lagrange polynomials
        dL = np.polyder(L)
        #Computation of the functions alpha
        alpha = (np.poly1d([1])-2*dL(x[i])*np.poly1d([1,-x[i]]))*L**2
        #Computation of the functions beta
        beta = np.poly1d([1,-x[i]])*L**2
        p = p + alpha*y[i]+beta*z[i]
    return p
```

The drawback of the Hermite interpolation is that it does not provide a solution to Runge's phenomenon. For example, using the previous implementation we compute the Hermite polynomial with $N = 9$ for Runge's function:

```
def f(x):
    return 1.0/(1+25*x**2)
def df(x):
    return -2*25*x/(1+25*x**2)**2
xi = np.linspace(-1,1,10)
yi = f(xi)
zi = df(xi)
x = np.linspace(-1,1,300)
y = f(x)
p = Hermite(xi,yi,zi)
y1 = p(x)
```

```python
plt.plot(x, y1, label="$H_{19}(x)$")
plt.plot(xi, yi, 'o', label='data')
plt.plot(x, y, '-.',label='$f(x)$')
plt.xlabel('$x$')
plt.ylabel('$y$',rotation=0)
plt.xlim(-1, 1)
plt.legend(loc='upper left', ncol=1)
plt.show()
```

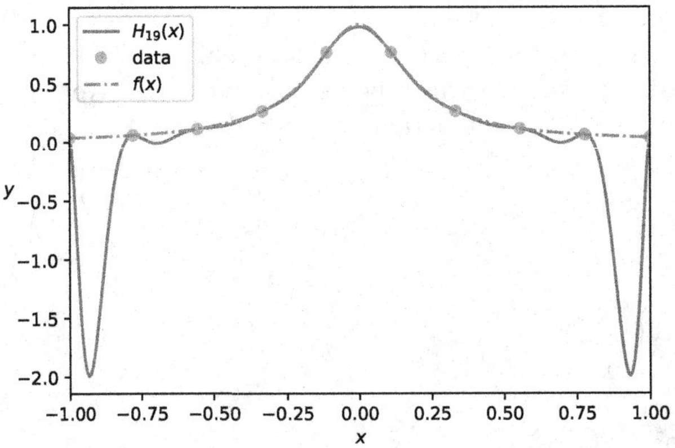

Comparing the result of the Hermite interpolation with Figure 6.2 we observe that the Hermite interpolation is worse closer to the endpoints of the interval $[-1, 1]$.

6.3.2 Error analysis

Now that we know what the Hermite interpolating polynomial is, it is rather useful to know its accuracy in approximating functions. The following result will appear quite useful later for theoretical estimate of the error of a particular method of numerical integration, the so-called Gaussian quadrature.

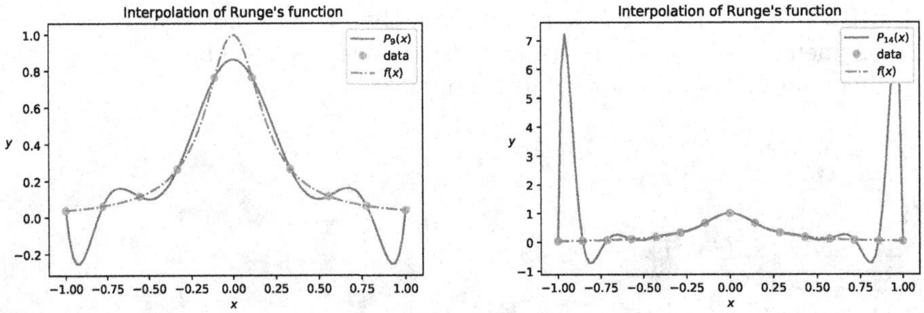

FIGURE 6.2
Lagrange interpolation of Runge's function with $N = 9$ and $N = 14$.

Theorem 6.4. *Let $N \in \mathbb{N}$, $f \in C^{2N+2}[a,b]$. If H_{2N+1} is the Hermite interpolant of $f(x)$ at the nodes $x_0, \dots, x_N \in [a,b]$, then for all $x \in [a,b]$ there is a $\xi = \xi(x) \in (a,b)$ such that*

$$f(x) - H_{2N+1}(x) = \frac{f^{2N+2}(\xi)}{(2N+2)!} \prod_{i=0}^{N} (x - x_i)^2 .$$

Proof. If x is one of $x_0, x_1, \dots, x_N$, then the result is obvious since $f(x_i) = H_{2N+1}(x_i)$ for all i.

Let $x \neq x_i$ and

$$Q(x) = \prod_{i=0}^{N} (x - x_i)^2 .$$

We define the function

$$h(t) := f(t) - H_{2N+1}(t) - \frac{f(x) - H_{2N+1}(x)}{Q(x)} Q(t) ,$$

with $t \in [a,b]$. Obviously, $h \in C^{2N+2}[a,b]$. Moreover, we have

$$h(x_i) = f(x_i) - H_{2N+1}(x_i) = 0, \quad i = 0, \dots, N ,$$

and for $t = x$ we have

$$h(x) = f(x) - H_{2N+1}(x) - \frac{f(x) - H_{2N+1}(x)}{Q(x)} Q(x) = 0 .$$

This means that $h(t)$ satisfies

$$h(x_0) = h(x_1) = \cdots = h(x_j) = h(x) = h(x_{j+1}) = \cdots = h(x_N) = 0 .$$

Therefore, because of Role's Theorem 4.2, there are $\xi_0, \xi_1, \dots, \xi_N$,

$$x_0 < \xi_0 < x_1 < \xi_1 < \cdots < x_j < \xi_j < x < \xi_{j+1} < x_{j+1} < \cdots < \xi_N < x_N ,$$

such that

$$h'(\xi_i) = 0, \quad i = 0, \dots, N .$$

The fact that $P'_{2N+1}(x_i) = f'(x_i)$ for $i = 0, \dots, N$ implies $h'(x_i) = 0$. Thus, the function h' has $2N+2$ roots. Similarly, h'' has at least $2N+1$ roots, h''' has at least $2N$ roots and finally for $h^{(2N+2)}$ we have that there is at least one $\xi \in (a,b)$ (which obviously depends on x) such that $h^{(2N+2)}(\xi) = 0$. But,

$$h^{(2N+2)}(t) = f^{(2N+2)}(t) - \frac{f(x) - H_{2N+1}(x)}{Q(x)} (2N+2)! ,$$

and for $t = \xi$ we get

$$0 = f^{(2N+2)}(\xi) - \frac{f(x) - H_{2N+1}(x)}{Q(x)} (2N+2)! .$$

Solving the last equation for $f(x) - H_{2N+1}(x)$ we obtain the error formula

$$f(x) - H_{2N+1}(x) = \frac{f^{2N+2}(\xi)}{(2N+2)!} Q(x) ,$$

which is the desired error formula. □

☞ Notice the differences between the two very similar Theorems 6.2 and 6.4. The error still depends on high-order derivatives of $f(x)$, but now the convergence as $N \to \infty$ is quadratic. Even if the convergence is quadratic, the presence of high-order derivatives can lead to large errors. For this reason Hermite interpolation is not the solution to Runge's problem.

6.3.3 Implementation details

Given the data $(x_i, f(x_i))$ and $(x_i, f'(x_i))$ for $i = 0, 1, \ldots, N$, the Hermite interpolation polynomial can be computed using divided differences and Newton's form of polynomials. Specifically, we can construct a table with the divided differences this time using the fact that $f[x_0, x_0] = f'(x_0)$, $f[x_1, x_1] = f'(x_1)$ etc. The idea is very similar to the Lagrange interpolation. The difference is that we use the data $f'(x_i)$ and the fact that we can write the Hermite interpolating polynomial in the form

$$H_{2N+1}(x) = \sum_{i=0}^{2N+1} f[z_0, z_1, \ldots, z_i] \prod_{j=0}^{i-1} (x - z_j) ,$$

where the nodes $z_0, z_1, \ldots, z_{2N+1}$ are defined such that $z_{2i} = z_{2i+1} = x_i$, for $i = 0, 1, \ldots, N$. Since $z_{2i} = z_{2i+1} = x_i$, for each i we have that $f[z_{2i}, z_{2i+1}] = f'(z_{2i}) = f'(x_i)$. The complete explanation of this property of the Hermite interpolating polynomial can be found in [107].

If we have for example the three triplets (x_i, y_i, y_i'), for $i = 0, 1, 2$, then we construct Table 6.5 similar to Table 6.3 using the z_i values instead of the x_i values. It is therefore straightforward to extend the algorithms and codes of Section 6.2.5 to compute the Hermite interpolation polynomial, and thus we leave it as an exercise for the reader.

The Hermite interpolation can also be extended to higher order polynomials provided data for high order derivatives of the function f. The extension of the Hermite interpolation in the case where certain derivatives have specified values at specified points is known as Hermite-Birkhoff interpolation, [10].

☞ While Hermite interpolation does not solve Runge's phenomenon, it can be more accurate than the Lagrange interpolation. The error of the Hermite interpolation depends on the values of high-order derivatives of $f(x)$. Lagrange interpolation is preferable because of its simplicity.

6.4 Spline Interpolation

One way to overcome the problems of Lagrange interpolation is by using spline interpolation. Spline interpolation is the interpolation of data using piecewise polynomial functions instead of one uniform polynomial. For this reason, spline interpolation is also known as piecewise-polynomial interpolation. The root of the word *spline* is the same as that of the word *splint*, which was originally a narrow piece of wood that could be used to join two boards. Later, the word was used to refer to a tool consisting of a long flexible piece of metal that could be

TABLE 6.5
Divided differences tableau for the computation of the coefficients of the Hermite polynomial

x_i	$f(x_i)$	1st order	2nd order	3rd order	4th order	5th order
$z_0 = x_0$	$\boxed{f[z_0] = f(x_0)}$					
		$\boxed{f[z_0, z_1] = f'(x_0)}$				
$z_1 = x_0$	$f[z_1] = f(x_0)$		$\boxed{f[z_0, z_1, z_2]}$			
		$f[z_1, z_2]$		$\boxed{f[z_0, z_1, z_2, z_3]}$		
$z_2 = x_1$	$f[z_2] = f(x_1)$		$f[z_1, z_2, z_3]$		$\boxed{f[z_0, z_1, z_2, z_3, z_4]}$	
		$f[z_2, z_3] = f'(x_1)$		$f[z_1, z_2, z_3, z_4]$		$\boxed{f[z_0, z_1, z_2, z_3, z_4, z_5]}$
$z_3 = x_1$	$f[z_3] = f(x_1)$		$f[z_2, z_3, z_4]$		$f[z_1, z_2, z_3, z_4, z_5]$	
		$f[z_3, z_4]$		$f[z_2, z_3, z_4, z_5]$		
$z_4 = x_2$	$f[z_4] = f(x_2)$		$f[z_3, z_4, z_5]$			
		$f[z_4, z_5] = f'(x_2)$				
$z_5 = x_2$	$f[z_5] = f(x_2)$					

used to draw smooth curves by forcing the tool to pass through specified points and tracing along the curve.

The simplest piecewise-polynomial approximation is the piecewise linear interpolation in which we connect the set of data points

$$\{(x_0, f(x_0)),\ (x_1, f(x_1)), \ldots, (x_N, f(x_N))\}\ ,$$

with straight lines.

6.4.1 Continuous piecewise linear interpolation

Piecewise linear interpolation, also known as linear spline interpolation, is basically the method we use to plot a function using the `plot` command of `matplotlib`. Instead of sketching a graph of the actual function $f(x)$, we first generate a grid of nodes $x_0, x_1, \ldots, x_N$, and then we compute the values $f(x_0), f(x_1), \ldots, f(x_N)$. In other words, we generate a set of data points $\{x_i, f(x_i)\}$ and we ask Python to make a plot using these points. The Python code in such a case can be the following:

```
x = np.linspace(0, 2.0*np.pi, 10)
y = np.sin(x)
plt.plot(x,y,'o-')
plt.xlabel('$x$'); plt.ylabel('$y$',rotation=0); plt.show()
```

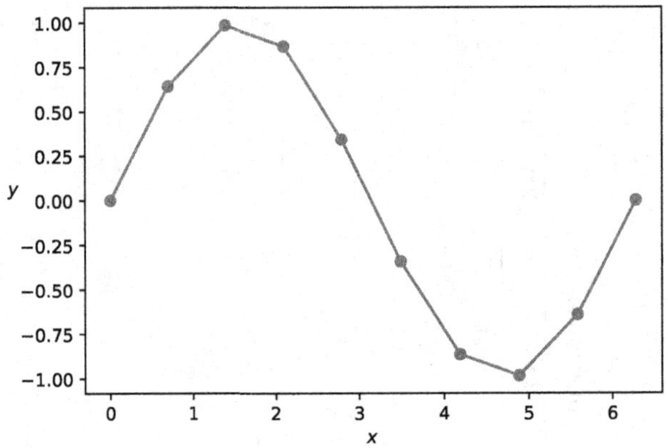

Python plots the function $\sin(x)$ using 10 nodes by connecting the points $\{x_i, f(x_i)\}$ with line segments. This piecewise linear function is in practical terms a linear interpolating spline.

> ☞ The resulting linear interpolating spline is continuous everywhere but not differentiable at the nodes x_i. In some circumstances we define a derivative of such piecewise polynomial function as a discontinuous function.

In the case of splines there is no interpolating polynomial. Instead, there is an interpolating function, which in the case of linear splines consists of different linear polynomials for each interval $[x_i, x_{i+1}]$. Specifically, the line-segment that connects the points $(x_i, f(x_i))$

and $(x_{i+1}, f(x_{i+1}))$ is given by the formula

$$s_i(x) = \frac{x_{i+1} - x}{x_{i+1} - x_i} f(x_i) + \frac{x - x_i}{x_{i+1} - x_i} f(x_{i+1}) , \tag{6.16}$$

for $x \in [x_i, x_{i+1}]$. It is easy to see that $s_i(x_i) = f(x_i)$ and $s_i(x_{i+1}) = f(x_{i+1})$. A piecewise linear function is assembled by linear functions s_i on each interval $[x_i, x_{i+1}]$, with $i = 0, \ldots, N$. A piecewise linear interpolant is a piecewise linear function that interpolates points $(x_i, f(x_i))$, $i = 0, 1, \ldots, N$. This is a function such as

$$P_l(x) = \begin{cases} s_0(x), & x \in [x_0, x_1) \\ s_1(x), & x \in [x_1, x_2) \\ \cdots & \\ s_{N-1}, & x \in [x_{N-1}, x_N] \end{cases} ,$$

where $s_i(x)$ is given by the formula (6.16).

Instead of computing all the linear functions $s_i(x)$, it is more convenient to follow the same technique we used in the case of the Lagrange interpolation and express the interpolating function as a linear combination of some basis functions $\phi_i(x)$:

$$P_l(x) = \sum_{i=0}^{N} c_i \phi_i(x) .$$

Let $a = x_0 < x_1 < x_2 < \cdots < x_N = b$ a partition of $[a, b]$ and $y_i = f(x_i)$, $i = 0, \ldots, N$ given points. The functions

$$\phi_i(x) = \begin{cases} \frac{x - x_{i-1}}{x_i - x_{i-1}}, & x_{i-1} \le x \le x_i \\ \frac{x_{i+1} - x}{x_{i+1} - x_i}, & x_i \le x \le x_{i+1} \\ 0, & x \in \bar{I} - (x_{i-1}, x_{i+1}) \end{cases} , \tag{6.17}$$

for $i = 1, \ldots, N-1$,

$$\phi_0(x) = \begin{cases} \frac{x_1 - x}{x_1 - x_0}, & x_0 \le x \le x_1 \\ 0, & x \in \bar{I} - (x_0, x_1) \end{cases} , \tag{6.18}$$

and

$$\phi_N(x) = \begin{cases} \frac{x - x_{N-1}}{x_N - x_{N-1}}, & x_{N-1} \le x \le x_N \\ 0, & x \in \bar{I} - (x_{N-1}, x_N) \end{cases} , \tag{6.19}$$

are the basis functions of the space of piecewise linear functions in which our new interpolants lie. The shape of $\phi_i(x)$ is depicted in Figure 6.3. These basis functions are also known to as hat functions because of their shape. They are also called linear B-splines. Observe that each basis function $\phi_i(x)$ for $i = 1, 2, \ldots, N-1$ is positive in only two intervals $[x_{i-1}, x_i] \cup [x_i, x_{i+1}]$ and zero anywhere else. Because of that, we say that they have support $[x_{i-1}, x_i] \cup [x_i, x_{i+1}]$. The two basis functions at the boundary intervals need to have support only one interval though. It is also important to note that the basis functions satisfy $\phi_i(x_i) = 1$ and $\phi_i(x_j) = 0$ if $j \ne i$. For this reason, we say that the basis functions $\phi_i(x)$ are normalized.

Having defined the basis functions of the space of piecewise linear functions $\mathbb{S}_1$, we write the piecewise linear interpolant as linear combination of the basis functions

$$P_l(x) = \sum_{i=0}^{N} f(x_i) \phi_i(x) .$$

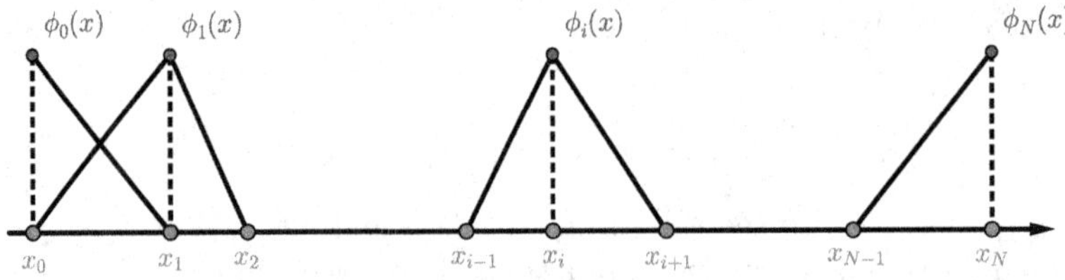

FIGURE 6.3
Linear spline basis functions with maximum value 1.

Since $\phi_i(x_j) = 0$ if $i \neq j$ and $\phi_i(x_i) = 1$ we have that $P_l(x_i) = f(x_i)$. It is noted that $P_l(x)$ is piecewise linear and continuous function in $[a, b]$. We will denote the space of continuous piecewise linear functions by

$$\mathbb{S}_1 = \{f : \ f \in C[a,b], f \text{ is linear polynomial on each } (x_i, x_{i+1}),\ 0 \leq i \leq N-1\} \ .$$

NumPy implementation

The piecewise linear interpolant can be computed easily using the function `numpy.interp`. The general call of this function is

```
interp(x, xp, fp, left, right, period)
```

where

`x`: An array with the x values at which we evaluate the interpolant

`xp`: An array with the values x_i, $i = 0, \ldots, N$

`fp`: An array with the values $y_i = f(x_i)$, $i = 0, \ldots, N$

`left`: Value to return if `x<xp[0]`. (Optional with default value `None`)

`right`: Value to return if `x>xp[-1]`. (Optional with default value `None`)

`period`: The period of the x values when f is periodic. In this case, the parameters `left` and `right` are ignored. (Optional with default value `None`)

The data $(x_i, f(x_i))$ or (x_i, y_i) must be stored in two arrays `xp` and `fp`. The points where we evaluate the interpolant are stored in an array `x`. The function `interp` returns the values of the interpolant at the nodes `x` stored in an array `y`.

As an example, we consider again the function $f(x) = \sin(x)$ in the interval $[0, 2\pi]$. We take $N = 9$ points $(x_i, f(x_i))$, $i = 0, 1, \ldots, N$ and we compute the linear spline interpolant evaluated on 50 points z_j, $j = 0, 1, \ldots, 49$.

```
# define the interpolation points
xp = np.linspace(0, 2*np.pi, 10)
yp = np.sin(xp)
# define the points for ploting
x = np.linspace(0, 2*np.pi, 50)
# generate the linear spline
y = np.interp(x, xp, yp)
# plot spline vs data
plt.plot(x, y, '-x', label='$S_1(x)$')
```

```
plt.plot(xp, yp, 'o', label='data')
plt.xlabel('$x$')
plt.ylabel('$y$',rotation=0)
plt.legend(loc='upper right', ncol=1)
```

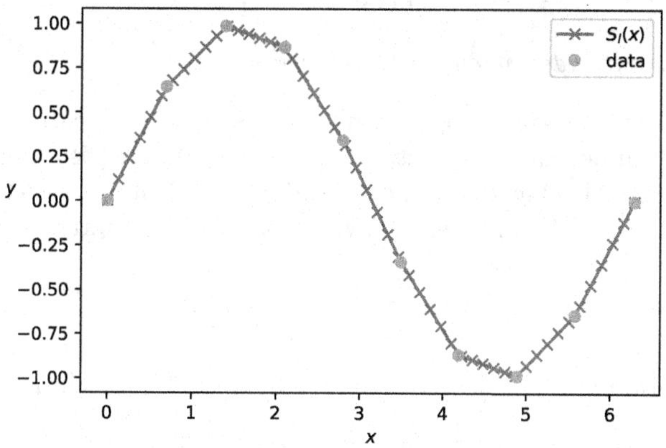

This figure is identical with the figure we generated by just ploting the same function using 10 nodes. With the interpolant, of course, we are able to compute approximation to $\sin(x)$ at any intermediate point x in $[0, 2\pi]$. Piecewise linear interpolation solves Runge's phenomenon. Specifically, the interpolant consists of straight lines connecting the interpolation points (x_i, y_i) and it cannot oscillate around Runge's function.

6.4.2 Some theoretical considerations

In piecewise linear interpolation, the function $f(x)$ is being interpolated by the function $P_\ell(x) = \sum_{i=0}^{N} f(x_i)\phi_i(x)$ at the points $(x_i, f(x_i))$, for $i = 0, 1, \dots, N$. In this notation $\phi_i(x)$ is the i-th basis function of the space of piecewise linear functions $\mathbb{S}_1$ given by the formulas (6.17)-(6.19). Therefore, the function P_l belongs to the space $\mathbb{S}_1$, since it is a linear combination of its basis functions. In order to see that the set $\{\phi_i\}$ spans the space $\mathbb{S}_1$, we need some elements from linear algebra.

Proposition 6.5. *The functions $\phi_i(x)$, $i = 0, 1, \dots, N$ span the space $\mathbb{S}_1$*

Proof. First observe that $\phi_i \in \mathbb{S}_1$. Moreover, if $\sum_{i=0}^{N} c_i\phi_i(x) = 0$ for all $x \in [a, b]$, then it is true that $\sum_{i=0}^{N} c_i\phi_i(x_j) = 0$ for $j = 0, 1, \dots, N$. This implies $c_j = 0$. Thus, $\phi_i(x)$ are linearly independent functions. Notice that every piecewise linear function $\phi(x) \in \mathbb{S}_1$ can be written as

$$\phi(x) = \sum_{i=0}^{N} c_i\phi_i(x) \ ,$$

with $c_i = \phi_i(x_i)$, which means that the functions ϕ_i span the space $\mathbb{S}_1$, and the proof is complete. □

6.4.3 Error analysis of piecewise linear interpolation

The question, again, is how well the piecewise linear function P_l approximates $f(x)$. In order to make things rigorous, we consider a grid of nodes $a = x_0 < x_1 < \cdots < x_N = b$.

The distance between two nodes x_i, x_{i+1} is denoted by $h_i = x_{i+1} - x_i$, for $i = 0, \dots, N-1$. We denote by h the maximum such distance, $h = \max_i h_i$. Then, the maximum distance between f and its interpolant P_l is $O(h^2)$. For this reason, we will use the L^∞-norm. This result is explained in the following theorem:

Theorem 6.6. *If $f \in C^2[a,b]$ and given the partition $a = x_0 < x_1 < \cdots < x_N = b$, then*

$$\|f - P_l\|_\infty \leq Ch^2 \ ,$$

where $C > 0$ is a constant independent of the choice of x_i.

Proof. Since we work with the maximum-norm we will search for the distance $|f(x) - P_l(x)|$ at a point $x \in [a,b]$. Assume that $x \in [x_i, x_{i+1}]$ for some i. Since P_l is a linear polynomial in $[x_i, x_{i+1}]$ that interpolates $f(x)$ at the points x_i and x_{i+1}, it coincides with the P_2 Lagrange interpolant for these points. By Theorem 6.2 we have that $x \in [x_i, x_{i+1}]$

$$f(x) - P_l(x) = \frac{f''(\xi)}{2}(x - x_i)(x - x_{i+1}) \ ,$$

where $\xi \in (x_i, x_{i+1})$. Taking absolute values on both sides of this relation we have

$$\begin{aligned}
|f(x) - P_l(x)| &= \frac{1}{2}|f''(\xi)| \; |x - x_i| \; |x - x_{i+1}| \\
&\leq \frac{1}{2}\max|f''(x)| \; |x_{i+1} - x_i| \; |x_i - x_{i+1}| \\
&= Ch^2 \ ,
\end{aligned}$$

where $C = \frac{1}{2}\|f''\|_\infty$. Since the last inequality holds for all x, then we conclude that the $\max|f(x) - P_l(x)|$ satisfy the same inequality. □

The estimated constant C in this proof is not the smallest constant we can have in the particular estimate. We can obtain a better estimate of C (implying smaller error) by assuming (without loss of generality) that x is closer to x_{i+1}. In such case we have that

$$\begin{aligned}
|f(x) - P_l(x)| &= \frac{1}{2}|f''(\xi)| \; |x - x_i| \; |x - x_{i+1}| \\
&= \frac{1}{2}\max|f''(x)| \; |x - x_{i+1} + x_{i+1} - x_i| \; |x - x_{i+1}| \\
&\leq \frac{1}{2}\max|f''(x)|(|x - x_{i+1}| + |x_{i+1} - x_i|)|x - x_{i+1}| \\
&= \frac{1}{2}\max|f''(x)|(|x - x_{i+1}|^2 + |x_{i+1} - x_i| \; |x - x_{i+1}|) \\
&\leq \frac{1}{2}\max|f''(x)|\left(\frac{|x_i - x_{i+1}|^2}{4} + \frac{|x_{i+1} - x_i|^2}{2}\right) \\
&= \frac{1}{8}\max|f''(x)|h_i^2 \\
&\leq Ch^2 \ .
\end{aligned}$$

Therefore, the actual constant is $C = \frac{1}{8}\|f''\|_\infty$.

We conclude that if the second derivative $f''(x)$ is bounded, then the piecewise linear interpolant $P_l(x)$ converges to $f(x)$ as $h \to 0$ for all $x \in [a,b]$.

6.4.4 Cubic spline interpolation

Studying the accuracy of piecewise-linear interpolation we showed that the distance between a function f and its linear spline interpolant is $O(h^2)$ with $h = \max_i |x_{i+1} - x_i|$. However, the piecewise linear interpolant is not differentiable. Sometimes, we require the interpolant to be smooth, and perhaps more accurate than the piecewise linear interpolant. In such cases we can use higher order piecewise polynomial interpolation, such as quadratic, cubic, quartic, quintic, etc.

Because the methodology for the construction of quadratic, cubic and other higher-order splines is the same, we study here the construction of a cubic spline. A cubic spline is a continuous piecewise cubic interpolant. A general cubic polynomial of the form $a_0 + a_1x + a_2x^2 + a_3x^3$ involves four constants. This fact gives us flexibility to ensure that the interpolant is continuously differentiable and it has a continuous second derivative. The construction of cubic splines does not, however, assume that the derivatives of the interpolant agree with those of the approximated function, even at the nodes.

Given a function f defined on $[a, b]$ and a set of nodes $a = x_0 < x_1 < \cdots < x_N = b$, a *cubic spline interpolant* P_c of f is a function equal to a different cubic polynomial $S_i(x)$ in each interval $[x_i, x_{i+1}]$ for every $i = 0, 1, \ldots, N-1$. Such a function can be expressed with the following formula of multiple cases (piecewise function)

$$P_c(x) = \begin{cases} S_0(x), & x \in [x_0, x_1)\ , \\ S_1(x), & x \in [x_1, x_2)\ , \\ \ldots & \\ S_{N-1}(x), & x \in [x_{N-1}, x_N]\ . \end{cases}$$

Since we require $P_c(x)$ to be twice continuously differentiable, the cubic polynomials S_i need to satisfy the following conditions:

(i) $S_i(x_i) = f(x_i)$ and $S_i(x_{i+1}) = f(x_{i+1})$ for each $i = 0, 1, \ldots, N-1$ (interpolates the data)

(ii) $S_{i+1}(x_{i+1}) = S_i(x_{i+1})$ for each $i = 0, 1, \ldots, N-2$ (is continuous)

(iii) $S'_{i+1}(x_{i+1}) = S'_i(x_{i+1})$ for each $i = 0, 1, \ldots, N-2$ (has continuous first derivative)

(iv) $S''_{i+1}(x_{i+1}) = S''_i(x_{i+1})$ for each $i = 0, 1, \ldots, N-2$ (has continuous second derivative)

Unfortunately, the continuity argument can only be used for the internal nodes and not for the boundary nodes. In order to be able to estimate all the coefficients, including those corresponding to boundary nodes, we impose boundary conditions. The most popular boundary conditions are the following:

(a) $P''_c(x_0) = P''_c(x_N) = 0$ (natural (or free) boundary)

(b) $P'_c(x_0) = f'(x_0)$ and $P'_c(x_N) = f'(x_N)$ (clamped boundary)

(c) $P'''_c(x)$ is continuous at x_1 and x_{N-1} ("not-a-knot" boundary)

(d) $P_c(x_0) = P_c(x_N)$ (Periodic boundary)

When natural boundary conditions are employed, we call the interpolating spline *natural spline*. The natural boundary conditions do not require any knowledge of the function f on the boundary. On the other hand, they impose conditions on the second derivatives of the interpolating spline at the boundary. These conditions are not essential for the shape of the function near the boundary, however, they introduce additional error. Clamped boundary conditions lead to more accurate approximations because they include more information about $f(x)$. However, for this type of boundary condition, it is necessary to have either the values of $f'(x)$ at the endpoints or an accurate approximation to those values. Finally, the "not-a-knot" condition does not imply any essential condition on the boundary, which is ideal when there is no information regarding the behavior of $f(x)$ at the endpoints x_0 and x_N.

One of the simplest examples for understanding the procedures behind cubic spline interpolation is the following: Assume that we want to find a natural spline that interpolates the set of points $(1,2)$, $(2,3)$ and $(4,6)$ in the interval $[1,4]$. In this case, $N=3$ and thus the spline consists of two ($N-1=2$) polynomials: The first in the interval $[1,2]$ can be written as

$$S_0(x) = a_0 + a_1(x-1) + a_2(x-1)^2 + a_3(x-1)^3 ,$$

and the other in the interval $[2,4]$ as

$$S_1(x) = b_0 + b_1(x-2) + b_2(x-2)^2 + b_3(x-2)^3 .$$

The polynomials S_0 and S_1 is written in such a way so as to minimize the computations later. In principle it is not required to express the polynomials in this form.

Therefore, there are 8 constants to be determined a_0, a_1, a_2, a_3 and b_0, b_1, b_2, b_3. To determine 8 unknowns requires 8 conditions: The first 4 conditions come from the fact that the spline must agree with the data at the nodes. Hence, from the conditions $S_0(1) = f(1)$, $S_0(2) = f(2) = S_1(2)$, $S_1(3) = f(3)$ we get

$$2 = f(1) = a_0 , \tag{6.20}$$

$$3 = f(2) = a_0 + a_1 + a_2 + a_3 , \tag{6.21}$$

$$3 = f(2) = b_0 , \tag{6.22}$$

$$6 = f(4) = b_0 + 2b_1 + 4b_2 + 8b_3 . \tag{6.23}$$

There are 2 conditions from the continuity of spline's first and second derivatives at the internal nodes, ($S_0'(2) = S_1'(2)$ and $S_0''(2) = S_1''(2)$):

$$S_0'(2) = S_1'(2) \Rightarrow a_1 + 2a_2 + 3a_3 = b_1 , \tag{6.24}$$

$$S_0''(2) = S_1''(2) \Rightarrow 2a_2 + 6a_3 = 2b_2 . \tag{6.25}$$

The last conditions come from the natural boundary conditions:

$$S_0''(1) = 0 \Rightarrow 2a_2 = 0 , \tag{6.26}$$

$$S_1''(3) = 0 \Rightarrow 2b_2 + 6b_3 = 0 . \tag{6.27}$$

Equations (6.20)–(6.27) form a system of 8 equations for the 8 unknown coefficients. Solving the corresponding 8×8 linear system is hard without the help of a computer. We obtained the coefficients $a_0 = 2$, $a_1 = 0$, $a_2 = 1/3$, $a_3 = 2/3$, $b_0 = 3$, $b_1 = 1$, $b_2 = -1/3$ and $b_3 = 7/24$. So the spline function can be written as

$$P_c(x) = \begin{cases} 2 + \frac{1}{3}(x-1)^2 + \frac{2}{4}(x-1)^3, & \text{for } x \in [1,2] \\ 3 + (x-2) - \frac{1}{3}(x-2)^2 + \frac{7}{24}(x-2)^3, & \text{for } x \in [2,4] \end{cases} .$$

6.4.5 An algorithm for cubic splines

The steps we followed in the previous example to find the coefficients of the cubic polynomials S_i can be followed systematically to formulate an algorithm for computing cubic splines with various boundary conditions. Here we will focus on the case of clamped boundary conditions. The algorithms for other boundary conditions can be formulated similarly with small adaptations.

Let $x_0 < x_1 < \cdots < x_N$ be a grid of an interval $[a, b]$ and f continuously differentiable function. We consider the problem of finding the cubic spline interpolant $P_c(x)$ so as to be a cubic polynomial $S_i(x)$ in each interval $[x_i, x_{i+1}]$ for $i = 0, 1, \ldots, N-1$. The function $P_c(x)$ interpolates f in the sense $P_c(x_i) = f(x_i)$, and satisfies the clamped boundary conditions $P_c'(x_0) = f'(x_0)$ and $P_c'(x_N) = f'(x_N)$. For convenience we will denote $f(x_i) = y_i$ and $P_c''(x_i) = z_i$. The values z_i are known as the *moments* of $f(x)$, and currently are unknown.

Because $P_c(x)$ is a cubic polynomial in each interval $[x_i, x_{i+1}]$, we have that $P_c''(x)$ is a linear polynomial in each $[x_i, x_{i+1}]$ and takes the values z_i and z_{i+1} at the endpoints x_i and x_{i+1}, respectively. Thus, for $x \in [x_i, x_{i+1}]$ we have

$$P_c''(x) = S_i''(x) = \frac{z_{i+1}}{h_i}(x - x_i) + \frac{z_i}{h_i}(x_{i+1} - x), \quad x \in [x_i, x_{i+1}] , \tag{6.28}$$

where $h_i = x_{i+1} - x_i$ for $i = 0, 1, \ldots, N-1$. If we integrate the formula (6.28) twice we obtain $S_i(x)$ as

$$P_c(x) = S_i(x) = \frac{z_{i+1}}{6h_i}(x - x_i)^3 + \frac{z_i}{6h_i}(x_{i+1} - x)^3 + c_i x + d_i, \quad x \in [x_i, x_{i+1}] ,$$

where c_i, d_i are integration constants. We can write the last equation in a more convenient form as

$$S_i(x) = \frac{z_{i+1}}{6h_i}(x - x_i)^3 + \frac{z_i}{6h_i}(x_{i+1} - x)^3 + C_i(x - x_i) + D_i(x_{i+1} - x) ,$$

where C_i and D_i are constants. We specify the constants C_i, D_i using the interpolation conditions

$$S_i(x_i) = y_i \quad \text{and} \quad S_i(x_{i+1}) = y_{i+1} ,$$

and we write $S_i(x)$ in the form

$$S_i(x) = \frac{z_{i+1}}{6h_i}(x-x_i)^3 + \frac{z_i}{6h_i}(x_{i+1}-x)^3 + \left(\frac{y_{i+1}}{h_i} - \frac{h_i}{6}z_{i+1}\right)(x-x_i) + \left(\frac{y_i}{h_i} - \frac{h_i}{6}z_i\right)(x_{i+1}-x) , \tag{6.29}$$

for $x \in [x_i, x_{i+1}]$. Given the moments z_i, we can use (6.29) to evaluate the cubic spline $P_c(x)$ for any $x \in [a, b]$. For the N intervals $[x_i, x_{i+1}]$, $i = 0, \ldots, N-1$ we define N cubic polynomials $S_0, S_1, \ldots S_{N-1}$ and $N + 1$ unknowns $z_0, z_1, \ldots, z_N$ to be specified.

To specify the z_i's we need to consider the continuity of $S'(x)$ at the interior nodes x_i, $i = 1, \ldots, N-1$. Specifically, we have the equations

$$S_{i-1}'(x_i) = S_i'(x_i), \quad \text{for} \quad i = 1, 2, \ldots, N-1 .$$

We compute the derivative S_i using (6.29) as

$$P_c'(x) = S_i'(x) = \frac{z_{i+1}}{2h_i}(x - x_i)^2 - \frac{z_i}{2h_i}(x_{i+1} - x)^2 + \frac{y_{i+1} - y_i}{h_i} - \frac{h_i}{6}(z_{i+1} - z_i) , \tag{6.30}$$

for $x \in [x_i, x_{i+1}]$. This yields

$$P_c'(x_i) = S_i(x_i) = -\frac{h_i}{6}z_{i+1} - \frac{h_i}{3}z_i + b_i ,$$

and also

$$P_c'(x_i) = S_{i-1}(x_i) = \frac{h_{i-1}}{6}z_i + \frac{h_{i-1}}{3}z_{i-1} + b_{i-1} ,$$

where

$$b_i = \frac{1}{h_i}(y_{i+1} - y_i) \ .$$

Taking into account the continuity of $P_c'(x)$ at every internal node x_i, we set $S_i(x_i) = S_{i-1}(x_i)$ in the last relationships to obtain the equations for the unknown moments

$$\frac{h_{i-1}}{h_{i-1}+h_i} z_{i-1} + 2z_i + \frac{h_i}{h_{i-1}+h_i} z_{i+1} = 6\frac{b_i - b_{i-1}}{h_{i-1}+h_i} \ , \tag{6.31}$$

for $i = 1, 2, \ldots, N-1$. These are $N-1$ equations for $N+1$ unknowns. In order to construct the two remaining equations required for a nonsingular system, we use the boundary conditions $P_c'(x_0) = f'(x_0)$ and $P_c'(x_N) = f'(x_N)$. By taking $x = x_0$ (and $i = 0$) and $x = x_N$ (and $i = N-1$) in (6.30) we obtain the additional equations

$$2z_0 + z_1 = 6\frac{b_0 - f'(x_0)}{h_0} \ , \tag{6.32}$$

and

$$z_{N-1} + 2z_N = 6\frac{b_{N-1} - f'(x_N)}{h_{N-1}} \ . \tag{6.33}$$

We define

$$u_i = \frac{h_i}{h_{i-1}+h_i}, \quad l_i = \frac{h_{i-1}}{h_{i-1}+h_i}, \qquad i = 1, 2, \ldots, N-1 \ ,$$

and $u_0 = 1$ and $l_N = 1$, and also

$$v_i = 6\frac{b_i - b_{i-1}}{h_{i-1}+h_i}, \quad i = 1, 2, \ldots, N-1 \ ,$$

and $v_0 = 6\frac{b_0 - f'(x_0)}{h_0}$ and $v_N = 6\frac{b_{N-1}-f'(x_N)}{h_{N-1}}$, and $c_i = 2$ for $i = 0, 1, \ldots, N$. Then we write equations (6.31), (6.32) and (6.33) in a form of $(N+1) \times (N+1)$ tridiagonal system of equations $\boldsymbol{A}\boldsymbol{z} = \boldsymbol{v}$

$$\begin{pmatrix} c_0 & u_0 & & & & \\ l_1 & c_1 & u_1 & & & \\ & l_2 & c_2 & u_2 & & \\ & & \ddots & \ddots & \ddots & \\ & & & l_{N-1} & c_{N-1} & u_{N-1} \\ & & & & l_N & c_N \end{pmatrix} \begin{pmatrix} z_0 \\ z_1 \\ z_2 \\ \vdots \\ z_{N-1} \\ z_N \end{pmatrix} = \begin{pmatrix} v_0 \\ v_1 \\ v_2 \\ \vdots \\ v_{N-1} \\ v_N \end{pmatrix} , \tag{6.34}$$

which is to be solved for z_i. Note that sometimes we write $\boldsymbol{A} = \text{tridiag}(\boldsymbol{l}, \boldsymbol{c}, \boldsymbol{u})$ to denote a tridiagonal matrix with sub-diagonal $\boldsymbol{l} = (l_1, \ldots, l_N)^T$, main diagonal $\boldsymbol{c} = (c_0, \ldots, c_N)^T$ and super-diagonal $\boldsymbol{u} = (u_0, \ldots, u_{N-1})^T$.

The first and last rows of the system $\boldsymbol{A}\boldsymbol{z} = \boldsymbol{v}$ are the only rows to be affected by the boundary conditions. For this reason, we only need to modify these particular rows in case of different boundary conditions. For example, in the case of natural boundary conditions we need to take $c_0 = c_N = 1$, $u_0 = l_N = 0$ and $v_0 = v_N = 0$.

The solution of a linear system with tridiagonal matrix $\boldsymbol{A}$ can be achieved efficiently using the tridiagonal version of the *LU* decomposition, forward and backward substitution Algorithms 37, 38 and 39 of Chapter 9. Because this is material we will study later, we will

Algorithm 21 Computation of the moments for clamped cubic splines

Given the points $y_0, y_1, \ldots, y_N$ at the nodes $x_0, x_1, \ldots, x_N$ and the boundary terms $f'(x_0)$ and $f'(x_N)$, compute the coefficients of the tridiagonal system (6.34)
for $i = 0 : N - 1$ **do**
  $h_i = x_{i+1} - x_i$
  $b_i = (y_{i+1} - y_i)/h_i$
end for
$u_0 = 1$, $c_0 = 2$, $v_0 = 6(b_0 - f'(x_0))/h_0$
for $k = 1 : N - 1$ **do**
  $c_i = 2$
  $u_i = h_i/(h_{i-1} + h_i)$
  $l_i = h_{i-1}/(h_{i-1} + h_i)$
  $v_i = 6(b_i - b_{i-1})/(h_{i-1} + h_i)$
end for
$l_N = 1$, $c_N = 2$, $v_N = 6(b_{n-1} - f'(x_N))/h_{N-1}$
Form the triadiagonal matrix $\boldsymbol{A} = \text{tridiag}(\boldsymbol{l}, \boldsymbol{c}, \boldsymbol{u})$
Solve the linear system $\boldsymbol{A}\boldsymbol{z} = \boldsymbol{v}$

use the generic NumPy function `solve` instead. The interested reader could employ banded matrix techniques of Section 2.3.1.

The algorithm for the computation of the moments $z_0, z_1, \ldots, z_N$ for cubic splines with clamped boundary conditions is summarized in Algorithm 21.

For the implementation of Algorithm 21 we will use the function `diags` of the module `scipy.sparse`. The particular function can create a NumPy array from its diagonals. Because the result will be a sparse matrix, we use the member function `todense` to convert the matrix to a usual NumPy array. We will discuss sparse matrices in Chapter 9. The rest of the implementation is included in the function `MyCubicSpline`.

```
import numpy as np
import numpy.linalg as npl
import scipy.sparse as sps

def MyCubicSpline(x, y, dy0, dyN):
    # Returns the moments of f given the data x, y
    # dy0 is the derivative at the left boundary
    # dyN is the derivative at the right boundary
    n = len(x)
    c = np.zeros(n); v = np.zeros(n); u = np.zeros(n-1)
    l = np.zeros(n-1); b = np.zeros(n-1); h = np.zeros(n-1)
    for i in range(n-1):
        h[i] = x[i+1] - x[i]
        b[i] = (y[i+1]-y[i])/h[i]
    u[0] = 1.0
    v[0] = 6.0*(b[0]-dy0)/h[0]
    c[0] = 2.0
    for i in range(1,n-1):
        c[i] = 2.0
        u[i] = h[i]/(h[i-1] + h[i])
        l[i-1] = h[i-1]/(h[i-1] + h[i])
        v[i] = 6.0*(b[i]-b[i-1])/(h[i-1] + h[i])
```

```
    l[n-2] = 1.0
    c[n-1] = 2.0
    v[n-1] = 6.0*(b[n-2]-dyN)/h[n-2]
    diagonals = [c, l, u]
    A = sps.diags(diagonals, [0, -1, 1]).todense()
    z = npl.solve(A,v)
    return z
```

To evaluate the cubic spline interpolant $P_c(x)$ at any point x given its moments z_i we can use the formula

$$S_i(x) = A_i + B_i(x - x_i) + C_i(x - x_i)^2 + D_i(x - x_i)^3 \,, \tag{6.35}$$

where

$$A_i = S_i(x_i), \quad B_i = S_i'(x_i), \quad C_i = \frac{1}{2}S_i''(x_i) \,.$$

The value D_i can be found using the continuity of the second derivative $S_i''(x_{i+1}) = S_{i+1}''(x_{i+1})$. Therefore,

$$A_i = y_i, \qquad B_i = -\frac{h_i}{6}z_{i+1} - \frac{h_i}{3}z_i + \frac{1}{h_i}(y_{i+1} - y_i),$$
$$C_i = \frac{z_i}{2}, \qquad D_i = \frac{1}{6h_i}(z_{i+1} - z_i),$$

which for $i = 0, 1, 2, \ldots, N-1$, and $x \in [x_i, x_{i+1}]$ we can write formula (6.35) in nested form as

$$S_i(x) = y_i + (x - x_i)\left(B_i + (x - x_i)\left(\frac{z_i}{2} + \frac{1}{6h_i}(x - x_i)(z_{i+1} - z_i)\right)\right) \,. \tag{6.36}$$

The algorithm for the evaluation of $P_c(x)$ using (6.36) and the computed moments from Algorithm 21 can be summarized in Algorithm 22. In this algorithm the loop condition $i = a : b : c$ means that the value a is the starting value, b is the stop value and c is the step.

Algorithm 22 Evaluation of the clamped cubic spline given its moments

Given the moments $z_1, \ldots, z_N$ and a point x
Find the value of i for which $x \in [x_i, x_{i+1}]$ is true
for $i = N-1 : 0 : -1$ **do**
  if $x \geq x_i$ **then**
    Break
  end if
end for
$h = x_{i+1} - x_i$
$B = -hz_{i+1}/6 - hz_i/3 + (y_{i+1} - y_i)/h$
Implement Formula (6.36) starting with the most inner brackets
$y = z_i/2 + (x - x_i)(z_{i+1} - z_i)/6h$
$y = B + (x - x_i)y$
$y = y_i + (x - x_i)y$
Variable y contains the value of the spline evaluated at x

The implementation of Algorithm 22 in Python can be as follows:

```
def EvalCubicSpline(x, y, z, xx):
    # Returns the cubic spline evaluated at xx
    # z = the moments of the cubic spline
    # xx = the vector with values of x on which we want the cubic spline
    n = len(x)
    m = len(xx)
    yy = np.zeros(m)
    for j in range(m):
        xvalue = xx[j]
        # First detect the index i
        for i in range(n-2, -1, -1):
            if (xvalue - x[i] >= 0.0):
                break
        # Implement formula (6.36)
        h = x[i+1] - x[i]
        B = -h*z[i+1]/6.0-h*z[i]/3.0+(y[i+1]-y[i])/h
        tmp =z[i]/2.0+(xvalue-x[i])*(z[i+1]-z[i])/6.0/h
        tmp = B+(xvalue-x[i])*tmp
        yy[j] = y[i] + (xvalue - x[i])*tmp
    return yy
```

To test the functions `MyCubicSpline` and `EvalCubicSpline` we can use the following script which results in the same figure as before and it is omitted.

```
def f(x):
    return 1.0/(1+25*x**2)
def df(x):
    return -2*25*x/(1+25*x**2)**2
x = np.linspace(-1,1,9)
y = f(x)
z = df(x)
xx = np.linspace(-1,1,300)
yy = f(xx)
ss = MyCubicSpline(x, y, z[0], z[-1])
zz=EvalCubicSpline(x, y, ss, xx)
plt.plot(xx,yy,xx,zz,x,y,'o')
```

We proceed with the analysis of the error for the cubic spline interpolation and other theoretical considerations following closely the theory presented in [127].

6.4.6 Theoretical considerations and error analysis

In addition to its practical significance, Algorithm 22 provides a method to prove that there is a unique natural cubic spline that interpolates a function f at a given set of nodes. In particular, by showing that $\boldsymbol{A}$ in (6.34) is invertible we show that there is a unique solution of the system $\boldsymbol{Az} = \boldsymbol{v}$ for the moments z_i, and thus the cubic spline is uniquely determined (since it can be constructed from these values).

In order to see that $\boldsymbol{A}$ is invertible, we first prove the following lemma:

Lemma 6.7. *The solution of the linear system* $\boldsymbol{Az} = \boldsymbol{v}$ *satisfies*

$$\max_i |z_i| \le \max_i |v_i| .$$

Proof. Let k be the index for which we have $|z_k| = \max_i |z_i|$. The k-th equation of the system $\boldsymbol{Az} = \boldsymbol{v}$ is then

$$l_k z_{k-1} + c_k z_k + u_k z_{k+1} = v_k .$$

Because

$$u_i = \frac{h_i}{h_{i-1} + h_i} \quad \text{and} \quad l_i = \frac{h_{i-1}}{h_{i-1} + h_i} ,$$

we have that $u_i + l_i = 1$, which implies $u_k + l_k = 1$. Then,

$$\begin{aligned} |v_k| &\ge 2|z_k| - l_k|z_{k-1}| - u_k|z_{k+1}| \\ &\ge 2|z_k| - l_k|z_k| - u_k|z_k| \\ &= (2 - l_k - u_k)|z_k| \\ &= |z_k| . \end{aligned}$$

On the other hand, we have that $\max_i |v_i| \ge |v_k|$, and thus $\max_i |v_i| \ge |z_k| = \max_i |z_i|$ which completes the proof. □

Lemma 6.7 implies that the solution of the system $\boldsymbol{Az} = \boldsymbol{v}$ is unique, and thus $\boldsymbol{A}$ is invertible. To see this assume for contradiction that $\boldsymbol{A}$ is singular. This means that there exists $\boldsymbol{z} \neq \boldsymbol{0}$ such that $\boldsymbol{Az} = \boldsymbol{0}$. By Lemma 6.7 we have that $0 < \max_i |z_i| \le 0$, which is impossible. Therefore, there is a unique cubic spline with clamped boundary conditions interpolating $N+1$ data (x_i, y_i), $i = 0, 1, \ldots, N$. Similar results hold for the other boundary conditions too. We continue with the error estimation in the case of clamped cubic spline interpolation.

Let the function f specifying the data points $y_i = f(x_i)$, $i = 0, 1, \ldots, N$ is a smooth function in $C^4[a, b]$. In the case of clamped spline interpolation, we can estimate the error between the interpolant and function f which is of optimal order. First we show optimal order approximation of the second derivatives. For this reason, we consider the system $\boldsymbol{Az} = \boldsymbol{v}$ (6.34) and we denote by $\boldsymbol{z}_f$ the vector of the second derivatives of f evaluated at the nodes x_i

$$\boldsymbol{z}_f = \begin{pmatrix} f''(x_0) \\ f''(x_1) \\ \vdots \\ f''(x_N) \end{pmatrix} .$$

We also define the residual vector

$$\boldsymbol{r} = \boldsymbol{Az} - \boldsymbol{Az}_f .$$

Recall, that the vector $\boldsymbol{z}$ contains the moments $P_c''(x_i)$. We estimate the maximum difference of the second derivatives at the nodes $\max_i |P_c''(x_i) - f''(x_i)|$. For convenience we introduce the norm notation $\|\boldsymbol{z} - \boldsymbol{z}_f\|_\infty = \max_i |z_i - z_{f_i}|$, where the quantity $\|\cdot\|_\infty$ is the usual maximum vector norm. (We discuss matrix norms in more detail in Chapter 9). As in Theorem 6.6, we define the grid-size $h = \max_i h_i$ where $h_i = |x_{i+1} - x_i|$. Once again the proof relies on Taylor's expansions of function f and its derivatives.

Proposition 6.8. *Let $f \in C^4[a,b]$ with $\max_x |f^{(4)}(x)| \le L$ for some $L > 0$. Then the following optimal error estimate holds*

$$\max_i |P_c''(x_i) - f''(x_i)| \le Ch^2 ,$$

where $C = \frac{3}{4}L$.

Proof. Using the residual $\boldsymbol{r} = \boldsymbol{Az} - \boldsymbol{Az}_f$ we obtain

$$\begin{aligned} r_0 &= c_0 z_0 + u_0 z_1 - c_0 f''(x_0) - u_0 f''(x_1) \\ &= v_0 - 2f''(x_0) - f''(x_1) \quad (\text{since } c_0 = 2,\ u_0 = 1) \\ &= \frac{6}{h_0}\left(\frac{y_1 - y_0}{h_0} - f'(x_0)\right) - 2f''(x_0) - f''(x_1) . \end{aligned}$$

Since $x_1 = x_0 + h_0$ the Taylor expansion of $f(x_1)$ and $f''(x_1)$ around x_0 yields the existence of ξ_1 and ξ_2 in $[x_0, x_1]$ such that

$$\begin{aligned} r_0 &= \frac{6}{h_0}\left(f'(x_0) + \frac{h_0}{2}f''(x_0) + \frac{h_0^2}{6}f'''(x_0) + \frac{h_0^3}{24}f^{(4)}(\xi_i) - f'(x_0)\right) \\ &\quad - 2f''(x_0) - \left(f''(x_0) + h_0 f'''(x_0) + \frac{h_0^2}{2}f^{(4)}(\xi_2)\right) \\ &= \frac{h_0^2}{4}f^{(4)}(\xi_1) - \frac{h_0^2}{2}f^{(4)}(\xi_2) . \end{aligned}$$

Taking the absolute value in the last relation we obtain $|r_0| \le Ch^2$, with $C = \frac{3}{4}L$. Similarly, we obtain $|r_N| \le Ch^2$, since the other boundary terms are almost identical. For the internal nodes x_i, $i = 1, \dots, N-1$ we have

$$\begin{aligned} r_i &= v_i - l_i f''(x_{i-1}) - c_i f''(x_i) - u_i f''(x_{i+1}) \\ &= 6\frac{b_i - b_{i-1}}{h_{i-1} + h_i} - \frac{h_{i-1}}{h_{i-1} + h_i}f''(x_{i+1}) - 2f''(x_i) - \frac{h_i}{h_{i-1} + h_i}f''(x_{i+1}) \\ &= \frac{6}{h_{i-1} + h_i}\left(\frac{f(x_{i+1}) - f(x_i)}{h_i} - \frac{f(x_i) - f(x_{i-1})}{h_{i-1}}\right) \\ &\quad - \frac{h_{i-1}}{h_{i-1} + h_i}f''(x_{i+1}) - 2f''(x_i) - \frac{h_i}{h_{i-1} + h_i}f''(x_{i+1}) . \end{aligned}$$

Since $x_{i-1} = x_i - h_i$ and $x_{i+1} = x_i + h_i$, Taylor's expansions of f' and f'' around x_i yields that there are $\xi_1, \xi_2, \xi_3, \xi_4 \in [x_{i-1}, x_{i+1}]$ such that

$$\begin{aligned} r_i &= \frac{1}{h_{i-1} + h_i}\left[6\left(f'(x_i) + \frac{h_i}{2}f''(x_i) + \frac{h_i^2}{6}f'''(x_i) + \frac{h_i^3}{24}f^{(4)}(\xi_1)\right.\right. \\ &\quad \left. -f'(x_i) + \frac{h_{i-1}}{2}f''(x_i) - \frac{h_{i-1}^2}{6}f'''(x_i) + \frac{h_{i-1}^3}{24}f^{(4)}(\xi_2)\right) \\ &\quad - h_{i-1}\left(f''(x_i) - h_{i-1}f''(x_i) + \frac{h_{i-1}^2}{2}f^{(4)}(\xi_3)\right) \\ &\quad \left. -2f''(x_i)(h_{i-1} + h_i) - h_i\left(f''(x_i) + h_i f'''(x_i) + \frac{h_i^2}{2}f^{(4)}(\xi_4)\right)\right] \\ &= \frac{1}{h_{i-1} + h_i}\left(\frac{h_i^3}{4}f^{(4)}(\xi_1) + \frac{h_{i-1}^3}{4}f^{(4)}(\xi_2) - \frac{h_{i-1}^3}{2}f^{(4)}(\xi_3) - \frac{h_i^3}{2}f^{(4)}(\xi_4)\right) . \end{aligned}$$

Taking again absolute values in the previous formulas we obtain $|r_i| \leq Ch^2$ with $C = \frac{3}{4}L$. Thus, the maximum over i satisfies

$$\|\boldsymbol{r}\|_\infty = \max_i |r_i| \leq Ch^2 \ .$$

Since $\boldsymbol{A}(\boldsymbol{z}-\boldsymbol{z}_f) = \boldsymbol{r}$, using Lemma 6.7 we have $\|\boldsymbol{z}-\boldsymbol{z}_f\|_\infty \leq \|\boldsymbol{r}\|_\infty$ and the result follows. □

The previous theorem states that the second derivative of the cubic spline at the nodes approximates the second derivative of the function f evaluated at the same nodes. Their maximum difference is of order h^2. This means that if we take $h \to 0$ the error becomes 0. Thus, we have convergence of the second derivatives at the nodes. Because the convergence is quadratic (since the error is $O(h^2)$), we say that the convergence is optimal. This is because we expect that the approximation of a function by polynomials of degree r will result to errors of $O(h^{r+1})$. In our case, the second derivative of the cubic spline is a piecewise linear function. The error of $O(h^2)$ follows the optimal error rule. We prove next that the error between the other derivatives are also optimal.

Theorem 6.9. *Let $f \in C^4[a,b]$ and $\max_x |f^{(4)}(x)| \leq L$ for some $L > 0$. If in addition $\frac{h}{h_i} \leq K$, for $i = 0, 1, \ldots, N-1$, then there exists a constant $C = C(k, K, L) > 0$ such that*

$$|f^{(k)}(x) - P_c^{(k)}(x)| \leq Ch^{4-k}, \quad k = 0, 1, 2, 3 \ .$$

Proof. We first prove the error estimate when $k = 3$. Let $x \in [x_i, x_{i+1}]$ for some $i = 0, 1, \ldots, N-1$. Then from (6.28) we have

$$\begin{aligned} f'''(x) - P_c'''(x) &= \frac{z_{i+1} - z_i}{h_i} - f'''(x) \\ &= \frac{z_{i+1} - f''(x_{i+1})}{h_i} - \frac{z_i - f''(i)}{h_i} + \\ &\quad + \frac{f''(x_{i+1}) - f''(x) - [f''(x_i) - f''(x)]}{h_i} - f'''(x) \ . \end{aligned}$$

Using Proposition 6.8 and the Taylor expansion of f'' around x we conclude that there are $\xi_1, \xi_2 \in [x_i, x_{i+1}]$ such that

$$\begin{aligned} |f'''(x) - P_c'''(x)| &\leq \frac{3}{2} L \frac{h^2}{h_i} + \frac{1}{h_i} \Big| (x_{i+1} - x) f'''(x) \\ &\quad + \frac{(x_{i+1} - x)^2}{2} f^{(4)}(\xi_1) - (x_i - x) f'''(x) - \frac{(x_i - x)^2}{2} f^{(4)}(\xi_2) - h_i f'''(x) \Big| \\ &\leq \frac{3}{2} L \frac{h^2}{h_i} + \frac{L}{2} \frac{h^2}{h_i} \\ &\leq Ch \ , \end{aligned}$$

where $C = 2KL$. In the case $k = 2$, we consider $x \in (a, b)$ and we choose the x_i, which is the closest node to x. For the specific node we will have $|x_i - x| \leq \frac{h}{2}$. From the fundamental theorem of calculus (see Theorem 4.4) we have

$$f''(x) - P_c''(x) = f''(x_i) - P_c''(x_i) + \int_{x_i}^{x} (f'''(s) - P_c'''(s)) \, ds \ .$$

Using Proposition 6.8 we can estimate an upper bound for the error in the second derivatives at the nodes, while the integrant at the right-hand side can be bounded using the previous estimate to obtain

$$\begin{aligned}|f''(x) - P_c''(x)| &\le |f''(x_i) - P_c''(x_i)| + \int_{x_i}^{x} |f'''(s) - P_c'''(s)|\, ds \\ &\le \frac{3}{4}Lh^2 + KLh^2 \\ &\le Ch^2\ ,\end{aligned}$$

where $C = \frac{7}{4}KL$.

We consider now the case $k = 1$. Because of the boundary conditions we have that $f'(x_0) = P_c'(x_0)$ and $f'(x_N) = P_c'(x_N)$. We thus define $\xi_0 = x_0$ and $\xi_{N+1} = x_N$. Since $f(x_i) = P_c(x_i)$ for all $i = 0, 1, \ldots, N$ we can deduce from Rolle's theorem in each interval $[x_i, x_{i+1}]$, $i = 0, 1, \ldots, N$ that there is a $\xi_{i+1} \in (x_i, x_{i+1})$ such that $f'(\xi_i) = P_c'(\xi_i)$ for $i = 1, \ldots, N$. Thus, we have $N + 2$ points ξ_i such that

$$f'(\xi_i) = P_c'(\xi_i),\ i = 0, 1, \ldots, N+1\ .$$

If now $x \in [a, b]$ we choose ξ_i such that $|x - \xi_i| \le h$. Without loss of generality we assume $\xi_i < x$ where the proof is the same as if $x < \xi_i$. Then since

$$f'(x) - P_c'(x) = \int_{\xi_i}^{x} (f''(s) - P_c''(s))\, ds\ ,$$

we have

$$\begin{aligned}|f'(x) - P_c'(x)| &\le \int_{\xi_i}^{x} |f''(s) - P_c''(s)|\, ds \\ &\le Ch^2|x - \xi_i| \\ &\le Ch^3\ ,\end{aligned}$$

with $C = \frac{7}{4}KL$.

Finally, in the case $k = 0$ we have

$$f(x) - P_c(x) = \int_{x_i}^{x} (f'(s) - P_c'(s))\, ds\ ,$$

and thus $|f(x) - P_c(x)| \le Ch^4$, which completes the proof. □

6.4.7 B-splines

Given a set of $N + 1$ nodes $x_0 < x_1 < \cdots < x_N$, we defined the space of continuous piecewise linear functions $\mathbb{S}_1$ with basis functions $\phi_i(x)$, $i = 0, 1, \ldots, N$ and the linear spline interpolant

$$P_1(x) = \sum_{i=0}^{N} f(x_i)\phi_i(x)\ , \tag{6.37}$$

where $\phi_i(x)$ are the hat functions defined in (6.17)–(6.19). This representation is very convenient and it does not practically require to impose any boundary conditions. This idea can be generalized for the general spline space of piecewise polynomial functions of degree m. Let $a = x_0 < x_1 < \cdots < x_N = b$ a grid of the interval $[a, b]$. We define the space of piecewise polynomial functions of degree m to be the linear space

$$\mathbb{S}_m = \{f \in C^{m-1}[a, b] :\ f \in \mathbb{P}_m[x_i, x_{i+1}]\ \text{ for }\ i = 0, 1, \ldots, N-1\}\ .$$

The elements of $\mathbb{S}_m$ are called splines of degree m (order $m+1$). We denote the basis functions of the space $\mathbb{S}_m$ by $\mathcal{B}_j^m(x)$, where m is the polynomial degree and j a counter running over the basis functions (usually from 0 to $M-1$). The basis functions $\mathcal{B}_j^m(x)$ are called B-splines as an abbreviation of the phrase "basis splines". Using the previous notation we have that $\mathcal{B}_j^1(x) = \phi_j(x)$, with support $[x_{j-1}, x_j] \cup [x_j, x_{j+1}]$. For the space $\mathbb{S}_1$ we also have $N+1$ basis functions, indicating that the dimension of the space is $N+1$. Because of these simple properties of the B-splines $\mathcal{B}_j^1$ we are able to represent the linear spline interpolant in the form (6.37).

For $m > 1$ the situation is more complicated compared to the case $m = 1$. For example, the dimension of the space (the number of basis functions) is greater than the number of nodes, and thus the coefficients a_j in a representation of the interpolant

$$P_m(x) = \sum_{j=0}^{M-1} a_j \mathcal{B}_j^m(x) ,$$

with $M = \dim(\mathbb{S}_m)$, are not necessarily equal to $f(x_j)$. In general the dimension of the space $\mathbb{S}_m$ is $N+m$. This is, briefly, because a spline is a polynomial of degree m in each of the N intervals $[x_i, x_{i+1}]$. Each polynomial has $m+1$ coefficients and thus we have in total $N(m+1)$ coefficients (degrees of freedom). On the other hand, the continuity of derivatives of order $0, \ldots, m-1$ at the $N-1$ internal nodes yields $(N-1)m$ constraints. The remaining degrees of freedom are $N(m+1)-(N-1)m = N+m$ set a requirement for equal number of basis functions.

There is a systematic way to generate the basis functions of the splines space $\mathbb{S}_m$ recursively, and we describe it in the sequel. For the sake of simplicity, we will do it in the case of uniform grid using convolutions. The convolution of two functions f, g is the function $f * g$ defined by the integral

$$(f * g)(x) = \int_{-\infty}^{\infty} f(y) g(x-y)\, dy .$$

The *cardinal* B-spline of degree 0 is the (characteristic) function

$$\varphi_0(x) = \begin{cases} 1, & \text{if } x \in [-1/2, 1/2) \\ 0, & \text{otherwise} \end{cases} . \tag{6.38}$$

For the notation of the cardinal splines we use the letter φ, which is different from ϕ used before for the basis functions of the space $\mathbb{S}_1$. We define the cardinal B-spline of degree m as the *convolution*

$$\varphi_m(x) = [\underbrace{\varphi_0 * \varphi_0 * \varphi_0 * \ldots * \varphi_0}_{m-\text{times}}](x) = [\varphi_{m-1} * \varphi_0](x) .$$

Thus,

$$\begin{aligned} \varphi_m(x) = [\varphi_{m-1} * \varphi_0](x) &= \int_{-\infty}^{\infty} \varphi_{m-1}(y)\varphi_0(x-y)\, dy \\ &= \int_{-\infty}^{\infty} \varphi_{m-1}(x-y)\varphi_0(y)\, dy = \int_{-1/2}^{1/2} \varphi_{m-1}(x-y)\, dy , \end{aligned}$$

which after change of integration variable we obtain the recursive formula

$$\varphi_m(x) = \int_{x-1/2}^{x+1/2} \varphi_{m-1}(z)\, dz . \tag{6.39}$$

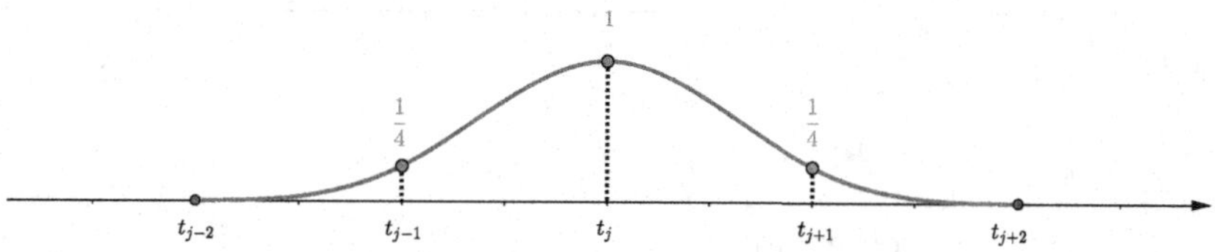

FIGURE 6.4
Cubic B-spline $\mathcal{B}_j^3$ generated by convolution.

All the cardinal B-splines are symmetric about the origin. To see this take

$$\begin{aligned}\varphi_m(-x) &= \int_{x-1/2}^{x+1/2} \varphi_{m-1}(z)\ dz \\ &= -\int_{x+1/2}^{x-1/2} \varphi_{m-1}(-\zeta)\ d\zeta \\ &= \int_{x-1/2}^{x+1/2} \varphi_{m-1}(\zeta)\ d\zeta \\ &= \varphi_m(x)\ .\end{aligned}$$

We can compute the derivatives of cardinal B-splines recursively with the formula

$$\frac{d}{dx}\varphi_m(x) = \frac{d}{dx}\int_{x-1/2}^{x+1/2} \varphi_{m-1}(z)\ dz = \varphi_{m-1}(x+1/2) - \varphi_{m-1}(x-1/2)\ .$$

The linear, quadratic, cubic, quartic and quintic cardinal B-splines obtained using the recursive formula (6.39) are presented in Table 6.6.

We define the nodes (also known as augmented knots)

$$\cdots < t_{-2} < t_{-1} < t_0 < t_1 < t_2 < \cdots\ ,$$

where here we assume that $t_{j+1} - t_j = h > 0$. The B-spline of degree m centered at the knot t_j can be found after appropriate translation of the cardinal B-spline. In particular, we define the B-spline of degree m as

$$\mathcal{B}_j^m(x) = \begin{cases} \varphi_m\left(\frac{x-t_j}{h}\right), & m = 2k+1 \\ \varphi_m\left(\frac{x-t_j}{h} - \frac{1}{2}\right), & m = 2k \end{cases}, k = 0, 1, 2, \dots\ .$$

The graph of a B-spline $\mathcal{B}_j^3$ is presented in Figure 6.4. We observe that $\mathcal{B}_j^3(t_j) = 1$, $\mathcal{B}_j^3(t_{j-1}) = \mathcal{B}_j^3(t_{j+1}) = 1/4$ and $\mathcal{B}_j^3(t_{j-2}) = \mathcal{B}_j^3(t_{j+2}) = 0$.

The B-splines of a general set of knots can be defined recursively by the Cox and de Boor recursion formula [31]. In particular, if $\cdots \le t_{-2} \le t_{-1} \le t_0 \le t_1 \le t_2 \le \cdots$, where we have extended the knots t_0 and t_N to the left and right, then the recursive relation defining the i-th B-spline of degree m, $\mathcal{B}_i^m$ is

$$\begin{aligned}&\mathcal{B}_i^0(x) = \begin{cases} 1, & \text{if } x \in [t_i, t_{i+1}) \\ 0, & \text{otherwise} \end{cases}, \\ &\mathcal{B}_i^k(x) = \frac{x - t_i}{t_{i+k-1} - t_i}\mathcal{B}_i^{k-1}(x) + \frac{t_{i+k} - x}{t_{i+k} - t_{i+1}}\mathcal{B}_{i+1}^{k-1}(x),\ \text{for } k = 1, 2, \dots, m\ ,\end{aligned} \tag{6.40}$$

TABLE 6.6
Cardinal B-splines up to degree 5

$$\varphi_0(x)=\begin{cases}1 & |x|\le 1/2\\ 0 & |x|>1/2\end{cases}$$

$$\varphi_1(x)=\begin{cases}1+x, & -1\le x\le 0\\ 1-x, & 0\le x\le 1\\ 0, & \text{otherwise}\end{cases}$$

$$\varphi_2(x)=\begin{cases}\frac{1}{3}\left(2x^2+6x+\frac{9}{2}\right), & -3/2\le x\le -1/2\\ \frac{4}{3}\left(\frac{3}{4}-x^2\right), & -1/2\le x\le 1/2\\ \frac{1}{3}\left(2x^2-6x+\frac{9}{2}\right), & 1/2\le x\le 3/2\\ 0, & \text{otherwise}\end{cases}$$

$$\varphi_3(x)=\begin{cases}\frac{1}{4}(x+2)^3, & -2\le x\le -1\\ \frac{1}{4}\left(4-6x^2-3x^3\right), & -1\le x\le 0\\ \frac{1}{4}\left(4-6x^2+3x^3\right), & 0\le x\le 1\\ \frac{1}{4}(2-x)^3, & 1\le x\le 2\\ 0, & \text{otherwise}\end{cases}$$

$$\varphi_4(x)=\begin{cases}\frac{8}{115}\left(\frac{5}{2}+x\right)^4, & -\frac{5}{2}\le x\le -\frac{3}{2}\\ \frac{2}{115}\left(55-20x-120x^2-80x^3-16x^4\right), & -\frac{3}{2}\le x\le -\frac{1}{2}\\ \frac{1}{115}\left(115-120x^2+48x^4\right), & -\frac{1}{2}\le x\le \frac{1}{2}\\ \frac{2}{115}\left(55+20x-120x^2+80x^3-16x^4\right), & \frac{1}{2}\le x\le \frac{3}{2}\\ \frac{8}{115}\left(\frac{5}{2}+x\right)^4, & \frac{3}{2}\le x\le \frac{5}{2}\\ 0, & \text{otherwise}\end{cases}$$

$$\varphi_5(x)=\begin{cases}\frac{1}{66}(3+x)^5, & -3\le x\le -2\\ \frac{1}{66}\left(51-75x-210x^2-150x^3-45x^4-5x^5\right), & -2\le x\le -1\\ \frac{1}{33}\left(33-30x^2+15x^4+5x^5\right), & -1\le x\le 0\\ \frac{1}{33}\left(33-30x^2+15x^4-5x^5\right), & 0\le x\le 1\\ \frac{1}{66}\left(51+75x-210x^2+150x^3-45x^4+5x^5\right), & 1\le x\le 2\\ \frac{1}{66}(3-x)^5, & 2\le x\le 3\\ 0, & \text{otherwise}\end{cases}$$

Note that this formula works even with repeated knots. We also use knots t_i instead of x_i because some of the B-splines will be centered around points t_i that are outside our domain. Contrary to the recursive formula (6.39), the Cox-de Boor formula will not necessarily produce normalized functions, but these can be normalized later if it is required.

Assume now that we want to compute the piecewise polynomial interpolant $P_m(x)$ such that $P_m(x_i)=y_i$ for $i=0,\dots,N+m-1$ given the B-splines $\mathcal{B}_i^m$. It is noted that in this case, the number of nodes must agree with the dimension of the space of splines. These points form $N+m$ equations with $N+m$ unknown coefficients

$$P_m(x_i)=\sum_{j=0}^{N+m-1} a_j\mathcal{B}_j^m(x_i)=y_i,\quad i=0,1,2,\dots,N+m-1\ ,$$

which can be written as a system of equations in the form $\boldsymbol{Aa}=\boldsymbol{y}$, where $A_{ij}=\mathcal{B}_{j-1}^m(x_{i-1})$. Due to the fact that $\mathcal{B}_i^m(x)\neq 0$ if $x\in[x_i,x_{i+m}]$, the system will be sparse and can be solved easily.

In the following code, we implement the Cox-de Boor recursive formula and we compute the cubic B-splines of the space $\mathbb{S}_3$ when $N = 10$, thus we compute $N + m = 13$ B-Splines. Aall the B-splines generated by the following code have maximum 2/3 and are not normalized.

```
# Define the B-splines using the Cox-de Boor recursive formula
def B(x, k, i, t):
    if k == 0:
        return 1.0 if t[i] < x <= t[i+1]  else 0.0
    else:
        c1 = (x - t[i])/(t[i+k] - t[i]) * B(x, k-1, i, t)
    c2 = (t[i+k+1] - x)/(t[i+k+1] - t[i+1]) * B(x, k-1, i+1, t)
    return c1 + c2
# Create uniform grid of nodes in [a,b]
a = -1.0; b = 1.0;
N=10
x = np.linspace(a,b,N+1)
#Compute all the cubic B-splines
m = 3
# Create uniform grid of knots
t = np.zeros(N+2*m+1)
for i in range(N+2*m+1):
    if i<=m:
        t[i] = x[0]-(m-i)*(x[1]-x[0])
    elif i>=N+m+1:
        t[i] = x[N]+(i-N-m)*(x[1]-x[0])
    else:
        t[i] = x[i-m]
# Plot the cubic B-splines
nn = 101
xx = np.linspace(a+1.e-10,b-1.e-10,nn)
yy = np.zeros(nn)
for j in range(N+m):
    for i in range (nn):
        yy[i]=B(xx[i], m, j, t)
    plt.plot(xx,yy)
plt.xlabel('$x$'); plt.ylabel('$y$',rotation=0); plt.show()
```

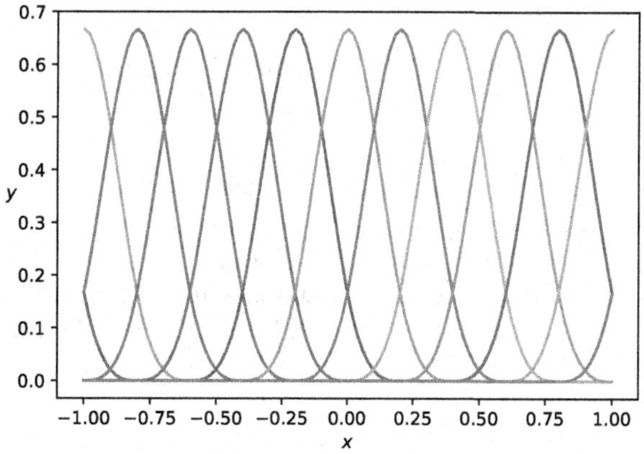

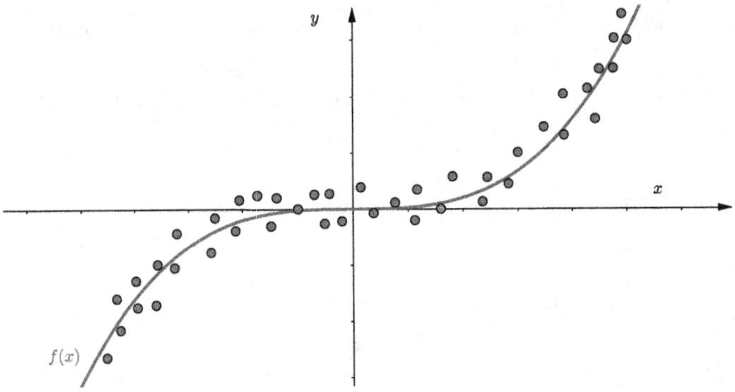

FIGURE 6.5
Least squares approximation of a dataset.

☞ The convolution formula for B-splines is related to the Cox-de Boor formula only for uniform grids. For non-uniform grids the basis functions will not be the same.

6.5 Method of Least Squares

We close this chapter with a different approach in approximation. Sometimes, the data points do not follow the graph of a smooth function, or they can be very "noisy", such as the data depicted in Figure 6.5. In such a case we prefer to approximate the data rather than interpolate them. The simplest approximation can be constructed using a straight line. In general, the approximation can be constructed using any function that can approximate the data reasonably well, and the method is known as the method of least squares. Figure 6.5 presents the situations where a dataset is quite noisy and chaotic but follows a certain pattern. The result of the least squares approximation is the function $f(x)$ represented by a continuous line that approximates the data points without interpolating them. The method for constructing such a smooth approximate function relies on the minimization of the error between the function and data. First we present the simple case of linear least squares.

6.5.1 Linear least squares

Suppose that we need to approximate some data $(x_0, y_0), (x_1, y_1), \ldots, (x_N, y_N)$ by a straight line of the form

$$y = ax + b \ .$$

The coefficients a and b have to be determined in such a way that the graph of the line is very close to these points.

If the point (x_i, y_i) lies on the graph of the line $y = ax + b$, then $ax_i + b - y_i = 0$. Otherwise, the error between the point (x_i, y_i) and the line can be defined as $|ax_i + b - y_i|$. If this is the case for all points (x_i, y_i) with $i = 0, 1, \ldots, N$, then the total absolute error of all $N+1$ points will be

$$\sum_{i=0}^{N} |ax_i + b - y_i| \ .$$

We want to find the values of the coefficients a and b to minimize such an error. We know from the first derivative test of calculus that the minimum of a smooth function is among its critical points; these can be the roots of its first derivatives. The way we defined the error as the absolute value of a function, we cannot use the first derivative test easily. For this reason, we will use the square of the absolute errors. This is the sum of the squared errors

$$S(a,b) = \sum_{i=0}^{N} (ax_i + b - y_i)^2 \ .$$

An extremum of this function is at a point where the partial derivatives are both zero

$$\frac{\partial}{\partial a} S(a,b) = 0 \quad \text{and} \quad \frac{\partial}{\partial b} S(a,b) = 0 \ .$$

These conditions lead to the following two equations

$$\sum_{i=0}^{N} 2(ax_i + b - y_i)x_i = 0 \quad \text{and} \quad \sum_{i=0}^{N} 2(ax_i + b - y_i) = 0 \ .$$

Rearranging these equations we get the so-called *normal equations*

$$\begin{cases} \left(\sum_{i=0}^{N} x_i^2\right) a + \left(\sum_{i=0}^{N} x_i\right) b = \sum_{i=0}^{N} y_i x_i \ , \\ \left(\sum_{i=0}^{N} x_i\right) a + (N+1)\ b = \sum_{i=0}^{N} y_i \ , \end{cases} \tag{6.41}$$

where we used the fact that $\sum_{i=0}^{N} i = N+1$. System (6.41) is a 2×2 system of linear equations with unknowns the coefficients a and b. Denoting

$$p = \sum_{i=0}^{N} x_i, \quad q = \sum_{i=0}^{N} y_i, \quad r = \sum_{i=0}^{N} y_i x_i, \quad s = \sum_{i=0}^{N} x_i^2 \ ,$$

we write the system (6.41) in matrix-vector form as

$$\begin{pmatrix} s & p \\ p & N+1 \end{pmatrix} \begin{pmatrix} a \\ b \end{pmatrix} = \begin{pmatrix} r \\ q \end{pmatrix} \ .$$

Using Cramer's rule (of Section 2.2.11) to solve it we compute the determinant

$$D = \det \begin{pmatrix} s & p \\ p & N+1 \end{pmatrix} = (N+1)s - p^2 \ ,$$

and then the solution becomes

$$a = \frac{1}{D} \det \begin{pmatrix} r & p \\ q & N+1 \end{pmatrix} = [(N+1)r - pq]/D \quad \text{and} \quad b = \frac{1}{D} \det \begin{pmatrix} s & r \\ p & q \end{pmatrix} = [sq - pr]/D \ .$$

Given the data points (x_i, y_i) for $i = 0, 1, \ldots, N$, we can express the previous process known as *linear least squares* (also known as linear regression) in the simple Algorithm 23.

Algorithm 23 Linear least squares method

Set $p = \sum_{i=0}^{N} x_i$, $q = \sum_{i=0}^{N} y_i$, $r = \sum_{i=0}^{N} y_i x_i$, $s = \sum_{i=0}^{N} x_i^2$, $D = (N+1)s - p^2$
Compute $a = [(N+1)r - pq]/D$
Compute $b = [sq - pr]/D$

Here is a simple implementation of a function that given the data x and y and a set of points stored in the variable xx returns the points yy such that yy=a*xx+b.

```
def linear_least_squares(x, y, xx):
# returns the vector yy with the linear least squares
# approximation of the input data (x,y)
    N = len(x)
    p = np.sum(x)
    q = np.sum(y)
    r = np.sum(np.multiply(x,y))
    s = np.sum(np.square(x))
    D = (N+1)*s-p**2
    a = [(N+1)*r - p*q]/D
    b = [s*q - p*r]/D
    yy = a*xx + b
    return yy
```

In order to test our code we generate some random data aligned with the line $y = 2x+1$. To generate a random vector with ten entries we use the NumPy function random.rand.

```
x = np.linspace(-1,2,10)
y = 2.0*x+1.0+np.random.rand(10)
xx = np.linspace(-1,2,20)
yy = linear_least_squares(x, y, xx)
plt.plot(xx,yy,'-',x,y,'o')
plt.xlabel('$x$')
plt.ylabel('$y$',rotation=0)
plt.show()
```

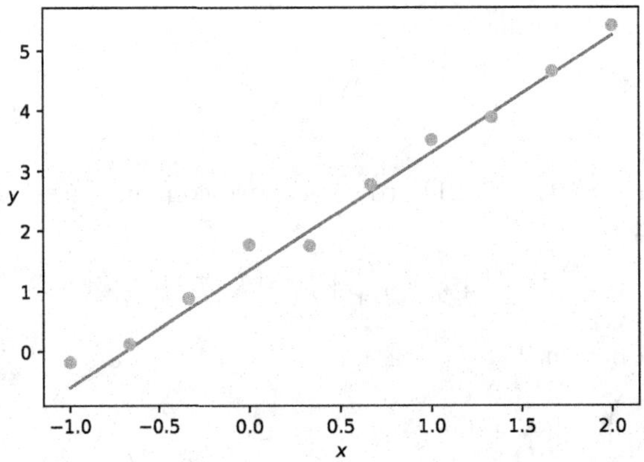

Printing the variables a and b you will observe that the resulting linear least squares approximation can be different from the line $y = 2x + 1$. This is because the points have been perturbed randomly and thus a different line minimizes the squared distance between the points and the line.

6.5.2 Polynomial fit

The linear least squares method can generate approximations to data aligned with a straight line. Usually, we observe this by plotting the data points in a graph. If our intuition says that the data points (x_i, y_i) are aligned with a non-linear curve, then we need to use a different method. Eventually, the method of linear least squares can be generalized to polynomials of degree m

$$P_m(x) = \sum_{k=0}^{m} a_k x^k \ .$$

The way to derive a linear system of equations with the unknown polynomial coefficients follows the exact same methodology we used to derive the 2×2 linear system in the case of linear least squares problem. For example, we define the error function

$$\begin{aligned} S(a_0, \dots, a_m) &= \sum_{i=0}^{N} |P_m(x_i) - y_i|^2 = \sum_{i=0}^{N} [(a_0 + a_1 x_i + \cdots + a_m x_i^m) - y_i]^2 \\ &= \sum_{i=0}^{N} \left[\sum_{k=0}^{m} a_k x_i^k - y_i \right]^2 \ . \end{aligned}$$

The critical points of the error function S can be found by estimating the roots of the partial derivatives

$$\frac{\partial}{\partial a_i} S(a_0, a_1, \dots, a_m) = 0, \quad \text{for} \quad i = 0, 1, \dots, m \ .$$

These equations can be written as

$$\begin{aligned} \frac{\partial}{\partial a_0} S(a_0, a_1, \dots, a_m) &= 2 \sum_{i=0}^{N} \left(\sum_{k=0}^{m} a_k x_i^k - y_i \right) = 0 \ , \\ \frac{\partial}{\partial a_1} S(a_0, a_1, \dots, a_m) &= 2 \sum_{i=0}^{N} \left(\sum_{k=0}^{m} a_k x_i^k - y_i \right) x_i = 0 \ , \\ \frac{\partial}{\partial a_2} S(a_0, a_1, \dots, a_m) &= 2 \sum_{i=0}^{N} \left(\sum_{k=0}^{m} a_k x_i^k - y_i \right) x_i^2 = 0 \ , \\ &\ \vdots \\ \frac{\partial}{\partial a_m} S(a_0, a_1, \dots, a_m) &= 2 \sum_{i=0}^{N} \left(\sum_{k=0}^{m} a_k x_i^k - y_i \right) x_i^m = 0 \ . \end{aligned}$$

The resulting linear system is of order $m+1$. Specifically, the coefficients a_j, $j = 0, 1, \dots, m$ can be found by solving the linear system

$$\boldsymbol{A z} = \boldsymbol{b} \ ,$$

where $\boldsymbol{z} = (a_0, a_1, \dots, a_m)^T$,

$$\boldsymbol{A} = \begin{pmatrix} N & \sum x_i & \sum x_i^2 & \cdots & \sum x_i^m \\ \sum x_i & \sum x_i^2 & \sum x_i^3 & \cdots & \sum x_i^{m+1} \\ \vdots & \vdots & \vdots & \ddots & \vdots \\ \sum x_i^{m-1} & \sum x_i^m & \sum x_i^{m+1} & \cdots & \sum x_i^{2m} \end{pmatrix} \quad \text{and} \quad \boldsymbol{b} = \begin{pmatrix} \sum y_i \\ \sum x_i y_i \\ \vdots \\ \sum x_i^m y_i \end{pmatrix} \ .$$

The solution of this linear system is not trivial, but in our implementation we solve it using the function `numpy.linalg.solve(A,b)`. The implementation of this method can be the following

```
def poly_least_squares(x, y, xx, m):
# Computes the least squares approximation of degree m
# for the data points (x,y) evaluated at xx
    N = len(x)
    A = np.zeros((m+1,m+1))
    b = np.zeros(m+1)
    s = np.zeros(2*m+1)
    # Assembly of A and b
    for i in range(N):
        temp = y[i]
        for j in range(m+1):
            b[j] = b[j] + temp
            temp = temp*x[i]
        temp = 1.0
        for j in range(2*m+1):
            s[j] = s[j] + temp
            temp = temp*x[i]
    for i in range(m+1):
        for j in range(m+1):
            A[i,j] = s[i+j]
    z = np.linalg.solve(A,b)
    # Evaluation of the polynomial at the nodes xx
    p = z[m]
    for j in range(m):
        p = p*xx+z[m-j-1]
    return p
```

We test our code by generating some random data aligned with the line $y = 2x^2 + 1$. Check that the least squares approximation is a polynomial close to $y = 2x^2 + 1$.

```
x = np.linspace(-1,2,10)
y = 2.0*np.square(x)+1.0+np.random.rand(10)
xx = np.linspace(-1,2,20)
yy = poly_least_squares(x, y, xx, 2)
plt.plot(xx,yy,'-',x,y,'o')
plt.xlabel('$x$'); plt.ylabel('$y$',rotation=0); plt.show()
```

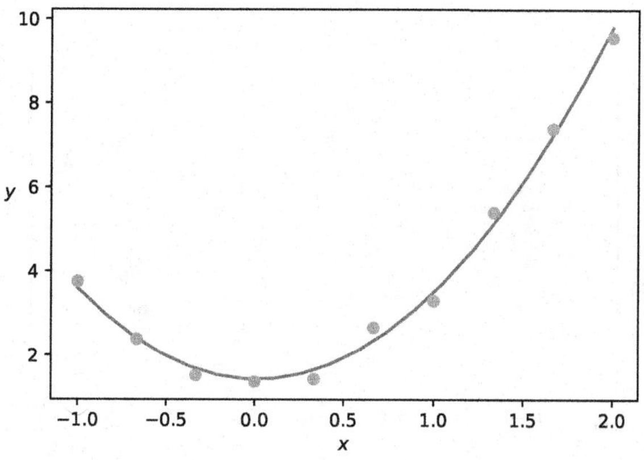

6.5.3 Non-polynomial fit

Sometimes, the data follow a law that isn't necessarily close to a polynomial. For example, their values might decay exponentially fast. This means that we would prefer to find a function of the form

$$f(x) = Ae^{Bx} \ ,$$

to approximate our data. This problem can be transformed to a linear one by using the points $(x_i, \ln y_i)$ instead of (x_i, y_i). In order to see that, consider the function

$$F(x) = \ln f(x) = \ln A + Bx \ ,$$

for which we need to minimize the residuals

$$R_i = \ln y_i - F(x_i) = \ln y_i - (\ln A + Bx_i) \ .$$

Setting $z_i = \ln y_i$ the least squares problem is now the problem of minimization of the function

$$S(a,b) = \sum_{i=0}^{N} (ax_i + b - z_i)^2 \ ,$$

with $a = B$ and $b = \ln A$.

As an example we consider a dataset aligned with the line $y = 3e^{2x}$.

```
x = np.linspace(0,2,10)
y = 3.0*np.exp(2.0*x)+0.1*np.random.rand(10)
z = np.log(y)
xx = np.linspace(0,2,20)
zz = linear_least_squares(x, z, xx)
yy = np.exp(zz)
plt.plot(xx,yy,'-',x,y,'o')
plt.xlabel('$x$')
plt.ylabel('$y$',rotation=0)
plt.show()
```

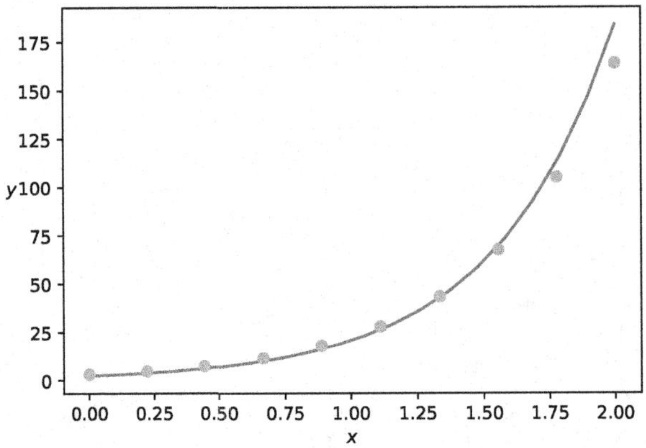

NumPy implementation

The least squares method is implemented in Python with the function `numpy.polyfit` that the error function $S(a_0, a_1, \ldots, a_m)$ and returns the coefficients of the approximating polynomial. Then the function `numpy.poly1d` can be used to obtain the polynomial as a callable function capable to be evaluated at any point x. The basic call of the function `polyfit` is

```
polyfit(x, y, deg, rcond, full, w, cov)
```

where

x: The x-coordinates of the dataset we want to approximate

y: The y-coordinates of the dataset we want to approximate

`deg`: The degree of the fitting polynomial

`rcond`: It is called *relative condition number* of the fit. Singular values smaller than this number, relative to the largest singular value (eigenvalues of the matrix $A^T A$) will be ignored. For more information please see Chapter 10. (Optional with default value `rcond=None`)

`full`: This is a boolean variable that determines the return value. When it is `False` the function returns only the coefficients of the polynomial. When it is `True` the function returns diagnostic information. (Optional with default value `full=False`

w: Is an array with the weights to apply to the y-coordinates of the sample points. (Optional with default value `w=None`)

cov: Is a boolean variables that returns an estimate of the covariance matrix of the approximation for statistical analysis. (Optional with default values `cov=False`)

The function `polyfit` returns an array `p` with the coefficients of the polynomial starting with the highest coefficient. Other returning variables are the **`residuals, rank, singular_values`** and **`rcond`** that are returned only if the parameter **`full=True`**. We do not present the use of these parameters here as they require knowledge of more advanced techniques of numerical analysis that we haven't discussed yet. One can return to the details of this function after studying the material of Chapter 10.

In order to compute the values of a polynomial at various values x we can use the function `numpy.poly1d`. The general call of this function is `poly1d(p, r, variable)` where `p` can be the output polynomial coefficients of the function `polyfit`, `r` is a boolean variable related with the computations of the roots of the polynomial, and `variable` is a string used when printing the polynomial `p`, and its default value is `None`.

An example of usage of the functions `polyfit` and `poly1d` is listed below, where we consider the case $y = 2x + 1$ again.

```
x = np.linspace(-1,2,10)
y = 2.0*np.square(x)+1.0+np.random.rand(10)
xx = np.linspace(-1,2,20)
yy = np.polyfit(x, y, 2)
zz = np.poly1d(yy)
plt.plot(xx,zz(xx),'-',x,y,'o')
plt.xlabel('$x$'); plt.ylabel('$y$',rotation=0); plt.show()
```

The specific code returns very similar results to our code for linear least squares approximations.

We will revisit the topic of the least squares problem later in this book when we explore relevant methods and present theoretical developments of approximation theory in Chapter 10.

6.6 The Module `scipy.interpolate`

SciPy provides the module `interpolate` with the implementation of several algorithms for interpolation problems. Here we present relevant routines for Lagrange and spline interpolation and their extensions in two dimensions, as well as an application in image processing.

6.6.1 Lagrange interpolation

The Lagrange interpolation is implemented in the function `scipy.interpolate.lagrange`. The `lagrange` function computes the Lagrange interpolation polynomial of the given data (x_i, y_i), for $i = 0, 1, \ldots, N$. The general call of this function is `lagrange(x,w)` where `x` is an array containing the values x_i (the array `x` in our code) and `w` an array with the corresponding values y_i (the array `y` in our code). The function `lagrange` returns the polynomial P_N as a function which can be evaluated at any point z.

```
import scipy.interpolate as spi

def f(x):
    return 1.0 / (1.0 + 25.0 * x**2)
# We evaluate the interpolant at the points z
z = numpy.linspace(-1, 1, 100)
# We use the data x,y
x = numpy.linspace(-1, 1, 10)
y = f(x)
# We compute the Lagrange polynomial p
p = spi.lagrange(x, y)
```

```
# We compare the graphs of the data and the interpolant
plt.plot(z,p(z),'-')
plt.plot(x,y,'o')
plt.plot(z,f(z),'-.')
plt.legend(["$P_9(z)$","data","$f(z)$"])
plt.xlabel('x'); plt.ylabel('y'); plt.show()
```

The results again will be the same as before and are omitted. The only important fact to note here is that using SciPy we can compute the interpolating polynomial with only one line of code.

6.6.2 Cubic spline interpolation

The cubic spline interpolant can be computed in Python with the function `CubicSpline` of the module `scipy.interpolate`. For example, if we want to interpolate the points (x_i, y_i) stored in the variables `x` and `y`, then we can simple call the function

```
cs = CubicSpline(x, y)
```

where the `cs` is the cubic spline function evaluated with "not-a-knot" boundary conditions, the default boundary conditions. We can evaluate the cubic spline at any point (or array) `xx` by typing `cs(xx)`. We can also compute the derivatives of the cubic spline using `cs(xx, 1)`, `cs(xx, 2)`, `cs(xx, 3)` for the first, second and third order derivatives respectively. The general call of the function `CubicSpline` is the following:

```
CubicSpline(x, y, axis, bc_type, extrapolate)
```

where here

`x`: An array with the values x_i, $i = 0, \dots, N$

`y`: An array with the values y_i, $i = 0, \dots, N$

`axis`: Axis along which y is assumed to be varying (if there are more than one indices). (Optional with default value `0`)

`bc_type`: Can be one of the following options:

- `'not-a-knot'`,
- `'periodic'`,
- `'clamped'`,
- `'natural'`.

It can also be a 2-tuple such as `bc_type=((order,value),(order,value))` where the first argument `order` is the order of the derivative at the boundary, and the second argument is its value. (Optional with default value `'not-a-not'`)

`extrapolate`: A boolean variable that determines whether to extrapolate to points outside the interval $[x_0, x_N]$. It can also take the values `'periodic'` or `None`. (Optional with default value `None`)

Here is an example where we interpolate the function $f(x) = \sin(x)$ using 10 points in $[0, 2\pi]$. In this code we compute the cubic spline along with its derivatives and we plot the results.

```
x = np.linspace(0.0, 2.0*np.pi, 10); y = np.sin(x)
cs = spi.CubicSpline(x, y)
xx = np.arange(0.0, 2*np.pi, 0.1)
plt.plot(x, y, 'o', label='data')
plt.plot(xx, cs(xx), label="$S$")
plt.plot(xx, cs(xx, 1), label="$S'$")
plt.plot(xx, cs(xx, 2), label="$S''$")
plt.plot(xx, cs(xx, 3), label="$S'''$")
plt.xlim(0, 2.0*np.pi); plt.legend(loc='lower left', ncol=2)
plt.xlabel('$x$'); plt.ylabel('$y$',rotation=0); plt.show()
```

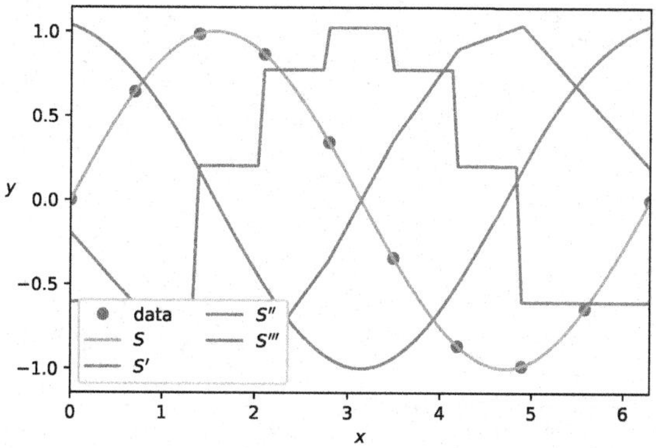

Observe that the third derivative of the cubic spline is a piecewise constant function but it was plotted as a piecewise linear function.

We already mentioned that the function `CubicSpline` returns a function as an object. The specific object (class) called `PPoly` is accompanied by additional methods. For example, if we store a cubic spline in a variable `cs` as before, then we can get its derivative as a new function using the method `cs.derivative(nu)`, where `nu` is the order of the derivative. This method returns just a piecewise polynomial object (`PPoly`). Similarly, one can obtain an antiderivative using the method `cs.antiderivative(nu)`. We can compute a definite integral of a cubic spline over an interval $[a, b]$ just by typing `cs.integrate(a,b)`. The function `integrate` can also has an additional argument `extrapolation` (which is `None` by default) with which we can specify whether to extrapolate the spline or not similarly to the function call `CubicSpline`.

Runge's example revisited

As we saw before, the Lagrange interpolation fails to accurately approximate Runge's function

$$f(x) = \frac{1}{1 + 25x^2} \ .$$

We try again Runge's example using cubic spline interpolation.

```
def f(x):
    return 1.0 / (1.0 + 25.0 * x**2)
num_points = 10
x = np.linspace(-1, 1, num_points)
y = f(x)
cs = spi.CubicSpline(x, y)
xx = np.linspace(-1, 1, 100)
plt.plot(x, y, 'o', label='data')
plt.plot(xx, f(xx), label='true')
plt.plot(xx, cs(xx), label='Spline')
plt.xlim(-1, 1)
plt.legend(loc='upper left', ncol=1)
plt.xlabel('$x$'); plt.ylabel('$y$',rotation=0); plt.show()
```

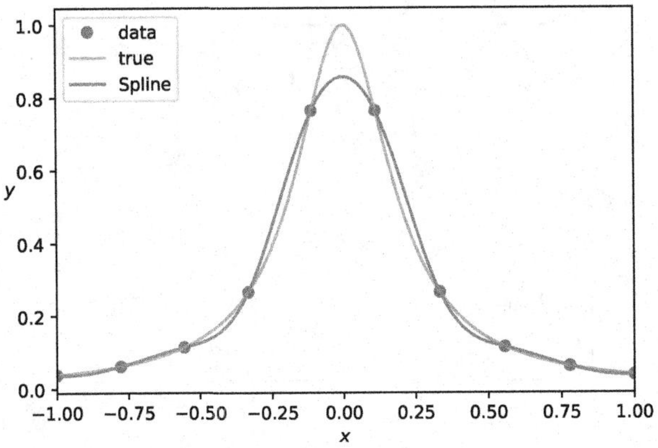

This time the cubic spline approximates Runge's function without problems. We encourage the reader to experiment with the specific example and try Runge's example with more and less nodes, and compare with analogous results using Lagrange interpolation.

6.6.3 Computations with B-splines

We can compute the B-spline representation of an interpolant in Python using the function `splrep` of the module `scipy.interpolate`. The general call of the function `splrep` is the following:

```
splrep(x, y, w, xb, xe, k, task, s, t,full_output, per, quite)
```

where here

`x`: An array with the values x_i, $i = 0, \ldots, N$

`y`: An array with the values y_i, $i = 0, \ldots, N$

`w`: A rank-1 array of weights with the same length as `x` and `y` for least squares fit. (Optional with default value `None`)

`xb, xe`: Floating point numbers that determine the endpoints of the interpolation interval. (Optional with default values `'None'`) `k`: The degree of the spline. (Optional with default value `3`)

`task`: A value between 1 for a subsequent call of the function with a new value of the smoothing factor `s`, 0 for evaluations of the spline coefficients using the specified smoothing factor `s` and -1 for weighted least squares approximation. (Optional with default value `0`)

`s`: A smoothing factor. (Optional with default value `None`)

`t`: The nodes needed for `task=-1`. (Optional with default value `None`)

`full_output`: If non-zero, then the function returns optional output. (Optional with default value `0`)

`per`: If non-zero then the data points are considered periodic. (Optional with default value `0`)

`quite`: A non-zero value suppress messages. (Optional with default value `1`)

The function returns a tuple `tck=(t,c,k)` where `t` is a vector with nodes, `c` a vector with the spline coefficients a_i, and k the degree of the spline. Optional arguments returned such as `fp` with the weighted sum of squared residuals of the spline approximation, `ier` as a flag of convergence, and `msg` a message corresponding to the `ier` flag.

After computing the spline coefficients a_i into the array `c` we can evaluate the spline function at any point x we want using the function `splev` of the module `scipy.interpolate`. The general call of the function `splev` is the following:

```
splev(x, tck, der, ext)
```

where

`x`: An array with the points x at which we need the values of the spline

`tck`: The tuple `tck` received in the output of `splrep`.

`der`: The order of the derivative to be computed less than or equal to k. (Optional with default value `0`)

`ext`: Controls the value returned for elements of `x` not in the interval defined by the node sequence. (Optional with default values `'0'`)

The function returns the values of the spline `y`.

As an example we consider again the Runge function and the following code:

```
def f(x):
    return 1.0/(1+25*x**2)
x = np.linspace(-1.0, 1.0, 10)
y = f(x)
spl = spi.splrep(x, y)
z = np.linspace(-1.0, 1.0, 200)
yz = spi.splev(z, spl)
yyz = f(z)
print(spl[2])
plt.plot(x, y, 'o', label='data')
plt.plot(z, yyz, label='true')
```

```
plt.plot(z, yz, label='Spline')
plt.xlim(-1, 1)
plt.legend(loc='upper left', ncol=1)
plt.xlabel('$x$')
plt.ylabel('$y$',rotation=0)
plt.show()
```

The result of this code is identical with the results obtained in Section 6.6.2. By printing also the value `spl[2]` we verify that the degree of the spline is 3.

6.6.4 General purpose interpolation and the function `interp1d`

There is a general purpose Python function that can be used with various different methods of interpolation. This is the function `interp1d` of the module `scipy.interpolate`. The general call of the function `interp1d` is

```
interp1d(x, y, kind, axis, copy, bounds_error, fill_value, assume_sorted)
```

where

`x`: A one-dimensional array with the values x_i

`y`: An n-dimensional array with length equal to the length of x containing the data y_i

`kind`: String specifying the kind of interpolation and it takes values `linear`, `nearest`, `nearest-up`, `zero`, `slinear`, `quadratic`, `cubic`, `previous`, and `next`. (Optional with default value `'linear'`)

`axis`: Specifies the axis of y along which to interpolate. (Optional with default value `-1`)

`copy`: Is a boolean variable which specifies if the variables `x,y` will be copied internally or not. (Optional with default value `True`)

`bounds_error`: Is a boolean variable which should be `True` when we use the function for extrapolation to compute data outside the region defined by the data (x_i, y_i). (Optional with default value `None`)

`fill_value`: This is the value to be used for points outside the interpolation domain. It can also take the value `"extrapolate"`. If this is `None` then nearest-neighbor extrapolation is used. (Optional with default value `NaN`)

`assume_sorted`: It determines whether the values of `x` are sorted or not. If they are not sorted (`False`) then the algorithm will sort them first. (Optional with default value `False`)

The `interp1d` function returns a class that can be used as an interpolation function for computations. As a simple example we consider again the interpolation of a data set produced using Runge's function in $[-1, 1]$.

```
def f(x):
    return 1.0/(1+25*x**2)
x = np.linspace(-1,1,10)
y = f(x)
p = spi.interp1d(x, y, kind = 'cubic' )
```

```
z = np.linspace(-1,1,100)
plt.plot(x, y, 'o', label='data')
plt.plot(z, f(z), label='true')
plt.plot(z, p(z), label='interpolant')
plt.xlim(-1, 1)
plt.legend(loc='upper left', ncol=1)
plt.xlabel('$x$'); plt.ylabel('$y$',rotation=0); plt.show()
```

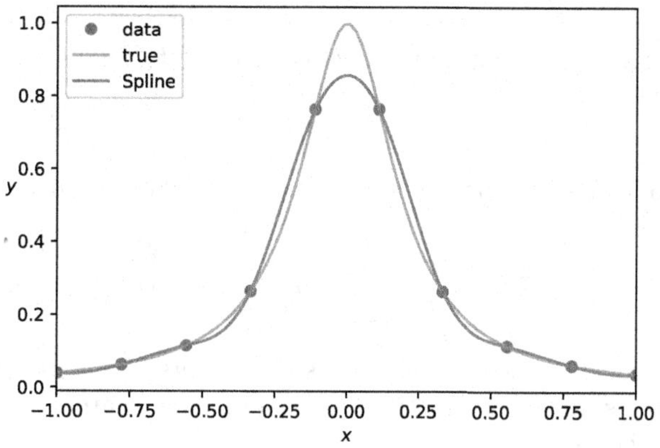

Using cubic spline interpolation with the function `intrp1d` we obtain the same results as in Section 6.6.2. We let the interested reader to experiment with the different kinds of interpolation methods offered with the function `interp1d`.

6.6.5 Interpolation in two dimensions

Two-dimensional interpolation is the interpolation of data points (x_i, y_i, z_i) for $i = 0, 1, \ldots, N$ by a polynomial $z = P(x, y)$ in two dimensions. Suppose that there is a function f such that $z_i = f(x_i, y_i)$ for all $i = 0, 1, \ldots, N$. This case is more complicated compared to the one-dimensional problem as the data points (x_i, y_i) can be scattered randomly in the xy-plane forming a non-uniform and unstructured grid of points. It is often the case that these data points can form a triangular grid like the one in Figure 6.6.

For the sake of simplicity, we consider only rectangular domains $[a, b] \times [c, d]$ and rectangular grids. Such grids consist of points (x_i, y_j), $i = 0, 1, \ldots, N$, $j = 0, 1, \ldots, M$ with $x_{i+1} = x_i + \Delta x$ for $i = 0, 1, \ldots, N-1$ and $y_{j+1} = y_j + \Delta y$ for $j = 0, 1, \ldots, M-1$. An example of a rectangular grid is depicted in Figure 6.6. For any of the previous grids we can apply Lagrange as well as Spline interpolation. We define the Lagrange polynomials for a rectangular grid as

$$\ell_i(x) = \prod_{k \neq i, k=0}^{N} \frac{x - x_k}{x_i - x_k}, \quad \text{for} \quad i = 0, 1, \ldots, N \ ,$$

and

$$\bar{\ell}_j(y) = \prod_{k \neq j, k=0}^{M} \frac{y - y_k}{y_j - y_k}, \quad \text{for} \quad j = 0, 1, \ldots, M \ .$$

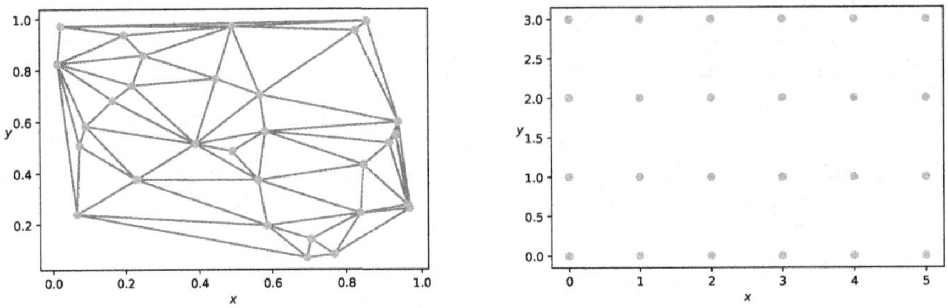

FIGURE 6.6
Examples of planar grids: unstructured triangular grid (left) versus structured rectangular grid (right).

The Lagrange interpolant of the data $(x_i, y_j, f(x_i, y_j))$ is the function

$$P(x,y) = \sum_{i=0}^{N}\sum_{j=0}^{M} f(x_i, y_j)\ell_i(x)\bar{\ell}_j(y) \ .$$

Bivariate spline functions can be defined as tensor products of one-dimensional splines. Let $\phi_i(x)$ and $\bar{\phi}_j(y)$ be the spine basis functions in x and y coordinates, respectively. Then, the spline interpolant is the function

$$S(x,y) = \sum_{i=0}^{\tilde{N}}\sum_{j=0}^{\tilde{M}} c_{i,j}\phi_i(x)\bar{\phi}_j(y) \ ,$$

where of course $c_{i,j}$ are equal to $f(x_i, y_j)$ only if the $\bar{\phi}_j$ and $\bar{\phi}_j$ are linear B-splines, and $\tilde{N}$, $\tilde{M}$ are the dimensions of the corresponding one-dimensional spline spaces. For more information about tensor product of splines, we refer to [117].

The module `scipy.interpolate` has the function `interp2d` which can interpolate data points (x_i, y_i, z_i) using spline interpolation in two dimensions. This function is a generalization of `interp1d` and its general call is the following:

```
interp2d(x, y, z, kind, copy, bounds_error, fill_value)
```

where

`x,y,z`: Arrays of values for the data points (x_i, y_i, z_i). In general, these arrays can be 1D with the values (x_i, y_i, z_i)

`kind`: String specifying the kind of spline interpolation. It takes values `linear`, `cubic` and `quintic`. (Optional with default value `'linear'`)

`copy`: Is a boolean variable which specifies whether the variables `x,y,z` will be copied internally or not. (Optional with default value `True`)

`bounds_error`: Is a boolean variable which should be `True` when we use the function for extrapolation purposes. (Optional with default value `'False'`)

`fill_value`: Is the value to be used for points outside the interpolation domain. If this is `None`, then nearest-neighbor extrapolation is used. (Optional with default value `None`)

The function returns a class that can be used as a spline function for computations. The minimum number of data points required for the spline interpolation is $(k+1)^2$ with $k=1$ for linear, $k=3$ for cubic and $k=5$ for quintic interpolation. Alternatives to the function `interp2d` are the functions `RectBivariateSpline` for splines on rectangular grids and `BivariateSpline`.

As an example we consider the interpolation of the Gaussian function $f(x,y) = e^{-(x^2+y^2)}$. First we plot the particular function on the rectangular domain $[-2,2]\times[-2,2]$ using the method `plot_surface` of MatPlotLib. The function `plot_surface` requires the transformation of the arrays `x` and `y` into two-dimensional rectangular grid arrays on which $Z_{ij} = f(X_{ij}, Y_{ij})$. We generate the one-dimensional grids using the function `arange` as explained in Section 1.4.2. Given the one-dimensional grids stored in the array `x` and `y` we generate the two-dimensional representations using the NumPy function `meshgrid` as `X, Y = np. meshgrid(x,y)`. Then we can generate the surface $z = f(x,y)$ using the function `plot_surface(X,Y,Z,cmap)` where `Z=f(X,Y)` and `cmap` defines the colormap we want to use. Executing the following code we observe that this function is like a Gaussian bell that decreases exponentially fast to 0 as x and y move away from the origin.

```
def f(x,y):
    z = np.exp(-(x**2+y**2))
    return z
# Specify that the graph is three-dimensional
fig = plt.figure()
ax = plt.axes(projection='3d')
# Generate the data
x = np.arange(-2, 2, 0.1)
y = np.arange(-2, 2, 0.1)
X, Y = np.meshgrid(x, y)
Z = f(X,Y)
# Plot the surface.
surf = ax.plot_surface(X, Y, Z, cmap='viridis')
plt.show()
```

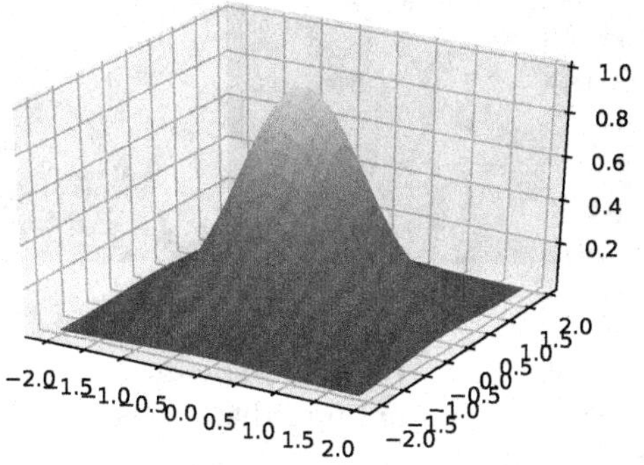

For the purposes of this example, we assume that we know the values of the function f on a sparse 10×10 grid of points of the domain $[-2,2]\times[-2,2]$. We interpolate the coarse 10×10 data in a dense 100×100 grid using the function `interp2d`. We will plot the original and interpolated data side by side. In the following code, we use the function `subplots` to plot two graphs side by side and also the function `pcolormesh` to plot the projection of the surface generated by the data on the xy-plane. The function **subplots** create a grid of plots with **nrows** rows and **ncols** columns. It also returns a figure **fig** and a class **axes** which determines the location of figures in the grid.

```
#define the function f
def f(x,y):
    z = np.exp(-(x**2+y**2))
    return z
# generate one-dimensional grids
x = np.linspace(-2, 2, 10)
y = x.copy()
z = f(x,y)
# generate two-dimensional grid arrays
X, Y = np.meshgrid(x, y)
Z = f(X,Y)
# interpolate the coarse data
z2 = spi.interp2d(x, y, Z, kind='cubic')
# evaluated the interpolant on dense grid
xx = np.linspace(-2, 2, 100)
yy = np.linspace(-2, 2, 100)
XX, YY = np.meshgrid(xx, yy)
ZZ = z2(xx,yy)
# plot the original and interpolated data
fig, axes = plt.subplots(nrows=1, ncols=2)
axes[0].pcolormesh(X, Y, Z)
axes[1].pcolormesh(XX, YY, ZZ)
plt.show()
```

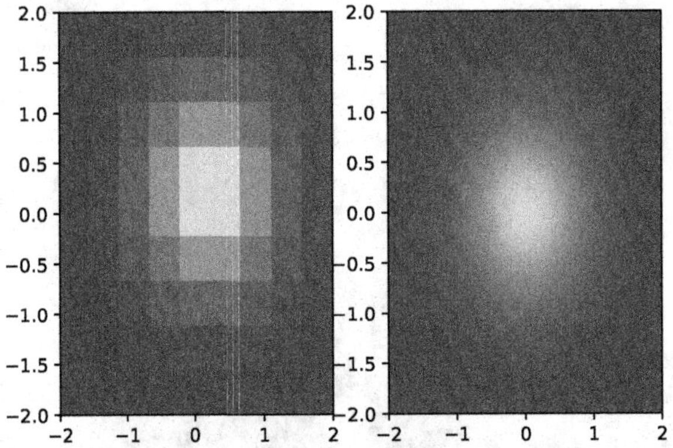

We observe that interpolating coarse data to dense data using cubic spline interpolation we obtain a smooth result. This has an immediate consequence in image processing which we present in the next section.

6.6.6 Application in image processing

Among the straightforward applications of interpolation from known data to unknown there are many applications in practical problems and especially in computer graphics and image processing. Here we present an application in image processing. Specifically, we show how we can improve the resolution of an image using spline interpolation.

Suppose that we have a digital photograph stored in our computer as a digital color image. A digital color image is an arranged set of pixels. Each pixel has each own color which can be represented by its values of the intensity of the basic colors red, green and blue (RGB). The color of each pixel is defined by a tuple of the form (R, G, B) where R stands for Red, G for Green and B for Blue. The colors can take values from 0 to 255 and usually are integer values. The value 0 indicates the complete absence of a specific color. For example, the blue color in the RGB scale will be $(0, 0, 255)$.

In this application we read a low resolution digital photograph of Lenna[4]. We assume that this photograph is stored in the file `Lenna.jpg`. In Python we can read an image using the command `imread` of the module `imageio`. The `imread` function in the following code reads the digital image stored in the file `Lenna.jpg` and stores it into the rectangular array `lenna`. Then we normalize its values and we store it in the array `photo1`. Practically, the matrix is a rectangular array with values the intensity of red, green and blue colors on each cell. In order to improve the resolution of this image, we will consider a denser grid (larger matrix) with the interpolated red, green and blue intensities.

After we have loaded the image in the computer memory, we print the shape of the array `lenna`. This matrix is $128 \times 128 \times 3$. This means that the picture consists of 128×128 pixels, with RGB values stored in three slices of the matrix `photo1`. Since we are working with functions, we consider the domain $[0, 128] \times [0, 128]$ and the grids in x and y axes with space $\Delta x = \Delta y = 1$. For the interpolation we generate a denser grid with $\Delta x = \Delta y = 0.1$. We can display the photo using the three dimensional array `photo1` and the command `imshow` of MatPlotLib.

After we separate the three colors using three different arrays, we interpolate the red, green, blue intensities using the function `interp2d` over a dense grid. We store the interpolated values in the arrays `better_red`, `better_green` and `better_blue`. We combine these arrays into a single array called `photo2`. The code of this application follows:

```python
import matplotlib.pyplot as plt
import imageio
import numpy as np
import scipy.interpolate as spi

# read the picture from file lena.jpg
lenna = imageio.imread("Lenna.jpg");
print(lenna.shape)
# scale the values to be between 0.0 and 1.0
photo1 = lenna.copy()/255.0
# get the intensity of red, green, blue
red = photo1[:,:,0]
green = photo1[:,:,1]
blue = photo1[:,:,2]
n = len(red)
# generate x and y coordinates
x = np.arange(0.0,n,1.0)
```

[4]Lenna is a standard test image used in computer science to demonstrate image processing techniques since 1973. It is a picture of the Swedish model Lena Forsén, shot by photographer Dwin Hooker.

```
y = x.copy()
# interpolated the red, green and blue intensities
better_red = spi.interp2d(x,y,red,kind='cubic')
better_green = spi.interp2d(x,y,green,kind='cubic')
better_blue = spi.interp2d(x,y,blue,kind='cubic')
# generate dense xx and yy coordinates
xx = np.arange(0.0,128.0,0.1)
yy = xx.copy()
row=len(xx)
col=len(yy)
# create array photo2 for the high-resolution picture
photo2 = np.zeros((row, col, 3))
# interpolate the data
photo2[:,:,0] = better_red(xx,yy)
photo2[:,:,1] = better_green(xx,yy)
photo2[:,:,2] = better_blue(xx,yy)
f, axs = plt.subplots(1,2,figsize=(15,15))
axs[0].imshow(photo1)
axs[1].imshow(photo2)
# choose a square of your choice such as [50,80]x[50,80]
zoom1 = photo1[50:80,50:80,:]
zoom2 = photo2[500:801,500:801,:]
# compare the different data
f, axs = plt.subplots(1,2,figsize=(15,15))
axs[0].imshow(zoom1)
axs[1].imshow(zoom2)
plt.show()
```

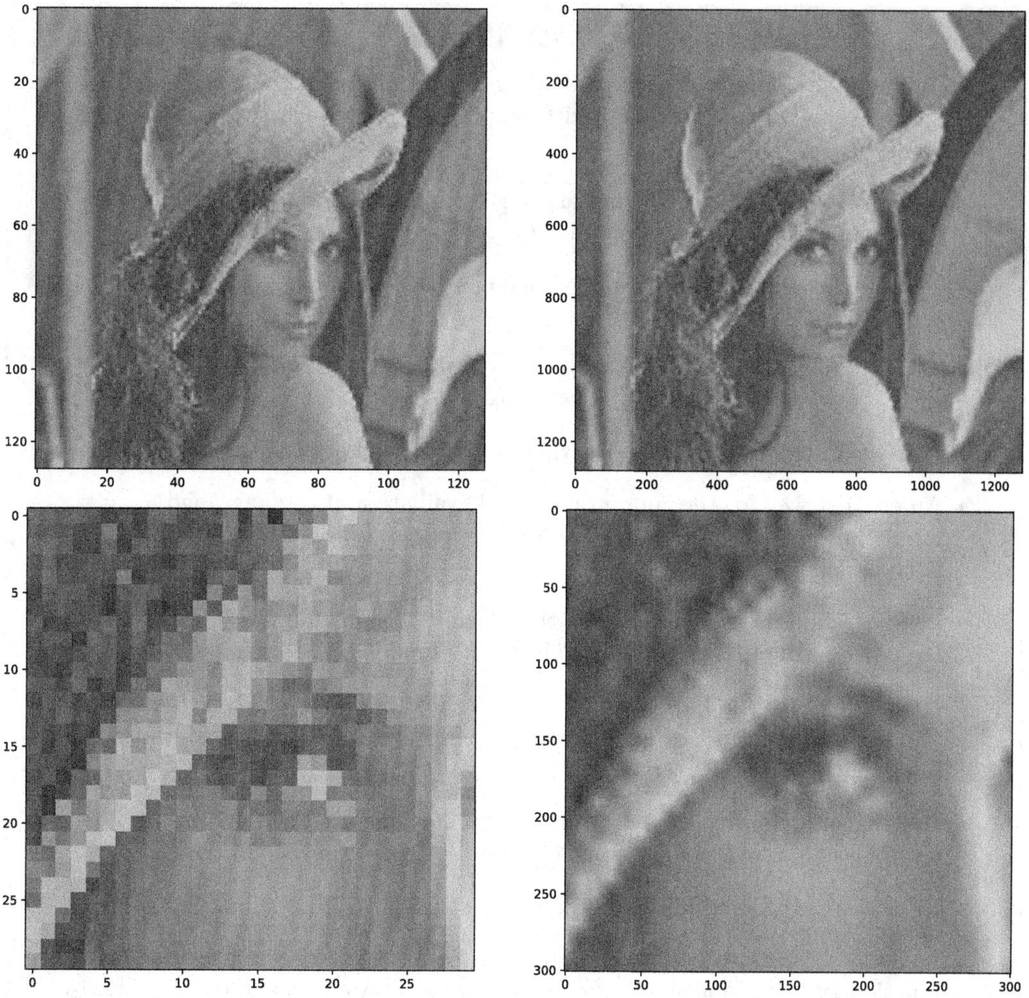

Magnifying the two images around the region $[50, 80] \times [50, 80]$ of the original image we observe that using interpolation we have improved the original resolution by smoothing out rough edges. It is noted that cubic splines smooth out oscillations and increase the regularity of the interpolating data (see Exercise 9 for a proof of this property). For this reason, using cubic splines especially in image processing we can obtain smoother images compared to the original one.

6.7 Further Reading

The theory of interpolation and least squares approximation is covered in all classic textbooks [25, 70, 76, 22, 26, 7, 108, 116]. In this book, we followed [127] in the development of the theory of splines. Specialized books on interpolation are, among many, the following [107, 31, 117, 118] while a practical introduction to two-dimensional interpolation can be found in [119]. A dedicated book to least squares problems is the book [12].

Chapter Highlights

- Interpolation is the procedure of finding a function (usually a polynomial or a piecewise polynomial function) with graph passing through a given set of points.
- Given $N+1$ points (x_i, y_i) the Lagrange interpolation is the set of methods for determining the unique polynomial $P_N(x)$ of degree N such that $P_N(x_i) = y_i$.
- The Lagrange interpolating polynomial can be written as

$$P_N(x) = \sum_{i=0}^{N} y_i \ell_i(x) ,$$

  where $\ell_i(x)$ are the Lagrange polynomial basis functions.
- An efficient way for the computation and evaluation of the Lagrange interpolation is via Newton's form for a representation of an interpolating polynomial.
- Lagrange interpolation fails when the magnitude of high-order derivatives is large. An example of such failure is Runge's function.
- If in addition to the interpolation data $y_i = P_N(x_i)$ we have the values of the first derivative, then Lagrange interpolation is generalized to the Hermite interpolation.
- Hermite interpolation might be more accurate compared to Lagrange interpolation, but it cannot solve the problem with Runge's function.
- Piecewise polynomial interpolation is the solution to Runge's problem.
- Spline is a piecewise polynomial function. A linear spline is a piecewise linear function and thus not differentiable at the nodes. A cubic spline is a piecewise cubic function, which is twice continuously differentiable.
- B-splines can be computed using recursive algorithms. One such algorithm is the Cox-de Boor algorithm.
- Cubic spline interpolation requires knowledge of the behavior of the interpolating polynomial at the boundary.
- Cubic splines are characterized by a smoothing property. For this reason, cubic splines are used in image processing.
- When the data points are too "noisy" we approximate them by a function or a polynomial without requiring interpolation.
- One such method is the method of least squares, which is based on the minimization of the error function

$$S = \sum_{i=0}^{N} |y_i - P(x_i)|^2 .$$

Exercises

1. Given the data $f(0) = 1$, $f(1) = 3$ and $f(3) = 5$:

   (a) Approximate the value $f(2)$ using an appropriate interpolating polynomial written in Lagrange's form.

   (b) Approximate the same value using Newton's divided difference formula.

   (c) Construct interpolating cubic spline and approximate the value $f(2)$.

2. Let x_i, $i = 1, 2, \ldots, n+1$ different nodes and let $y_i \in \mathbb{R}$, $i = 1, 2, \ldots, n+1$. The interpolating polynomial is written in Newton's form as:

$$p_n(x) = a_1 + a_2(x - x_1) + a_3(x - x_1)(x - x_2) + \cdots + a_{n+1}(x - x_1)\cdots(x - x_{n+1}),$$

   where the coefficients a_i, $i = 1, \ldots, n+1$ can be computed using the following algorithm:

Algorithm 24 Newton's polynomial

$a_i = y_i$, $i = 1, 2, \ldots, n+1$
for $k = 2 : n+1$ **do**
  for $i = 1 : k-1$ **do**
    $a_k = (a_k - a_i)/(x_k - x_i)$
  end for
end for

If the coefficients a_i, $i = 1, \ldots, n+1$ are known, then the value of the interpolating polynomial at the point z can be computed using Horner's formula:

Algorithm 25 Horner's formula

$s = a_{n+1}$
for $i = n : -1 : 1$ **do**
  $s = a_i + (z - x_i)s$
end for
$p_n(z) = s$

   (a) Write Python functions `coefs` and `evalp` implementing the previously described algorithms for the coefficients of the interpolating polynomial and its evaluation at values z using Horner's formula. Write a function `newtinterp` with input arguments (x, y, z), where (x_i, y_i), $i = 1, \ldots, n+1$, the interpolation data points and $\mathbf{z} = [z_1, \ldots, z_m]$ the vector with the m points at which we want to evaluate the interpolating polynomial. This will compute the coefficients of the interpolating polynomial a_i using your function `coefs` and will return the values $u_i = p_n(z_i)$, $i = 1, 2, \ldots, m$ using your function `evalp`.

   (b) Consider the function $f(x) = \sin x$, $x \in [0, 2\pi]$, and a uniform partition of $[0, 2\pi]$ with 6 points x_i, $i = 1, \ldots, 6$. Using your function `newtinterp.m` compute the interpolating polynomial p_5, that interpolates function f at the nodes x_i, at 101 equidistributed points $z_i \in [0, 2\pi]$.

   (c) Verify your results and compare them with those obtained by using the Python functions `polyfit`, `polyval` and `CubicSpline`.

3. Consider the following data:

t_i	1991	1996	2001	2006	2011	2016
y_i	3,516,000	3,762,300	3,916,200	4,209,100	4,399,400	4,747,200

where y_i is the population of New Zealand in the year t_i.

(a) Complete the divided difference tableau and compute by hand the Newton's form of the interpolating polynomial of the given data.

(b) Using the function `newtinterp` of Problem 2, compute the interpolating polynomial $p \in \mathbb{P}_5$, that interpolates the values (t_i, y_i).

(c) Compare the numerical solution with your theoretical computations.

(d) Estimate the population of New Zealand for all the years between 1991 until 2016.

(e) Compare the results using the appropriate Python function with a cubic spline.

(f) Using your interpolating polynomial find an approximation of the population of New Zealand for the years 2018 and 2034.

(g) Repeat the computations with integer values t_i and y_i instead of floats. What do you observe?

4. Let

$$f(x) = \frac{1}{1+25x^2}, \quad -1 \le x \le 1.$$

(a) Using your function `newtinterp` compute the interpolating polynomial $p \in \mathbb{P}_n$, which interpolates the points $(x_i, f(x_i))$, $i = 1, 2, \ldots, N+1$ of a uniform partition of $[-1, 1]$ for $N = 5, 10, 20, 100$. Let z_i, $i = 1, 2, \ldots, 201$, equidistributed points of $[-1, 1]$. Plot your interpolants and comment on the results.

(b) Compute the values of the interpolating polynomial $p(z_i)$ and compare the results with the Python cubic spline function. Comment on the accuracy of the two methods.

5. Consider the polynomial $f(x) = x^5 + x^4$ in the interval $[-2, 2]$. Using five equidistributed nodes in $[-2, 2]$:

(a) Compute the Lagrange interpolating polynomial.

(b) Compute the cubic spline interpolant with natural boundary conditions.

(c) Compute the cubic spline interpolant with clamped boundary conditions.

(d) Sketch the graphs of the function and the three approximating functions and compare the results.

6. Consider the function

$$f(x) = \begin{cases} 0, & x \in [0,1]\ , \\ (x-1)^4, & x \in (1,2]\ . \end{cases}$$

(a) Find the coefficients a, b, c, d in the following piecewise and continuously differentiable polynomial function

$$p(x) = \begin{cases} 0, & x \in [0,1]\ , \\ a + b(x-1) + c(x-1)^2 + d(x-1)^2, & x \in (1,2]\ . \end{cases}$$

in order to interpolate the function f such as

$$p(0) = f(0),\ p'(0) = f'(0),\ p(1) = f(1),\ p(2) = f(2),\ p'(2) = f'(2)\ .$$

(b) Compute the cubic spline interpolant with clamped boundary conditions at the nodes $\{0, 1, 2\}$.

(c) Modify the definition of the Hermite interpolating polynomial to interpolate the specific function f at the same nodes.

(d) Sketch the graphs of the function f and its approximations.

7. Consider the function $f(x) = \cos(k\pi x)$ for $x \in [0, 1]$ and $k \in \mathbb{N}$.

(a) Estimate the maximum norm of the error in the Lagrange interpolation for uniform grid of N nodes.

(b) What is the relationship between k and N so as the error converges to 0 while $k, N \to \infty$?

(c) Estimate the maximum norm of the error of the linear spline interpolant for the same grid.

(d) Find again the relationship between k and N so as the error converges to 0 while $k, N \to \infty$.

(e) In the case where $k = 500$, which method would you prefer to use for the interpolation problem in order to get an error less than 10^{-2}?

8. A F1 car is racing in the Albert Park Circuit for the Australian Grand Prix. The particular circuit has length 5.303 km. The lap times in minutes:seconds and the corresponding recorded speeds in km/h are given in the following table:

Time ($minutes : seconds$)	$0 : 0.000$	$1 : 30.245$	$1 : 25.533$	$1 : 20.259$	$1 : 22.453$
Lap	0	4	8	12	16
Speed (km/h)	212	236	255	224	240

(a) Use Lagrange interpolation to predict the location of the car when $t = 3 : 50.000$.

(b) Repeat the calculation using Hermite interpolation.

(c) Use the derivative of the Hermite interpolant to estimate whether the car ever exceeds the speed of 255 km/h.

(d) Predict the maximum speed of the car.

9. Let $f \in C^2[a, b]$ and $P_c(x)$ the cubic spline interpolant with clamped boundary conditions of the function f at the nodes x_i, $i = 0, 1, \ldots, N$ of $[a, b]$.

(a) Prove that

$$\int_a^b (f''(x) - P_c''(x))\, P_c''(x)\; dx = 0\ .$$

(b) Prove that

$$\int_a^b \left(P_c''(x)\right)^2\, dx \le \int_a^b \left(f''(x)\right)^2\, dx\ .$$

(c) Explain the geometric interpretation of the previous results and explain why it is called the cubic spline smoothness theorem.

10. Consider the following dataset:

x_i	0.0	0.1	0.2	0.3	0.4	0.5	0.6
y_i	2.8	2.9	2.7	2.4	2.3	2.1	1.8

(a) Compute a piecewise linear interpolant for the points (x_i, y_i) using appropriate Python function.

(b) Compute the linear least squares approximation of the same set of points using appropriate Python function.

(c) Sketch the graph of the dataset and the approximations and comment on the results.

11. Consider the (nonlinear) function $f(x) = \sqrt{x}$ in $[0, 1]$.

(a) Generalize the linear least squares method to approximate data that follow the nonlinear rule $f(x)$.

(b) Write Python function to implement the specific nonlinear least squares method.

(c) Generate a set of points (x_i, y_i), $i = 0, 1, \ldots, 9$ using random perturbations of the points $(x_i, f(x_i))$, $i = 0, 1, \ldots, 9$ and test your computer code.

12. Consider the function $f(x) = 1/(x^2 + 1)$ in the interval $[0, 1]$.

(a) Compute the Lagrange interpolant $P_2(x)$ on a uniform grid with 3 nodes without the help of Python.

(b) Compute the linear spline (piecewise linear interpolant) $S_1(x)$ using the same grid.

(c) Approximate the integrals like

$$\int_0^1 f(x)\ dx \approx \int_0^1 P_2(x)\ dx\ ,$$

and

$$\int_0^1 f(x)\ dx \approx \int_0^1 S_1(x)\ dx\ .$$

(d) Which method gives more accurate results?

13. Consider the function $f(x) = 1/(x^2 + 2)$ in the interval $[-10, 10]$.

(a) Use Lagrange interpolation to compute the values $f(0)$ and $f(9.76)$ using $10, 11, 20, 21, 100$ and 101 nodes.

(b) Use linear and cubic spline interpolation to approximate the same value and using the same grids.

(c) Compare your results with results obtained using the Python function `interp1d`.

14. Consider the function $f(x, y) = 1/(x^2 + y^2 + 1)$ in the interval $[-5, 5] \times [-5, 5]$.

(a) Write a Python program to implement the two-dimensional Lagrange interpolation using rectangular grids as described in Section 6.6.5.

(b) Use your code to compute the Lagrange interpolant and estimate the values $f(0, 0)$, $f(4.76, 0)$ and $f(4.76, 4, 76)$ using 100 and 101 nodes.

(c) Compare your results with results obtained using the Python function `interp2d` using linear, cubic and quintic spline interpolation.

(d) Use appropriate functions from the module MatPlotLib to present the results.

7

Numerical Integration

Integrals of functions often appear in applications as well as in various computations. For example, the energy of a system is usually expressed as an integral. The issue is that most integrals do not have closed form solutions. For instance, the integral

$$\int_a^b \sqrt{1+\cos^2(x)}\, dx ,$$

is a case where its computation is extremely hard if not impossible. In such cases we need automated methods to approximate integrals. In this chapter, we discuss numerical methods for the approximation of definite integrals of single-variable functions. Numerical integration is also known as *quadrature* and the approximate formulas for the integrals are called *quadrature rules.*

7.1 Introduction to Numerical Quadrature and Midpoint Rule

Quadrature is based on the approximation of an integral

$$I[f] = \int_a^b f(x)\, dx ,$$

by a finite sum I_N of the form

$$I_N[f] = \sum_{i=0}^{N} w_i f(z_i) .$$

The constants w_i are called *quadrature weights* while the z_i's are the *quadrature points* or *nodes.* Usually, a quadrature rule is specified by the points z_i and a particular set of weights w_i. Ideally, we would like the quadrature to converge to the exact integral as the number of quadrature points tends to infinity

$$\lim_{N\to\infty} I_N[f] = I[f] .$$

The roots of numerical integration can be found in the definition of definite integral: Given a continuous function $f(x)$ and a partition of the interval $[a,b]$ with $\{x_i\}_{i=0}^{N+1}$ such that $a = x_0 < x_1 < \cdots < x_{N+1} = b$ we choose $z_i \in [x_i, x_{i+1}]$ for $i = 0,1,\ldots,N$. Then, we define the Riemann integral as

$$\int_a^b f(x)\, dx = \lim_{N\to\infty} \sum_{i=0}^{N} (x_{i+1}-x_i) f(z_i) .$$

DOI: 10.1201/9781003287292-7

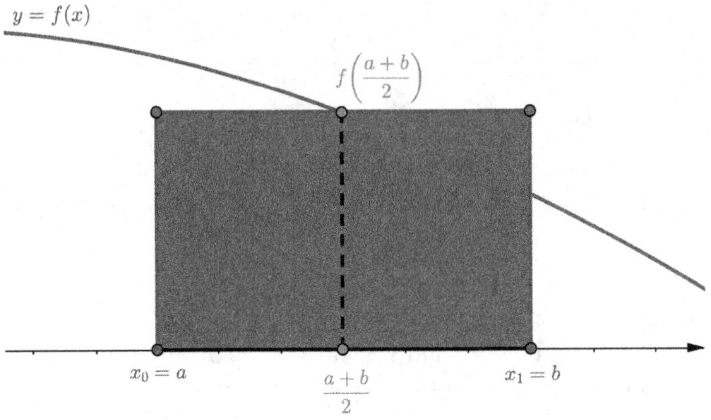

FIGURE 7.1
The midpoint rule.

Choosing a large value of N we approximate the integral with a finite sum such as

$$\int_a^b f(x)\, dx \approx \sum_{i=0}^{N} (x_{i+1} - x_i) f(z_i) .$$

This definition suggests that the weights are $w_i = x_{i+1} - x_i$ and the nodes z_i, and leads to a number of quadrature rules based on how we choose the values $z_i \in [x_i, x_{i+1}]$.

The simplest of all quadrature rules is the midpoint rule with one interval $[a, b] = [x_0, x_1]$. We take the point z_0 to be the midpoint $z_0 = \frac{a+b}{2}$. This quadrature rule approximates the area prescribed by the graph of f, the lines $x = a$, $x = b$ and the x-axis with the area of the parallelogram with sides z_0 and $x_1 - x_0$. This is depicted in Figure 7.1. The approximate formula of an integral $I[f]$ is given by

$$I[f] \approx I_0[f] = w_0 f(z_0) = (b - a) f\left(\frac{a+b}{2}\right) .$$

When we use only one interval, $N = 1$, the error between the exact value of the integral and its approximation can be significantly large, especially when the interval $[a, b]$ is large. For this reason, we divide the interval $[a, b]$ into N subintervals of the form $[x_i, x_{i+1}]$ for $i = 0, 1, \ldots, N$, such that $a = x_0 < x_1 < \cdots < x_{N+1} = b$. For the sake of simplicity we assume that the intervals $[x_i, x_{i+1}]$ have uniform length $h = x_{i+1} - x_i$. The integral $I[f]$ can be written as

$$I[f] = \int_a^b f(x)\, dx = \sum_{i=0}^{N} \int_{x_i}^{x_{i+1}} f(x)\, dx .$$

We now apply the midpoint rule to the smaller intervals $[x_i, x_{i+1}]$. We choose z_i to be the midpoint of the interval $[x_i, x_{i+1}]$, that is

$$z_i = \frac{x_i + x_{i+1}}{2} .$$

The application of midpoint rule in each of these intervals gives

$$\int_{x_i}^{x_{i+1}} f(x)\, dx \approx (x_{i+1} - x_i) f\left(\frac{x_i + x_{i+1}}{2}\right) = hf\left(\frac{x_i + x_{i+1}}{2}\right) ,$$

which leads to the composite midpoint rule

$$I_N[f] = \sum_{i=0}^{N}(x_{i+1} - x_i) f\left(\frac{x_i + x_{i+1}}{2}\right) = h \sum_{i=0}^{N} f\left(\frac{x_i + x_{i+1}}{2}\right) .$$

This suggests that given a subdivision $a = x_0 < x_1 < \cdots < x_{N+1} = b$ of $[a, b]$, the weights of the midpoint rule are $w_i = (x_{i+1} - x_i) = h$ and the quadrature nodes are $z_i = (x_i + x_{i+1})/2$.

> ☞ When we divide an interval of integration $[a, b]$ into subintervals $[x_i, x_{i+1}]$ and we apply a quadrature rule on each subinterval, the resulting quadrature rule is called *composite* rule.

In order to illustrate the use of the midpoint rule for the approximation of integrals we consider the integral

$$\int_0^{2\pi} \sin(x)\, dx = [-\cos(x)]_0^{2\pi} = 0 ,$$

and we apply the midpoint rule with 10 quadrature nodes in the following Python code:

```
1 f = lambda x: np.sin(x)
2 N = 10 # intervals
3 a = 0.0; b= 2.0*np.pi; h=(b-a)/N
4 x = np.linspace(a, b, N+1)
5 z = (x[1:] + x[:-1]) / 2.0
6 w = h*np.ones_like(z)
7 quad = np.inner(w, f(z))
8 print('Result =', quad)
```

```
Result = -2.7755575615628914e-16
```

Line 4 contains the definition of a partition of $[a, b]$, line 5 the definition of the quadrature nodes using the midpoints of each interval $[x_i, x_{i+1}]$, and line 6 the definition of the weights w_i. The operation $z_i = (x_{i+1} + x_i)/2$ could have been introduced in a `for` loop but we used the vectorized form `z=(x[1:]+x[:-1])/2.0` instead. An alternative way to vectorize this operation is by using the indices explicitly using the command `z=(x[1:N]+x[0:N-1])/2.0`. Note that the midpoint rule has been used widely in many situations due to its simplicity. We continue with two different classes of quadrature rules, namely the Newton-Cotes and the Gaussian quadrature rules. These are among the most sophisticated classes of quadrature and are the most popular in practice. These methods have been introduced by Sir Isaac Newton (1642–1726), Roger Cotes (1682–1716) and Johann Carl Friedrich Gauss (1777–1855), respectively (since the integrand is a polynomial, see Figure 7.2).

7.2 Newton-Cotes Quadrature Rules

Newton-Cotes rules are based on the approximation of the integrant $f(x)$ by an interpolating polynomial. The integral of f is then approximated by the integral of the interpolating polynomial, which we can compute exactly.

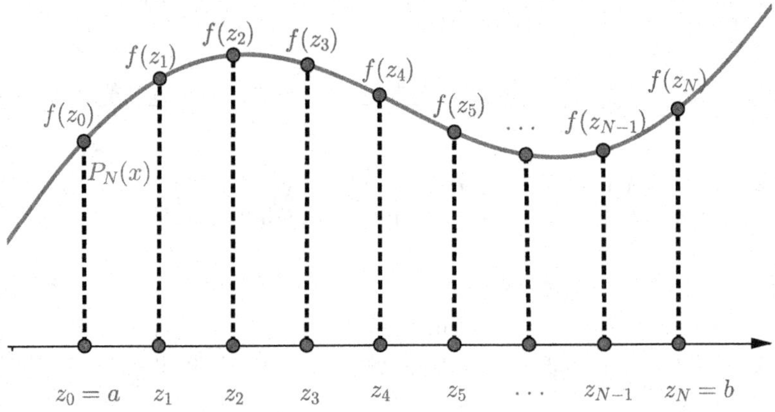

FIGURE 7.2
Illustration of the interpolating polynomial at the quadrature nodes z_i.

Using $N+1$ equally-spaced points $a = x_0 < x_1 < \ldots < x_N = b$, the quadrature nodes are defined as $z_i = x_i$ for $i = 0, 1, \ldots, N$. This situation is depicted in Figure 7.2. We evaluate $f(x)$ at these points and take the interpolating polynomial

$$P_N(x) = \sum_{i=0}^{N} f(z_i)\ell_i(x) \quad \text{where} \quad \ell_i(x) = \prod_{\substack{j=0 \\ j\neq i}}^{N} \frac{x - x_j}{x_i - x_j} .$$

The quadrature formula for $I_N[f]$, which approximates $I[f]$, is defined as

$$I_N[f] = \int_a^b P_N(x)\, dx = \int_a^b \sum_{i=0}^{N} f(z_i)\ell_i(x)\, dx = \sum_{i=0}^{N} \left[f(z_i) \int_a^b \ell_i(x)\, dx \right] = \sum_{i=0}^{N} w_i f(z_i) ,$$

where the weights are given as

$$w_i = \int_a^b \ell_i(x)\, dx, \quad i = 0, 1, \ldots, N .$$

The general Newton-Cotes quadrature rule has the form

$$I_N[f] = \sum_{i=0}^{N} w_i f(z_i) . \tag{7.1}$$

A classical example for numerical quadrature is the *trapezoidal rule*. In the trapezoidal rule only two quadrature nodes are needed. These are the endpoints of the interval $[a, b]$, $z_0 = x_0 = a$ and $z_1 = x_1 = b$, and the linear interpolant $P_1(x)$.

☞ Newton-Cotes quadrature rules are based on the idea of approximating an integral of a function $f(x)$ by the integral of an interpolating polynomial of $f(x)$.

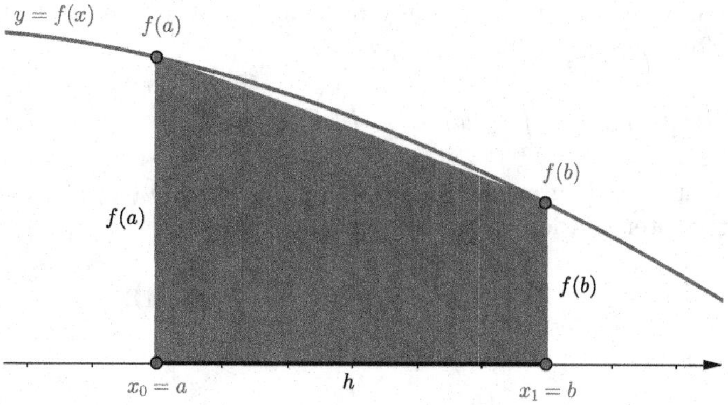

FIGURE 7.3
The trapezoidal rule.

7.2.1 Trapezoidal rule

The trapezoidal rule uses linear polynomials (i.e. $N = 1$) to approximate the function $f(x)$. For $N = 1$ we have

$$\ell_0(x) = \frac{x - x_1}{x_0 - x_1} = -\frac{x - b}{h} \quad \text{and} \quad \ell_1(x) = \frac{x - x_0}{x_0 - x_1} = \frac{x - a}{h} ,$$

where $h = b - a$ is the length of the interval $[a, b]$. In this case, we have

$$w_0 = \int_a^b \ell_0(x)\, dx = \frac{1}{h}\int_a^b (x - b)\, dx = \frac{1}{2h}(b - a)^2 = \frac{h}{2} ,$$

and

$$w_1 = \int_a^b \ell_1(x)\, dx = \frac{1}{h}\int_a^b (x - a)\, dx = \frac{1}{2h}(b - a)^2 = \frac{h}{2} .$$

Substitution of the weights w_i into the formula (7.1) yealds

$$I_1[f] = \frac{h}{2}\,[f(a) + f(b)] \ . \tag{7.2}$$

After approximating the function $f(x)$ with a linear interpolant, the integral is approximated by the area enclosed inside the trapezium formed by the linear interpolant, and the lines $y = 0$, $x = a$ and $x = b$ (see Figure 7.3). Recall that the area of a trapezium is given by the formula Area $= [(\text{large base}) + (\text{small base})] \times \frac{(\text{height})}{2}$. This is the same with formula (7.2).

> ☞ The trapezoidal rule requires only two quadrature nodes and approximates the actual integral by the area of the trapezium bounded by the straight line joining the two endpoints, the x-axis and the lines $x = a$ and $x = b$.

7.2.2 Error estimate of the trapezoidal rule

To find out the accuracy of the trapezoidal rule, we define the error between the exact and approximate values of an integral as

$$E_1[f] = I[f] - I_1[f] \ .$$

The subscript 1 indicates that we use only one interval to approximate the integral, *viz.* $[a,b]$. Using the formulas for $I[f]$ and $I_1[f]$ we have

$$E_1[f] = I[f] - I_1[f] = \int_a^b f(x)\,dx - \int_a^b P_1(x)\,dx = \int_a^b (f(x) - P_1(x))\,dx\;,$$

where P_1 is the first order Lagrange interpolating polynomial of $(a, f(a))$ and $(b, f(b))$. From the formula of the interpolation error (6.7) we have that

$$f(x) - P_1(x) = \frac{f''(\xi)}{2!}(x - x_0)(x - x_1)\;,$$

for some $\xi \in (x_0, x_1) = (a, b)$. Thus,

$$\begin{aligned} E_1[f] &= \frac{1}{2!}\int_a^b (x - x_0)(x - x_1)f''(\xi)\,dx = \frac{1}{2}f''(\xi)\int_a^b (x-a)(x-b)\,dx \\ &= -\frac{1}{12}(b-a)^3 f''(\xi)\;. \end{aligned} \tag{7.3}$$

Taking the absolute value of the error $E_1[f]$ we obtain

$$|E_1[f]| = \left|\frac{1}{12}(b-a)^3 f''(\xi)\right| = \frac{1}{12}(b-a)^3\,|f''(\xi)| \le Ch^3\;,$$

where $h = b - a$ and $C = \frac{1}{12}\max|f''(\xi)|$.

The shaded area in Figure 7.3 represents the area of the trapezium. We observe that when the length of the interval $[a,b]$ is large (for example if $h > 1$) then the interpolation error can be quite large since it is proportional to the length h^3.

☞ Integrating in large intervals $[a,b]$ with $h = b - a > 1$ requires composite rules for high accuracy. Composite rules rely on the use of a grid of points $a = x_0 < x_1 < \cdots < x_N = b$ to define a partition of $[a,b]$.

7.2.3 Composite trapezoidal rule

To increase the accuracy in computations we use the composite trapezoidal rule. Consider a uniform partition of $[a,b]$ with nodes $x_0 = a < x_1 < \cdots < x_N = b$. We integrate using the property

$$\int_a^b f(x)\,dx = \sum_{i=0}^{N-1}\int_{x_i}^{x_{i+1}} f(x)\,dx\;.$$

To derive the composite trapezoidal rule, we apply the trapezoidal rule in each interval $[x_i, x_{i+1}]$. In particular, we have

$$\int_{x_i}^{x_{i+1}} f(x)\,dx \approx \frac{h}{2}\,[f(x_i) + f(x_{i+1})]\;,$$

where $h = x_{i+1} - x_i$, for $i = 0, 1, 2, \ldots, N-1$. Hence, the composite trapezoidal rule can be written as

$$\begin{aligned} I_N[f] &= \sum_{i=0}^{N-1}\frac{h}{2}\,[f(x_i) + f(x_{i+1})] \\ &= \frac{h}{2}\,[f(x_0) + 2f(x_1) + 2f(x_2) + \cdots + 2f(x_{N-1}) + f(x_N)]\;. \end{aligned} \tag{7.4}$$

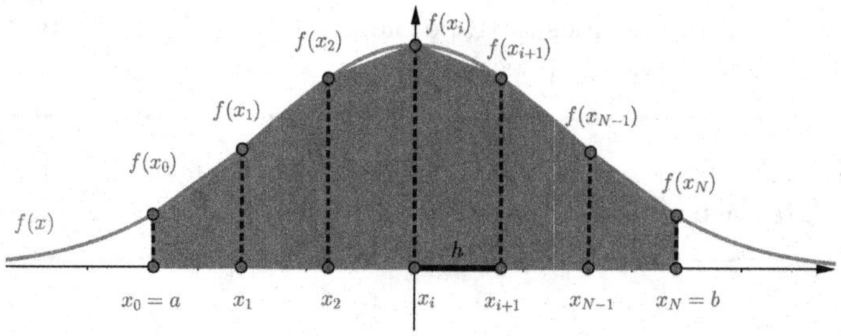

FIGURE 7.4
The composite trapezoidal rule.

Figure 7.4 shows the approximation of the integral with the composite trapezoidal rule. The composite trapezoidal rule can be viewed as the exact integration of the piecewise linear interpolant of $f(x)$ at the nodes x_i, $i = 0, 1, \ldots, N$.

Implementation of the composite trapezoidal rule in Python is easy as well. The following Python function computes an approximation of the integral of `f` from `a` to `b` using `N` subintervals.

```
1  def trapezoidal(f, a, b, N):
2      # Composite trapezoidal rule with N subintervals
3      x = np.linspace(a, b, N+1)
4      y = f(x)
5      h = (b - a)/N
6      # First add the internal nodes
7      sum = 0.0
8      for i in range(1,N):
9          sum += 2.0*y[i]
10     # Add the two boundary nodes
11     sum = 0.5*h*(f(a) + sum + f(b))
12     return sum
```

In the function `trapezoidal` we first define the $N+1$ quadrature nodes in the variable `x`. We compute the interpolation points `y=f(x)` and we define a stepsize `h`. Then we compute the sum over all the internal nodes in a `for` loop. The command of line 11 computes the result taking into account the contribution of the boundary nodes. It is left as an exercise for the reader to replace the `for` loop of the previous code using the NumPy function `sum`.

To test the function `trapezoidal` we estimate the integral

$$\int_0^{\pi/2} \sin(x)\, dx = 1 \ .$$

For testing purposes we use $N = 10$ subintervals.

```
f = lambda x: np.sin(x)
trapezoidal(f, 0.0, np.pi/2.0, 10)
```

```
0.9979429863543572
```

We observe that the result is correct up to two decimal digits or in other words the error is $O(10^{-2})$. This is because $h = 2\pi/10 \leq C \cdot 10^{-1}$ for some constant $C > 0$. On the other hand, the error of the composite trapezoidal rule is expected to be $O(h^2)$. The error is estimated analytically in the next section.

☞ The composite trapezoidal rule is more efficient than the standard trapezoidal rule in large intervals, and can be used even for general functions f.

7.2.4 Error estimate of the composite trapezoidal rule

Denote $h = x_{i+1} - x_i$. Then the truncation error $E_i[f]$ in each interval $[x_i, x_{i+1}]$ is given by the formula (7.3) as

$$E_i[f] = -\frac{h^3}{12} f''(\xi_i) ,$$

with $\xi_i \in (x_i, x_{i+1})$. The truncation error of a composite rule is the sum of the individual truncation errors of the simple rule applied in each interval $[x_i, x_{i+1}]$

$$E_N[f] = \sum_{i=0}^{N-1} E_i[f] = -\frac{h^3}{12} \sum_{i=0}^{N-1} f''(\xi_i) . \tag{7.5}$$

Applying the intermediate value theorem to the second derivative and the fact that

$$\min_{a\leq x\leq b} f''(x) \leq \frac{1}{N} \sum_{i=0}^{N-1} f''(\xi_i) \leq \max_{a\leq x\leq b} f''(x) ,$$

we conclude that there is a $\xi \in (a, b)$ such that

$$f''(\xi) = \frac{1}{N} \sum_{i=0}^{N-1} f''(\xi_i) .$$

Thus,

$$\sum_{i=0}^{N-1} f''(\xi_i) = N f''(\xi) = \frac{b-a}{h} f''(\xi) .$$

From (7.5) we obtain

$$E_N[f] = -\frac{(b-a)h^2}{12} f''(\xi) \leq Ch^2 ,$$

where $h = x_{i+1} - x_i$. This means that the error $E_N[f] \to 0$ as $N \to \infty$, and the error of the composite trapezoidal rule is $O(h^2)$. Thus, the numerical approximation converges to the exact value of the integral as $N \to \infty$,

$$\lim_{N\to\infty} I_N[f] = I[f] .$$

7.2.5 Simpson's rule

Simpson's rule (named after Thomas Simpson (1710–1761) and also known as the 1/3 rule) is a Newton-Cotes rule for the quadratic interpolation polynomial $P_2(x)$. So for Simpson's rule we have to use 3 nodes and $N = 2$.

We take $x_0 = a$, $x_1 = (a+b)/2$ and $x_2 = b$, and also $z_i = x_i$ for $i = 0,1,2$. The length of each interval $[x_i, x_{i+1}]$ is $h = (b-a)/2$. The quadratic Lagrange basis functions given by the formula (6.6) are

$$\ell_0(x) = \frac{(x-x_1)(x-x_2)}{(x_0-x_1)(x_0-x_2)}, \quad \ell_1(x) = \frac{(x-x_0)(x-x_2)}{(x_1-x_0)(x_1-x_2)}, \quad \ell_2(x) = \frac{(x-x_0)(x-x_1)}{(x_2-x_0)(x_2-x_1)} .$$

The integration of these functions is easier if we introduce a variable $\xi = x - x_1$ with origin at x_1. In the new variable, the coordinates of the nodes are $\xi_0 = -h$, $\xi_1 = 0$ and $\xi_2 = h$. The quadrature weights become

$$w_i = \int_a^b \ell_i(x)\, dx = \int_{-h}^{h} \ell_i(\xi)\, d\xi, \quad \text{for} \quad i = 0,1,2 .$$

Specifically, we have

$$w_0 = \int_{-h}^{h} \frac{(\xi-0)(\xi-h)}{(-h)(-2h)} d\xi = \frac{1}{2h^2}\int_{-h}^{h} (\xi^2 - h\xi) d\xi = \frac{h}{3} ,$$

$$w_1 = \int_{-h}^{h} \frac{(\xi+h)(\xi-h)}{(h)(-h)} d\xi = -\frac{1}{h^2}\int_{-h}^{h} (\xi^2 - h^2) d\xi = \frac{4h}{3} ,$$

and

$$w_2 = \int_{-h}^{h} \frac{(\xi+h)(\xi-0)}{(2h)(h)} d\xi = \frac{1}{2h^2}\int_{-h}^{h} (\xi^2 + h\xi) d\xi = \frac{h}{3} .$$

Thus, Simpson's rule for the approximation of an integral is

$$I_2[f] = \sum_{i=0}^{2} w_i f(z_i) = \frac{h}{3}\left[f(a) + 4f\left(\frac{a+b}{2}\right) + f(b)\right] .$$

Similarly to the trapezoidal rule, Simpson's rule can be inaccurate if the length h of the interval $[a,b]$ is large. For this reason, we consider the composite Simpson's rule by dividing the interval $[a,b]$ into smaller intervals and applying Simpson's rule in each of the subintervals.

7.2.6 Composite Simpson's rule

Let $N = 2m$ be an even number. We define $h = (b-a)/N$ and $x_i = a + ih$ for $i = 0,1,\ldots,N$. Because Simpson's rule has three nodes, we consider the intervals $[x_{2k}, x_{2k+2}]$ for $k = 0,1,\ldots,\frac{N}{2}-1$. Then the integral $I[f]$ can be written as

$$\int_a^b f(x)\, dx = \sum_{k=0}^{N/2-1} \int_{x_{2k}}^{x_{2k+2}} f(x)\, dx .$$

We apply Simpson's rule in the intervals $[x_0, x_2]$, $[x_2, x_4]$, $\ldots$, $[x_{N-2}, x_N]$ to obtain

$$\int_{x_{2k}}^{x_{2k+2}} f(x)\, dx \approx \frac{h}{3}\left[f(x_{2k}) + 4f(x_{2k+1}) + f(x_{2k+2})\right] .$$

The composite Simpson's rule is given by the sum

$$\begin{aligned} I_N[f] &= \sum_{k=0}^{N/2-1} \frac{h}{3}\left[f(x_{2k}) + 4f(x_{2k+1}) + f(x_{2k+2})\right] \\ &= \frac{h}{3}\left[f(x_0) + 4f(x_1) + 2f(x_2) + 4f(x_3) + \cdots + 2f(x_{N-2}) + 4f(x_{N-1}) + f(x_N)\right] . \end{aligned}$$

> ☞ Note that the composite Simpson's rule requires $N = 2m$ to be an even number.

The implementation of Simpson's rule in Python is very similar to the implementation of the trapezoidal rule. The difference is that the values of the integrant evaluated at the internal quadrature nodes x_i for $i = 1, 2, \ldots N-1$ are multiplied by 2 when i is odd and by 4 when i is even. For this reason, an implementation of the composite Simpson rule could compute the two sums in two different `for` loops. In our implementation we avoid the use of `for` loops and we just use a vectorized algorithm. In particular we store all the values $f(x_i)$ in a vector `s`. We then multiply first all the entries by 2 except for the first and the last entries (line 6). Then we multiply all the even entries again by 2 so as to form the coefficients 4 (line 7). Finally we compute the sum and we multiply by $h/3$ in the last command of the code.

```
1  def Simpson(f, a, b, N):
2  # Composite Simpson rule with N subintervals
3      h = (b - a)/float(N)
4      x=np.linspace(a, b, N + 1)
5      s = f(x)
6      s[1:N]=2*s[1:N]
7      s[1:N:2]=2*s[1:N:2]
8      return h/3.0*np.sum(s)
```

Testing the code for the same integral as in the case of the trapezoidal rule

$$\int_0^{\pi/2} \sin(x)\, dx = 1 ,$$

using $N = 10$ intervals we observe that Simpson's method is more accurate compared to the trapezoidal rule.

```
f = lambda x: np.sin(x)
Simpson(f, 0.0, np.pi/2.0, 10)
```

```
1.0000033922209004
```

The result is correct with 5 decimal digits indicating that the composite Simpson rule is accurate with error of order h^4.

7.2.7 Error of Simpson's rule

Similar to the trapezoidal rule error, the error of Simpson's rule will be

$$E_2[f] = \int_a^b [f(x) - P_2(x)]\, dx ,$$

where P_2 is the unique interpolating polynomial of the function $f(x)$ at the nodes $x_0 = a$, x_1 and $x_2 = b$. Thus,

$$E_2[f] = -\frac{1}{24}\int_a^b (x-a)\left(x - \frac{a+b}{2}\right)^2 (b-x)\, f^{(4)}(\xi(x))\, dx ,$$

where due to the continuity of the function $f^{(4)}(x)$ we have

$$E_2[f] = -\frac{1}{24} f^{(4)}(\xi_1) \int_a^b (x-a)\left(x - \frac{a+b}{2}\right)^2 (b-x)\, dx ,$$

with $\xi_1 \in (a,b)$. Thus, if $f \in C^4[a,b]$, there is a $\xi \in (a,b)$ such that

$$E_2[f] = -\frac{(b-a)^5}{2^4 \cdot 180} f^{(4)}(\xi) .$$

Working similarly for the composite Simpson's rule we obtain

$$\begin{aligned}
E_N[f] &= \int_a^b f(x)dx - I_N[f] \\
&= -\frac{1}{2^4 \cdot 180}\left\{(x_2 - x_0)^5 f^{(4)}(\xi_0) + (x_4 - x_2)^5 f^{(4)}(\xi_1) + \cdots + (x_N - x_{N-2})^5 f^{(4)}(\xi_{N/2-1})\right\} \\
&= -\frac{1}{2^4 \cdot 180}(2h)^5 \sum_{k=0}^{N/2-1} f^{(4)}(\xi_i) = -\frac{2^4 h^4}{2^4 \cdot 180}(2h) \sum_{k=0}^{N/2-1} f^{(4)}(\xi_i) \\
&= -\frac{h^4}{180}(Nh)\frac{2}{N} \sum_{k=0}^{N/2-1} f^{(4)}(\xi_i) = -\frac{h^4}{180}(b-a) f^{(4)}(\xi) ,
\end{aligned}$$

with $\xi_k \in (x_{2k}, x_{2k+2})$, for $k = 0, 1, \ldots, N/2-1$. The existence of $\xi \in (a,b)$ for which

$$f^{(4)}(\xi) = \frac{2}{N} \sum_{k=0}^{N/2-1} f^{(4)}(\xi_i) ,$$

is guaranteed by the intermediate value theorem and the assumption $f \in C^4[a,b]$. Therefore, for all $f \in C^4[a,b]$ there is a $\xi \in (a,b)$ such that

$$E_N[f] = -\frac{b-a}{180} h^4 f^{(4)}(\xi) .$$

☞ Simpson's rule is fifth order accurate while its composite rule is fourth order. Interestingly we have gained two orders of accuracy from Trapezoidal rule by increasing the degree of the polynomial only by one degree!

7.2.8 Degree of exactness

The degree of exactness of a quadrature rule $I_N[f]$ is the largest degree of polynomial for which the quadrature rule can integrate without making any error. In other words the degree of exactness is the integer n such that the quadrature error $E[x^k] = 0$ for all $k = 0, 1, \ldots, n$ and $E[x^{n+1}] \neq 0$. Since the integral is a linear operator we conclude that a quadrature rule of degree of exactness n can integrate exactly (up to the machine precision) polynomials of maximum degree n.

Obviously the trapezoidal rule has degree of exactness 1. Although Simpson's rule was designed to be exact for polynomials of degree 2, it happens to be exact also for polynomials of degree 3. In order to see this, we compute the errors

$$E[f] = \int_a^b f(x)\, dx - \frac{b-a}{6}\left(f(a) + 4f\left(\frac{a+b}{2}\right) + f(b)\right) ,$$

for $f(x) = 1, x, x^2$ and x^3:

$$
\begin{aligned}
E[1] &= (b-a) - \frac{b-a}{6}(1+4+1) = 0 \,, \\
E[x] &= \frac{1}{2}[x^2]_a^b - \frac{b-a}{6}\left(a + 4\frac{a+b}{2} + b\right) = \frac{1}{2}(b^2-a^2) - \frac{(b-a)(b+a)}{2} = 0 \,, \\
E[x^2] &= \frac{1}{3}[x^3]_a^b - \frac{b-a}{6}\left(a^2 + 4\frac{(a+b)^2}{4} + b^2\right) \\
&= \frac{1}{3}(b^3-a^3) - \frac{b-a}{3}(a^2+ab+b^2) \\
&= \frac{1}{3}\left(b^3 - a^3 - (a^2b + ab^2 + b^3 - a^3 - a^2b - ab^2)\right) \\
&= 0 \,, \\
E[x^3] &= \frac{1}{4}[x^4]_a^b - \frac{b-a}{6}\left(a^3 + 4\frac{(a+b)^3}{8} + b^3\right) \\
&= \frac{1}{4}(b^4-a^4) - \frac{b-a}{6}\left(a^3 + \frac{a^3 + 3a^3b + 3ab^2 + b^3}{2} + b^3\right) \\
&= \frac{1}{4}(b^4-a^4) - \frac{b-a}{6}\cdot\frac{3}{2}(a^3 + a^2b + ab^2 + b^3) \\
&= 0 \,.
\end{aligned}
$$

The conclusion is implied by the fact that $E[1] = E[x] = E[x^2] = E[x^3] = 0$ while $E[x^4] = (a-b)^5/120 \neq 0$. Thus, the degree of exactness for Simpson's rule is 3.

☞ The knowledge of the degree of exactness of a quadrature rule helps in choosing appropriate quadrature for the exact computation of integrals involving polynomials.

7.3 Gaussian Quadrature Rules

Various Newton-Cotes formulas have been derived by integrating interpolating polynomials. The error term in the interpolating polynomial of degree N involves the $(N+1)$-th order derivative of the function being approximated. So simple Newton-Cotes formulas with $N+1$ nodes are expected to be exact for polynomials of degree at least N.

In this section we study Gaussian quadrature rules. Gaussian rules have been designed to be simple and integrate exactly polynomials of degree greater than N. The key ingredient in Gaussian quadrature is the choice of the quadrature nodes, which are not equally spaced and do not include the end-points of the integration interval. In the approximation of an integral using the general quadrature formula

$$
\int_a^b f(x)\,dx \approx \sum_{i=0}^{N} w_i f(z_i) \,, \tag{7.6}
$$

the nodes $z_0, z_1, \ldots, z_N$ in the interval $[a,b]$ and coefficients $w_0, w_1, \ldots, w_N$, need to be chosen appropriately. This means that a quadrature rule can be defined by determining these $2(N+1)$ parameters. The coefficients $w_0, w_1, \ldots, w_N$ in the approximation formula

(7.6) can take any value, and the nodes $z_0, z_1, \ldots, z_N$ are restricted only by the fact that they must lie in the interval of integration $[a, b]$. We will determine all these parameters such that the resulting method integrates exactly polynomials of degree $2N+1$. The space of polynomials of degree $2N+1$ can be generated by the $2(N+1)$ monomials x^i, for $i = 0, 1, \ldots, 2N+1$, and this fact will help us to determine the $2(N+1)$ parameters of the quadrature.

7.3.1 Choice of nodes and weights

For simplicity and convenience we will consider the interval $[a, b] = [-1, 1]$. Suppose that we want to approximate the integral

$$\int_{-1}^{1} f(x)\, dx\ ,$$

using formula (7.6) for $N = 1$. In this case, we need to determine the weights w_0, w_1, and the nodes z_0, z_1 so that the integration formula $w_0 f(z_0) + w_1 f(z_1)$ is exact whenever $f(x)$ is a polynomial of degree $2N + 1 = 2 \cdot 1 + 1 = 3$ or less. This means that we need

$$\int_{-1}^{1} f(x)\, dx = w_0 f(z_0) + w_1 f(z_1)\ , \tag{7.7}$$

for polynomial functions

$$f(x) = a_0 + a_1 x + a_2 x^2 + a_3 x^3\ ,$$

for some constants a_0, a_1, a_2 and a_3. Due to the linearity of the integral we have

$$\int \left(a_0 + a_1 x + a_2 x^2 + a_3 x^3\right)\, dx = a_0 \int 1\, dx + a_1 \int x\, dx + a_2 \int x^2\, dx + a_3 \int x^3\, dx\ .$$

Thus, it suffices to show that the quadrature rule is exact when $f(x)$ is 1, x, x^2 and x^3. Then the formula will be exact for any linear combination of these monomials. Applying the quadrature formula (7.7) to these monomials we observe that the parameters w_0, w_1, z_0, and z_1 satisfy the equations

$$\begin{aligned}
w_0 \cdot 1 + w_1 \cdot 1 &= \int_{-1}^{1} dx = 2\ ,\\
w_0 \cdot z_0 + w_1 \cdot z_1 &= \int_{-1}^{1} x\, dx = 0\ ,\\
w_0 \cdot z_0^2 + w_1 \cdot z_1^2 &= \int_{-1}^{1} x^2\, dx = \frac{2}{3}\ ,\\
w_0 \cdot z_0^3 + w_1 \cdot z_1^3 &= \int_{-1}^{1} x^3\, dx = 0\ .
\end{aligned}$$

Solving the above system of four equations with four unknowns, we find the unique solution

$$w_0 = 1, \quad w_1 = 1, \quad z_0 = -\frac{\sqrt{3}}{3}, \quad \text{and} \quad z_1 = \frac{\sqrt{3}}{3}\ .$$

Substituting into the integral formula (7.7), we obtain the quadrature formula

$$\int_{-1}^{1} f(x)\, dx \approx f\left(-\frac{\sqrt{3}}{3}\right) + f\left(\frac{\sqrt{3}}{3}\right)\ .$$

By construction, this quadrature rule is exact for polynomials of degree at most 3 and thus it has degree of exactness 3.

The question now is whether there is a generic way to determine the nodes and weights of the Gaussian quadrature for any value of N. In fact, there are many ways to determine the parameters of Gaussian quadrature. In the next section we present one such method based on the Legendre polynomials. Because of its construction this quadrature rule is known to as the Gauss-Legendre quadrature.

7.3.2 Gauss-Legendre quadrature rules

In this section, we present how to find simple quadrature formulas that integrate exactly polynomials of degree $2N+1$. In particular, we consider a special class of methods based on the Legendre polynomials. Legendre polynomials are named after Adrian-Marie Legendre (1752–1833) and are solutions to Legendre's differential equation

$$\frac{d}{dx}\left[(1-x^2)\frac{d}{dx}P_n(x)\right]+n(n+1)P_n(x)=0 ,$$

and they form a basis for the set of polynomials of degree $2N+1$

$$\{P_0(x), P_1(x), \ldots, P_n(x), \ldots\} .$$

Legendre polynomials can also be defined recursively by the formula

$$(n+1)P_{n+1}(x)=(2n+1)xP_n(x)-nP_{n-1}(x) \quad \text{for } n=1,2,\ldots ,$$

with $P_0(x)=1$ and $P_1(x)=x$. Their most important property is that they are "orthogonal" in the sense that

$$\int_{-1}^{1} P_m(x)P_n(x)\,dx=\frac{2}{2n+1}\delta_{mn} ,$$

where

$$\delta_{mn}=\begin{cases}1, & \text{if } m=n\\ 0, & \text{if } m\neq n\end{cases} ,$$

denotes the Kronecker delta.

The first five normalized Legendre polynomials are

$$P_0(x)=1, \quad P_1(x)=x, \quad P_2(x)=x^2-\frac{1}{3} ,$$

$$P_3(x)=x^3-\frac{3}{5}x, \quad \text{and} \quad P_4(x)=x^4-\frac{6}{7}x^2+\frac{3}{35} .$$

The roots of these polynomials are distinct and lie in the interval $(-1,1)$. Moreover, they are symmetric around the origin. In our case, the roots of the $(N+1)$-th degree Legendre polynomial are taken to be the nodes $z_0, z_1, \ldots, z_N$ and these produce the quadrature formula that is exact for any polynomial of degree at most $2N+1$.

If $z_0, z_1, \ldots, z_N$ are the roots of the $(N+1)$-th Legendre polynomial $P_{N+1}(x)$, then the weights w_i for $i=0,1,\ldots,N$ are defined by

$$w_i=\int_{-1}^{1}\prod_{\substack{j=0\\ j\neq i}}^{N}\frac{x-z_j}{z_i-z_j}\,dx . \tag{7.8}$$

We first show that using these nodes and weights for the Gaussian quadrature the formula

$$\sum_{i=0}^{N} w_i f(z_i) ,$$

is exact for polynomials of degree $2N+1$. This fact is summarized in the following theorem:

Theorem 7.1. *For any polynomial $P(x)$ of degree at most $2N+1$, we have*

$$\int_{-1}^{1} P(x)\, dx = \sum_{i=0}^{N} w_i P(z_i) ,$$

where z_i are the roots of the $(N+1)$-th Legendre polynomial and the weights w_i are given by (7.8).

Proof. Initially we prove this result for polynomials $P(x)$ of degree less than or equal to N. The Lagrange polynomial that interpolates the polynomial $P(x)$ at the nodes z_i, $i = 0, 1, \ldots, N$ is

$$P(x) = \sum_{i=0}^{N} P(z_i)\ell_i(x) = \sum_{i=0}^{N} \prod_{\substack{j=0\\ j\neq i}}^{N} \frac{x - z_j}{z_i - z_j} P(z_i) .$$

This representation of the polynomial $P(x)$ using Lagrange polynomial basis functions is exact due to the uniqueness of the interpolating polynomial. Integrating the last formula yields

$$\int_{-1}^{1} P(x)\, dx = \int_{-1}^{1} \left[\sum_{i=0}^{N} \prod_{\substack{j=0\\ j\neq i}}^{N} \frac{x - z_j}{z_i - z_j} P(z_i) \right] dx = \sum_{i=0}^{N} \left[\int_{-1}^{1} \prod_{\substack{j=1\\ j\neq i}}^{N} \frac{x - z_j}{z_i - z_j}\, dx \right] P(z_i)$$
$$= \sum_{i=0}^{N} w_i P(z_i) .$$

Hence the quadrature is exact for polynomials of degree less than or equal to N.

For polynomials $P(x)$ of degree at least $N+1$ but less than or equal to $2N+1$, the long division of $P(x)$ by the $(N+1)$-th Legendre polynomial $P_{N+1}(x)$ will have quotient $Q(x)$ and residual $R(x)$ of degree less than or equal to N. This long division can be written as

$$P(x) = Q(x)P_{N+1}(x) + R(x) ,$$

and implies that if z_i is a root of $P_{N+1}(x)$, then $P(z_i) = R(z_i)$. Since the degree of the polynomial $Q(x)$ is less than $N+1$, then the orthogonality property of the Legendre polynomials implies

$$\int_{-1}^{1} Q(x)P_{N+1}(x)\, dx = 0 ,$$

which means that the integral of $P(x)$ depends only on the residual $R(x)$. On the other hand, since $R(x)$ is a polynomial of degree less than or equal to N, we have

$$\int_{-1}^{1} R(x)\, dx = \sum_{i=0}^{N} w_i R(z_i) .$$

TABLE 7.1
Gaussian nodes and weights for $N = 0, 1, 2, 3, 4$

N	z_i	w_i
0	0	2
1	$\pm\sqrt{\frac{1}{3}}$	1
2	$\pm\sqrt{\frac{3}{5}}$	$\frac{5}{9}$
	0	$\frac{8}{9}$
3	$\pm\sqrt{\frac{3}{7}-\frac{2}{7}\sqrt{\frac{6}{5}}}$	$\frac{18+\sqrt{30}}{36}$
	$\pm\sqrt{\frac{3}{7}+\frac{2}{7}\sqrt{\frac{6}{5}}}$	$\frac{18-\sqrt{30}}{36}$
4	0	$\frac{128}{225}$
	$\pm\frac{1}{3}\sqrt{5-2\sqrt{\frac{10}{7}}}$	$\frac{322+13\sqrt{70}}{900}$
	$\pm\frac{1}{3}\sqrt{5+2\sqrt{\frac{10}{7}}}$	$\frac{322-13\sqrt{70}}{900}$

So we conclude that

$$\int_{-1}^{1} P(x)\, dx = \int_{-1}^{1} [Q(x)P_{N+1}(x) + R(x)]\, dx = \int_{-1}^{1} R(x)\, dx = \sum_{i=0}^{N} w_i R(z_i)$$
$$= \sum_{i=0}^{N} w_i P(z_i)\ ,$$

which shows that the Gauss-Legendre quadrature is exact for polynomials of degree at most $2N + 1$. □

In order to implement the Gauss-Legendre quadrature we need to know the roots of the Legendre polynomials and the respective weights. These nodes and weights are not easy to be determined analytically. Table 7.1 presents the nodes and weights of Gauss-Legendre quadrature for $N = 1, 2, 3$ and 4. For large values of N we usually employ numerical methods for their computation.

7.3.3 Gaussian quadrature on general intervals

While we developed the Gaussian quadrature in the interval $[-1, 1]$, it is usually required to approximate integrals over general intervals $[a, b]$. This can be done by changing the variable of integration x. Specifically, the general integral

$$\int_a^b f(x)\, dx\ ,$$

can be transformed into an integral over $[-1, 1]$ by utilizing the change of variables

$$t = \frac{2x - a - b}{b - a} \quad \text{or equivalently} \quad x = \frac{b-a}{2}t + \frac{a+b}{2}\ .$$

This change of variable transforms the original integral into

$$\int_a^b f(x)\,dx = \int_{-1}^1 f\left(\frac{b-a}{2}t + \frac{a+b}{2}\right)\cdot\frac{b-a}{2}\,dt\ .$$

Approximating the last integral using the Gaussian quadrature gives

$$\int_a^b f(x)\,dx \approx \frac{b-a}{2}\sum_{i=0}^N w_i f\left(\frac{b-a}{2}t_i + \frac{a+b}{2}\right)\ . \tag{7.9}$$

☞ To derive composite Gaussian quadrature formulas we divide the interval $[a,b]$ into sub-intervals $[x_i, x_{i+1}]$ and we apply Gaussian quadrature in each subinterval.

It is worth mentioning that the midpoint rule can be derived as the Gauss-Legendre rule with $N=0$. This is a direct consequence of (7.9) for $N=0$, and with values $w_0=2$ and $x_0=0$ from the Table 7.1.

An implementation of a Gauss-Legendre quadrature rule with $N=2$ can be the following:

```
def GaussLegendre3(f, a, b):
    # Implementation of Gauss-Legendre quadrature with
    # 3 nodes in general interval [a,b]
    N = 2
    # define quadrature weights and nodes
    w = np.array([5.0/9.0, 8.0/9.0, 5.0/9.0])
    z = np.array([-np.sqrt(3.0/5.0), 0.0, np.sqrt(3.0/5.0)])
    # implement formula (7.9)
    c1 = (b-a)/2.0
    c2 = (a+b)/2.0
    s = c1*np.inner(w, f( c1*z + c2 ))
    return s
```

In order to test this function we consider the approximation of the integral

$$\int_{-1}^1 e^x \cos x\,dx\ ,$$

using Gauss-Legendre quadrature with $N=3$. Doing the computations analytically we have

$$\begin{aligned}\int_{-1}^1 e^x\cos x\,dx &\approx 0.\bar{5}e^{0.774596692}\cos(0.774596692) + 0.\bar{8}\ \cos 0\\ &+ 0.\bar{5}e^{-0.774596692}\cos(-0.774596692)\\ &= 1.9333904\ .\end{aligned}$$

One can evaluate the exact value of the previous integral using integration by parts. In particular, its exact value is $[(1+e^2)\sin(1)-(1-e^2)\cos(1)]/2e \approx 1.9334214$. So the absolute error of the Gauss-Legendre quadrature in this case is about 3.2×10^{-5}. Using our function we get the following result:

```
f = lambda x: 0.5*np.exp(x/2.0)*np.cos(x/2.0)
GaussLegendre3(f, -2.0, 2.0)
```

```
1.9333904692642978
```

This result agrees with our theoretical computation. If we need to use more nodes and especially if $N \geq 4$, then we need to approximate the quadrature nodes and weights using appropriate computational techniques. For example, the quadrature nodes can be approximated using Newton's method applied to the Legendre polynomials, while the computation of the quadrature weights can be implemented using the composite trapezoidal or Simpson rule.

Approximations of the Gauss-Legendre quadrature nodes and weights can be obtained in Python using the function `leggauss` of the module `numpy.polynomial.lagendre`. The simple call `leggauss(degree)` will return two one-dimensional arrays `x` and `y` containing the nodes and weights for the Gauss-Legendre quadrature with N points, respectively. The parameter `degree` should be equal to $N + 1$.

7.3.4 Error analysis of the Gaussian quadrature

The orthogonality property of Legendre polynomials was found to be very useful in the construction of a Gaussian quadrature rule. Here we consider general orthogonal polynomials of degree at most $N + 1$ (instead of the Legendre polynomials) and a Gaussian quadrature rule with $N + 1$ nodes. In what follows, we estimate the error between the actual integral and the result of the quadrature. In general, we can derive Gaussian quadrature rules using any kind of orthogonal polynomials, and thus the proof here applies to every Gaussian quadrature rule.

Theorem 7.2. *Consider a function $f \in C^{2N+2}[a, b]$ and the quadrature nodes $z_0, z_1, \ldots, z_N$ in $[a, b]$. Then for the Gaussian quadrature rule*

$$I_N[f] = \sum_{i=0}^{N} w_i f(z_i) \ ,$$

the following error estimate holds true:

$$I[f] - I_N[f] = \frac{f^{(2N+2)}(\zeta)}{(2N+2)!} \int_a^b Q^2(x)\, dx \ ,$$

where

$$Q(x) = \prod_{i=0}^{N} (x - z_i) \ ,$$

and $\zeta \in (a, b)$.

Proof. In this proof we will make use of the Hermite interpolation polynomial H_{2N+1}. From Theorem 6.4 we know that the error between $f(x)$ and its Hermite interpolant $H_{2N+1}(x)$ at the nodes $z_0, z_1, \ldots, z_N$ is

$$f(x) - H_{2N+1}(x) = \frac{f^{(2N+2)}(\xi(x))}{(2N+2)!} Q^2(x) \ .$$

Since the Gaussian quadrature with $N+1$ nodes is exact for polynomials of degree at most $2N+1$, then

$$\int_a^b H_{2N+1}\, dx = I_N[H_{2N+1}] = I_N[f]\ .$$

The last equality is a consequence of the fact the $f(z_i) = H_{2N+1}(z_i)$ since H_{2N+1} interpolates f at z_i.

Integrating the error formula between f and H_{2N+1} we have that

$$\int_a^b f(x)\, dx - I_N[f] = \int_a^b [f(x) - H_{2N+1}]\, dx = \frac{f^{(2N+2)}(\zeta)}{(2N+2)!}\int_a^b Q^2(x)\, dx\ .$$

The last equality follows from the mean value theorem of integral calculus (Theorem 4.6), and the proof is complete. □

This theorem ensures the convergence of the Gaussian quadrature to the correct integral, in the sense that

$$\lim_{N\to\infty} I_N[f] = \int_a^b f(x)\, dx\ .$$

It also indicates that the error in the Gaussian quadrature is proportional to $[(2N+2)!]^{-1}$, which can be remarkably small.

Remark 7.3. *All the analysis and the derivations of this chapter can be generalized to the case of weighted integrals of the form*

$$\int_a^b \omega(x) f(x)\, dx\ ,$$

which can be found in various applications.

7.4 The Module `scipy.integrate`

Each of the quadrature rules we have discussed so far is implemented in the module `scipy.integrate`.

7.4.1 Trapezoidal rule

The composite trapezoidal rule is implemented in the module `scipy.integrate` with the function `trapezoid`. The general call of the function `trapezoid` is the following:

```
trapezoid(y, x, dx, axis)
```

where

y: An array with the values $y_i = f(x_i)$, $i = 0, \ldots, N$

x: An array with the nodes x_i, $i = 0, \ldots, N$. (Optional with default value `None`)

dx: Spacing of quadrature nodes along axis of `y` only when `x=None`. (Optional with default value `1.0`)

`axis`: Axis along which to integrate. (Optional with default value `-1`)

For example, we show here the use of the function **trapezoid** for the integral

$$\int_0^{\pi/2} \sin(x)\; dx \; .$$

After we create a grid of $N+1$ points x_i and the values $f(x_i)$ we call the function **trapezoid** in its simplest form.

```
from scipy.integrate import trapezoid
f = lambda x: np.sin(x)
x = np.linspace(0.0, np.pi/2.0, 11)
y = f(x)
print( trapezoid(y,x) )
```

```
0.9979429863543573
```

Observe that the command **linspace** generates 11 quadrature nodes which means that $N = 10$, and the result is identical with our implementation.

7.4.2 Simpson's rule

Composite Simpson's rule is implemented in the module `scipy.integrate` with the function **simpson**. This function's general call is almost the same with that of the function **trapezoid** with the difference that in this case, we need to specify how to treat the case, where N is odd number. The general call of the function **simpson** is the following:

```
simpson(y, x, dx, axis, even)
```

where

y: An array with the values $y_i = f(x_i)$, $i = 0, \ldots, N$

x: An array with the nodes x_i, $i = 0, \ldots, N$. (Optional with default value `None`)

dx: Spacing of quadrature nodes along axis of `y` only when `x=None`. (Optional with default value `1`)

axis: Axis along which to integrate. (Optional with default value `-1`)

even: A string that specifies how to treat the case if N is not even. It can take the value `'first'` to use Simpson's rule for the first $N-2$ intervals and the trapezoidal rule on the last interval, and `'last'` so as to use trapezoidal rule on the first interval and Simpson's rule for the rest of the intervals, and `avg` to return the average of the previous two case. (Optional with default value `'avg'`)

For example, we present here the use of the function **simpson** using the same integral as before

$$\int_0^{\pi/2} \sin(x)\; dx \; .$$

After we create a grid of odd number of points x_i (N even) and the values $f(x_i)$, we call the function **simpson** in its simplest form.

```python
from scipy.integrate import simpson
f = lambda x: np.sin(x)
x = np.linspace(0.0, np.pi/2.0, 11)
y = f(x)
print( simpson(y,x) )
```

```
1.0000033922209006
```

The result is identical with our implementation, and the accuracy is better than the accuracy of the trapezoidal rule with the same number of quadrature nodes.

7.4.3 Gaussian quadrature

A Gaussian quadrature rule is implemented in SciPy in the function `fixed_quad` of the module `scipy.integrate`. The general call of the function `fixed_quad` is

```python
fixed_quad(func,a,b,args,n)
```

where `func` is the integrant, `a` and `b` are the integration limits, `args` is an optional argument with default value `None` containing additional arguments needed in the function `func`, if any. Finally, the argument `n` is again optional with default value 5 and is the order of the quadrature rule (i.e. `n`= $N+1$). The previous example can be implemented using the function `fixed_quad` in the following way:

```python
from scipy.integrate import fixed_quad
f = lambda x: 0.5*np.exp(x/2.0)*np.cos(x/2.0)
print( fixed_quad(f,-2.0,2.0,n=3) )
```

```
(1.9333904692642971, None)
```

We observe that the function returns the approximate value of the integral (which agrees with our previous computations) and also a second result (`None`) which consists of statistical information if there are any.

SciPy has also general purpose routines that can control the error, such as the functions `quads` and `quadrature`. The function `quadrature` is an adaptive Gaussian quadrature rule. The basic call of the function `quadrature` is

```python
quadrature(func,a,b,args,tol,rtol,maxiter,vec_func,miniter)
```

where

`func`: Is the name of the function $f(x)$

`a` and `b`: The end-points of the interval $[a,b]$

`args`: Additional arguments and parameters for the functions f. (Optional with default value `args=()`)

`xtol` and `rtol`: The absolute and relevant error tolerances for the termination of the method. (Options with default value for both tolerances `1.49e − 8`)

`maxiter`: The maximum order of Gaussian quadrature. (Optional with default value `maxiter=50`)

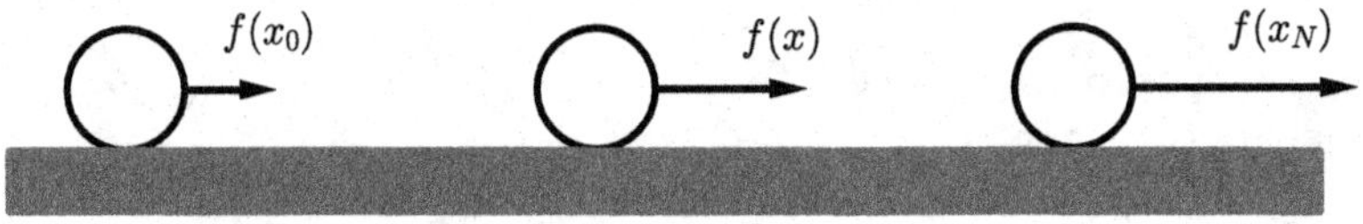

FIGURE 7.5
Object moves under the influence of the force $f(x)$.

`vec_funct`: Boolean variable which is `True` or `False` if `func` is a multivariable function. (Optional with default value `vec_func=True`)

`miniter`: Is the minimum order of the Gaussian quadrature. (Optional with default value is `miniter`)

The function `quadrature` returns the approximation of the integral and the difference between the last two estimates. For example, the approximation of the integral

$$\int_{-1}^{1} e^x \cos x \, dx$$

can be done using the following code

```
from scipy.integrate import quadrature
f = lambda x: 0.5*np.exp(x/2.0)*np.cos(x/2.0)
print( quadrature(f,-2.0,2.0,tol=1.e-15,maxiter=100) )
```

```
(1.933421496299271, 9.692333602373537e-10)
```

We observe that the last two estimates of the integral differ by an amount of $O(10^{-10})$.

☞ Adaptive quadrature rules are useful in cases where the integrant function consists of different scales or varies irregularly in different areas of the integration interval.

The module `scipy.integrate` also contains general purpose functions for the evaluation of definite integrals in one, two and general multiple dimensions. The functions `quad`, `dblquad`, `tplquad` and `nquad` can be used for the computation of integrals with one, two, three or multiple variables respectively.

7.4.4 Application in classical mechanics

Work in physics measures the energy transfer during the motion of an object under the influence of external force and it is measured in Joules (J) (1 Joule = 1 Newton · meter). For example, the object in Figure 7.5 changes location x due to a force $f(x)$. The work done by the force $f(x)$ to move the object from the location x_0 to x_N is

$$W = \int_{x_0}^{x_N} f(x) \, dx ,$$

and is practically the area enclosed by the graph of the function $f(x)$, the x-axis and the lines $x = x_0$ and $x = x_N$. Assume that $N = 14$. We measure the force applied to the object of Figure 7.5 at the locations $x_0, x_1, \ldots, x_{15}$ and we report its magnitude in Table 7.2.

TABLE 7.2
The magnitude of the force $f(x_i)$ measured in Newtons at the locations x_i, $i = 0, 1, \ldots, 15$ measured in meters

x_i	0.0	0.1	0.2	0.3	0.4	0.5	0.6	0.7	0.8	0.9	1.0	1.1	1.2	1.3	1.4
$f(x_i)$	0.0	0.45	1.45	2.3	3.1	3.1	3.1	2.5	1.1	1.1	1.1	0.8	0.6	0.3	0.0

The most straightforward way to estimate the work done by such a force is by approximating the work integral using the composite trapezoidal rule. Here, we employ the function `trapezoid` of the module `scipy.integrate` with quadrature nodes the values x_i for $i = 0, 1, \ldots, 15$.

```
from scipy.integrate import trapezoid
import numpy as np

x = np.arange(0.0,1.5,0.1)
y = np.array([0.0, 0.45, 1.45, 2.3, 3.1, 3.1, 3.1,
              2.5, 1.1, 1.1, 1.1, 0.8, 0.6, 0.3, 0.0])
print( trapezoid(y,x) )
```

```
2.1
```

Using the previous code we obtained the result which is 2.1 J. If we want to use different quadrature nodes, then we must first interpolate the data, and then integrate the interpolant. If for example we take the cubic spline interpolant, then we could use Gaussian quadrature such as the formula (7.7) in each interval. Because the interpolant is a cubic polynomial in each interval, the result of Gaussian quadrature with at least two nodes will contain only the interpolation error.

7.5 Further Reading

Numerical quadrature is a topic contained in every textbook related to numerical analysis [25, 70, 76, 22, 26, 7, 127, 108]. Specialized books on numerical integration includes the classical book by Davis and Rabinowitz [30], and other more recent books such as [14, 38]. A specialized book on Gaussian quadrature is [131]. Modern computer codes for adaptive quadrature rules are usually based on the techniques described in [101]. An interesting book, especially for postgraduate students, is that by Stewart [124], which is structured in individual lectures.

Chapter Highlights

- Quadrature or numerical integration is the approximation of integrals by finite sums of the form
$$\int_a^b f(x)\,dx \approx \sum_{i=0}^{N} w_i f(z_i)\ ,$$
where z_i are called quadrature nodes and w_i weights.
- The simplest quadrature rule is the midpoint method with $N = 0$, $w_i = b - a$ and $z_i = (a + b)/2$.
- Composite quadrature rules are obtained if we divide the integration domain into multiple subintervals and apply the respective quadrature rule in each of these subintervals.
- The weights and nodes of the composite midpoint rule given a grid $a = x_0 < x_1 < \cdots < x_N = b$ are defined as $z_i = (x_i + x_{i+1})/2$ and $w_i = x_{i+1} - x_i$, respectively.
- Newton-Cotes quadrature rules come from the integration of an interpolating polynomial of f at the $N + 1$ nodes x_i, $i = 0, 1, \ldots, N$. When $N = 1$ the method is called trapezoidal rule, and when $N = 2$ Simpson's rule.
- The error of the trapezoidal rule is $O(h^3)$ and the respective error of Simpson's rule is $O(h^5)$ with $h = b - a$. The composite rules have errors of order one less than the simple rules.
- The estimation of the discretization error in a numerical method can provide a proof of its convergence.
- Composite Simpson's rule requires $N = 2m$ to be even number.
- The degree of exactness is the largest degree of polynomial for which the quadrature rule can integrate exactly.
- The degree of exactness of Newton-Cotes rules with $N + 1$ nodes is at least N.
- Gaussian quadrature rules with $N + 1$ nodes have degree of exactness $2N + 1$, and error $O([(2N + 2)!]^{-1})$.
- Gauss-Legendre rule employs the $N + 1$ zeros of the $(N + 1)$-th Legendre polynomial $z_0, z_1, \ldots, z_N$ as nodes and weights given by the formula (7.8).
- The midpoint rule can be considered as a Gauss-Legendre rule with $N = 0$. Thus, it has degree of exactness 1.
- Usually the nodes and weights of Gauss-Legendre rule are computed numerically.
- The module `scipy.integrate` contains the implementation of all classical quadrature rules as well as general purpose automatic integration routines.

Exercises

1. Consider the integrals

$$I_1 = \int_0^1 x^2 \, dx, \quad I_2 = \int_0^1 x^2 e^{-x} \, dx, \quad \text{and} \quad I_3 = \int_e^{e+1} \frac{1}{x \ln x} \, dx \; .$$

   (a) Compute approximations (without the help of Python) of the following integrals using the Midpoint, Trapezoidal and Simpson rules.

   (b) Compute the exact values of the integral I_1, I_2 and I_3.

   (c) In each case estimate the errors between your approximations and the exact values, and compare with the formula of Section 7.2.2.

   (d) Verify the results using appropriate Python functions.

2. Consider the integral

$$I[f] = \int_a^b f(x) \, dx \; .$$

   (a) Using Lagrange polynomial interpolation on appropriate grids derive the following Newton-Cotes-type quadrature rules known to as Adams-Bashforth-Moulton quadrature formulas:

$$Q_1[f] = \frac{h}{24}[55f(a) - 59f(a-h) + 37f(a-2h) - 9f(a-3h)] \; ,$$

$$Q_2[f] = \frac{h}{24}[9f(a+h) + 19f(a) - 5f(a-h) + f(a-2h)] \; .$$

   (b) Write Python programs to implement the quadrature rules Q_1 and Q_2.

   (c) Estimate experimentally the degree of exactness of these formulas.

   (d) Describe their advantages and disadvantages compared to other Newton-Cotes methods.

   (e) Use these formulas to estimate the integrals of Problem 1.

3. Write a Python function with name `midpointquad` and arguments as in the following code to implement the composite midpoint rule:

```
def midpointquad(func, a, b, n):
# quad = midpointquad(func, a, b, n)
# comments

return quad
```

   where `func` indicates the name of a function, `a` and `b` are the lower and upper limits of integration, and `n` is the number of points (not the number of intervals).

   (a) Debug your `midpointquad` routine by computing the integral $\int_0^1 2x \, dx = 1$. Even if you use only one interval (`n=2`) you should get the exact answer because the midpoint rule integrates linear functions exactly.

(b) Use your midpoint routine to estimate the integral of Runge's function, $f(x) = 1/(1+x^2)$ over the interval $[-5, 5]$. The exact answer is $2\arctan(5)$. Fill in the following table, using scientific notation for the error values so you can see the pattern.

n	h	Midpoint Result	Error
11	1.0		
101	0.1		
1001	0.01		
10001	0.001		

(c) Estimate experimentally the order of accuracy (an integer power of h). This can be done by the errors and the values of h of the previous question.

(d) Study the degree of exactness of the midpoint rule by computing the integrals of the following polynomials using the single interval $[0, 1]$ and `n = 2`.

`func`	Midpoint Result	Error
1		
$2x$		
$3x^2$		
$4x^3$		

(e) For some methods, but not all, the degree of exactness is one less than the order of accuracy. Is that the case for the midpoint rule?

4. Implement the trapezoidal and Simpson's rules and repeat the study of the previous problem.

5. Implement the Gauss-Legendre Quadrature rule with N nodes. For the computation of the quadrature nodes and weights of the Gauss-Legendre quadrature rule use the Python function `leggauss` of the module `numpy.polynomial.legendre`.

(a) Debug your function by showing that its exactness is at least 1 for $N = 1$ using the integral $\int_0^1 2x\, dx = 1$.

(b) Fill in the following table by computing with your code the integrals over $[0, 1]$ of the indicated integrands.

f	Error for $N = 2$	Error for $N = 3$
$3x^2$		
$4x^3$		
$5x^4$		
$6x^5$		
$7x^6$		
Degree		

(c) Compute accuracy estimates of the integral of Runge's function in the interval $[-5, 5]$. Recall that the exact answer is $2\arctan(5)$. Fill in the following table with the computed errors:

N	Gauss-Legendre Result	Error
3		
7		
11		
15		

6. Consider the (rectangular) quadrature rule defined for any $f \in C[a, b]$ by

$$I[f] = (b - a)f(a) ,$$

with error

$$E[f] = \int_a^b f(x)\,dx - I[f]\ .$$

(a) Prove that the quadrature rule $I[f]$ integrates exactly polynomials of degree 0. In other words prove that $E[p] = 0$ for all $p \in \mathbb{P}_0$.

(b) Prove that for $f \in C^1[a,b]$ there is a $\xi \in (a,b)$ such that

$$E[f] = \frac{(b-a)^2}{2} f'(\xi)\ .$$

(c) Let $N \in \mathbb{N}$, $h = (b-a)/N$ and $x_i = a+ih$, $i = 0,1,\ldots,N$. Prove that for $f \in C^1[a,b]$ there is a $\xi \in (a,b)$ such that

$$\int_a^b f(x)\,dx - \sum_{i=0}^{N-1} f(x_i) = \frac{b-a}{2} h f'(\xi)\ .$$

7. Let $f \in C^1[a,b]$. Consider the quadrature rule

$$I[f] = (b-a)f(a) + \frac{(b-a)^2}{2} f'(a)\ ,$$

with error

$$E[f] = \int_a^b f(x)\,dx - I[f]\ .$$

(a) Prove that the quadrature rule $I[f]$ integrates exactly polynomials of degree less than or equal to 1. In other words prove that $E[p] = 0$ for all $p \in \mathbb{P}_1$.

(b) Prove that for $f \in C^2[a,b]$ there is a $\xi \in (a,b)$ such that

$$E[f] = \frac{(b-a)^3}{6} f''(\xi)\ .$$

(c) If $N \in \mathbb{N}$, $h = (b-a)/N$ and $x_i = a+ih$, $i = 0,1,\ldots,N$. Prove that for $f \in C^2[a,b]$ there is a $\xi \in (a,b)$ such that

$$\int_a^b f(x)\,dx - \sum_{i=0}^{N-1} \left[h f(x_i) + \frac{h^2}{2} f'(x_i) \right] = \frac{b-a}{6} h^2 f''(\xi)\ .$$

8. Consider a function f on $[a,b]$ and a partition $\{x_0, x_1, x_2\}$ such that $x_0 = a$, $x_1 = a+h$ and $x_2 = b$ with $h = (b-a)/3$. We approximate the integral

$$\int_a^b f(x)\,dx\ ,$$

using the quadrature rule defined by

$$I[f] = \frac{9}{4} h f(x_1) + \frac{3}{4} h f(x_2)\ .$$

(a) Find the degree of exactness of $I[f]$.

(b) Apply this formula to estimate the integral

$$\int_0^1 x^2 e^{-x}\,dx\ .$$

(c) Estimate the error between the numerical and the exact solution and compare with Simpson's rule.

9. Consider $f \in C^1[a,b]$ and a uniform grid of nodes x_i, $i = 0, 1, \ldots, N$. Define the composite quadrature rule

$$I_N[f] = h\left[\frac{1}{2}f(x_0) + \sum_{i=1}^{N-1} f(x_i) + \frac{1}{2}f(x_N)\right] - \frac{h^2}{12}\left[f'(x_N) - f'(x_0)\right] ,$$

with $h = (b-a)/N$.

(a) Derive a quadrature rule with one interval which applied to each interval $[x_i, x_{i+1}]$ leads to the composite rule $I_N[f]$.

(b) Prove that for $f \in C^4[a,b]$ there is $\xi \in (a,b)$ such that

$$\int_a^b f(x)\, dx - I_N[f] = \frac{b-a}{720} h^4 f^{(4)}(\xi) .$$

10. Gaussian quadrature can be used to approximate integrals of functions that suffer from singularities at the endpoints of the integration domain. Verify that Gaussian quadrature can compute the following singular integrals

(a) $\int_0^1 \frac{\log(1-x)}{x}\, dx = -\frac{\pi^2}{6}$,

(b) $\int_0^1 \frac{\log(1+x)}{x}\, dx = -\frac{\pi^2}{12}$,

(c) $\int_0^1 \frac{\log(1+x^2)}{x}\, dx = -\frac{\pi^2}{24}$.

11. A car is traveling on a straight road for $20sec$ with speeds recorded at the time instances $t_0, t_1, \ldots, t_N$ as in the following table:

t_i	0.0	4.5	6.1	8.4	10.0	12.7	15.3	17.2	18.8	20.0
$v(t_i)$	0.0	12.0	31.3	55.1	51.0	45.2	23.2	12.6	5.8	0.0

The speeds $v(t_i)$ are in km/h while the times t_i in seconds.

(a) Determine the distance covered by the car using the data of the previous table.

(b) Determine the average speed of the car.

(c) Compute a cubic spline interpolant and use it to find the total distance covered by the car. Employ adaptive quadrature.

8

Numerical Differentiation and Applications to Differential Equations

There are many numerical methods as well as applications that rely on the approximation of derivatives. For example, the Secant method required the approximation of the first derivative for the approximation of roots of functions. Approximation of derivatives is of significant importance also for the derivation of efficient methods for the numerical solution of differential equations. In this chapter, we turn our attention to the numerical approximation of derivatives in a systematic way. We also discuss applications of the approximation of solutions of ordinary differential equations. We close this chapter with an application in epidemiology.

8.1 Numerical Differentiation

Recall that the derivative can be defined by any of the following limits

$$
\begin{aligned}
f'(x_0) &= \lim_{x\to x_0} \frac{f(x)-f(x_0)}{x-x_0}\ ,\\
f'(x_0) &= \lim_{h\to 0} \frac{f(x_0+h)-f(x_0)}{h}\ ,\\
f'(x_0) &= \lim_{h\to 0} \frac{f(x_0)-f(x_0-h)}{h}\ .
\end{aligned}
$$

Using these limits we can obtain even more formulas for the first derivative. For example, adding the last two limits and dividing by 2 we obtain the formula

$$f'(x_0) = \lim_{h\to 0} \frac{f(x_0+h)-f(x_0-h)}{2h}\ .$$

The derivative $f'(x_0)$ is the slope of the tangent line of $f(x)$ passing through the point $(x_0, f(x_0))$. Each limit in the previous definitions of the first derivative is the limit of the slope of a secant as $h \to 0$. Since the derivative $f'(x_0)$ is a limit of approximating secant slopes, a first way to find an approximation for $f'(x_0)$ is to choose a small value for the parameter h and evaluate the fraction without the limit for the particular value of h.

8.1.1 First-order accurate finite difference approximations

In the formula

$$f'(x_0) = \lim_{h\to 0} \frac{f(x_0+h)-f(x_0)}{h}\ ,$$

DOI: 10.1201/9781003287292-8

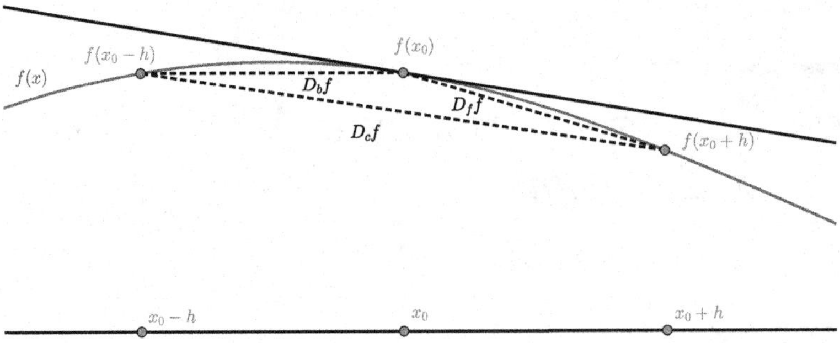

FIGURE 8.1
Finite difference approximations of the first derivative.

we take h to be sufficiently small that $x_0 + h \approx x_0$. For such h we expect that

$$f'(x_0) \approx \frac{f(x_0 + h) - f(x_0)}{h} .$$

This difference quotient is known to as the *forward finite difference approximation* of the derivative $f'(x_0)$ and is denoted by

$$D_f f(x_0) = \frac{f(x_0 + h) - f(x_0)}{h} . \tag{8.1}$$

This is the slope of the secant crossing the graph of f at the points $(x_0, f(x_0))$ and $(x_0 + h, f(x_0 + h))$ and is presented in Figure 8.1.

For example, consider the function $f(x) = x \sin(x)$ and assume that we want to find an approximation of the derivative $f'(\pi/2)$. The derivative of $f(x)$ is $f'(x) = \sin(x) + x \cos(x)$, and thus $f'(\pi/2) = 1$. We approximate this derivative using the formula (8.1) with $h = 0.01$ and $x_0 = \pi/2$. This can be implemented in Python in the following way:

```
def f(x):
    y = x*np.sin(x)
    return y
x0 = np.pi/2
h = 0.01
# approximation of the derivative using (8.1)
derivative = ( f(x0+h) - f(x0) ) / h
print('df(x) =',derivative, 'Error =',np.abs(derivative-1.0))
```

```
df(x) = 0.992096084232319 Error = 0.007903915767681013
```

Our first observation is that the discretization error should be $O(h)$: Using $h = 0.01$ we obtain an approximation of the derivative with two decimal digits correct $f'(x_0) \approx 0.99$. This is an indication of first order convergence rate. By taking smaller values of h we can achieve better accuracy.

Similarly, fixing $0 < h < 1$ in the limit

$$f'(x_0) = \lim_{h \to 0} \frac{f(x_0) - f(x_0 - h)}{h} ,$$

we obtain the finite difference approximation

$$f'(x_0) \approx \frac{f(x_0) - f(x_0 - h)}{h} .$$

This approximation is called *backward difference approximation* and it is denoted by

$$D_b f(x_0) = \frac{f(x_0) - f(x_0 - h)}{h} . \tag{8.2}$$

This formula can be implemented with the following simple code

```
x0 = np.pi/2
h = 0.01
# approximation of the derivative using (8.2)
derivative = ( f(x0) - f(x0-h) ) / h
print('df(x) =',derivative, 'Error =',np.abs(derivative-1.0))
```

```
df(x) = 1.0078039166010244 Error = 0.007803916601024419
```

This result has the same accuracy as the forward finite difference approximation indicating that the backward finite difference approximation is again first-order accurate.

8.1.2 Second-order accurate finite difference approximations

Consider the following formulas for the first derivative

$$f'(x_0) = \lim_{h\to 0} \frac{f(x_0 + h) - f(x_0)}{h} \quad \text{and} \quad f'(x_0) = \lim_{h\to 0} \frac{f(x_0) - f(x_0 - h)}{h} .$$

Adding these two expression we get

$$2f'(x_0) = \lim_{h\to 0} \frac{f(x_0 + h) - f(x_0 - h)}{h} ,$$

and dividing by 2 yields

$$f'(x_0) = \lim_{h\to 0} \frac{f(x_0 + h) - f(x_0 - h)}{2h} .$$

Taking $0 < h < 1$ we obtain the *central finite difference approximation*

$$f'(x_0) \approx \frac{f(x_0 + h) - f(x_0 - h)}{2h} .$$

This approximation of the first derivative will be denoted by

$$D_c f(x_0) = \frac{f(x_0 + h) - f(x_0 - h)}{2h} .$$

An implementation of the central finite difference approximation in Python for the same f as before can be the following

```
x0 = np.pi/2
h = 0.01
derivative = (f(x0+h) - f(x0-h))/(2.0*h)
print('df(x) =',derivative, 'Error =',np.abs(derivative-1.0))
```

```
df(x) = 0.9999500004166717 Error = 4.999958332829735e-05
```

Observe here that contrary to the forward and backward finite differences, the central finite difference approximation is more accurate; the error is $O(h^2)$ since it has four digits correct.

☞ Forward and backward differences to the first derivative are first-order accurate, while the central difference approximation has order of accuracy 2.

8.1.3 Error estimation

In order to derive finite difference formulas in rigorous way and at the same time estimate their accuracy[1] we use Taylor polynomials.

Forward finite difference approximation

Suppose that f has bounded second derivative and $0 < h < 1$ is fixed. We consider the first-order Taylor expansion

$$f(x_0 + h) = f(x_0) + hf'(x_0) + \frac{h^2}{2!}f''(c) ,$$

where $c \in (x_0, x_0 + h)$. Solving for the first derivative we obtain

$$f'(x_0) = \frac{f(x_0 + h) - f(x_0)}{h} + \frac{h}{2}f''(c) .$$

Let $|f''(c)| \leq 2C$ for some constant $C > 0$, then

$$f'(x_0) = \frac{f(x_0 + h) - f(x_0)}{h} + O(h) . \tag{8.3}$$

Discarding the terms of $O(h)$ we obtain the forward difference approximation

$$D_f f(x_0) = \frac{f(x_0 + h) - f(x_0)}{h} .$$

Rearranging (8.3) we deduce that there is a constant $C > 0$ such that

$$|f'(x_0) - D_f f(x_0)| \leq Ch .$$

This means that the forward finite difference formula has first-order convergence rate $O(h)$, and if we take the limit $h \to 0$, then the approximation $D_f f(x_0) \to f'(x_0)$ linearly.

Backward finite difference approximation

Similarly, for the derivation of the backward finite difference formula and the estimation of its convergence rate we consider the Taylor polynomial

$$f(x_0 - h) = f(x_0) - hf'(x_0) + \frac{h^2}{2!}f''(c) ,$$

[1]Numerical accuracy and convergence rate are equivalent notions.

where $c \in (x_0 - h, x_0)$. Solving again for the first derivative we obtain

$$f'(x_0) = \frac{f(x_0) - f(x_0 - h)}{h} + \frac{h}{2} f''(c) ,$$

or in other words

$$f'(x_0) = D_b f(x_0) + O(h) .$$

This means that

$$|f'(x_0) - D_b f(x_0)| \leq Ch ,$$

for some $C > 0$, and thus the convergence is linear in this case as well.

Central finite difference approximation

In order to derive a second-order finite difference approximation to $f'(x_0)$ we consider the following second-order Taylor expansions

$$f(x_0 + h) = f(x_0) + hf'(x_0) + \frac{h^2}{2!} f''(x_0) + \frac{h^3}{3!} f'''(c_1) ,$$

and

$$f(x_0 - h) = f(x_0) - hf'(x_0) + \frac{h^2}{2!} f''(x_0) - \frac{h^3}{3!} f'''(c_2) ,$$

where c_1, c_2 in the intervals $(x_0, x_0 + h)$ and $(x_0 - h, x_0)$, respectively. We can now eliminate the second derivatives by subtracting the two Taylor polynomials to obtain

$$f(x_0 + h) - f(x_0 - h) = 2hf'(x_0) + \frac{h^3}{6}(f'''(c_1) + f'''(c_2)) .$$

Solving for the first derivative we obtain

$$f'(x_0) = \frac{f(x_0 + h) - f(x_0 - h)}{2h} + \frac{h^2}{3}(f'''(c_1) + f'''(c_2)) .$$

Let $|f'''(x)| \leq 3C/2$ for some positive constant $C > 0$ and for all x in the domain of f. Then,

$$f'(x_0) = \frac{f(x_0 + h) - f(x_0 - h)}{2h} + O(h^2) .$$

This means that absolute error is

$$|f'(x_0) - D_c f(x_0)| \leq Ch^2 ,$$

for some $C > 0$, and the convergence is quadratic.

8.1.4 Approximation of second-order derivatives

In order to approximate high-order derivatives by finite differences, we follow the exact same procedure as in the case of first derivatives. Suppose that we want to approximate the second derivative $f''(x_0)$. We consider the following third-order Taylor polynomials

$$f(x_0 + h) = f(x_0) + hf'(x_0) + \frac{h^2}{2!} f''(x_0) + \frac{h^3}{3!} f'''(x) + \frac{h^4}{4!} f^{(4)}(c_1) ,$$

and

$$f(x_0 - h) = f(x_0) - hf'(x_0) + \frac{h^2}{2!} f''(x_0) - \frac{h^3}{3!} f'''(x) + \frac{h^4}{4!} f^{(4)}(c_2) ,$$

where c_1, c_2 in the intervals $(x_0, x_0 + h)$ and $(x_0 - h, x_0)$. Since we want to find the second derivative, we add these Taylor expansions to simplify the first and third derivatives

$$f(x_0 + h) + f(x_0 - h) = 2f(x_0) + h^2 f''(x_0) + \frac{h^4}{24}(f^{(4)}(c_1) + f^{(4)}(c_2)) .$$

Solving the last relationship for $f''(x_0)$ we have

$$f''(x_0) = \frac{f(x_0 + h) - 2f(x_0) + f(x_0 - h)}{h^2} - \frac{h^2}{24}(f^{(4)}(c_1) + f^{(4)}(c_2)) .$$

Let $|f^{(4)}(x)| \leq 12C$ for some constant $C > 0$. Thus,

$$f''(x_0) = \frac{f(x_0 + h) - 2f(x_0) + f(x_0 - h)}{h^2} + O(h^2) .$$

The particular finite difference approximation is denoted by D^2 and

$$D^2 f(x_0) = \frac{f(x_0 + h) - 2f(x_0) + f(x_0 - h)}{h^2} .$$

We conclude that there is $C > 0$ such that

$$|f''(x_0) - D^2 f(x)| < Ch^2 ,$$

and thus $D^2 f(x_0)$ is 2nd order accurate.

As an example we consider the function $f(x) = x\sin(x)$ and $x_0 = \pi/2$. Now $f''(x) = 2\cos(x) - x\sin(x)$ which implies $f''(\pi/2) = -\pi/2$. An implementation in Python could be

```
x0 = np.pi/2
h = 0.01
d2f = (f(x0+h) - 2.0*f(x0) + f(x0-h))/(h**2)
print('d2f(x) =',derivative2, 'Error =',np.abs(d2f-(-np.pi/2)))
```

```
d2f(x) = -1.5707832368705432 Error = 1.3089924353337778e-05
```

In this example we see once again that the practice can verify the theory since for $h = 0.01$ we get an error of $O(h^2)$.

> ☞ Finite difference approximations of any order and for any derivative can be obtained using appropriate combinations of Taylor expansions.

8.1.5 Richardson's extrapolation

We can derive high-order, accurate approximations of the derivatives using high-order Taylor polynomials, like in the case of central finite difference formula for the first derivative. On the other hand, there is a general methodology that can improve the accuracy of an approximation. This methodology is known to as the *Richardson extrapolation.*

Assume that we want to compute a quantity D and $d(h)$ is an approximation of D with error $E(h)$,

$$D = d(h) + E(h) .$$

The error function is expected to be $E(h) = Ch^p$ for some power p. For example, if $D = f'(x_0)$ and $d(h) = (f(x_0 + h) - f(x_0))/h$, then $E(h) = Ch$ for some constant C.

The idea of Richardson's extrapolation is based on the elimination of the error $E(h)$ by taking two different values of the parameter h, namely h_1 and h_2. Then we have

$$D = d(h_1) + Ch_1^p ,$$

and

$$D = d(h_2) + Ch_2^p ,$$

where we included the error terms to obtain the exact value of the quantity D. Eliminating the constant C we obtain a new formula for the quantity D

$$D = \frac{(h_1/h_2)^p d(h_2) - d(h_1)}{(h_1/h_2)^p - 1} ,$$

which is exact, and if we take for simplicity $h_2 = h_1/2$ we obtain

$$D = \frac{2^p d(h_1/2) - d(h_1)}{2^p - 1} .$$

In the case of forward finite differences with $p = 1$, $h_1 = h$ and $h_2 = h/2$ we have

$$d(h/2) = \frac{f(x_0 + h/2) - f(x_0)}{h/2} ,$$

and therefore

$$\begin{aligned} D = f'(x_0) &\approx \frac{2\frac{f(x_0+h/2)-f(x_0)}{h/2} - \frac{f(x_0+h)-f(x_0)}{h}}{2-1} \\ &= \frac{4f(x_0 + h/2) - f(x_0 + h) - 3f(x_0)}{h} . \end{aligned}$$

Implementing the Richardson extrapolation for the forward difference approximation of the first derivative in Python we can estimate the error of the new method.

```
x0 = np.pi/2
h = 0.01
d1 = (f(x0+h) - f(x0))/h
d2 = (f(x0+h/2.0) - f(x0))/(h/2.0)
derivative = 2.0*d2-d1
print('df(x) =',derivative, 'Error =',np.abs(derivative-1.0))
```

```
df(x) = 1.000024950548184 Error = 2.4950548183966248e-05
```

We observe that the result has been improved, and practically we approximated the derivative with second-order accuracy. On the other hand, we didn't obtain the exact value of the derivative. The reason is that the error $E(h)$ is not exactly Ch^p but $|E(h)| \leq Ch^p$. More precisely, the error is $E(h) = c_1 h^p + c_2 h^{p+1} + \cdots$. Richardson's extrapolation method eliminates only the error term $c_1 h^p$ while the rest of the terms remain intact. For this reason, the error after proceeding with Richardson's extrapolation is expected to be $O(h^{p+1})$. To verify the inequality in the error formula, estimate numerically the error of the forward, backward and central finite differences for the function $f(x) = \sin(x)$ and $x = \pi$.

☞ Richardson extrapolation can improve the accuracy of numerical methods. We saw how we can improve finite difference formulas but one can repeat the same procedure to improve quadrature formulas too. The resulting quadrature formulas are called Romberg integration formulas.

8.2 Applications to Ordinary Differential Equations

In this section we introduce fundamental techniques for the numerical approximation of solutions of ordinary differential equations. Ordinary differential equations are equations where the unknowns are functions and the equations themselves contain derivatives of the unknown functions. One idea is to employ finite difference methods to transform differential equations into algebraic equations which can be solved numerically.

8.2.1 First-order ordinary differential equations

Because this section serves as an introduction to the numerical solution of differential equations we consider only first-order scalar differential equations posed in a finite interval. Moreover, we consider initial values of the unknown quantities that serve as current states of the problem described by the differential equation. This is because differential equations are the manifestation of deterministic problems and, given the current state of a physical system, we should be able to determine uniquely any future state of the system by solving the corresponding differential equations. An ordinary differential equation along with appropriate initial values is called *initial value problem.*

A general initial value problem for first-order ordinary differential equations can be written as

$$\frac{d}{dt}y(t) = f(t, y(t)), \quad a \leq t \leq b, \quad y(a) = y_0 \ , \tag{8.4}$$

where the independent variable is denoted by t since it usually represents time. It is reasonable to assume that we know the initial value[2] of the unknown quantity $y(t)$ at the present time $y(a) = y_0$, and that we want to predict its value in latter time.

☞ The major difficulty in solving ordinary differential equations is the presence of the derivative of the unknown function. We will try to eliminate this derivative, and approximate the continuous problem with a discrete.

The function $f(t, y)$ is a function of two independent variables and is very important for the solution of the problem. The regularity of this function specifies the solvability and usefulness of the initial value problem. Specifically, it is meaningful to solve initial value problems that have a unique solution, in the interval $[a, b]$. If the initial value problem has no solution, then there is no point in searching for approximate solutions; or if the solution is not unique, then we will not be able to speak with great certainty about any future prediction. For problems with a unique solution given some initial data we say that are well-posed in the Hadamard sense or that respect Newton's principle of determinacy.

[2]This can be an actual measurement.

We know from the theory of ordinary differential equations that if $f(t, y)$ is continuous, $f \in C([a, b] \times \mathbb{R})$ and satisfies a *Lipschitz* condition in terms of y uniformly in t, then for any given initial value $y_0 \in \mathbb{R}$ the initial value problem (8.4) has a unique solution.

The Lipschitz condition is perhaps something that can generate some difficulties if you are not familiar with the definition. In Section 4.2 we saw that a function f satisfies a Lipschitz conditions if there is a constant $L \geq 0$ such that for all $t \in [a, b]$ and $y_1, y_2 \in \mathbb{R}$ then $|f(t, y_1) - f(t, y_2)| \leq L|y_1 - y_2|$. This condition is satisfied by differentiable functions with bounded derivative. For example, any function $f(t, y)$ with $\left|\frac{\partial}{\partial y} f(t, y)\right| \leq M$ satisfies a Lipschitz condition with $L = M$. In this book, we will consider smooth enough functions such that the respective initial value problem can be solved uniquely, and the focus will be on the numerical discretization of the differential equations. For more information on the theory of ordinary differential equations, we refer to [11].

8.2.2 Euler method

The Euler method is perhaps the simplest numerical method for the solution of initial value problems of first-order ordinary differential equations studied by Leonard Euler (1707–1783). Assume that we are looking for approximations to the solution $y(t)$ of the initial-value problem

$$\frac{d}{dt} y(t) = f(t, y(t)), \quad a \leq t \leq b, \quad y(a) = y_0 \ .$$

The function $y(t)$ is a differentiable function, obviously, since the differential equation involves its first derivative. This means that $y(t)$ is smooth and is defined for all values of $t \in [a, b]$. To simplify the problem, we try to find approximations to $y(t)$ not for every value of $t \in [a, b]$ but for a finite (discrete) set of nodes $t_0, t_1, \ldots, t_N$ in the interval $[a, b]$. These nodes are known as *grid points* or *mesh points*. Once the approximate solution is obtained at the grid points, we can estimate the solution at any node other than t_i via interpolation.

Setup of a uniform mesh

We first make the assumption that the grid points $a = t_0 < t_1 < \cdots < t_N = b$ are distributed uniformly throughout the interval $[a, b]$. This condition is ensured by choosing a positive integer N and selecting the common distance between the nodes to be

$$h = \frac{b - a}{N} = t_{i+1} - t_i \ .$$

This distance is called *stepsize* or *timestep*. The grid points then can be given by the formula

$$t_i = a + i \cdot h, \quad \text{for each } i = 0, 1, 2, \ldots, N \ .$$

Discretization of the differential equation

The main difficulty when solving differential equations is that these equations contain derivatives of the unknown function $y(t)$. In the particular case of a first-order differential equation we only have a first derivative. For example, let's say that we take the differential equation at the node t_i,

$$\frac{d}{dt} y(t_i) = f(t_i, y(t_i)) \ . \tag{8.5}$$

We simplify the problem by substituting the derivative $dy(t_i)/dt$ with an approximation at t_i using forward finite differences

$$\frac{d}{dt} y(t_i) \approx \frac{y(t_{i+1}) - y(t_i)}{t_{i+1} - t_i} = \frac{y(t_{i+1}) - y(t_i)}{h} \ .$$

Substituting the previous approximation of the derivative into the differential equation (8.5) leads to the algebraic approximation

$$\frac{y(t_{i+1})-y(t_i)}{h} \approx f(t_i, y(t_i)) \ .$$

For $i=0$ we have that $y(t_1) \approx y_0 + hf(t_0, y_0)$. Let us denote

$$y_1 = y_0 + hf(t_0, y_0) \ ,$$

so we have $y(t_1) \approx y_1$ and

$$\frac{y_1 - y_0}{h} = f(t_0, y_0) \ .$$

We will denote the approximation of the value $y(t_i)$ by $y_i \approx y(t_i)$. Suppose that we have computed the approximation y_i for some $i > 0$. The next step is to find the new approximation $y_{i+1} \approx y(t_{i+1})$. The previous approximations suggest that the ordinary differential equation at $t = t_i$ can be approximated by the finite difference formula

$$\frac{y_{i+1} - y_i}{h} = f(t_i, y_i) \ . \tag{8.6}$$

Thus, the solution at the next step $y(t_{i+1}) \approx y_{i+1}$ can be obtained by solving (8.6) with respect to y_{i+1}

$$\begin{aligned} y_{i+1} &= y_i + hf(t_i, y_i), \quad \text{for } i = 0, 1, \ldots, N-1 \ , \\ y_0 &= \text{given} \ . \end{aligned} \tag{8.7}$$

This forms an algorithm for the approximations $y_i \approx y(t_i)$, $i = 1, 2, \ldots, N$. Given now the initial condition $y(t_0) = y_0$, we can compute the approximation of $y(t_1)$

$$y(t_1) \approx y_1 = y_0 + hf(t_0, y_0) \ ,$$

and then given y_1 we can compute the approximation of $y(t_2)$

$$y(t_2) \approx y_2 = y_1 + hf(t_1, y_1) \ ,$$

and so on. In this way we compute the approximations $y_0, y_1, \ldots, y_N$ at the grid points $t_0, t_1, \ldots, t_N$, respectively. This method of finding approximations to the solution $y(t)$ using the formula $y_{i+1} = y_i + hf(t_i, y_i)$ is known to as the Euler method.

As an example, we apply Euler's method to approximate the solution of the initial value problem

$$y' = y, \quad 0 \le t \le 2, \quad y(0) = 1 \ ,$$

at $t = 2$ with stepsize $h = 0.5$. In this problem $f(t, y) = y$. When the function f does not depend explicitly on t, like this case here, we say that the ordinary differential equation is *autonomous*. Euler's method for the specific initial value problem is written as $y_{i+1} = y_i + hf(t_i, y_i) = y_i + hy_i$ or better $y_{i+1} = 1.5y_i$ for $i = 0, 1, \ldots N$ with $N = 2/0.5 = 4$. So, given the initial condition $y(0) = y_0 = 1$ we have

$$\begin{aligned} y(0.5) &\approx y_1 = 1.5y_0 = 1.5 \cdot 1 = 1.5, \qquad y(1) \approx y_2 = 1.5y_1 = 2.25 \\ y(1.5) &\approx y_3 = 1.5y_2 = 3.375, \qquad y(2) \approx y_4 = 1.5y_3 = 5.0625 \ . \end{aligned}$$

It is easy to verify that the exact solution to this problem is $y(t) = e^t$. So the exact solution at $t = 2$ is $y(2) \approx 7.389$ while the value we found using Euler's method is 5.0625. Our approximation obviously is not accurate but this is the tribute we pay for using such a large stepsize $h = 0.5$. With the help of a computer we can take much smaller stepsize h

and obtain more accurate approximations. We proceed with an implementation of Euler's method. We define the function `f(t,y)` for the function $f(t, y)$ and also the function `euler` that implements Euler's method with input arguments the time instances t_i, $i = 0, 1, \ldots, N$ in a vector `t`, the function `f` and the initial condition `y0`. The function `euler` returns the solution vector `y` with the approximation of the solution at the time instances t_i.

```
def f(t, y):
    return y
def euler(t, f, y0):
    n = len(t)
    y = np.zeros(n)
    h = (t[-1]-t[0])/(n-1)
    y[0]=y0
    for i in range(n-1):
        y[i+1] = y[i]+h*f(t[i], y[i])
    return y
a = 0.0; b = 2.0; N = 20
t = np.linspace(a, b, N+1)
y0 = 1.0
y = euler(t,f,y0)
# Plot the results
fig = plt.figure()
axes = fig.add_subplot(1, 1, 1)
axes.plot(t, y, '-o', label="Euler")
axes.plot(t, np.exp(t), '-d', label="Exact solution")
axes.set_xlabel("t"); axes.set_ylabel("y"); axes.legend(loc=2); plt.show()
```

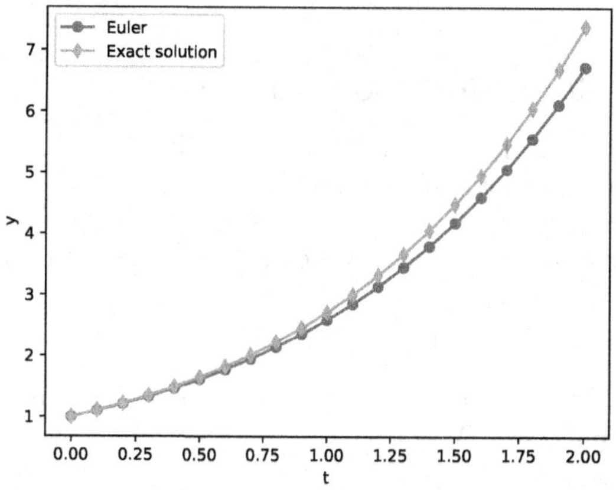

In this code we took $N = 20$ nodes. We observe that the value $y(2)$ is approximated better now with the value $y_{20} = 6.7275$. By taking $N = 200$ we obtain even better results. In the graph produced by MatPlotLib we observe that the error in the approximations of y_i increases with the time t_i. There is a simple explanation behind this phenomenon which we explain below.

To interpret Euler's method geometrically note that for the first step

$$y_1 = y_0 + hy'(t_0) \ ,$$

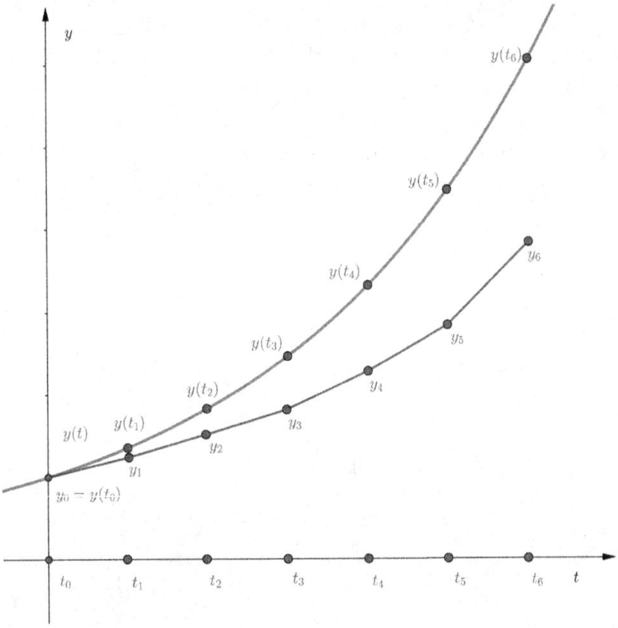

FIGURE 8.2
Computation of the approximations y_i on the tangent lines to $y(t_i)$ translated to the points (t_i, y_i).

since $f(t_0, y_0) = f(t_0, y(t_0)) = y'(t_0)$. This means that y_1 lies on the tangent line of y at t_0. Continuing like that, the value y_{i+1} lies on the tangent line at t_i but translated to the point (t_i, y_i). And so, the distance between y_i and $y(t_{i+1})$ is increasing with t as it is depicted in Figure 8.2.

8.2.3 Alternative derivation and error estimates

Following the same methodology for the derivation of Euler's method as in the derivation of finite difference approximations of derivatives, we estimate the error and also show the convergence of the numerical solution as the stepsize h tends to 0.

Suppose that $y(t)$ is the unique solution of the initial-value problem

$$\frac{d}{dt}y(t) = f(t, y), \quad a \le t \le b, \quad y(a) = y_0 \ .$$

Moreover, assume that $y \in C^2[a, b]$. Then the Taylor expansion of the solution $y(t_{i+1})$ around t_i for each $i = 0, 1, 2, \ldots, N-1$ is

$$y(t_{i+1}) = y(t_i) + (t_{i+1} - t_i)y'(t_i) + \frac{(t_{i+1} - t_i)^2}{2}y''(\xi_i) \ ,$$

for some $\xi_i \in (t_i, t_{i+1})$. Because $h = t_{i+1} - t_i$, we have

$$y(t_{i+1}) = y(t_i) + hy'(t_i) + \frac{h^2}{2}y''(\xi_i) \ . \tag{8.8}$$

Discarding the residual of Taylor's polynomial and denoting the approximation $y_i \approx y(t_i)$ we define

$$y_{i+1} = y_i + hf(t_i, y_i) \quad \text{for } i = 0, 1, \ldots, N-1 \ , \tag{8.9}$$

where y_0 is the given initial value $y(t_0)$.

Local truncation error (or one-step error)

Local truncation error is the error in the computation of the numerical solution y_{i+1} at t_{i+1} as if the solution at the previous step t_i was known and equal to the exact solution $y_i = y(t_i)$. We denote the *local truncation error* of the i-th step with τ_i.

Suppose that $y(t_i)$ is given, then the approximation of $y(t_{i+1})$ by y_{i+1} with Euler's method is

$$y_{i+1} = y(t_i) + h \cdot f(t_i, y(t_i)) \ . \tag{8.10}$$

The local truncation error is defined as

$$\tau_i = y(t_{i+1}) - y_{i+1} \ , \tag{8.11}$$

where y_{i+1} is defined recursively by (8.9). Substituting (8.10) into (8.11) we obtain the expression for the local truncation error

$$\tau_i = y(t_{i+1}) - [y(t_i) + h \cdot f(t_i, y(t_i))] \ .$$

Let $|y''(t)| \leq M$ for all $t \in [a, b]$. The Taylor expansion of $y(t_{i+1}) = y(t_i + h)$ around t_i implies

$$\tau_i = y(t_i) + hy'(t_i) + \frac{h^2}{2}y''(\xi_i) - [y(t_i) + hy'(t_i)] = \frac{h^2}{2}y''(\xi_i) \ ,$$

where $\xi_i \in (t_i, t_{i+1})$. Taking absolute values on both sides of the last relation, and using the bound $|y''(t)| \leq M$ for $t \in [a, b]$ we obtain

$$|\tau_i| \leq Mh^2/2 \ .$$

This means that the local truncation error is of order h^2, which seems very satisfactory at first glance. On the other hand, we always compute the solution at t_{i+1} using an approximation y_i rather than the exact solution $y(t_i)$ (except for $i = 0$). Thus, the actual error at every step will be larger as it accumulates local errors from the previous steps. The actual error is called *global error.*

Global error estimation

In order to find an estimate of the global error we need two lemmata. The first lemma expresses the fact that the exponential growth is larger than the polynomial one.

Lemma 8.1. *For all $x \geq -1$ and any positive integer m, we have:*

$$0 \leq (1 + x)^m \leq e^{mx} \ .$$

Proof. Using Taylor's expansion (see Section 4.3.2) with $f(x) = e^x$, $c = 0$ and $n = 1$ we have

$$e^x = 1 + x + \frac{1}{2}x^2 e^{\xi} \ ,$$

where ξ is between x and 0. Thus,

$$0 \leq 1 + x \leq 1 + x + \frac{1}{2}x^2 e^{\xi} = e^x \ ,$$

and because $1 + x \geq 0$, we obtain

$$0 \leq (1 + x)^m \leq (e^x)^m = e^{mx} \ .$$

□

The second lemma is actually a summation formula of a geometric sum. The global error is practically a geometric sum formed by the local errors. For this reason, we use the next summation formula to find the global error in closed form.

Lemma 8.2 (Geometric sum). *The following summation formula holds for any $i \in \mathbb{N}$ and $s \in \mathbb{R}$:*

$$\sum_{j=0}^{i}(1+s)^j = \frac{1-(1+s)^{i+1}}{1-(1+s)} = \frac{(1+s)^{i+1}-1}{s} .$$

Proof. Let $r = 1+s$, then we have

$$S = 1 + r + r^2 + \ldots + r^i ,$$

which after multiplication with r gives

$$S \cdot r = r + r^2 + \ldots + r^{i+1} .$$

Subtracting the previous expressions we have

$$S \cdot r - S = r^{i+1} - 1 ,$$

and solving for S we obtain the summation formula

$$S = \frac{r^{i+1}-1}{r-1} = \frac{(1+s)^{i+1}-1}{s} .$$

□

We are ready now to prove a global error estimate.

Theorem 8.3 (Error estimate for Euler's method). *Suppose that $f(t, y)$ is continuous and satisfies a Lipschitz condition with Lipschitz constant L on*

$$D = \{ (t, y) |\ a \leq t \leq b \quad and \quad -\infty < y < \infty \} .$$

Moreover, assume that there is a constant $M > 0$ such that

$$|y''(t)| \leq M, \quad for\ all \quad t \in [a, b] ,$$

where $y(t)$ denotes the unique solution of the initial-value problem

$$y' = f(t, y), \quad a \leq t \leq b, \quad y(a) = y_0 .$$

Let $y_0, y_1, \ldots, y_N$ be the approximations obtained by Euler's method for some integer $N > 0$. Then for each $i = 0, 1, 2, \ldots, N-1$,

$$|y(t_{i+1}) - y_{i+1}| \leq \frac{hM}{2L}\left[e^{L(t_{i+1}-a)} - 1\right] .$$

Proof. We will estimate the difference $|y(t_{i+1}) - y_{i+1}|$. Taking Taylor's expansion to $y(t_{i+1})$ we have that for $i = 0, 1, 2, \ldots, N-1$,

$$y(t_{i+1}) = y(t_i) + h \cdot f(t_i, y(t_i)) + \frac{h^2}{2} y''(\xi_i) ,$$

where $\xi \in (t_i, t_{i+1})$. On the other hand, Euler's method gives

$$y_{i+1} = y_i + h \cdot f(t_i, y_i) .$$

Subtracting the last two equations we have

$$y(t_{i+1}) - y_{i+1} = y(t_i) - y_i + h \cdot [f(t_i, y(t_i)) - f(t_i, y_i)] + \frac{h^2}{2} y''(\xi_i) \ .$$

Taking absolute values on both sides yields

$$|y(t_{i+1}) - y_{i+1}| \leq |y(t_i) - y_i| + h \cdot |f(t_i, y(t_i)) - f(t_i, y_i)| + \frac{h^2}{2} |y''(\xi_i)| \ .$$

Since f satisfies a Lipschitz condition in the second variable with constant L, and $|y''(t)| \leq M$ we have

$$\begin{aligned}
|y(t_{i+1}) - y_{i+1}| &\leq (1 + hL)|y(t_i) - y_i| + \frac{h^2 M}{2} \\
&\leq (1 + hL)^2 |y(t_{i-1}) - y_{i-1}| + (1 + (1 + hL)) \frac{h^2 M}{2} \\
&\ \ \vdots \\
&\leq (1 + hL)^{i+1} |y(0) - y_0| + \left(1 + (1 + hL) + (1 + hL)^2 + \ldots + (1 + hL)^i\right) \frac{h^2 M}{2} \\
&= \frac{(1 + hL)^{i+1} - 1}{hL} \frac{h^2 M}{2} .
\end{aligned}$$

Using Lemma 8.1 (and in particular the inequality $(1 + hL)^i \leq e^{hLi}$) we have

$$|y(t_{i+1}) - y_{i+1}| \leq \frac{e^{Lh(i+1)} - 1}{hL} \frac{h^2 M}{2} \ .$$

Since $t_{i+1} = a + (i + 1)h$, we have that $h(i + 1) = t_{i+1} - a$ and thus, the last inequality can be written as

$$|y(t_{i+1}) - y_{i+1}| \leq \frac{hM}{2L} \left[e^{(t_{i+1} - a)L} - 1 \right] \ ,$$

for each $i = 0, 1, 2, \ldots, N - 1$, which is the desired estimate. □

In order to find the error estimate in the previous theorem, we assumed that the solution $y(t)$ has bounded second derivative $|y''(t)| \leq M$. In fact this is not unreasonable. Since y is a solution of a differential equation, we can estimate the second derivative by taking the first derivative on both sides of the differential equation $y' = f(t, y)$ to obtain

$$y'' = \frac{d}{dt}(y') = \frac{d}{dt} f(t, y(t)) = f_t + f_y \cdot f \ . \tag{8.12}$$

One of the main reasons for studying the error estimates is to have a theoretical justification for the convergence of the numerical solution to the exact solution. This means that when the discretization parameter h tends to zero, then the numerical solution tends to the exact solution. The error estimate also provides with information about the speed of convergence. The speed of convergence (like in Newton's method and so forth) is described again by the rate of convergence. In Theorem 8.3 we proved that $|y(t_i) - y_i| \leq Ch$ for some constant C (independent of h). This implies that the numerical solution converges to the exact solution since

$$\lim_{h \to 0} |y(t_i) - y_i| \leq \lim_{h \to 0} Ch = 0 \ ,$$

and that the convergence is linear. In principle, we cannot be excited about a method that converges linearly. A higher-order alternative shall be derived in the following section.

8.2.4 Implicit variants of Euler's method

Given the approximation y_i, we can compute the approximation y_{i+1} using Euler's method explicitly by the formula $y_{i+1} = y_i + hf(t_i, y_i)$. For this reason, Euler's method is characterized as *explicit*[3]. In general, numerical methods of the form $y_{i+1} = F(y_i)$ for $i = 0, 1, \ldots$ are called explicit. Numerical methods that require the solution of a system of the form $y_{i+1} = F(y_{i+1})$, for $i = 0, 1, \ldots$ are called *implicit methods*. Implicit methods perform better than the explicit ones, especially when the solution of the differential equation varies rapidly in small periods of time. Problems with such behavior of their solution are called *stiff*. An example of implicit method is the implicit Euler method. The implicit Euler method can be derived from the differential equation

$$y'(t_{i+1}) = f(t_{i+1}, y(t_{i+1})) ,$$

using the backward finite difference approximation to the first derivative[4]

$$y'(t_{i+1}) \approx \frac{y(t_{i+1}) - y(t_i)}{h} .$$

This leads to the definition of the implicit Euler method

$$\begin{aligned} y_{i+1} &= y_i + hf(t_i, y_{i+1}) \quad \text{for } i = 0, 1, \ldots, N-1 , \\ y_0 &= \text{given} . \end{aligned} \tag{8.13}$$

If $f(t, y)$ is a nonlinear function in terms of y, then we cannot solve explicitly the recursive relation of (8.13) with respect to y_{i+1}. Equation (8.13) can be seen as a fixed point equation $y_{i+1} = F(y_{i+1})$, where techniques of Chapter 5 can be applied. We are able to solve the problem explicitly in the special case where the function $f(t, y)$ is linear in terms of y.

For example, if $y' = y$, the implicit Euler method is

$$y_{i+1} = y_i + hy_{i+1} ,$$

which after solving for y_{i+1} gives the explicit solution

$$y_{i+1} = \frac{1}{1-h} y_i, \quad i = 0, 1, 2, \ldots, N .$$

Another implicit Euler-type method is the trapezoidal method. Also known as the Crank-Nicolson method, the implicit trapezoidal method can be derived as follows: First integrate the differential equation $y'(t) = f(t, y(t))$ on the interval $[t_i, t_{i+1}]$ to obtain

$$y(t_{i+1}) = y(t_i) + \int_{t_i}^{t_{i+1}} f(t, y(t))\, dt . \tag{8.14}$$

We then use the trapezoidal quadrature rule (see Section 7.2.1) to approximate the integral on the right-hand side of (8.14). This approximation yields the formula

$$y_{i+1} = y_i + \frac{h}{2} \left[f(t_i, y_i) + f(t_{i+1}, y_{i+1}) \right] ,$$

where again y_{i+1} is an approximation of $y(t_{i+1})$. The trapezoidal method stands between the explicit and implicit Euler's methods, and has order of convergence $O(h^2)$, while the implicit Euler method has $O(h)$.

[3]The explicit Euler method is also known to as the forward Euler method.

[4]The implicit Euler method is also known to as the backward Euler method.

8.2.5 Improved Euler method

The improved Euler method, is also known as Heun's method (named after Karl Heun 1859–1929). It can be seen as an extension of Euler's method to achieve second-order convergence rate. To make this somewhat clearer, we first present the improved Euler method. Given the initial value problem $y' = f(t, y)$ with $y(0) = y_0$, and the mesh $y_0 < y_1 < \cdots < y_N$, then the approximations y_i of the values $y(t_i)$ with the improved Euler method can be determined using the formulas

$$\begin{aligned} \tilde{y}_{i+1} &= y_i + hf(t_i, y_i) , \\ y_{i+1} &= y_i + \frac{h}{2}\left(f(t_i, y_i) + f(t_{i+1}, \tilde{y}_{i+1})\right) , \end{aligned} \qquad \text{for } i = 0, 1, \ldots, N-1 .$$

The first step of the improved Euler method is Euler's method and serves as an intermediate stage to obtain an initial approximation of the solution. The second step uses the approximation $\tilde{y}_{i+1}$ in the trapezoidal rule to obtain an improved approximation y_{i+1} explicitly.

The derivation of the improved Euler method might sound impossible at first glance. The truth is that it can be derived using Taylor's expansions (like what we did so far to eliminate derivatives), and also the estimate (8.12) for the second derivative, $y'' = f_t + f_y \cdot f$.

Consider the general first-order ordinary differential equation

$$y'(t) = f(t, y(t)) ,$$

and the Taylor expansion

$$y(t + h) = y(t) + hy'(t) + \frac{h^2}{2} y''(t) + O(h^3) .$$

The first derivative in this expansion can be replaced by the right-hand side of the differential equation, and the second derivative is obtained by (8.12). Therefore, the Taylor expansion becomes

$$\begin{aligned} y(t + h) &= y(t) + hf(t, y) + \frac{h^2}{2}\left[f_t(t, y) + f_y(t, y)f(t, y)\right] + O(h^3) \\ &= y(t) + \frac{h}{2} f(t, y) + \frac{h}{2}\left[f(t, y) + hf_t(t, y) + hf_y(t, y)f(t, y)\right] + O(h^3) . \end{aligned}$$

Recalling the multivariate Taylor expansion

$$f(t + h, y + k) = f(t, y) + hf_t(t, y) + f_y(t, y)k + \ldots ,$$

we see that the quantity in the square brackets can be interpreted as

$$f(t + h, y + hf(t, y)) = f(t, y) + hf_t(t, y) + hf_y(t, y)f(t, y) + O(h^2) .$$

Therefore, we get

$$y(t + h) = y(t) + \frac{h}{2} f(t, y) + \frac{h}{2} f(t + h, y + hf(t, y)) + O(h^3) .$$

If we take $t = t_i$ and discard the high-order terms, then we get

$$y_{i+1} = y_i + \frac{h}{2} f(t_i, y_i) + \frac{h}{2} f(t_{i+1}, y_i + hf(t_i, y_i)) . \qquad (8.15)$$

Setting $\tilde{y}_{i+1} = y_i + hf(t_i, y_i)$ we have

$$y_{i+1} = y_i + \frac{h}{2} f(t_i, y_i) + \frac{h}{2} f(t_{i+1}, \tilde{y}_{i+1}) .$$

An alternative way to formulate the improved Euler method is by setting $k_1 = f(t_i, y_i)$ and $k_2 = f(t_i + h, y_i + hk_1)$. Then for given $y_0 = y(t_0)$, the formula (8.15) can be written as

$$y_{i+1} = y_i + h\left(\frac{1}{2}k_1 + \frac{1}{2}k_2\right) \quad \text{for } i = 0, 1, \dots ,$$

where

$$k_1 = f(t_i, y_i) ,$$
$$k_2 = f(t_i + h, y_i + hk_1) .$$

The values k_1 and k_2 are called *intermediate stages*. In order to compute the solution y_{i+1} we need to compute first the two intermediate stages k_1 and k_2. The improved Euler method is a particular example of a class of numerical methods known as Runge-Kutta methods.

From the previous derivation we see that the local error is $O(h^3)$. Thus, the global error is expected to be $O(h^2)$. We proceed now with the implementation of the numerical method and a computational demonstration of its accuracy. In the following implementation, we modify the function `euler` by adding the new intermediate step $\tilde{y}_{i+1}$ in the variable `z`.

```
def f(t, y):
    return y
def improved_euler(t, f, y0):
    n = len(t)
    y = np.zeros(n)
    h = (t[-1]-t[0])/(n-1)
    y[0]=y0
    for i in range(n-1):
        k1 = f(t[i], y[i])
        k2 = f(t[i+1], y[i] + h*k1)
        y[i+1] = y[i] + 0.5*h*(k1+k2)
    return y
a = 0.0; b = 2.0; N = 10
t = np.linspace(a, b, N+1)
y0 = 1.0
y = improved_euler(t, f, y0)
# Plot the results
fig = plt.figure()
axes = fig.add_subplot(1, 1, 1)
axes.plot(t, y, '-o', label="Improved Euler")
axes.plot(t, np.exp(t), '-d', label="Exact solution")
axes.set_xlabel("t"); axes.set_ylabel("y"); axes.legend(loc=2); plt.show()
```

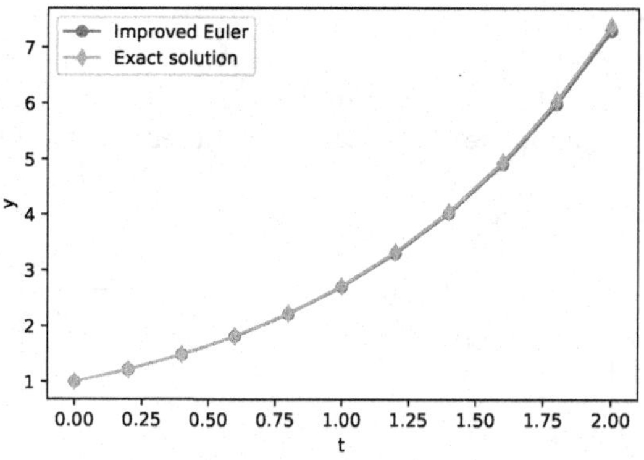

In this code we used only 10 nodes for the computations (stepsize $h = 0.2$) and we observe in the output plot that the exact and numerical solution almost coincide. This is the benefit of using high-order methods as we achieve faster convergence and the required stepsize can be taken large.

☞ The higher the order of convergence of a numerical method, the faster the method converges. This leads to better approximations with larger stepsizes. We say that such methods are efficient.

8.2.6 The notion of stability

When it comes to the use of numerical methods for the approximation of solutions of differential equations one common problem is the choice of the stepsize h. In principle, we want to take h as large as possible without having any discount in the accuracy of the solution. By taking h quite large sometimes can lead to unpleasant surprises[5]. In such situations, we need to take the value of h smaller and try again. There is a question that arises whether or not we can have an indication of how small we can take the stepsize h before we start spending time trying to solve our problem.

Let's see first a problematic situation. Consider the very simple ordinary differential equation

$$y'(t) = \lambda y(t), \quad \lambda < 0 ,$$

with initial condition $y(0) = 1$. The exact solution of this problem is the function $y(t) = e^{\lambda t}$. Let's try using our code, the case with $\lambda = -10$ and with $N = 5$. In this case, the stepsize $h = 0.4$ which is large.

```
def f(t, y):
    return -10.0*y
a = 0.0; b = 2.0; N = 5
h = (b-a)/N
t = np.linspace(a, b, N+1)
y0 = 1.0
y = euler(t, f, y0)
# Plot the results
```

[5]The numerical solution converges as $h \to 0$. If $h > 0$ is large enough the error might be surprisingly large

```
fig = plt.figure()
axes = fig.add_subplot(1, 1, 1)
axes.plot(t, y, '-o', label="Euler")
axes.plot(t, np.exp(-10*t), '-d', label="Exact solution")
axes.set_xlabel("t"); axes.set_ylabel("y"); axes.legend(loc=2); plt.show()
```

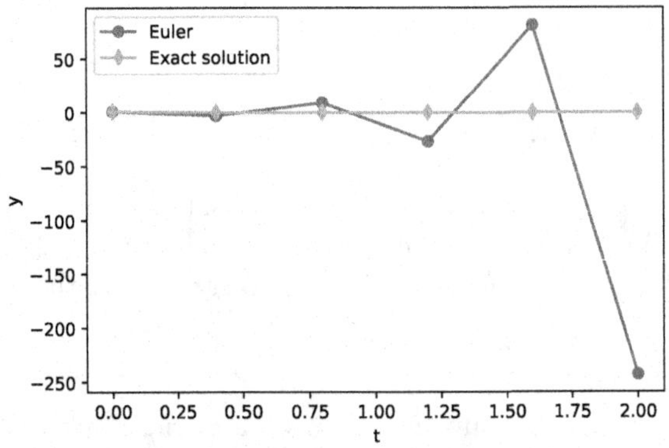

We observe that by taking $h = 0.4$ the Euler method diverges from the exact solution of the problem. More precisely, we observe that the solution grows with time instead of converging to 0 as it should. By taking $h = 0.1$ ($N = 20$) we see in the graph that the numerical solution converges to 0 very quickly. Not that we should be totally happy about this result, but at least the solution seems to converge. Taking even smaller values of h we observe convergence eventually.

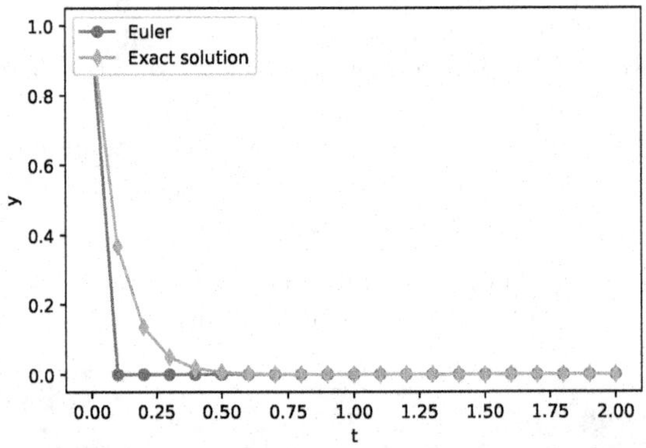

To understand why this is happening, we apply Euler's method to the equation $y'(t) = \lambda y(t)$ for $\lambda < 0$. This gives $y_{n+1} = y_n + h\lambda y_n$, or even better

$$y_{n+1} = (1 + h\lambda)y_n, \quad n = 0, 1, 2, \ldots, \quad \text{and} \quad y_0 = 1 \ .$$

It can easily be seen that the previous recursive sequence can be expressed explicitly as

$$y_n = (1 + h\lambda)^n, \quad n = 0, 1, \ldots \ .$$

So, we have that

$$\begin{aligned}
|y_n| &\to 0, && \text{if } |1+h\lambda| < 1 \ , \\
|y_n| &= 1, && \text{if } |1+h\lambda| = 1 \ , \\
|y_n| &\to \infty, && \text{if } |1+h\lambda| > 1 \ .
\end{aligned}$$

Therefore, for stability we need $|1 + h\lambda| < 1$. Solving this inequality for $h\lambda$ yields $-2 < h\lambda < 2$. Since $h > 0$ and $\lambda < 0$, we need $h\lambda \in (-2, 0)$ or else $h < -2/\lambda$ for stability. In our previous example where $\lambda = -10$ this means $h < 0.2$ and this explains the unwanted behavior for $h = 0.4 > 0.2$.

☞ Stability depends always on the initial-value problem. An unconditionally stable method has no restrictions in the choice of the stepsize h. If conditions apply in the stepsize in order to achieve meaningful solutions, then we say that the method has stability restrictions.

Applying the implicit Euler method to the same problem, we have that

$$y_{n+1} = y_n + h\lambda y_{n+1}, \quad n = 0, 1, 2, \ldots, \quad \text{and} \quad y_0 = 1$$

which yields

$$y_{n+1} = \frac{y_n}{1 - h\lambda}, \quad n = 0, 1, 2, \ldots, \quad \text{and} \quad y_0 = 1 \ .$$

This recursive relation leads to the analytical formula of the numerical solution

$$y_n = \frac{1}{(1 - h\lambda)^{-n}}, \quad n = 0, 1, 2, \ldots \ .$$

For $\lambda < 0$ the quantity $1 - h\lambda > 1$. Therefore, $|y_n| \to 0$ for all values of h. This means that the implicit Euler method is unconditionally stable. Such methods, without stability limitations on h, are called A-stable methods. Usually, implicit methods are A-stable.

8.3 Runge-Kutta Methods

So far, we have studied numerical methods for the solution of ordinary differential equations as a consequence of the approximation of the first derivative with finite differences. This lead us to the formulation of a method that belongs in, perhaps, the most important class of numerical methods for ordinary differential equations, the so-called Runge-Kutta methods. Runge-Kutta methods where first introduced by Carl Runge (1856-1927) and Martin Kutta (1867-1944), while the more recent contributions of John Butcher evolved these methods in their current form. Runge-Kutta methods can be derived following many different methodologies and also by approximating integrals with numerical quadrature. As an introduction to this topic, we present briefly explicit, implicit and adaptive Runge-Kutta methods.

8.3.1 Explicit Runge-Kutta methods

Consider the initial value problem in the interval $[0, T]$

$$\begin{aligned}
y'(t) &= f(t, y(t)), \quad t \in (0, T] \ , \\
y(0) &= y_0 \ ,
\end{aligned}$$

and a uniform partition $0 = t_0 < t_1 < \cdots < t_N = T$ of the interval $[0, T]$ with stepsize $h = T/N$. The improved Euler method, described by the formula

$$y_{i+1} = y_i + h\left(\frac{1}{2}k_1 + \frac{1}{2}k_2\right), \quad i = 0, 1, \dots ,$$

where

$$k_1 = f(t_i, y_i) ,$$
$$k_2 = f(t_i + h, y_i + hk_1) .$$

is a specific example of a general class of methods known to as Runge-Kutta methods. Runge-Kutta methods for the numerical solution of ordinary differential equations usually consist of multiple stages k_i. For example, the most commonly used Runge-Kutta method with four stages can be described as follows,

$$y_{i+1} = y_i + h\left(\frac{1}{6}k_1 + \frac{1}{3}k_2 + \frac{1}{3}k_3 + \frac{1}{6}k_4\right), \quad i = 0, 1, \dots, N-1 ,$$

where

$$k_1 = f(t_i, y_i) ,$$
$$k_2 = f\left(t_i + \frac{h}{2}, y_i + \frac{k_1}{2}\right) ,$$
$$k_3 = f\left(t_i + \frac{h}{2}, y_i + \frac{k_2}{2}\right) ,$$
$$k_4 = f(t_i + h, y_i + k_3) .$$

This method is known to as the classical, explicit four-stage and fourth order Runge-Kutta method. It is called explicit because the computation of any intermediate stage k_i requires the knowledge of previous stages only. (Otherwise the method is called implicit.) It is a four-stage method because it consists of four intermediate stages k_i, for $i = 1, 2, 3, 4$, and fourth-order because the global error can be proved to be $|y(t_i) - y_i| = O(h^4)$.

All Runge-Kutta methods can be formulated in the same way. The general explicit Runge-Kutta method with s intermediate stages is the following:

$$y_{i+1} = y_i + h\sum_{j=1}^{s} b_j k_j, \quad i = 0, 1, \dots ,$$

where

$$k_1 = f(t_i, y_i) ,$$
$$k_2 = f(t_i + c_2 h, y_i + h(a_{21}k_1)) ,$$
$$k_3 = f(t_i + c_3 h, y_i + h(a_{31}k_1 + a_{32}k_2)) ,$$
$$\vdots$$
$$k_s = f(t_i + c_s h, y_i + h(a_{s1}k_1 + a_{s2}k_2 + \cdots + a_{s,s-1}k_{s-1})) ,$$

for appropriate coefficients a_{ij}, b_i, c_i, for $i, j = 1, 2, \dots, s$. The choice of these parameters determine uniquely the Runge-Kutta method. Note that in the previous definition $c_1 = 0$ and $a_{ij} = 0$ for all $j \geq i$. Thus, the parameters a_{ij}, b_i and c_i form the lower triangular matrix

$$\boldsymbol{A} = \begin{pmatrix} 0 & 0 & \cdots & 0 \\ a_{21} & 0 & & 0 \\ a_{31} & a_{32} & \ddots & \vdots \\ \vdots & & \ddots & 0 \\ a_{s1} & a_{s2} & \cdots & a_{s,s-1} \end{pmatrix} ,$$

TABLE 8.1
Butcher tableau for the general explicit Runge-Kutta method with s stages

$$\begin{array}{c|c} \boldsymbol{c} & \boldsymbol{A} \\ \hline & \boldsymbol{b}^T \end{array} = \begin{array}{c|ccccc} 0 & & & & & \\ c_2 & a_{21} & & & & \\ c_3 & a_{31} & a_{32} & & & \\ \vdots & \vdots & & \ddots & & \\ c_s & a_{s1} & a_{s2} & \cdots & a_{s,s-1} & \\ \hline & b_1 & b_2 & \cdots & b_{s-1} & b_s \end{array}$$

and the vectors

$$\boldsymbol{b} = \begin{pmatrix} b_1 \\ b_2 \\ \vdots \\ b_s \end{pmatrix} \quad \text{and} \quad \boldsymbol{c} = \begin{pmatrix} 0 \\ c_2 \\ c_3 \\ \vdots \\ c_s \end{pmatrix} .$$

As we will see later, $\boldsymbol{A}$ is a lower triangular matrix only for explicit Runge-Kutta methods, where the computation of each new step requires only the knowledge of the previous steps. We usually do not write the 0's in the upper triangular part of matrix $\boldsymbol{A}$ but we leave their places empty. In principle, any consistent Runge-Kutta method must satisfy the condition

$$\sum_{j=1}^{i-1} a_{ij} = c_i \quad \text{for} \quad i = 2, 3, \dots, s .$$

This information is usually incorporated in a table known as *Butcher tableau*, which is described in Table 8.1.

Interpreting the previous tables, if we want to compute the solution $y_{i+1} \approx y(t_{i+1})$ given the solution $y_i \approx y(t_i)$ using an explicit Runge-Kutta method with s intermediate stages we need first to compute the intermediate stages

$$k_\ell = f\left(t_i + c_\ell h,\ y_i + h \sum_{j=1}^{\ell-1} a_{\ell,j} k_j\right), \quad \ell = 1, 2, \dots, s , \tag{8.16}$$

and then the approximation of the solution at $t = t_{i+1}$ given by

$$y_{i+1} = y_i + h \sum_{j=1}^{s} b_j k_j, \quad i = 0, 1, \dots . \tag{8.17}$$

The computation of k_ℓ in (8.16) requires all the intermediate stages $k_1, k_2, \dots, k_{\ell-1}$.

☞ The computation of an intermediate stage k_ℓ of any explicit Runge-Kutta method with s stages requires the knowledge of the previous stages k_j for $j < \ell$. Thus, in the worst case scenario, in order to compute the intermediate stages we need to solve a linear system. If the matrix $\boldsymbol{A}$ in the Butcher tableau consists of one diagonal only, then the linear system can be solved in a trivial way as the resulting equations will be decoupled.

TABLE 8.2
Butcher tableau for the Euler method

0	0
	1

TABLE 8.3
Butcher tableau for the improved Euler method

0		
1	1	
	1/2	1/2

TABLE 8.4
Butcher tableau for second-order methods with two stages

0		
α	α	
	$1 - 1/2\alpha$	$1/2\alpha$

Examples of explicit Runge-Kutta methods

The first example is the Euler method which can be described by the tableau in Table 8.2. According to the general formulas (8.16)–(8.21) the Euler method is the following 1-stage Runge-Kutta method

$$\begin{aligned} k_1 &= f(t_i, y_i), \\ y_{i+1} &= y_i + hk_1, \end{aligned} \qquad i = 0, 1, 2, \ldots \ . \tag{8.18}$$

A second example is the improved Euler method with two stages. The method as it was described in Section 8.2.5 is the following two-stage method

$$\begin{aligned} k_1 &= f(t_i, y_i), \\ k_2 &= f(t_i + h, y_i + hk_1), \\ y_{i+1} &= y_i + h\left(\frac{1}{2}k_1 + \frac{1}{2}k_2\right), \end{aligned} \qquad i = 0, 1, 2, \ldots \ , \tag{8.19}$$

and thus the corresponding Butcher tableau is given in Table 8.3. The improved Euler method is not the only second-order Runge-Kutta method with two stages. All methods described by the tableau of Table 8.4 are also second-order accurate.

The last example we provide in Table 8.5 is the classical four-stage, fourth-order Runge-Kutta method, which is perhaps the most popular explicit Runge-Kutta method.

8.3.2 Implicit and diagonally implicit Runge-Kutta methods

As previously discussed, the matrix $\boldsymbol{A}$ is strictly lower triangular for all explicit Runge-Kutta methods. One result of this property is that the computation of an intermediate

TABLE 8.5
Butcher tableau classical four-stage fourth-order Runge-Kutta method

0				
1/2	1/2			
1/2	0	1/2		
1	0	0	1	
	1/6	1/3	1/3	1/6

TABLE 8.6
Butcher tableau for the general Runge-Kutta method with s stages

$$\begin{array}{c|c} \mathbf{c} & \boldsymbol{A} \\ \hline & \boldsymbol{b}^T \end{array} = \begin{array}{c|cccc} c_1 & a_{11} & a_{12} & \cdots & a_{1s} \\ c_2 & a_{21} & a_{21} & \cdots & a_{2s} \\ \vdots & \vdots & \vdots & \ddots & \vdots \\ c_s & a_{s1} & a_{s2} & \cdots & a_{ss} \\ \hline & b_1 & b_2 & \cdots & b_s \end{array}$$

stage requires only the knowledge of the previous intermediates stages. On the other hand, the most general Runge-Kutta method is described by the tableau of Table 8.6 with full matrix $\boldsymbol{A}$.

This tableau corresponds to the general implicit Runge-Kutta method with s stages

$$k_\ell = f\left(t_i + c_\ell h,\ y_i + h\sum_{j=1}^{s} a_{\ell,j}k_j\right), \quad \ell = 1, 2, \dots, s\ , \tag{8.20}$$

and the corresponding formula for the approximate solution at $t = t_{i+1}$

$$y_{i+1} = y_i + h\sum_{j=1}^{s} b_j k_j, \quad i = 0, 1, \dots\ . \tag{8.21}$$

It is easy to see that for the computation of the stage k_ℓ the knowledge of all the previous and the future stages is required. This formulates an s-dimensional nonlinear system of equations which can be solved using Newton's or any other fixed point method. If $\boldsymbol{A}$ is lower triangular but with non-zero diagonal entries, then the method is called *diagonally implicit* Runge-Kutta method (DIRK). A DIRK method does not require the solution of a nonlinear system but only scalar nonlinear equations for the computation of each stage k_ℓ.

Examples of implicit Runge-Kutta methods

The simplest implicit Runge-Kutta method is the implicit Euler method, which is described by the tableau of Table 8.7. Although it is implicit method, it can be seen also as diagonally implicit since there is only one stage. This unique intermediate stage is

$$k_1 = f(t_i + h, y_i + hk_1)\ , \tag{8.22}$$

TABLE 8.7
Butcher tableau for the implicit Euler method

1	1
	1

TABLE 8.8
Butcher tableau for the two-stage, fourth-order Gauss-Legendre Runge-Kutta method

$\frac{1}{2} - \frac{1}{6}\sqrt{3}$	$\frac{1}{4}$	$\frac{1}{4} - \frac{1}{6}\sqrt{3}$
$\frac{1}{2} + \frac{1}{6}\sqrt{3}$	$\frac{1}{4} + \frac{1}{6}\sqrt{3}$	$\frac{1}{4}$
	$\frac{1}{2} + \frac{1}{2}\sqrt{3}$	$\frac{1}{2} - \frac{1}{2}\sqrt{3}$

which is a nonlinear equation formulated directly as a fixed point equation with fixed point the unknown intermediate stage k_1. For the computation of the unknown k_1 we can introduce the iteration

$$k_1^{(p+1)} = f\left(t_i + h, y_i + hk_1^{(p)}\right), \quad p = 0, 1, 2, \dots ,$$

with a given initial guess $k_1^{(0)}$. The value of the initial guess $k_1^{(0)}$ could be the value $f(t_i, y_i)$, which is the intermediate stage of the Euler method. After computing the intermediate stage k_1, we compute the actual approximation of the solution

$$y_{i+1} = y_i + hk_1 . \tag{8.23}$$

This main step of the implicit Euler method is the same as the main step of the explicit Euler method but they differ in the value of k_1.

Other very important implicit Runge-Kutta methods are the so-called Gauss-Legendre methods which can be derived using Gaussian quadrature. Among the many advantages of Gauss-Legendre Runge-Kutta methods is their accuracy of s-stage methods, which is $2s$. Thus, we can achieve fourth order accuracy by using only two stages. The two-stage, fourth-order Gauss-Legendre Runge-Kutta method is given by the Butcher tableau of Table 8.8.

Formulating the intermediate stages using the Table 8.8 we observe that the equations for k_1 and k_2 are

$$\begin{aligned} k_1 &= f(t_i + c_1 h,\ y_i + h(a_{11}k_1 + a_{12}k_2)) , \\ k_2 &= f(t_i + c_2 h,\ y_i + h(a_{21}k_1 + a_{22}k_2)) , \end{aligned} \tag{8.24}$$

and they form a nonlinear system of two equations with two unknowns (k_1, k_2). The methods of Section 5.5.6 can be applied for the solution of this system, or the following simple fixed point iteration also known to as the Gauss-Seidel method for nonlinear systems: Given the initial guesses $k_1^{(0)}$, $k_2^{(0)}$, we approximate the values of k_1 and k_2 by the iterations

$$\begin{aligned} k_1^{(p+1)} &= f(t_i + c_1 h,\ y_i + h(a_{11}k_1^{(p)} + a_{12}k_2^{(p)})), \\ k_2^{(p+1)} &= f(t_i + c_2 h,\ y_i + h(a_{21}k_1^{(p+1)} + a_{22}k_2^{(p)})), \end{aligned} \quad p = 0, 1, 2, \dots . \tag{8.25}$$

Because of its simplicity, the specific iterative method can converge very slowly or it may not converge at all. In case of divergence, the use of Newton-type methods can be unavoidable.

Using the approximations of k_1 and k_2 the main step of the two-stage, fourth-order Runge-Kutta method is then

$$y_{i+1} = y_i + h(b_1 k_1 + b_2 k_2) \;, \tag{8.26}$$

which has not practical implications in its computation.

8.3.3 Adaptive Runge-Kutta methods

So far, we have considered uniform partition of the temporal interval of numerical integration using fixed $h = (b-a)/N = t_{i+1} - t_i$ for all $i = 0, 1, \ldots$. On the other hand, especially in cases where the time integration of a differential equation is time-consuming, it would have been preferable to consider a variable stepsize so as to improve the speed of the computations without compromising their accuracy. Here, we present a technique of using explicit Runge-Kutta methods with variable stepsize h_i. One can use also implicit Runge-Kutta methods in a similar manner. The technique is based on the application of two different Runge-Kutta methods. For economy, we choose methods with the same matrix $\boldsymbol{A}$ and vector $\boldsymbol{c}$ in their Butcher tableaux, while we keep the vector $\boldsymbol{b}$ different. In principle, we use Runge-Kutta methods with different order of accuracy. The idea is to pretend that the solution of the high-order method y_i is very accurate (like being exact) and compare it with the one obtained using the low-order method y_i^*. If the error between the two approximations y_i and y_i^* of the same solution at t_i is greater than a prescribed tolerance, then we repeat the computation of the specific step with smaller h. Consider two s-stage Runge-Kutta methods, let's say, given by the tableaux

$$\begin{array}{c|c} \boldsymbol{c} & \boldsymbol{A} \\ \hline & \boldsymbol{b}^T \end{array} \quad \text{and} \quad \begin{array}{c|c} \boldsymbol{c} & \boldsymbol{A} \\ \hline & (\boldsymbol{b}^*)^T \end{array} \;.$$

Usually, we write the two Runge-Kutta methods in a combined Tableau as in Table 8.9.

Since the matrices $\boldsymbol{A}$ and $\boldsymbol{c}$ are common for both methods, the intermediate stages are the same, and they are computed once for both methods using the formula (8.16)

$$k_\ell = f\left(t_i + c_\ell h,\; y_i + h\sum_{j=1}^{\ell-1} a_{\ell,j} k_j\right) \;, \quad \ell = 1, 2, \ldots, s \;. \tag{8.27}$$

Because the vectors $\boldsymbol{b}$ and $\boldsymbol{b}^*$ are different, the two methods will result in two different approximations to the solution

$$y_{i+1} = y_i + h\sum_{j=1}^{s} b_j k_j \quad \text{and} \quad y_{i+1}^* = y_i + h\sum_{j=1}^{s} b_j^* k_j \;. \tag{8.28}$$

The absolute error between the two solutions is then

$$e_i = |y_{i+1} - y_{i+1}^*| = h\left|\sum_{j=1}^{s} (b_j - b_j^*) k_j\right| \;. \tag{8.29}$$

If the error e_i is greater than a prescribed tolerance, then we repeat the computation for the step i by using stepsize h equal to the half of the previous stepsize.

The algorithm for a general adaptive Runge-Kutta method with s stages is summarized in Algorithm 26. In that algorithm, we assume that we want to solve the problem $y' = f(t, y)$ in the temporal interval $[a, b]$ with $y(a) = y_0$. We assume also that $\boldsymbol{A}$, $\boldsymbol{c}$, $\boldsymbol{b}$ and $\boldsymbol{b}^*$ of the Runge-Kutta method are all given.

TABLE 8.9
Butcher tableau for the general explicit adaptive Runge-Kutta with s stages

$$\begin{array}{c|c} \mathbf{c} & \boldsymbol{A} \\ \hline & \boldsymbol{b}^T \\ & (\boldsymbol{b}^*)^T \end{array} = \begin{array}{c|ccccc} 0 & & & & & \\ c_2 & a_{21} & & & & \\ c_3 & a_{31} & a_{32} & & & \\ \vdots & \vdots & & \ddots & & \\ c_s & a_{s1} & a_{s2} & \cdots & a_{s,s-1} & \\ \hline & b_1 & b_2 & \cdots & b_{s-1} & b_s \\ & b_1^* & b_2^* & \cdots & b_{s-1}^* & b_s^* \end{array}$$

Algorithm 26 Adaptive Runge-Kutta method

Initialize the stepsize $h < b - a$ and time $t_i = 0$
Set $y(a) = y_0$
while $t_i < b$ **do**
  $error = 1$
  while $error > TOL$ **do**
    Set $h = h/2$
    for $l = 1 : s$ **do**
      Compute the intermediate stages

$$k_l = f\left(t_i + c_l h,\ y_i + h\sum_{j=1}^{l-1} A_{lj}k_j\right)$$

    end for
    Compute

$$y_{i+1} = y_i + h\sum_{j=1}^{s} b_j k_j \quad \text{and} \quad y_{i+1}^* = y_i + h\sum_{j=1}^{s} b_j^* k_j \ ,$$

    Compute the $error = |y_{i+1} - y_{i+1}^*|$
  end while
  Set $t_i = t_i + h$
end while

The simplest example of an adaptive Runge-Kutta method is the combination of the Euler and improved Euler methods given by the tableau in Table 8.10. In this case, the Euler method is of order 1 while the improved Euler method of order 2, and that is why we call it RK12.

The most popular adaptive Runge-Kutta method is the Runge-Kutta-Fehlberg method of order 5 given by the Table 8.11 This method was introduced by Erwin Fehlberg (1911-1990) and is of order 5 since one of the methods is of order 4 and the other of order 5. This particular method is referred to as RK45, and can be found implemented in every scientific package dedicated to the numerical solution of ordinary differential equations, including SciPy.

TABLE 8.10
Butcher tableau for the adaptive Runge-Kutta 12 method

0		
1	1	
	1/2	1/2
	1	0

TABLE 8.11
Butcher tableau for the RK45, Runge-Kutta-Fehlberg method of order 5

0						
1/4	1/4					
3/8	3/32	9/32				
12/13	1932/2197	−7200/2197	7296/2197			
1	439/216	−8	3680/513	−845/4104		
1/2	−8/27	2	−3544/2565	1859/4104	−11/40	
	16/135	0	6656/12825	28561/56430	−9/50	2/55
	25/216	0	1408/2565	2197/4104	−1/5	0

TABLE 8.12
Butcher tableau for the RK Radau IIA method of order 5

$\frac{2}{5}-\frac{\sqrt{6}}{10}$	$\frac{11}{45}-\frac{7\sqrt{6}}{360}$	$\frac{37}{225}-\frac{169\sqrt{6}}{1800}$	$-\frac{2}{225}+\frac{\sqrt{6}}{75}$
$\frac{2}{5}+\frac{\sqrt{6}}{10}$	$\frac{37}{225}+\frac{169\sqrt{6}}{1800}$	$\frac{11}{45}+\frac{7\sqrt{6}}{360}$	$-\frac{2}{225}-\frac{\sqrt{6}}{75}$
1	$\frac{4}{9}-\frac{\sqrt{6}}{36}$	$\frac{4}{9}+\frac{\sqrt{6}}{36}$	$\frac{1}{9}$
	$\frac{4}{9}-\frac{\sqrt{6}}{36}$	$\frac{4}{9}+\frac{\sqrt{6}}{36}$	$\frac{1}{9}$

8.4 The Module `scipy.integrate` Again

Several functions for the numerical solution of general initial value problems of ordinary differential equations are available in `scipy.integrate` module of SciPy. Specifically, among the many solvers of the module `integrate` one can find the explicit Runge-Kutta-Fehlberg methods 23 and 45 implemented in the functions `RK23` and `RK45`, and the implicit Radau IIA method of order 5 with the tableau as in Table 8.12, implemented in the function `Radau`. In general, Radau methods with s stages have order of convergence $2s-1$ and are very effective when applied to stiff equations. All these methods can also be called via the function `solve_ivp` which we present next.

8.4.1 Generic integration of initial value problems

We present here the generic function `solve_ivp` for the numerical solution of initial-value problems of ordinary differential equations. This function is designed to integrate numerically initial value problems of scalar and systems of ordinary differential equations of the form

$$y'(t) = f(t,y), \quad t \in (a,b] ,$$
$$y(a) = y_0 .$$

The general call of this function is the following:

```
solve_ivp(fun, t_span, y0, method, t_eval, args, options)
```

where the variables hidden in the `options` parameter can be ignored as all of them are optional and can be used in sophisticated problems. The most useful parameters for the purposes of this book are the following:

`fun`: Is the name of the function $f(t,y)$

`t_span`: The end-points of the interval $[a,b]$. Is a tuple `(t0,tf)` defining the (temporal) interval of integration (t_0, t_f). The solver starts with `t=t0` and ends with the computation of the solution at `t=tf`

`y0`: Is the initial condition which can be an one-dimensional array

`method`: Is the preferred numerical method. It can take the values `RK45`, `RK23`, `DOP853` and for stiff problems suggested values are `Radau`, `BDF` and `LSODA`. (Optional with default value `method=RK45`)

`t_eval`: An array with values of the times t_i on which we want the solution to be computed. (Optional with default value `t_eval=None`)

`args`: If the function f has additional arguments such as $f(t,y,a,b,c)$, then we pass the extra arguments in the tuple `args=(a,b,c)`. (Optional with default value `args=None`)

The additional `options` are summarized below:

`dense_output`: Is a boolean variable to specify whether to compute a continuous solution or not. (Optional with default value `dense_output=False`)

`vectorized`: Is a boolean variable that specifies if the function `fun` is implemented in a vectorized fashion. (Optional with default value `vectorized=False`)

`first_step`: Is a float that specifies initial stepsize h. (Optional with default value `first_step=None`)

`max_step`: Is a float that specifies the maximum allowed stepsize. (Optional with default value `max_step=np.inf`)

`rtol` and `atol`: Are floats or arrays specifying the relative and absolute tolerances. (Optional with default values `rtol=1.e-3` and `atol=1.e-6`)

`jac`: Is the Jacobian matrix necessary for the methods `Radau`, `BDF` and `LSODA`. When `None` then appropriate finite difference approximation is considered. (Optional with default value `jac=None`)

`jac_sparsity`: Is an `array_like` variable or a sparse matrix or `None`. This variable specifies a sparsity structure of the $n \times n$ Jacobian matrix for a finite difference approximation. If `None` then the Jacobian matrix is considered dense. Sparse matrix usually has more zero than non-zero entries. This variable is not supported by the `LSODA` method. (Optional with default value `jac_sparsity=None`).

lband and **uband**: These are integer variables specifying the bandwidth of the Jacobian for the `LSODA` method. If `None` then the matrix is considered full. (Optional with default value `None`)

min_step: Is a float specifying the minimum allowed stepsize for the `LSODA` method. (Optional with default value `0`)

And the function returns a class with the members the following variables:

t: Is an array containing the times at which the solution is computed

y: Is an array with the values of the approximated solution at `t`

sol: Is an instance of a class containing the solution as in the previous argument unless the argument `dense_output` was set to `False` where the return value is `None`

t_events: Contains for each event type a list of arrays at which an event of that type event was detected. It is `None` if the a events was None

t_events: A list of arrays (or `None`) containing for each value of `t_events`, the corresponding values of the solution. `None` is returned if the value of `events` is `None`

nfev: The number of evaluations of the right-hand side

njev: The number of evaluations of the Jacobian

nlu: The number of the LU decompositions required

status: Is -1 if the integration failed, 0 if the solver successfully computed the solution for the prescribed time interval, and 1 if a termination event occurred

message: Is a string with a description of the termination reason

success: Is a boolean variable which is `True` if the `status` is greater or equal to zero

In order to demonstrate the usage of the function `solve_ivp` we consider again the initial value problem

$$y'(t) = y(t), \quad t \in (0, 2] ,$$
$$y(0) = 1 ,$$

and we approximate its solution using the method RK45. For accurate results we consider the tolerance `rtol=1.e-6`.

```python
from scipy.integrate import solve_ivp

def f(t, y):
    return y
a = 0.0; b = 2.0
tspan = [a, b]
y0 = [1.0]
sol = solve_ivp(f, tspan, y0, 'RK45', rtol=1.e-6)
# Plot the results
fig = plt.figure()
axes = fig.add_subplot(1, 1, 1)
axes.plot(sol.t, sol.y[0], '-o', label="RK45")
axes.plot(sol.t, np.exp(sol.t), '-d', label="Exact solution")
axes.set_xlabel("t"); axes.set_ylabel("y"); axes.legend(loc=2); plt.show()
```

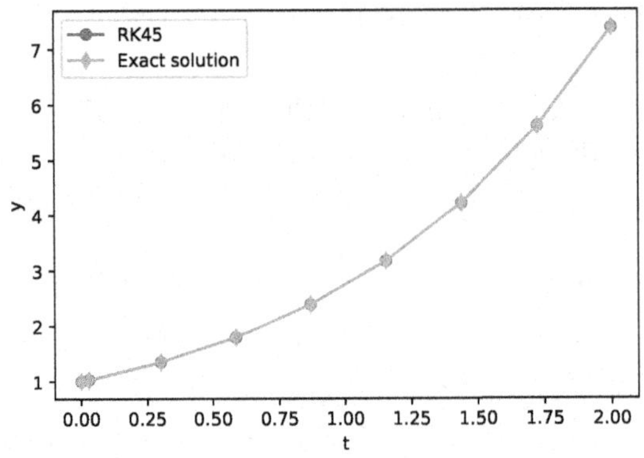

In order to plot the solution obtained by the function `solve_ivp`, we use the member `sol.y[0]`. The `[0]` is necessary since the variable `y` is at least a two-dimensional array but we need to plot only a one-dimensional array. The numerical and exact solutions agree as expected and it is impossible to observe differences within the scale of the specific graph.

8.4.2 Systems of ordinary differential equations

All the aforementioned numerical methods can be generalized for systems of ordinary differential equations in a straightforward manner. For simplicity, consider the following coupled system with two equations and two unknowns $y_1(t)$ and $y_2(t)$

$$\begin{cases} y_1'(t) = f_1(t, y_1(t), y_2(t)) \\ y_2'(t) = f_2(t, y_1(t), y_2(t)) \end{cases}, \quad t \in (a, b] , \tag{8.30}$$

with initial conditions $y_1(a) = y_1^0$ and $y_2(a) = y_2^0$. The system (8.30), like any other system of ordinary differential equations, can be written in vector form

$$\begin{aligned} \boldsymbol{y}'(t) &= \boldsymbol{f}(t, \boldsymbol{y}(t)), \quad t \in (a, b] , \\ \boldsymbol{y}(a) &= \boldsymbol{y}_0 , \end{aligned} \tag{8.31}$$

where $\boldsymbol{y}_0 = (y_1^0, y_2^0)$,

$$\boldsymbol{y}(t) = \begin{pmatrix} y_1(t) \\ y_2(t) \end{pmatrix} \quad \text{and} \quad \boldsymbol{f}(t, \boldsymbol{y}(t)) = \begin{pmatrix} f_1(t, y_1(t), y_2(t)) \\ f_2(t, y_1(t), y_2(t)) \end{pmatrix} .$$

The application of the numerical methods to (8.31) follows the same formalism as in the case of scalar equations. For example, given a uniform, temporal grid $a = t_0 < t_1 < \cdots < t_N = b$ the recursive algorithm of Euler's method for system (8.31) is

$$\begin{aligned} \boldsymbol{y}_{i+1} &= \boldsymbol{y}_i + h\ \boldsymbol{f}(t, \boldsymbol{y}_i), \quad i = 0, 1, \ldots, N-1 , \\ \boldsymbol{y}_0 &= \text{given} , \end{aligned} \tag{8.32}$$

which is the same as the algorithm (8.7) with the difference that here instead of computing scalar variables we compute vectors $\boldsymbol{y}_i = (y_1^i, y_2^i)$ for the approximations $y_1^i \approx y_1(t_i)$ and $y_2^i \approx y_2(t_i)$. One may want to write Euler's method without using vectors. In such case Euler's method is written as

$$\begin{aligned} y_1^{i+1} &= y_1^i + h f_1(t_i, y_1^i, y_2^i) , \\ y_2^{i+1} &= y_2^i + h f_2(t_i, y_1^i, y_2^i) , \\ y_1^0, y_2^0 &= \text{given} . \end{aligned} \tag{8.33}$$

In each step, each equation of (8.33) provides with a new approximation for each of the two variables y_1, y_2 separately, while the vector form (8.32) will provide a vector containing the same approximations of y_1, y_2.

The function `solve_ivp` as one may observe from its description it can solve also systems of ordinary differential equations. For more information, we refer the interested reader to more specialized books on the subject. Some of them are listed in the further reading section.

8.4.3 Application in epidemiology

As an application of the previous derivations we present a mathematical model that describes to some extend the spread of an infectious disease such as the common flu or the virus COVID-19 (SARS-CoV-2) in a population of N individuals. The particular model consists of three ordinary differential equations and is known as *compartmental model* in epidemiology, because the population is assigned to compartments with labels. In the simple case we study here, we consider three compartments with labels S, I and R representing the number of *Susceptible, Infectious* and *Recovered* parts of the population. This model is called SIR model to emphasize the three different compartments.

In the SIR model, t is the elapsed time in days, $S(t)$ is the number of susceptible individuals, $I(t)$ is the number of infectious individuals that can infect susceptible individuals, and $R(t)$ is the number of recovered individuals that they are either immunized or deceased. For sure the recovered population cannot get infected again. When, the susceptible individuals $S(t)$ come in contact with an infectious individual, then they become infectious. In our simple example, we assume that the population remains constant, which can be translated in mathematical language into

$$S(t) + I(t) + R(t) = \text{constant} = N \ .$$

If β is the average number of contacts per person per time, then the transition rate between susceptible S and infectious I individuals is expressed as

$$S' = -\frac{\beta}{N} SI \ .$$

The transition rate between infectious I and immunized R individuals is assumed to be proportional to the number of infectious individuals, and in particular

$$R' = \gamma I \ ,$$

where $\gamma = 1/D$ where D is the average time period in which an individual is infectious. The requirement of the constancy of the population $S + I + R = N$ leads to the law

$$S' + I' + R' = 0 \ ,$$

which gives us the third ordinary differential equation

$$I' = \frac{\beta}{N} SI - \gamma I \ .$$

Defining the solution vector $\boldsymbol{y}(t) = (S(t), I(t), R(t))^T$, we write the SIR model as a system

$$\boldsymbol{y}' = \boldsymbol{f}(t, \boldsymbol{y}) \quad \text{where} \quad \boldsymbol{f}(t, \boldsymbol{y}) = \begin{pmatrix} -\frac{\beta}{N} SI \ , \\ \frac{\beta}{N} SI - \gamma I \\ \gamma I \end{pmatrix} \ . \tag{8.34}$$

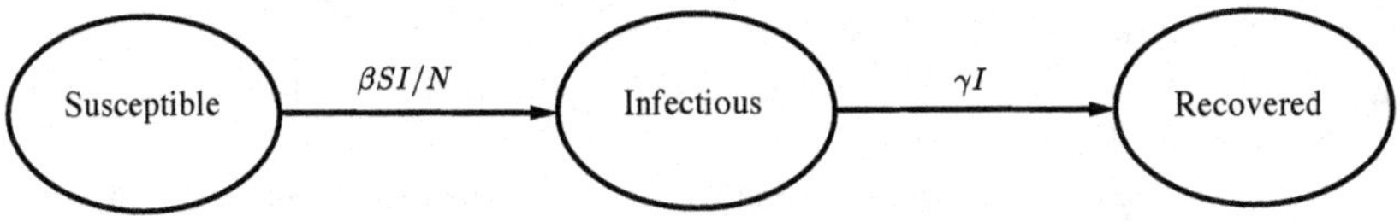

FIGURE 8.3
Schematic representation of the simplest SIR model.

The relation between the unknowns S, I and R is depicted schematically in Figure 8.3.

Note that the third equation is not necessary for the computation of the quantities S and I as the first two equations are not coupled with the third. The reason for keeping it into the equations is for simplicity, because otherwise we would need to solve this equation separately after the computation of the unknowns S and I.

The system (8.34) is nonlinear, and in order to solve it we will use the function `solve_ivp` of the module `scipy.integrate`. To test our code we consider the population of an isolated country with $N = 4,000,000$, and an infectious disease that is characterized by the parameters $\beta = 0.350$ and $\gamma = 0.035$. As initial conditions to our initial-value problem we consider $S(0) = N - 1$, $I(0) = 1$ and $R(0) = 0$, in the sense that initially there is one case of an infection in the whole population.

```
# Define the right-hand side and input parameters
def f(t, y, beta, gamma, N):
    z = np.zeros(3)
    z[0]=-beta/N * y[0] * y[1]
    z[1]= beta/N * y[0] * y[1] - gamma * y[1]
    z[2]= gamma * y[1]
    return z
beta = 0.35; gamma = 0.035
N = 4.e+6; a = 0.0; b = 200.0; tspan = [a, b]
y0 = [N-1, 1, 0]
args1=(beta, gamma, N)
# Solve the ODE
sol1 = solve_ivp(f, tspan, y0, 'RK45', args=args1, rtol=1.e-6)
# Plot the results
fig = plt.figure()
axes = fig.add_subplot(1, 1, 1)
axes.plot(sol1.t, sol1.y[0], label="S(t)")
axes.plot(sol1.t, sol1.y[1], label="I(t)")
axes.plot(sol1.t, sol1.y[2], label="R(t)")
axes.set_xlabel("t"); axes.set_ylabel("y"); axes.legend(loc=1); plt.show()
```

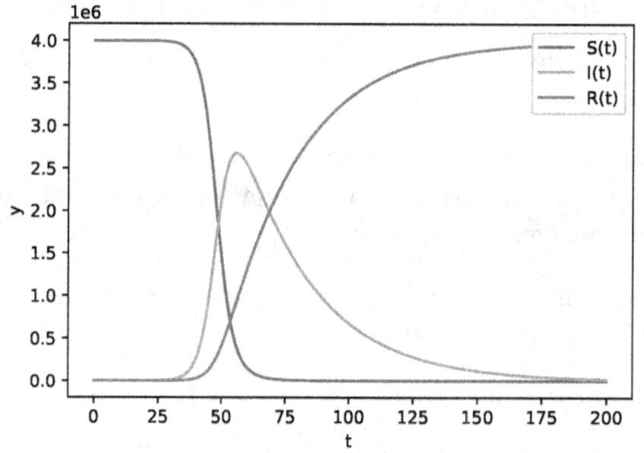

Using the simplified SIR model (8.34), and with the particular parameters, we can predict that the population will be immunized by the disease in less than a year. Additional parameters and factors must be considered for accurate predictions. An improvement of the SIR model known as $SIRD$ model is presented in the exercises.

8.5 Further Reading

The approximation of derivatives with finite differences is covered by almost all classic textbooks in numerical analysis [25, 70, 76, 22, 26, 7, 108, 127]. A list of specialized books on finite differences and on the numerical solution of differential equations includes the following [44, 19, 121, 51, 36, 71, 55, 56, 54, 114, 83, 2, 82, 39]. An introduction to the mathematical modeling of infectious diseases can be found in [15, 61].

Chapter Highlights

- First derivatives can be approximated using forward, backward and central finite differences.
- The forward and backward finite differences are first-order accurate with error $O(h)$. The central finite difference approximation of the first derivative is second-order accurate with error $O(h^2)$.
- Finite difference formulas can be obtained using Taylor expansions of functions around appropriately chosen points.
- We can approximate second-order derivatives with finite differences such as

$$f''(x_0) \approx \frac{f(x_0+h) - 2f(x_0) + f(x_0-h)}{h^2} ,$$

which is second-order accurate with error $O(h^2)$.
- Given an approximation formula that depends on h, one can obtain improved approximations using Richardson's extrapolation.
- Given an initial value problem $y'(t) = f(t, y(t))$, $t \in (a, b]$ with initial condition $y(a) = y_0$, a numerical approximation $y_i \approx y(t_i)$ at the nodes $a = t_0 < t_1 < \cdots < t_N = b$ with uniform distance $h = t_{i+1} - t_i$ can be computed using Euler's method

$$y_{i+1} = y_i + hf(t_i, y_i), \quad i = 0, 1, \ldots, N-1 .$$

- The local truncation error is defined as the error between the approximate solution y_{i+1} and the exact solution of the differential equation $y(t_{i+1})$ when $y_i = y(t_i)$ is exact.
- The global error is the error between the numerical solution y_{i+1} and the exact solution $y(t_{i+1})$.
- The local error in Euler's method is $O(h^2)$ while the global error is $O(h)$.
- More sophisticated methods such as Runge-Kutta methods can result in more accurate numerical approximations of ordinary differential equations.
- If the solution of the differential equation varies rapidly, then the system is characterized as stiff.
- We usually solve stiff ordinary differential equations using implicit numerical methods.
- Mathematical modeling of infectious diseases relies on ordinary differential equations. The SIR model is a simple compartmental model that describes the spread of an infectious disease.
- An SIR model divides the population in three compartments, Susceptible (S), Infectious (I) and Recovered (R) individuals and is a coupled system of three first order ordinary differential equations.

Exercises

1. Consider a smooth function $f(x)$ defined on $\mathbb{R}$ and $x_0 \in \mathbb{R}$.

   (a) Using appropriate Taylor expansions of the function f and a parameter h, find a finite difference approximation of the first derivative $f'(x_0)$ in the form

   $$f'(x_0) \approx \frac{f(x_0 - 2h) - 8f(x_0 - h) + 8f(x_0 + h) + f(x_0 + 2h)}{12h} .$$

   (b) Determine the order of accuracy of your finite difference approximation.

2. Consider a smooth function $f(x)$ defined on $\mathbb{R}$ and $x_0 \in \mathbb{R}$.

   (a) Using appropriate Taylor expansions of the function f and a parameter h, find a finite difference approximation of the first derivative $f'(x_0)$ in the form

   $$f'(x_0) \approx \frac{3f(x_0) - 4f(x_0 - h) + f(x_0 - 2h)}{2h} .$$

   (b) Determine the order of accuracy of your finite difference approximation.

   (c) Derive the same formula using Richardson's extrapolation method and the backward finite difference approximation of first order.

3. Consider a smooth function $f(x)$ defined on $\mathbb{R}$ and $x_0 \in \mathbb{R}$.

   (a) Using appropriate Taylor expansions of the function f and a parameter h, find a finite difference approximation of the first derivative $f'(x_0)$ in the form

   $$f'(x_0) \approx \frac{-3f(x_0) + 4f(x_0 + h) - f(x_0 + 2h)}{2h} .$$

   (b) Determine the order of accuracy of your finite difference approximation.

   (c) Derive the same formula using Richardson's extrapolation method and the forward finite difference approximation of first order.

4. Consider the function $f(x) = \sin(\pi x)$ and the values $h = 0.2, 0.1, 0.05, 0.01, 0.005, 0.001, 0.0005, 0.0001$.

   (a) Write a Python program to approximate the derivative $f'(x_0)$ for $x_0 = 0.4$ with central, forward and backward differences for all values of the parameter h.

   (b) In addition to the previous approximations consider the finite difference approximation

   $$D_\ell f(x_0) = \frac{f(x_0 - 2h) - 8f(x_0 - h) + 8f(x_0 + h) - f(x_0 + 2h)}{12h} ,$$

   (c) Let p be the convergence rate of such that $|D_\ell f(x_0) - f'(x_0)| \leq Ch^p$. Estimate numerically the convergence rate p for each of the previous finite difference approximations: Consider a given function $f(x)$ and values $h = h_1, h_2, \dots h_n$ with $h_{i+1} < h_i$. Compute the approximation $D_\ell f(x)$ of the derivative $f'(x)$ and then compute the error $E_i = |f'(x) - D_\ell f(x)|$ for each h_i. This error must be $E_i = Ch_i^p$. Dividing the values E_i and E_{i+1} we compute the fraction

   $$E_i/E_{i+1} = (h_i/h_{i+1})^p$$

and then with the help of logarithms we solve for p to obtain

$$p = \frac{\log(E_i/E_{i+1})}{\log(h_i/h_{i+1})} \ .$$

Estimate the convergence rate p with $x_0 = 0.4$ and $x_0 = 1.0$.

(d) Repeat the previous question to estimate the convergence rate of the forward, backward and central finite differences.

(e) What do you observe?

5. Repeat the procedure of the previous question to approximate the second derivative of the function $f(x) = \sin(\pi x)$ at $x_0 = 0.4$ using the finite difference approximation

$$f''(x_0) \approx \frac{f(x_0 + h) - 2f(x_0) + f(x_0 - h)}{h^2} \ .$$

6. Consider a finite difference approximation of the first derivative $f'(x_0)$ of the form

$$f'(x_0) \approx a\ f(x_0) + b\ f(x_0 + h) + c\ f(x_0 + 2h) \ ,$$

for a small value of h. We say that a finite difference formula is exact for polynomials $p(x)$ for degree r if the previous approximation is exact

$$p'(x_0) = a\ p(x_0) + b\ p(x_0 + h) + c\ p(x_0 + 2h) \ .$$

(a) Show that the previous finite difference approximation is exact for quadratic polynomials if and only if it is exact for the monomials 1, x and x^2.

(b) Find the values a, b and c so that the specific finite difference approximation is exact for quadratic polynomials.

7. In this problem you will practice complex number arithmetic in Python. Consider a real function $f \in C^3[a, b]$ and assume that we want to approximate the derivative $f'(x_0)$ with high accuracy. For this reason, we consider the small step parameter $h_i = 10^{-i}$ for $i = 1, 2, \ldots, 300$.

(a) Given $0 < h < 1$ define an approximation of the first derivative to be

$$D_{\mathrm{i}} f(x_0) = \frac{\mathrm{Im}[f(x_0 + \mathrm{i}h)]}{h} \ ,$$

where i denotes the imaginary unit, and Im is the imaginary part of a complex number. Prove that there is a $\xi \in [a, b]$ such that for $x_0 \in [a, b]$

$$|f'(x_0) - D_{\mathrm{i}} f(x_0)| \leq \frac{|f^{(3)}(\xi)|}{6} h^2 \ .$$

(b) Write a Python function to implement the approximate formula $D_{\mathrm{i}} f$ of the first derivative.

(c) Consider the function

$$f(x) = \frac{e^x}{\sin^3 x + \cos^3 x} \ ,$$

and its derivative $f'(x)$. Compute the approximations of $f'(1.5)$ for all values of $h = h_i$, $i = 1, 2, \ldots, 300$ and plot a graph of the error as a function of h. Determine what is the optimal value h_i which can be used in this method.

(d) For the same values of h, compute the central approximation of the derivative $D_c f$ and compare the errors. Determine again what is the optimal value h_i that can be used in the central difference formula.

(e) Explain the differences between the two methods, and also the various phenomena you observe.

[The approximation of the derivative in this problem is known to as the *complex-step* derivative approximation. For more information, we refer to the publications [123, 85]]

8. Write the composite trapezoidal rule for the integration of a function $f(x)$ over an interval $[a, b]$ as

$$I[f] = I_N[f] + E_N[f] ,$$

where $I[f]$ is the exact value of the integral, $I_N[f]$ is the quadrature rule and $E_N[f]$ is the error given by the formula

$$E_N[f] = -\frac{(b-a)h^2}{12} f''(\xi) ,$$

where $h = (b-a)/N$.

(a) Using Richardson's extrapolation consider two values N_1 and N_2 such that $N_1 = 2N_2$ and eliminate the error to obtain the Romberg's quadrature formula

$$I[f] = \frac{4}{3} I_{N_2}[f] - \frac{1}{3} I_{N_1}[f] .$$

(b) Implement this improved quadrature formula in a Python function and test your code using the integral

$$\int_0^1 e^{-x^2}\, dx = \sqrt{\pi}/2 .$$

(c) Estimate by numerical means the order of convergence of Romberg's quadrature formula.

9. This exercise is based on the work [52]. Let $x, h \in \mathbb{R}$, and

$$g(x) = \int_{-h}^{h} t f(t+x)\, dt ,$$

with $f \in C^4[x-h, x+h]$.

(a) Using Simpson's rule show that

$$\int_{-h}^{h} t f(t+x)\, dt = \frac{h}{3}[-hf(x-h) + hf(x+h)] + O(h^5) .$$

(b) Using the central finite difference approximation formula

$$f'(x) = \frac{f(x+h) - f(x-h)}{2h} + O(h^2) ,$$

show that

$$f'(x) \approx \frac{3}{2h^3} \int_{-h}^{h} t f(x+t)\, dt ,$$

and estimate the error in the approximation.

(c) Write a Python code to approximate the function $g(x)$ for $x \in [-1, 1]$ when $f(x) = e^{x^2}$ implementing the previous approximations.

(d) Show that

$$\int_{-h}^{h} tf(t+x)\, dt = h^2 \int_{-1}^{1} uf(hu+x)\, du \ .$$

(e) Use Gauss-Legendre quadrature to approximate the same function $g(x)$ as before.

10. Consider the initial value problem

$$\begin{cases} \frac{d}{dt}y(t) = f(t, y(t)), & a \leq t \leq b\ , \\ y(a) = y_0\ , \end{cases} \tag{8.35}$$

(a) Prove that the initial value problem with $f(t, y) = y$ and $y_0 = 1$ admits the exact solution $y(t) = e^t$.

(b) Write a function `Euler` implementing Euler's method to solve the specific initial-value problem. The function should require as input the function f, the endpoints a and b, the initial condition y_0 and the temporal stepsize Δt. It will return the approximation of the solution at the points of the temporal partition in a vector y.

(c) Write similar functions to implement the improved and implicit Euler methods.

(d) Estimate the convergence rates of these methods experimentally. For this reason, consider different values N_i and stepsize $h = h_1, h_2, \ldots, h_n$ with $h_{i+1} < h_i$, such that $h_i = (b-a)/N_i$. Compute the solution at $t_{N_i} = b$ for all values of h_i and also the errors $E_i = |y_{N_i} - y(t_{N_i})|$. Estimate the convergence rate using the formula

$$p = \frac{\log(E_i/E_{i+1})}{\log(h_i/h_{i+1})}\ .$$

11. For the initial value problem (8.35) consider a uniform grid $a = t_0 < t_1 < \cdots < t_N = b$ with $h = t_{i+1} - t_i$, and the trapezoidal method for the computation of the approximations $y_i \approx y(t_i)$:

$$y_{i+1} = y_i + \frac{h}{2}\left[f(t_{i+1}, y_{i+1}) + f(t_i, y_i)\right], \quad i = 0, 1, \ldots, N-1\ .$$

(a) Implement this method in Python to solve the problem with $f(t, y) = y$ in $[0, 2]$ and $y_0 = 1$.

(b) Estimate experimentally the convergence rate of the specific numerical method.

(c) Formulate the trapezoidal method as a Runge-Kutta method.

(d) Study the stability of the trapezoidal method and compare with that of implicit Euler's method.

(e) Devise an algorithm using Newton's method for the solution of the resulting non-linear equation in the trapezoidal method and compare with the improved Euler (Heun) method.

12. Consider the initial value problem for the following system of ordinary differential equations

$$\begin{cases} y_1'(t) = -y_2(t)\ , \\ y_2'(t) = y_1(t)\ , \end{cases} \quad t \geq 0\ ,$$

given that $y_1(0) = 1, \quad y_2(0) = 0$.

(a) Show that the following conservation of energy holds

$$(y_1(t))^2 + (y_2(t))^2 = 1, \quad \text{for all} \quad t \geq 0 .$$

(b) Write the approximations (y_1^i, y_2^i), $i = 1, 2, \ldots$ of (explicit) Euler's method and show that $(y_1^i)^2 + (y_2^i)^2 \to \infty$ as $i \to \infty$.

(c) Show that the analogous approximations (y_1^i, y_2^i) of the trapezoidal method (of the Exercise 11) with constant stepsize h respects the conservation of energy $(y_1^i)^2 + (y_2^i)^2 = 1$ for all $i \geq 1$. This is an example of a symplectic method that preserves quadratic invariants.

(d) What is the conservation properties of implicit Euler's method for the same problem?

(e) Use the function `solve_ivp` to solve the particular system experimenting with the available methods for non-stiff problems. In all cases comment on the conservation properties of these methods.

13. A metallic rod of length $L = 1\ m$ is heated by a heating source located at its center of gravity. The endpoints of the rod are kept at $0°C$. If $u(x)$ denotes the temperature of the rod at location x, then it satisfies the ordinary differential equation

$$-u''(x) + u(x) = (4\pi^2 + 1)\sin(2\pi x) ,$$

which accompanied with the boundary conditions

$$u(0) = u(1) = 0 ,$$

forms a two-point boundary-value problem.

(a) To find approximations to this problem first consider a uniform grid of the interval $[0, 1]$ of points $x_{i+1} = x_i + h$ for $i = 0, 1, \ldots, N - 1$ with $x_0 = 0$ and $h = 1/N$. Then using the second order formula for the second derivative

$$u''(x_i) \approx \frac{u(x_{i-1}) - 2u(x_i) + u(x_{i+1})}{h^2} ,$$

and given that $u(x_0) = u(x_N) = 0$ derive a linear systems of equations $\boldsymbol{Au} = \boldsymbol{b}$ where

$$\boldsymbol{u} = \begin{pmatrix} U_1 & U_2 & \cdots & U_{N-1} \end{pmatrix}^T ,$$

for the approximation $U_i \approx u(x_i)$.

(b) Write a Python function to solve this problem and perform experiments with various values of N.

(c) Verify that the function $u(x) = \sin(2\pi x)$ satisfies the two-point boundary value problem.

(d) Compute maximum error between the numerical and analytical solution at the nodes x_i such as

$$E_N = \max_{0 \leq i \leq N} |U_i - u(x_i)| .$$

(e) Estimate the convergence rate and show that $E_N \leq Ch^2$ for some constant $C > 0$ independent of h.

14. A significant drawback of the simple SIR model is that we consider the deceased individuals as recovered. An extension of SIR model that distinguishes the two cases is called $SIRD$ model and includes an additional compartment D for the *deceased* individuals. The $SIRD$ system reads

$$\begin{aligned} S' &= -\frac{\beta}{N} IS \ , \\ I' &= \frac{\beta}{N} IS - \gamma I - \mu I \ , \\ R' &= \gamma I \ , \\ D' &= \mu I \ , \end{aligned} \tag{8.36}$$

where β, γ, μ are the infection, recovered and deceased rates, respectively.

(a) Write a computer code in Python that solves system (8.36).

(b) Using $\beta = 0.35$, $\gamma = 0.035$ and $\mu = 0.1$ and the initial conditions

$$(S, I, R, D) = (N, 1, 0, 0) \ ,$$

solve system (8.36) and predict the death rate of the diseased in a population of $N = 4,000,000$ individuals.

(c) Try different parameters β and μ to minimize the death rate.

9

Numerical Linear Algebra

Numerical linear algebra is the branch of numerical analysis that focuses on the development of methods and algorithms for solving problems with matrices such as the solution of linear systems. It is perhaps the most useful branch of numerical analysis since the solution of linear systems is a major issue in most engineering applications. It is remarkable that even the solution of nonlinear systems of equations may require the solution of linear systems. For example, as we saw in Section 5.5.6, Newton's method for nonlinear systems requires the solution of a linear system of equations at every iteration. In this chapter, we will limit the presentation to some direct and iterative methods for the numerical solution of linear systems. We start with the most popular and basic method, the Gaussian elimination.

9.1 Numerical Solution of Linear Systems

A general system of linear equations (simultaneous equations) can be expressed as

$$\begin{aligned}
a_{11}\,x_1 + a_{12}\,x_2 + \cdots + a_{1n}\,x_n &= b_1 \ , \\
a_{21}\,x_1 + a_{22}\,x_2 + \cdots + a_{2n}\,x_n &= b_2 \ , \\
&\ \vdots \\
a_{n1}\,x_1 + a_{n2}\,x_2 + \cdots + a_{nn}\,x_n &= b_n \ .
\end{aligned}$$

The constant coefficients a_{ij} for each $i, j = 1, 2, \dots, n$, and b_j for each $j = 1, 2, \dots, n$ of this system are assumed to be given, and we need to find the unknowns $x_1, x_2, \dots, x_n$. Computer languages, though, have not been equipped with structures to store a linear system other than matrices and vectors. Fortunately, a linear system can be expressed in matrix-vector form $\boldsymbol{Ax} = \boldsymbol{b}$ where

$$\boldsymbol{A} = \begin{pmatrix} a_{11} & a_{12} & \cdots & a_{1n} \\ a_{21} & a_{22} & \cdots & a_{2n} \\ \vdots & \vdots & & \vdots \\ a_{n1} & a_{n2} & \cdots & a_{nn} \end{pmatrix} ,$$

is the coefficient matrix, $\boldsymbol{x} = \begin{pmatrix} x_1 & x_2 & \cdots & x_n \end{pmatrix}^T$ is the solution of the linear system, and the vector $\boldsymbol{b} = \begin{pmatrix} b_1 & b_2 & \cdots & b_n \end{pmatrix}^T$ is the right-hand side.

In order to develop a numerical method to solve linear systems we must first revise the method of Gaussian[1] elimination. Gaussian elimination (also known as forward elimination or row reduction) is a *direct method* used to transform a linear system into an upper triangular system, in the sense that the resulting coefficient matrix $\boldsymbol{A}$ is upper triangular. A direct

[1]The Gaussian elimination method is named after the German mathematician and physicist Johann Carl Friedrich Gauss (1777–1855)

DOI: 10.1201/9781003287292-9

method is a numerical method to compute the exact solution (not an approximation like Newton's method) when exact arithmetic is being used. On the contrary, iterative methods (such as Newton's method) generate an approximation to the solution of the linear system.

Performing Gaussian elimination to a linear system $\boldsymbol{A}\boldsymbol{x} = \boldsymbol{b}$ with a non-singular matrix $\boldsymbol{A}$, $(\det(\boldsymbol{A}) \neq 0)$, we transform the system into a new system $\boldsymbol{U}\boldsymbol{x} = \boldsymbol{c}$ where

$$\boldsymbol{U} = \begin{pmatrix} u_{11} & u_{12} & \cdots & u_{1n} \\ 0 & u_{22} & \cdots & u_{2n} \\ \vdots & \vdots & & \vdots \\ 0 & 0 & \cdots & u_{nn} \end{pmatrix} ,$$

with $u_{ii} \neq 0$ for $i = 1, 2, \dots, n$. The new system $\boldsymbol{U}\boldsymbol{x} = \boldsymbol{c}$ apparently has the same solution as the system $\boldsymbol{A}\boldsymbol{x} = \boldsymbol{b}$ regardless that $\boldsymbol{U}$ is not the same as $\boldsymbol{A}$. It is noted that the right-hand side $\boldsymbol{b}$ is also transformed in a new right-hand side $\boldsymbol{c}$. In the end, we can solve the system $\boldsymbol{U}\boldsymbol{x} = \boldsymbol{c}$ easily using backward substitution. The best way to understand this method is with an example, but first we present the methodology in its full generality. Therefore, we first describe the steps of the Gaussian elimination applied to the following linear system of equations

$$\begin{aligned} E_1 :&\quad a_{11}\, x_1 + a_{12}\, x_2 + \cdots + a_{1n}\, x_n = b_1 \ , \\ E_2 :&\quad a_{21}\, x_1 + a_{22}\, x_2 + \cdots + a_{2n}\, x_n = b_2 \ , \\ &\vdots \\ E_n :&\quad a_{n1}\, x_1 + a_{n2}\, x_2 + \cdots + a_{nn}\, x_n = b_n \ , \end{aligned}$$

where we denote the i-th equation by E_i for simplicity.

Gaussian elimination is based on three linear operations that lead to equivalent systems of equations, in the sense that the resulting system has the same solution as the original one. We call these operations *basic*. These basic operations are the following:

- We can multiply any equation with a non-zero number. For example, equation E_i can be multiplied by any non-zero constant λ with the resulting equation used in place of E_i. This operation is denoted $(\lambda E_i) \to (E_i)$.
- We can add a multiple of one equation to another equation. For example, equation E_j can be multiplied with any constant λ and added to equation E_i. This operation is denoted $(E_i + \lambda E_j) \to (E_i)$.
- We can swap two equations. For example, equation E_i and E_j can exchange places in the system. This operation is denoted $(E_i) \leftrightarrow (E_j)$.

We use the basic operations to eliminate columns and transform the coefficients of the low-trangular part of the original system into zeros. In particular, we use these operations to eliminate all the entries a_{ij} with $j < i$, i.e. all the entries under the main diagonal of the corresponding coefficient matrix. For example, in order to eliminate the entry a_{21} we need to multiply the first equation with a_{21}/a_{11} and subtract the result from the second equation. The resulting equation replaces the second equation E_2 and we say that we have eliminated the entry a_{21}. This operation is expressed as

$$(E_2 - (a_{21}/a_{11})\ E_1) \to (E_2) \ .$$

For example, consider the linear system

$$\begin{aligned} E_1 :&\quad x_1 + x_2 \phantom{{}- x_3} = 2 \ , \\ E_2 :&\quad 2x_1 + x_2 - x_3 = 2 \ , \\ E_3 :&\quad 3x_1 - x_2 - x_3 = 1 \ . \end{aligned}$$

First we use equation E_1 to eliminate the unknown x_1 from equations E_2 and E_3 by performing the basic operations

$$(E_2 - 2E_1) \to (E_2) ,$$
$$(E_3 - 3E_1) \to (E_3) .$$

The first operation eliminates the unknown x_1 from equation E_2, and the second from equation E_3. This step will eliminate only the coefficients of the unknown x_1 stored in the first column of the coefficient matrix A. This leads to the equivalent new system

$$\begin{array}{llllll} E_1: & x_1 & +x_2 & & = & 2 \\ E_2: & & -x_2 & -x_3 & = & -2 \\ E_3: & & -4x_2 & -x_3 & = & -5 \end{array} \tag{9.1}$$

After eliminating the first column we do not use E_1 again to avoid destroying what we have achieved so far. Subsequently, E_2 is used to eliminate the unknown x_2 from E_3 by performing the basic operation

$$(E_3 - 4E_2) \to (E_3) ,$$

to obtain the equivalent system

$$\begin{array}{llllll} E_1: & x_1 & +x_2 & & = & 2 \\ E_2: & & -x_2 & -x_3 & = & -2 \\ E_3: & & & 3x_3 & = & 3 \end{array} \tag{9.2}$$

After applying the permitted linear transformations, the new system is in the form of an upper triangular system. The later system of equations is written in *triangular* (or *reduced*) form and can be solved for the unknowns by a *backward substitution process*: We solve the equation E_3 for x_3 dividing by 3 to obtain $x_3 = 1$. Then we substitute the solution $x_3 = 1$ into E_2 and we solve for x_2 to obtain $x_2 = 1$. We substitute the values of x_2 and x_3 into E_1 and solve for $x_1 = 1$. The solution vector we found is the vector

$$\boldsymbol{x} = \begin{pmatrix} 1 \\ 1 \\ 1 \end{pmatrix} .$$

By performing all the previous linear transformations to the original linear system we didn't alter the unknowns x_i, and the solution we found is the correct one.

☞ Observe that each entry of the right-hand side of the previous linear system is the sum of the coefficients of the corresponding equation. Such a linear system has solution 1 for each unknown. To see this consider its matrix-vector form and think of the multiplication $\boldsymbol{Ax}$ with $\boldsymbol{x}$ a vector full of ones.

The same linear transformations, and therefore Gaussian elimination, can be applied to the coefficient matrix $\boldsymbol{A}$ and the right hand side $\boldsymbol{b}$ of the linear system. This is important because the implementation of any linear algebra algorithm in any programming language will involve matrices and not linear equations. For theoretical purposes, and when we work with pen and paper, we usually combine the coefficient matrix $\boldsymbol{A}$ and right-hand side $\boldsymbol{b}$ together into the *augmented* matrix $(\boldsymbol{A}|\boldsymbol{b})$. This is done by merging the right-hand side to a new $(n+1)$-th column in the matrix $\boldsymbol{A}$ separated with a vertical line. For example, the augmented matrix of the previous linear system is the matrix

$$(\boldsymbol{A}|\boldsymbol{b}) = \left(\begin{array}{rrr|r} 1 & 1 & 0 & 2 \\ 2 & 1 & -1 & 2 \\ 3 & -1 & -1 & 1 \end{array}\right) .$$

When we apply the linear transformations to the augmented matrix, we use the rows of the augmented matrix instead of equations. For this reason, we call the basic operations of the Gaussian elimination *row operations.* If the i-th row of the augmented matrix is denoted by R_i, then the basic row operations to eliminate the first column in the previous example will be

$$(R_2 - 2R_1) \rightarrow (R_2) \ ,$$
$$(R_3 - 3R_1) \rightarrow (R_3) \ .$$

The resulting augmented matrix (with eliminated first column) will contain the coefficients of the reduced linear system (9.1)

$$\left(\begin{array}{ccc|c} 1 & 1 & 0 & 2 \\ 0 & -1 & -1 & -2 \\ 0 & -4 & -1 & -5 \end{array}\right) .$$

The corresponding row operations for eliminating the second column of the new augmented matrix will be

$$(R_3 - 4R_2) \rightarrow (R_3) \ ,$$

and the reduced augmented matrix that describes system (9.2) will be

$$\left(\begin{array}{ccc|c} 1 & 1 & 0 & 2 \\ 0 & -1 & -1 & -2 \\ 0 & 0 & 3 & 3 \end{array}\right) .$$

We can denote the last matrix by $(\boldsymbol{U}|\boldsymbol{c})$ to indicate that $\boldsymbol{A}$ and $\boldsymbol{b}$ have been transformed to an upper triangular matrix $\boldsymbol{U}$ and new right-hand side vector $\boldsymbol{c}$.

Now we need to perform back substitution to the (transformed) upper triangular system $\boldsymbol{U}\boldsymbol{x} = \boldsymbol{c}$ to compute the unknown values x_i. This can be done by writing the system $\boldsymbol{U}\boldsymbol{x} = \boldsymbol{c}$ in equations form

$$\begin{array}{cccc} x_1 & +x_2 & & = & 2 \\ & -x_2 & -x_3 & = & -2 \\ & & 3x_3 & = & 3 \end{array}$$

and then first solve the last equation x_3 which gives $x_3 = 1$. Solving the second equation for x_2 using $x_3 = 1$ we obtain $x_2 = 1$ and finally solving the first equations for x_1 we obtain $x_1 = 1$.

9.1.1 Algorithm for the naive Gaussian elimination

Suppose that we store the coefficients of a linear system into a matrix $\boldsymbol{A}$ and vector $\boldsymbol{b}$. The previous elimination procedure can be formulated in the following steps: For eliminating the k-th column for $k = 1, \ldots, n-1$ (the n-th column is used to compute x_n and is not eliminated) we define the *multipliers*

$$m_{ik} = \frac{a_{ik}}{a_{kk}} \quad \text{for} \quad i = k+1, \ldots, n \ ,$$

and we perform the substitutions

$$\begin{array}{ll} a_{ij} \leftarrow a_{ij} - m_{ik}a_{kj}, & j = k, \ldots, n, \\ b_i \leftarrow b_i - m_{ik}b_k, & \end{array} \quad \text{for} \quad i = k+1, \ldots, n \ ,$$

where we modify the entries of $\boldsymbol{A}$ instead of using a new matrix $\boldsymbol{U}$, and the same with vector $\boldsymbol{c}$, which is stored in vector $\boldsymbol{b}$. This can save memory space, especially when the matrices are large. These steps are summarized using vectorized format in Algorithm 27. The multipliers m_{ik} in Algorithm 27 are stored in a matrix $\boldsymbol{M}$. This is not necessary though and one can use a simple variable to store the individual multipliers locally in memory unless they are required for later use. We will see that these multipliers can play an important role in computations.

Algorithm 27 Naive Gaussian elimination

Given the arrays $\boldsymbol{A}$ and $\boldsymbol{b}$ the solution $\boldsymbol{x}$ replaces $\boldsymbol{b}$
for $k = 1 : n-1$ **do**
  for $i = k+1 : n$ **do**
    $\boldsymbol{M}(i,k) = \boldsymbol{A}(i,k)/\boldsymbol{A}(k,k)$
    $\boldsymbol{A}(i,k:n) = \boldsymbol{A}(i,k:n) - \boldsymbol{M}(i,k)\cdot\boldsymbol{A}(k,k:n)$
    $\boldsymbol{b}(i) = \boldsymbol{b}(i) - \boldsymbol{M}(i,k)\cdot\boldsymbol{b}(k)$
  end for
end for

In order to formulate the backward substitution for upper triangular systems, we write the reduced system $\boldsymbol{U}\boldsymbol{x} = \boldsymbol{c}$ in equations form

$$\begin{aligned} a_{11}x_1 + a_{12}x_2 + \cdots + a_{k1}x_k + \cdots + a_{n1}x_n &= b_1 \ , \\ &\vdots \\ a_{k,k}x_k + a_{k,k+1}x_{k+1} + \cdots + a_{kn}x_n &= b_k \ , \\ &\vdots \\ a_{nn}x_n &= b_n \ . \end{aligned}$$

The back substitution can be performed first by solving the last equation of the reduced system for x_n

$$x_n = \frac{b_n}{a_{nn}} \ ,$$

and then by computing the rest of the unknowns in a `for` loop using the formula

$$x_k = \frac{1}{a_{kk}}\left(b_k - \sum_{j=k+1}^{n} a_{kj}x_j\right) \quad \text{for} \quad k = n-1, n-2, \ldots, 1 \ .$$

The last formula can be derived by solving the k-th equation for x_k. For the computation of x_k we only used the value b_k for the current k. Since we do not use previous values b_{k-1}, then we can replace them with the values x_k for memory economy. Taking into account this observation we summarize the backward substitution in Algorithm 28.

Algorithm 28 Backward substitution for solving upper triangular systems

Given an upper triangular array $\boldsymbol{A}$, the right-hand side $\boldsymbol{b}$ is replaced by the solution $\boldsymbol{x}$
$\boldsymbol{b}(n) = \boldsymbol{b}(n)/\boldsymbol{A}(n,n)$
for $k = n-1 : 1 : -1$ **do**
  $\boldsymbol{b}(k) = (\boldsymbol{b}(k) - \boldsymbol{A}(k,k+1:n)\cdot\boldsymbol{b}(k+1:n)) \ / \ \boldsymbol{A}(k,k)$
end for

☞ We use backward substitution to solve upper triangular systems. The corresponding algorithm for lower triangular systems is called forward substitution. The forward substitution algorithm is presented in detail in Algorithm 30.

We combine Algorithms 27 and 28 in a Python function `Gauss_elimination`. Note that algorithms are usually presented with natural indices while the implementation takes into account the Python indexing conventions. Taking as input the arrays `A` and `b`, the function `Gauss_elimination` returns the solution of the corresponding linear system stored in the variable `b`.

```
def Gauss_elimination(A,b):
    n = len(b)
    # Gaussian elimination
    for k in range(n-1):
        for i in range(k+1,n):
            if A[k,k]!=0.0:
                mik=A[i,k]/A[k,k]
                A[i,k+1:n] = A[i,k+1:n] - mik * A[k,k+1:n]
                b[i] = b[i] - mik * b[k]
    # backward substitution
    for k in range(n-1,-1,-1):
        b[k] = (b[k]-np.dot(A[k,k+1:n],b[k+1:n]))/A[k,k]
    return b
```

We test our code in solving the linear system $\boldsymbol{Ax} = \boldsymbol{b}$ of the previous section. Because the function `Gauss_elimination` alters the arrays `A` and `b` we call it with arguments `A.copy()` and `b.copy()`, respectively. This will leave the arrays `A` and `b` unaltered for further use, while inside the function new unrelated copies will be modified. In other situations we would prefer to overate `A` and `b` for memory economy.

```
A = np.array([[ 1., 1., 0.],
              [ 2., 1.,-1.],
              [ 3.,-1.,-1.]])
b = np.array([ 2., 2., 1.])
x = Gauss_elimination( A.copy(), b.copy() )
print(x)
```

```
[1. 1. 1.]
```

It is worth mentioning that there are variants of Gaussian elimination for banded or tridiagonal systems. An example of a tridiagonal algorithm can be found in the exercises.

Operations count

In Algorithm 27 we first compute the multipliers. Every time we compute a multiplier we perform one division, and thus in total we need

$$\sum_{k=1}^{n-1}(n-k) = \sum_{k=1}^{n-1} k = \frac{n(n-1)}{2} ,$$

TABLE 9.1
Operations count for the Gaussian elimination and back substitution

Gaussian elimination	Multiplications	Additions	Divisions
Multipliers	$\frac{n(n-1)}{2}$		$n-1$
Elimination of $\boldsymbol{A}$	$\frac{n(n-1)(2n-1)}{6}$	$\frac{n(n-1)(2n-1)}{6}$	
Elimination of $\boldsymbol{b}$	$\frac{n(n-1)}{2}$	$\frac{n(n-1)}{2}$	
Back Substitution	$\frac{n(n-1)}{2}$	$\frac{n(n-1)}{2}$	n
Total	$\frac{n(n-1)(n+4)}{3}$	$\frac{n(n-1)(2n+5)}{6}$	$2n-1$

divisions. Because the denominator a_{kk} is common in $n-k$ multipliers for $k = 1, \ldots, n-1$, it is faster to first compute the ratios $1/a_{kk}$ and then multiply them with a_{ik}. Thus, we can have $(n-1)$ divisions and $n(n-1)/2$ multiplications.

The second stage of the elimination algorithm requires

$$\sum_{k=1}^{n-1}(n-k)^2 = \sum_{k=1}^{n-1} k^2 = \frac{n(n-1)(2n-1)}{6} ,$$

multiplications and the same amount of additions/subtractions. This is because of the three nested loops. The computations of the right-hand sides b_i are included in two nested loops. Therefore, we require

$$\sum_{k=1}^{n-1}(n-k) = \sum_{k=1}^{n-1} k = \frac{n(n-1)}{2} ,$$

multiplications and the same amount of additions (subtractions), as in the case of the computation of multipliers.

The last stage of Gaussian elimination consists of the back substitution Algorithm 28, which has two nested loops. The total number of multiplications is $n(n-1)/2$ with n divisions and $n(n-1)/2$ subtractions. In the divisions we include also the computations of the value b_n, which is outside of the `for` loop.

The total number of the operations required by the Gaussian elimination and back substitution is summarized in Table 9.1.

☞ The total number of operations for a dense $n \times n$ matrix is $O(n^3)$. From these operations $O(n^2)$ operations is the back substitution indicating that the most significant load is on the elimination process. Cramer's rule for the solution of the same linear system would require $O((n+1)!)$ operations. Therefore, Gaussian elimination is considered one of the most efficient algorithms for the solution of general linear systems.

9.1.2 *LU* factorization

Assume that we want to solve n linear systems with multiple right hand sides $\boldsymbol{b}$ but the same matrix $\boldsymbol{A}$. This situation is not rare. If we knew the various vectors $\boldsymbol{b}$ *a priori*, then we could adjust the Algorithms 27 and 28 to perform elimination and back substitution with matrix $\boldsymbol{b}$ instead of vector. But in the case where the various vectors $\boldsymbol{b}$ are not known from the beginning, then we need to perform Gauss elimination on matrix $\boldsymbol{A}$ multiple times. This doesn't sound really efficient! We will spend $O(n^3)$ operations every time we call the elimination function to perform the same thing. For this reason, we usually use the so-called

LU factorization (or *decomposition*). In general, factorizations of matrices help improving the efficiency of algorithms as well as to understand matrix properties.

The LU factorization is an algorithm that factorizes a matrix $\boldsymbol{A}$ into a product of two matrices using the elimination procedure. In particular, the result of the LU factorization is a lower triangular matrix $\boldsymbol{L}$ and an upper triangular matrix $\boldsymbol{U}$ such that $\boldsymbol{A} = \boldsymbol{LU}$. In the LU factorization the matrix $\boldsymbol{U}$ is the usual upper triangular matrix we receive as output of the Gauss elimination, and $\boldsymbol{L}$ is lower triangular with the multipliers m_{ik} in its entries. Below we show how we factorize $\boldsymbol{A}$ into a product $\boldsymbol{LU}$.

To construct the LU factorization algorithm, we express every step $k = 1, \dots, n-1$ of the elimination algorithm as

$$a_{ij}^{(k+1)} = a_{ij}^{(k)} - m_{ik} a_{kj}^{(k)}, \quad \text{for} \quad i = k, \dots, n, \quad j = k+1, \dots, n \ ,$$

where the multipliers m_{ik} are defined as

$$m_{ik} = \frac{a_{ik}^{(k)}}{a_{kk}^{(k)}}, \quad \text{for} \quad i = k+1, \dots, n \ ,$$

and we do not compute the right-hand side entries b_i. Here $a_{ij}^{(1)} = a_{ij}$. We define the upper triangular matrix $\boldsymbol{U}$ via

$$u_{ij} = a_{ij}^{(n)} \quad \text{for} \quad i = 1, \dots, n \quad \text{and} \quad j = i, \dots, n \ ,$$

and zero elsewhere. This is the usual upper triangular matrix we construct with Gauss elimination. The lower triangular matrix $\boldsymbol{L}$ is used to store the multipliers m_{ik} in its lower triangular part

$$\boldsymbol{L} = \begin{pmatrix} 1 & 0 & 0 & \cdots & 0 & 0 \\ m_{21} & 1 & 0 & \cdots & 0 & 0 \\ m_{31} & m_{32} & 1 & \cdots & 0 & 0 \\ \vdots & \vdots & \vdots & & \vdots & \vdots \\ m_{n1} & m_{n2} & m_{n3} & \cdots & m_{n,n-1} & 1 \end{pmatrix} .$$

The matrices $\boldsymbol{L}$ and $\boldsymbol{U}$ defined via the elimination process form a multiplicative factorization of $\boldsymbol{A} = \boldsymbol{LU}$.

☞ Performing the steps of Gaussian elimination we can derive the LU factorization of a matrix $\boldsymbol{A}$. Matrix $\boldsymbol{L}$ is a lower triangular matrix with 1 in the principal diagonal and the multipliers stored in its lower triangular part $l_{ij} = m_{ij}$ for $j = 1, \dots, n-1$ and $i = j+1, \dots, n$. Matrix $\boldsymbol{U}$ is the usual upper triangular matrix resulting in the Gaussian elimination. A lower (or upper) triangular matrix with ones in the main diagonal is called unit lower (or upper) triangular matrix.

Now that we know how to write $\boldsymbol{A}$ as a product $\boldsymbol{LU}$, we repeat the method of solving a linear system of the form $\boldsymbol{Ax} = \boldsymbol{b}$ given its LU factorization:

- Suppose that the factorization $\boldsymbol{A} = \boldsymbol{LU}$, where $\boldsymbol{L}$ and $\boldsymbol{U}$ are given lower and upper triangular matrices, respectively.
- We can solve the system $\boldsymbol{Ax} = \boldsymbol{b}$ for $\boldsymbol{x}$ as many times as we want by using the following two steps:
  - Set $\boldsymbol{y} = \boldsymbol{Ux}$ and solve the lower triangular system $\boldsymbol{Ly} = \boldsymbol{b}$ for $\boldsymbol{y}$ using forward substitution. Since $\boldsymbol{L}$ is triangular, determining $\boldsymbol{y}$ requires only $O(n^2)$ operations.

  - Once $\boldsymbol{y}$ is known, solve the upper triangular system $\boldsymbol{Ux} = \boldsymbol{y}$ using back substitution. This requires only an additional $O(n^2)$ operations to determine the solution $\boldsymbol{x}$.

- The number of operations needed to solve the system $\boldsymbol{Ax} = \boldsymbol{b}$ using its LU factorization[2] is reduced from $O(n^3)$ to $O(n^2)$.

For example, let us compute the LU factorization for the matrix $\boldsymbol{A}$ and solve the linear system $\boldsymbol{Ax} = \boldsymbol{b}$ where

$$\boldsymbol{A} = \begin{pmatrix} 1 & 1 & 0 \\ 2 & 1 & -1 \\ 3 & -1 & -1 \end{pmatrix} \quad \text{and} \quad \boldsymbol{b} = \begin{pmatrix} 2 \\ 2 \\ 1 \end{pmatrix} .$$

The Gaussian elimination consists of the row operations

$$\begin{aligned} &(R_2 - 2R_1) \to (R_2), \\ &(R_3 - 3R_1) \to (R_3), \qquad (R_3 - 4R_2) \to (R_3) . \end{aligned}$$

The multipliers m_{ij} are stored in a matrix $\boldsymbol{L}$, which combined with the upper triangular matrix $\boldsymbol{U}$ produced by the Gaussian elimination comprise the LU factorization of $\boldsymbol{A}$ with

$$\boldsymbol{L} = \begin{pmatrix} 1 & 0 & 0 \\ 2 & 1 & 0 \\ 3 & 4 & 1 \end{pmatrix} \quad \text{and} \quad \boldsymbol{U} = \begin{pmatrix} 1 & 1 & 0 \\ 0 & -1 & -1 \\ 0 & 0 & 3 \end{pmatrix} .$$

For economy in computer memory we can store matrices $\boldsymbol{L}$ and $\boldsymbol{U}$ in the entries of matrix $\boldsymbol{A}$ excluding the main diagonal of matrix $\boldsymbol{L}$ which has only 1. For example, we can store the previous matrices $\boldsymbol{L}$ and $\boldsymbol{U}$ in $\boldsymbol{A}$ as

$$\boldsymbol{A} = (\boldsymbol{L}\ \boldsymbol{U}) = \begin{pmatrix} 1 & 1 & 0 \\ 2 & -1 & -1 \\ 3 & 4 & 3 \end{pmatrix} .$$

To find the solution of the original system $\boldsymbol{Ax} = \boldsymbol{b}$, we first use forward substitution to solve the system $\boldsymbol{Ly} = \boldsymbol{b}$

$$\begin{pmatrix} 1 & 0 & 0 \\ 2 & 1 & 0 \\ 3 & 4 & 1 \end{pmatrix} \begin{pmatrix} y_1 \\ y_2 \\ y_3 \end{pmatrix} = \begin{pmatrix} 2 \\ 2 \\ 1 \end{pmatrix} .$$

Solving the first row for y_1 and then moving to the second row for y_2 and so on we have

$$\begin{aligned} &y_1 = 2 , \\ &2y_1 + y_2 = 2 \Rightarrow y_2 = 2 - 2y_1 = -2 , \\ &3y_1 + 4y_2 + y_3 = 1 \Rightarrow y_3 = 1 - 3y_1 - 4y_2 = 3 . \end{aligned}$$

Finally, we solve system $\boldsymbol{Ux} = \boldsymbol{y}$ for the solution $\boldsymbol{x}$. The system $\boldsymbol{Ux} = \boldsymbol{y}$ can be written as

$$\begin{pmatrix} 1 & 1 & 0 \\ 0 & -1 & -1 \\ 0 & 0 & 3 \end{pmatrix} \begin{pmatrix} x_1 \\ x_2 \\ x_3 \end{pmatrix} = \begin{pmatrix} 2 \\ -2 \\ 3 \end{pmatrix} .$$

Using backward substitution we obtain $x_3 = 1$, $x_2 = 1$ and $x_1 = 1$.

---

[2] Of course to obtain the LU factorization it requires $O(n^3)$ operations but we do it only once and for all right-hand sides $\boldsymbol{b}$.

LU factorization expressed with matrices

If we use the previous algorithm with input

$$\boldsymbol{A} = \begin{pmatrix} 1 & 1 & 0 \\ 2 & 1 & -1 \\ 3 & -1 & -1 \end{pmatrix} ,$$

this gives

$$\boldsymbol{L} = \begin{pmatrix} 1 & 0 & 0 \\ 2 & 1 & 0 \\ 3 & 4 & 1 \end{pmatrix} \quad \text{and} \quad \boldsymbol{U} = \begin{pmatrix} 1 & 1 & 0 \\ 0 & -1 & -1 \\ 0 & 0 & 3 \end{pmatrix} .$$

We show how to express the *LU* factorization using matrices. We start with the elimination of the first column. We store the multipliers with minus sign in its first column of a matrix $\boldsymbol{M}_1$ with 1 in the main diagonal and 0 elsewhere

$$\boldsymbol{M}_1 = \begin{pmatrix} 1 & 0 & 0 \\ -2 & 1 & 0 \\ -3 & 0 & 1 \end{pmatrix} .$$

Then the product $\boldsymbol{M}_1\boldsymbol{A}$ describes the elimination of the first column

$$\boldsymbol{M}_1\boldsymbol{A} = \begin{pmatrix} 1 & 4 & 0 \\ 0 & -1 & -1 \\ 0 & -4 & -1 \end{pmatrix} .$$

Likewise, we store the multipliers we used to eliminate the second column in a matrix $\boldsymbol{M}_2$

$$\boldsymbol{M}_2 = \begin{pmatrix} 1 & 0 & 0 \\ 0 & 1 & 0 \\ 0 & -4 & 1 \end{pmatrix} .$$

Then the matrix $\boldsymbol{M}_2(\boldsymbol{M}_1\boldsymbol{A})$ describes the elimination of the second column of the matrix $\boldsymbol{M}_1\boldsymbol{A}$, which is the second stage of the Gaussian elimination with

$$\boldsymbol{M}_2(\boldsymbol{M}_1\boldsymbol{A}) = \begin{pmatrix} 1 & 1 & 0 \\ 0 & -1 & -1 \\ 0 & 0 & 3 \end{pmatrix} .$$

Now is obvious that $\boldsymbol{U} = \boldsymbol{M}_2\boldsymbol{M}_1\boldsymbol{A}$ and $\boldsymbol{L} = (\boldsymbol{M}_2\boldsymbol{M}_1)^{-1} = \boldsymbol{M}_1^{-1}\boldsymbol{M}_2^{-1}$. That is

$$\boldsymbol{M}_1^{-1} = \begin{pmatrix} 1 & 0 & 0 \\ 2 & 1 & 0 \\ 3 & 0 & 1 \end{pmatrix} \quad \text{and} \quad \boldsymbol{M}_2^{-1} = \begin{pmatrix} 1 & 0 & 0 \\ 0 & 1 & 0 \\ 0 & 4 & 1 \end{pmatrix} ,$$

and therefore

$$\boldsymbol{L} = \boldsymbol{M}_1^{-1}\boldsymbol{M}_2^{-1} = \begin{pmatrix} 1 & 0 & 0 \\ 2 & 1 & 0 \\ 3 & 4 & 1 \end{pmatrix} .$$

Taking this into consideration we have the following theorem:

Theorem 9.1 (Existence and uniqueness of ***LU*** factorization). *If an upper triangular matrix* $\boldsymbol{U}$ *can be found by Gaussian elimination from a matrix* $\boldsymbol{A}$ *without encountering zero diagonal elements, then* $\boldsymbol{A}$ *has a unique LU factorization* $\boldsymbol{A} = \boldsymbol{L}\boldsymbol{U}$*, where* $\boldsymbol{L}$ *is a low triangular matrix with ones in the main diagonal.*

Proof. The Gaussian elimination is expressed as

$$\boldsymbol{U} = \boldsymbol{M}_{n-1}\boldsymbol{M}_{n-2}\cdots\boldsymbol{M}_1\boldsymbol{A} ,$$

where $\boldsymbol{M}_i$ is lower triangular matrix with ones in its main diagonal for all $i = 1, \dots, n-1$. All these matrices $\boldsymbol{M}_i$ are invertible since all their diagonal entries of $\boldsymbol{M}_i$ are nonzero, and thus $\det(\boldsymbol{M}_i) = 1$. Therefore, the matrix $\boldsymbol{A}$ can be written as

$$\boldsymbol{A} = \boldsymbol{M}_1^{-1}\boldsymbol{M}_2^{-1}\cdots\boldsymbol{M}_{n-1}^{-1}\boldsymbol{U} = \boldsymbol{L}\boldsymbol{U} ,$$

which shows the existence of *LU* decomposition. To prove that this is unique, we assume that there are two different *LU* decompositions of $\boldsymbol{A}$

$$\boldsymbol{A} = \boldsymbol{L}_1\boldsymbol{U}_1 = \boldsymbol{L}_2\boldsymbol{U}_2 .$$

Separating the lower and upper triangular matrices we have that $\boldsymbol{L}_2^{-1}\boldsymbol{L}_1 = \boldsymbol{U}_2\boldsymbol{U}_1^{-1}$. Since the left-hand side is lower triangular and the right-hand side upper triangular (see Sections 2.2.8 and 2.2.9), we conclude that both products are equal to a diagonal matrix, let's say $\boldsymbol{D}$. Moreover, the matrix $\boldsymbol{D}$ must be the $n \times n$ identity matrix $\boldsymbol{I}$ because the product $\boldsymbol{L}_2^{-1}\boldsymbol{L}_1$ must have all the entries of the main diagonal equal to 1. We conclude that

$$\boldsymbol{L}_2^{-1}\boldsymbol{L}_1 = \boldsymbol{U}_2\boldsymbol{U}_1^{-1} = \boldsymbol{I} ,$$

or equivalently that $\boldsymbol{L}_1 = \boldsymbol{L}_2$ and $\boldsymbol{U}_1 = \boldsymbol{U}_2$. □

While this procedure can be generalized for any matrix $\boldsymbol{A}$, apart from theoretical purposes it serves no other purpose, and thus we move on with the implementation of the *LU* factorization algorithm.

9.1.3 Implementation of *LU* factorization

The implementation of the *LU* factorization follows the same steps with the Gaussian elimination of Algorithm 27 and is summarized in Algorithm 29.

Algorithm 29 *LU* factorization of a matrix $\boldsymbol{A}$

```
for k = 1 : n − 1 do
    for i = k + 1 : n do
        if A(k, k) ≠ 0 then
            m_ik = A(i, k)/A(k, k)
            A(i, k : n) = A(i, k : n) − m_ik · A(k, k : n)
            A(i, k) = m_ik
        end if
    end for
end for
```

During the elimination process the upper triangular matrix $\boldsymbol{U}$ replaces the upper triangular part of $\boldsymbol{A}$, while the multipliers are stored in the lower triangular part. Specifically, in the following implementation of the *LU* factorization we receive in the output the matrix

$$\boldsymbol{A} = (\boldsymbol{L}\ \boldsymbol{U}) = \begin{pmatrix} u_{11} & u_{12} & u_{13} \\ l_{21} & u_{22} & u_{23} \\ l_{31} & l_{32} & u_{33} \end{pmatrix} .$$

This matrix contains the upper triangular matrix $\boldsymbol{U}$ and the lower triangular matrix $\boldsymbol{L}$ but not the principal diagonal of $\boldsymbol{L}$, which we know anyway that consists of ones.

The implementation of Algorithm 29 in Python could be the following:

```
def LU(A):
    # LU factorization of matrix A
    n = len(A)
    for k in range(n-1):
        for i in range(k+1,n):
            if A[k,k] != 0.0:
                mik = A[i,k]/A[k,k]
                A[i,k+1:n] = A[i,k+1:n] - mik*A[k,k+1:n]
                A[i,k] = mik
    return A
```

Solving linear systems with *LU* factorization

We continue with the implementation of the algorithm for the solution of a linear system $\boldsymbol{Ax}=\boldsymbol{b}$ given the $\boldsymbol{LU}$ factorization of $\boldsymbol{A}$. As we discussed before, in order to solve a system $\boldsymbol{Ax}=\boldsymbol{b}$ given that $\boldsymbol{A}=\boldsymbol{LU}$, we set $\boldsymbol{Ux}=\boldsymbol{y}$. Then, we first solve the system $\boldsymbol{Ly}=\boldsymbol{b}$ and then the system $\boldsymbol{Ux}=\boldsymbol{y}$. The solution of the system $\boldsymbol{Ux}=\boldsymbol{y}$ is solved with backward substitution (Algorithm 28).

The system $\boldsymbol{Ly}=\boldsymbol{b}$ can be solved easily using forward substitution. We first write the system $\boldsymbol{Ly}=\boldsymbol{b}$ in the form of equations

$$
\begin{aligned}
&l_{11}y_1 = b_1 \ ,\\
&l_{21}y_1 + l_{22}y_2 = b_2 \ ,\\
&\quad\vdots\\
&l_{k1}y_1 + l_{k2}y_2 + \cdots + l_{k,k-1}y_{k-1} + l_{kk}y_k = b_k \ ,\\
&\quad\vdots\\
&l_{n1}y_1 + l_{n2}y_2 + \cdots + l_{n,k-1}y_{k-1} + \cdots + l_{n,n-1}y_{n-1} + l_{nn}y_n = b_n \ ,
\end{aligned}
$$

Solving the k-th equation for y_k yields the formula of forward substitution:

$$
y_k = \left(b_k - \sum_{j=1}^{k-1} l_{kj}y_j\right)/l_{kk}, \quad k = 1,2,\ldots,n \ .
$$

The algorithm for the forward substitution for a general lower triangular matrix $\boldsymbol{L}$ is summarized in Algorithm 30.

Algorithm 30 Forward substitution for solving lower triangular systems

---

Given the lower triangular array $\boldsymbol{L}$, the right-hand side $\boldsymbol{b}$ of the system $\boldsymbol{Ly}=\boldsymbol{b}$ the solution $\boldsymbol{y}$ replaces $\boldsymbol{b}$
$\boldsymbol{b}(1) = \boldsymbol{b}(1)/\boldsymbol{L}(1,1)$
for $k = 2:n$ **do**
    $\boldsymbol{b}(k) = (\boldsymbol{b}(k) - \boldsymbol{L}(k,1:k-1)\cdot\boldsymbol{b}(1:k-1))\ /\ \boldsymbol{L}(k,k)$
end for

In practical implementations matrix $\boldsymbol{L}$ is stored in the lower triangular part of $\boldsymbol{A}$, while $\boldsymbol{U}$ in the upper triangular part of $\boldsymbol{A}$. The diagonal entries l_{kk} of $\boldsymbol{L}$ can be ignored because they are all 1. A code for the solution of the system $\boldsymbol{L}\boldsymbol{y} = \boldsymbol{b}$ is the following:

```
def solveLU(A,b):
    # Solution of the linear system LUx=b
    # matrix A contains the LU factorization of A
    n = len(A)
    # Solve the low triangular system Ly=b, note that L_ii = 1.0
    for k in range(1,n):
        b[k] = b[k] - np.dot(A[k,0:k],b[0:k])
    # Solve the upper triangular system Ux=b
    b[n-1] = b[n-1]/A[n-1,n-1]
    for k in range(n-2,-1,-1):
        b[k] = (b[k] - np.dot(A[k,k+1:n],b[k+1:n]))/A[k,k]
    return b
```

In the next piece of code we use the above functions to generate the LU factorization of the matrix

$$\boldsymbol{A} = \begin{pmatrix} 1 & 1 & 0 \\ 2 & 1 & -1 \\ 3 & -1 & -1 \end{pmatrix} ,$$

and then solve two systems with different right-hand sides $\boldsymbol{b}_1 = (2,2,1)^T$ and $\boldsymbol{b}_2 = (4,4,2)^T$. This is done by performing the LU factorization once and then forward and backward substitutions twice for the solution of the two systems $\boldsymbol{L}\boldsymbol{U}\boldsymbol{x} = \boldsymbol{b}_1$ and $\boldsymbol{L}\boldsymbol{U}\boldsymbol{x} = \boldsymbol{b}_2$.

```
A = np.array([[ 1., 1., 0.],
              [ 2., 1.,-1.],
              [ 3.,-1.,-1.]])
b1 = np.array([ 2., 2., 1.])
b2 = np.array([ 4., 4., 2.])
# Perform LU factorization once
A = LU(A)
# Solve the first system using forward/backward substitution
x1 = solveLU(A,b1)
print(x1)
#Solve the second system using forward/backward substitution
x2 = solveLU(A,b2)
print(x2)
```

```
[1. 1. 1.]
[2. 2. 2.]
```

9.2 Pivoting Strategies

In the previous discussions we assumed that the diagonal entries $a_{kk}^{(k)}$ in the elimination of $\boldsymbol{A}$ are not zero. Thus, we were able to compute the multipliers without issues. These diagonal entries are called *pivots* and play an important role in achieving stable and accurate results in Gaussian elimination. There are cases where the naive Gaussian elimination fails because $\boldsymbol{A}$ is nearly singular (nearly non-invertible with a tiny determinant) or $\boldsymbol{A}$ is such that floating point arithmetic leads to wrong solutions[3]. For example consider a linear system with one of the matrices

$$\boldsymbol{A}_1 = \begin{pmatrix} 0 & 1 \\ 1 & 1 \end{pmatrix} \quad \text{or} \quad \boldsymbol{A}_2 = \begin{pmatrix} 10^{-17} & 1 \\ 1 & 1 \end{pmatrix} .$$

In the case of $\boldsymbol{A}_1$ since $a_{11} = 0$ we cannot perform even the first step of the Gaussian elimination. In the second case, let us try to solve using our `Gauss_elimination` function the system $\boldsymbol{A}_2\boldsymbol{x} = \boldsymbol{b}$ with $\boldsymbol{b} = (1, 2)^T$. The specific system has exact solution $\boldsymbol{x} \approx (1, 1)$, but our code returns `x=[0., 1.]`, which contains a huge error of $O(1)$.

```
A = np.array([[ 1.e-17, 1.],
              [ 1., 1.]])
b = np.array([ 1., 2.])
x = Gauss_elimination(A,b)
print(x)
```

```
[0. 1.]
```

To correct these problems we change the order of the equations. In the case of the system $\boldsymbol{A}_2\boldsymbol{x} = \boldsymbol{b}$ instead of solving the original system we can solve the system where the first equation becomes second, and the second first

$$\begin{pmatrix} 1 & 1 \\ 10^{-17} & 1 \end{pmatrix} \begin{pmatrix} x_1 \\ x_2 \end{pmatrix} = \begin{pmatrix} 2 \\ 1 \end{pmatrix} ,$$

```
A = np.array([[ 1., 1.],
              [ 1.e-17, 1.]])
b = np.array([ 2., 1.])
x = Gauss_elimination(A,b)
print(x)
```

```
[1. 1.]
```

Observe that after changing the order of the rows in the matrix $\boldsymbol{A}_2$ and vector $\boldsymbol{b}$ the solution is more accurate. This is the way to solve systems with matrix $\boldsymbol{A}_1$ as well. This technique is called pivoting and is described briefly in the next section.

☞ Gaussian elimination is an unstable algorithm in the sense that small perturbation of the data (entries of $\boldsymbol{A}$ or $\boldsymbol{b}$) can lead to a totally different solution. Thus, it is very dangerous to perform the naive Gaussian elimination algorithm when we use computers because of errors due to finite precision arithmetic. On the other hand, the forward and backward substitution algorithms are stable [64].

[3]For more information see [42].

9.2.1 Gaussian elimination with partial pivoting

The simplest pivoting strategy at stage k of the Gaussian elimination is to select the entry of the k-th column of the augmented matrix with the largest absolute value below the main diagonal (let's say p-th row), and swap rows k and p. This can be described algorithmically as: At stage k of the outer loop of the Gaussian elimination

- We determine the smallest $p \geq k$ such that

$$|a_{pk}| = \max_{k \leq i \leq n} |a_{ik}|$$

- Interchange rows k and p

$$(R_k) \leftrightarrow (R_p)$$

No interchange of columns is used in this case. This is the reason that the specific pivoting strategy is called partial pivoting. There is a column pivoting strategy where we proceed as in the row pivoting but interchanging columns instead.

A Python code in vector form for swapping rows fast can be implemented as follows,

```
def rowSwap(v,i,j):
    if len(v.shape) == 1:
        v[i],v[j] = v[j],v[i]
    else:
        v[[i,j],:] = v[[j,i],:]
```

A modification of the Gaussian elimination with partial pivoting in Python can be the following:

```
def pGauss_elimination(A,b,tol=1.0e-15):
    n = len(b)
    for k in range(n-1):
        # swap rows if necessary
        p = np.argmax(np.abs(A[k:n,k])) + k
        if np.abs(A[p,k]) < tol:
            error.err('singular matrix has been detected')
        if p != k:
            rowSwap(b,k,p)
            rowSwap(A,k,p)
        # perform Gauss elimination
        for i in range(k+1,n):
            if A[k,k] != 0.0:
                mik = A[i,k]/A[k,k]
                A[i,k+1:n] = A[i,k+1:n] - mik*A[k,k+1:n]
                b[i] = b[i] - mik*b[k]
    # check if the matrix is singular
    if np.abs(A[n-1,n-1]) < tol:
        error.err('singular matrix has been detected')
    # perform backward substitution
    b[n-1] = b[n-1]/A[n-1,n-1]
    for k in range(n-2,-1,-1):
        b[k] = (b[k] -np.dot(A[k,k+1:n],b[k+1:n]))/A[k,k]
    return b
```

Note that the previous algorithm uses a tolerance `tol` to detect tiny entries on the main diagonal. Now using the Gaussian elimination with partial pivoting to the system $\boldsymbol{A}_1\boldsymbol{x} = \boldsymbol{b}$ with $\boldsymbol{b} = (1, 2)^T$ we get

```
A = np.array([[ 0., 1.],
              [ 1., 1.]])
b = np.array([ 1., 2.])
x = pGauss_elimination(A,b)
print(x)
```

```
[1. 1.]
```

This is the correct answer, and the row interchange happened automatically inside the function `pGauss_elimination`.

Scaled partial pivoting

In addition to partial pivoting there is a *scaled partial pivoting* (or scaled-column pivoting) algorithm. Scaled partial pivoting is needed when equations and unknowns are scaled differently. An example is the previous system with matrix $\boldsymbol{A}_2$. In this case, you may select row pivots relative to the scale factors:

- Before factorization select scale factors

$$s_i = \max_{1 \le j \le n} |a_{ij}|, \quad i = 1, \ldots, n$$

- At stage i of the factorization, select p such that

$$\left| \frac{a_{pk}}{s_p} \right| = \max_{k \le i \le n} \left| \frac{a_{ik}}{s_i} \right|$$

- Interchange rows k and p

$$(R_k) \leftrightarrow (R_p).$$

We leave the implementation of this as an exercise to the reader. This can also be found in the book [78]. In addition to partial pivoting, the complete pivoting technique can be considered as the ultimate pivoting technique.

Complete pivoting

Complete pivoting is a more exhaustive procedure and less common than partial pivoting, in which the largest entry of the entire unreduced matrix is transferred into a pivoting position. In this case, we perform interchanges in columns and rows during the elimination process. This can be described briefly by the following steps: Choose row r and column c as follows:

- Find r, c such as

$$|a_{rc}| = \max_{i \le k, l \le n} |a_{kl}|$$

- Interchange rows i and r and columns i and c

☞ Complete pivoting technique is avoided in practice because it can slow down the elimination process. On the other hand, partial pivoting is so common that it is a default procedure in Gaussian elimination and LU factorization.

9.2.2 *LU* factorization with pivoting

Further to the failure of Gaussian elimination due to problematic pivots, *LU* factorization fails in the same way due to its equivalence with the Gaussian elimination. For this reason, we incorporate partial pivoting in the *LU* factorization too. First we define *permutation* matrices to express row (or column) interchanges. A permutation matrix $\boldsymbol{P}$ is an identity matrix with its rows or columns interchanged. Multiplication of a matrix $\boldsymbol{A}$ with a permutation matrix $\boldsymbol{P}$ is equivalent with interchanging rows (or columns) of $\boldsymbol{A}$. For example, if we want to interchange rows 1 and 3 of a matrix

$$\boldsymbol{A} = \begin{pmatrix} a_{11} & a_{12} & a_{13} \\ a_{21} & a_{22} & a_{23} \\ a_{31} & a_{32} & a_{33} \end{pmatrix} ,$$

then we can multiply $\boldsymbol{A}$ with the permutation matrix

$$\boldsymbol{P} = \begin{pmatrix} 0 & 0 & 1 \\ 0 & 1 & 0 \\ 1 & 0 & 0 \end{pmatrix} ,$$

obtained from the identity matrix by swapping the first and third rows. This is

$$\boldsymbol{PA} = \begin{pmatrix} 0 & 0 & 1 \\ 0 & 1 & 0 \\ 1 & 0 & 0 \end{pmatrix} \cdot \begin{pmatrix} a_{11} & a_{12} & a_{13} \\ a_{21} & a_{22} & a_{23} \\ a_{31} & a_{32} & a_{33} \end{pmatrix} = \begin{pmatrix} a_{31} & a_{32} & a_{33} \\ a_{21} & a_{22} & a_{23} \\ a_{11} & a_{12} & a_{13} \end{pmatrix} .$$

☞ A permutation matrix is a row-permutation of the identity matrix, and $\boldsymbol{P}^{-1} = \boldsymbol{P}^T$.

The Gaussian elimination algorithm was modified appropriately (with minor changes) to accomodate pivoting techniques. Based on the following theorem we can do the same with the *LU* factorization.

Theorem 9.2. *For any $n \times n$ matrix $\boldsymbol{A}$ of rank n, there is a reordering of rows such that*

$$\boldsymbol{PA} = \boldsymbol{LU} ,$$

where $\boldsymbol{P}$ is a permutation matrix that reorders the rows of $\boldsymbol{A}$.

We demonstrate the *LU* factorization with partial pivoting for the matrix

$$\boldsymbol{A} = \begin{pmatrix} 3 & 17 & 10 \\ 2 & 4 & -2 \\ 6 & 18 & -12 \end{pmatrix} .$$

As it is required by the partial pivoting technique, we choose for the first step the entry of the first column with the largest magnitude. We thus need to swap rows 1 and 3 since the entry with the maximum magnitude in first column is the entry 6. This row interchange can be represented by the matrix

$$\boldsymbol{P}_1 = \begin{pmatrix} 0 & 0 & 1 \\ 0 & 1 & 0 \\ 1 & 0 & 0 \end{pmatrix} .$$

Thus, we eliminate initially the first column of matrix

$$P_1A = \begin{pmatrix} 0 & 0 & 1 \\ 0 & 1 & 0 \\ 1 & 0 & 0 \end{pmatrix} \cdot \begin{pmatrix} 3 & 17 & 10 \\ 2 & 4 & -2 \\ 6 & 18 & -12 \end{pmatrix} = \begin{pmatrix} 6 & 18 & -12 \\ 2 & 4 & -2 \\ 3 & 17 & 10 \end{pmatrix} ,$$

using the row operations $(R_2 - 1/3\ R_1) \to (R_2)$ and $(R_3 - 1/2\ R_1) \to (R_3)$ to obtain the reduced matrix $A^{(1)} = M_1P_1A$ where

$$A^{(1)} = \begin{pmatrix} 6 & 18 & -12 \\ 0 & -2 & 2 \\ 0 & 8 & 16 \end{pmatrix} \quad \text{and} \quad M_1 = \begin{pmatrix} 1 & 0 & 0 \\ -1/3 & 1 & 0 \\ -1/2 & 0 & 1 \end{pmatrix} .$$

Next, we need to eliminate the second column of this matrix, but using partial pivoting we perform the interchange of rows 2 and 3 described by the permutation matrix

$$P_2 = \begin{pmatrix} 1 & 0 & 0 \\ 0 & 0 & 1 \\ 0 & 1 & 0 \end{pmatrix} .$$

We perform the elimination of the second column of the matrix

$$P_2A^{(1)} = \begin{pmatrix} 1 & 0 & 0 \\ 0 & 0 & 1 \\ 0 & 1 & 0 \end{pmatrix} \cdot \begin{pmatrix} 6 & 18 & -12 \\ 0 & 8 & 16 \\ 0 & -2 & 2 \end{pmatrix} ,$$

using the row operation $(R_3 + 1/4\ R_2) \to (R_3)$ to obtain at the end of the procedure $A^{(2)} = M_2P_2A^{(1)}$ where

$$A^{(2)} = \begin{pmatrix} 6 & 18 & -12 \\ 0 & 8 & 16 \\ 0 & 0 & 6 \end{pmatrix} \quad \text{and} \quad M_2 = \begin{pmatrix} 1 & 0 & 0 \\ 0 & 1 & 0 \\ 0 & 1/4 & 1 \end{pmatrix} .$$

If we write $M = M_2P_2M_1P_1$, then $U = A^{(2)} = MA$ and the LU factorization with partial pivoting returns the matrices

$$L = \begin{pmatrix} 1 & 0 & 0 \\ 1/2 & 1 & 0 \\ 1/3 & -1/4 & 1 \end{pmatrix} \quad \text{and} \quad U = \begin{pmatrix} 6 & 18 & -12 \\ 0 & 8 & 16 \\ 0 & 0 & 6 \end{pmatrix} .$$

Taking

$$P = P_2P_1 = \begin{pmatrix} 0 & 0 & 1 \\ 1 & 0 & 0 \\ 0 & 1 & 0 \end{pmatrix} ,$$

we have the factorization $PA = LU$

$$\begin{pmatrix} 0 & 0 & 1 \\ 1 & 0 & 0 \\ 0 & 1 & 0 \end{pmatrix} \begin{pmatrix} 3 & 17 & 10 \\ 2 & 4 & -2 \\ 6 & 18 & -12 \end{pmatrix} = \begin{pmatrix} 1 & 0 & 0 \\ 1/2 & 1 & 0 \\ 1/3 & -1/4 & 1 \end{pmatrix} \begin{pmatrix} 6 & 18 & -12 \\ 0 & 8 & 16 \\ 0 & 0 & 6 \end{pmatrix} .$$

In general, L contains the permutations of the rows and is the matrix

$$L = P(M_{n-1}P_{n-1} \cdots M_2P_2M_1P_1)^{-1} ,$$

where

$$P = P_{n-1} \cdots P_2P_1 .$$

☞ The matrices $\boldsymbol{L}$ and $\boldsymbol{U}$ in the $\boldsymbol{PA} = \boldsymbol{LU}$ factorization with pivoting are not the same as in the $\boldsymbol{A} = \boldsymbol{LU}$ factorization without pivoting.

It is important to keep a record of the row interchanges in order to formulate the permutation matrix $\boldsymbol{P}$. The permutation matrix $\boldsymbol{P}$ is sparse and it is not good practice to store its zero entries. In practice we store only the interchanges using a vector `perm`. Initially, the vector `perm` is $(0, 1, 2, \ldots, n)^T$ (in Python indices). Whenever two rows are interchanged, the corresponding interchange is also carried out in the vector `perm`. Thus, the vector `perm` contains the order in which the original rows have been rearranged. This information is passed on to the solution phase (`solveLU`), by rearranging the entries of the constant vector in the same order before proceeding to forward and backward substitutions.

The algorithm of the *LU* factorization with partial pivoting is summarized in Algorithm 31

Algorithm 31 *LU* factorization with partial pivoting

Given the array $\boldsymbol{A}$
for $k = 1 : n-1$ **do**
  Find p such that $|\boldsymbol{A}(p,k)| = \max_{k \le p \le n} |\boldsymbol{A}(k:n,k)|$
  Swap rows $\boldsymbol{A}(k,:) \leftrightarrow \boldsymbol{A}(p,:)$
  Swap rows $perm(k) \leftrightarrow perm(p)$
  for $i = k+1 : n$ **do**
    if $\boldsymbol{A}(i,k) \neq 0$ **then**
      $m_{ik} = \boldsymbol{A}(i,k)/\boldsymbol{A}(k,k)$
      $\boldsymbol{A}(i,k+1:n) = \boldsymbol{A}(i,k+1,n) - m_{ik} \cdot \boldsymbol{A}(k,k+1:n)$
      $\boldsymbol{A}(i,k) = m_{ik}$
    end if
  end for
end for

The implementation of the *LU* factorization with partial pivoting in Python is the same as the Gaussian elimination with partial pivoting with the only difference that we keep a record of the permutations.

```
def LUpivot(A,tol=1.0e-15):
    n = len(A)
    # define permutation vector
    perm = np.array(range(n))
    for k in range(0,n-1):
        # perform row interchange if necessary
        p = np.argmax(np.abs(A[k:n,k])) + k
        if np.abs(A[p,k]) < tol:
            error.err('singular matrix has been detected')
        if p != k:
            swapRows(A,k,p)
            swapRows(perm,k,p)
        # perform Gauss elimination
        for i in range(k+1,n):
            if A[i,k] != 0.0:
```

```
                mik = A[i,k]/A[k,k]
                A[i,k+1:n] = A[i,k+1:n] - mik*A[k,k+1:n]
                A[i,k] = mik
    return A, perm
```

In order to solve a linear system $\boldsymbol{Ax} = \boldsymbol{b}$ using LU factorization with partial pivoting, we multiply both sides of the system with $\boldsymbol{P}$ to obtain $\boldsymbol{LUx} = \boldsymbol{Pb}$ since $\boldsymbol{PA} = \boldsymbol{LU}$. Then we perform the usual steps of forward and backward substitution to the vector $\boldsymbol{Pb}$ instead of $\boldsymbol{b}$. The implementation could be the following:

```
def solveLUpivot(A,b,perm):
    n = len(A)
    # Store right-hand side in solution vector x
    x = b.copy()
    for i in range(n):
        x[i]=b[perm[i]]
    # Forward-backward substitution
    for k in range(1,n):
        x[k] = x[k] - np.dot(A[k,0:k],x[0:k])
    x[n-1] = x[n-1]/A[n-1,n-1]
    for k in range(n-2,-1,-1):
        x[k] = (x[k] - np.dot(A[k,k+1:n],x[k+1:n]))/A[k,k]
    return x
```

Testing these programs using the system $\boldsymbol{Ax} = \boldsymbol{b}$ with

$$\boldsymbol{A} = \begin{pmatrix} 3 & 17 & 10 \\ 2 & 4 & -2 \\ 6 & 18 & -12 \end{pmatrix} \quad \text{and} \quad \boldsymbol{b} = \begin{pmatrix} 30 \\ 4 \\ 12 \end{pmatrix},$$

we have:

```
A = np.array([[ 3., 17., 10.],
              [ 2., 4.,-2.],
              [ 6.,18.,-12.]])
b = np.array([ 30., 4., 12.])
[A,perm] = LUPivot(A)
x = solveLUpivot(A,b,perm)
print(x)
```

```
[1. 1. 1.]
```

In order to verify the result we perform some extra computations to retrieve the matrices $\boldsymbol{P}$, $\boldsymbol{L}$ and $\boldsymbol{U}$ from the output of our functions.

```
# construct lower triangular matrix L
L = np.tril(A,-1)+np.eye(3)
# construct upper triangular matrix U
U = np.triu(A)
# construct permutation matrix P
P = np.eye(3)
P = P[perm,:]
```

```
print('P='); print(P)
print('L='); print(L)
print('U='); print(U)
```

```
P=
[[0. 0. 1.]
 [1. 0. 0.]
 [0. 1. 0.]]
L=
[[ 1.          0.          0.        ]
 [ 0.5         1.          0.        ]
 [ 0.33333333 -0.25        1.        ]]
U=
[[  6.  18. -12.]
 [  0.   8.  16.]
 [  0.   0.   6.]]
```

Observe that the output agrees with the example presented above.

9.3 Condition Number of a Matrix

We saw that Gaussian elimination can be unstable for some linear systems due to finite precision arithmetic. It would have been very useful if we had a tool to detect matrices that can lead to failure of Gaussian elimination. The sensitivity on matrices to small errors can be expressed with the condition number of a matrix. We first define rigorously the notion of norm to compare vectors and matrices.

9.3.1 Vector and matrix norms

A vector norm is a function that maps vectors to positive real numbers. A norm of a vector $\boldsymbol{x}$ usually is denoted by $\|\boldsymbol{x}\|$, and it can be interpreted as the length of $\boldsymbol{x}$. Formally, a vector norm is the mapping $\|\cdot\| : \mathbb{R}^n(\text{or } \mathbb{C}^n) \to \mathbb{R}$ and satisfies the following properties:

1. $\|\boldsymbol{x}\| \geq 0$, $\boldsymbol{x} \in \mathbb{R}^n$, $(\|\boldsymbol{x}\| = 0 \Leftrightarrow \boldsymbol{x} = \boldsymbol{0})$
2. $\|\boldsymbol{x} + \boldsymbol{y}\| \leq \|\boldsymbol{x}\| + \|\boldsymbol{y}\|$, $\boldsymbol{x}, \boldsymbol{y} \in \mathbb{R}^n$ (Triangle inequality)
3. $\|a\boldsymbol{x}\| = |a|\|\boldsymbol{x}\|$ for $a \in \mathbb{R}$, $\boldsymbol{x} \in \mathbb{R}^n$

The most common norms are the ℓ_1-norm, ℓ_2-norm and ℓ_∞-norm defined as

$$\begin{aligned}
\|\boldsymbol{x}\|_1 &= |x_1| + \cdots + |x_n|, && (\ell_1\text{-norm})\\
\|\boldsymbol{x}\|_2 &= \sqrt{|x_1|^2 + \cdots + |x_n|^2}, && (\ell_2\text{-norm})\\
\|\boldsymbol{x}\|_\infty &= \max_{1\leq i\leq n} |x_i|, && (\ell_\infty\text{-norm})
\end{aligned}$$

where the absolute value is used when the vectors have complex entries ($x_i \in \mathbb{C}$).

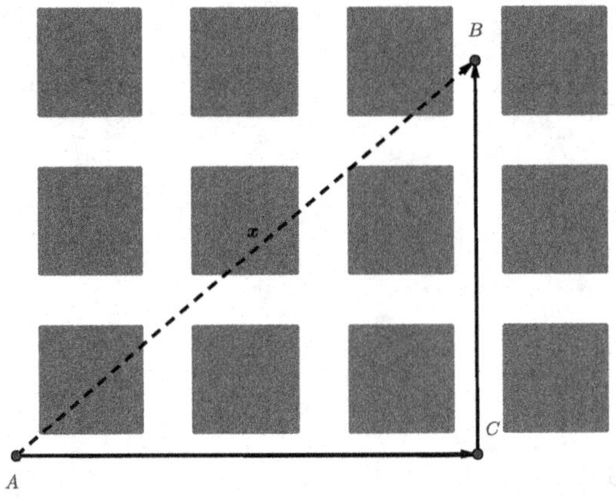

FIGURE 9.1
Measuring distances in Manhattan.

Sometimes, we find it useful to use the more general norm (ℓ_p-norm)

$$\|\boldsymbol{x}\|_p = \left(\sum_{i=1}^{n} |x_i|^p\right)^{1/p} , \quad p > 1 ,$$

which for $p = \infty$ is $\|\boldsymbol{x}\|_\infty = \lim_{p\to\infty} \|\boldsymbol{x}\|_p$. The $\|\boldsymbol{x}\|_2$ is usually referred to as the *Euclidean norm*.

To understand the difference between the various norms we present in Figure 9.1 a top view of a hypothetical city, let's say Manhattan with the squares representing buildings. Suppose that we want to move from point A to point B. The distance between A and B is measured by the ℓ_2-norm $\|\boldsymbol{x}\|_2$ and is the modulus of the vector $\boldsymbol{x}$ depicted with broken line. This distance though is not an accurate estimate of the distance we need to cover. On the contrary, the distance we need to travel is the distance from A to C and from C to B. This distance is the ℓ_1-norm $\|\boldsymbol{x}\|_1 = |x_1| + |x_2|$. So norms measure distances in various topologies. Distances measured with different norms have the same order of magnitude and thus the choice of a norm will depend on the application[4].

Norms estimate also distances between vectors by their difference $\|\boldsymbol{a} - \boldsymbol{b}\|$. This is very useful computation as, for example, in the computation of an approximation $\tilde{\boldsymbol{x}}$ of a vector $\boldsymbol{x}$, the norm of their difference $\|\boldsymbol{x} - \tilde{\boldsymbol{x}}\|$ or even the normalized quotient $\|\boldsymbol{x} - \tilde{\boldsymbol{x}}\|/\|\boldsymbol{x}\|$ gives us information about the error or the relative error.

Some vector norms are related to the inner product between vectors. The inner (or dot) product of two vectors is defined as the product of a row with a column vector $\boldsymbol{x}^T \cdot \boldsymbol{y}$. (Sometimes we will omit the dot in the notation like we do in matrix multiplication.) In linear algebra we usually consider vectors as column vectors. So the inner product of the vectors $\boldsymbol{x}, \boldsymbol{y} \in \mathbb{R}^n$ is the sum of the products of their entries

$$\boldsymbol{x}^T \cdot \boldsymbol{y} = \sum_{i=1}^{n} \boldsymbol{x}_i \boldsymbol{y}_i .$$

In this book, the inner product will be denoted by $\langle \boldsymbol{x}, \boldsymbol{y}\rangle = \boldsymbol{x}^T \cdot \boldsymbol{y}$ to emphasize the relationship between $\boldsymbol{x}$ and $\boldsymbol{y}$. It is easy to see that

$$\boldsymbol{x}^T \cdot \boldsymbol{x} = \langle \boldsymbol{x}, \boldsymbol{x}\rangle = \|\boldsymbol{x}\|_2^2 ,$$

[4]In finite-dimensional spaces.

which implies

$$\|\boldsymbol{x}\|_2 = \sqrt{\langle \boldsymbol{x}, \boldsymbol{x} \rangle} \ .$$

If the vectors $\boldsymbol{x}$ and $\boldsymbol{y}$ are complex, then the definition of the transpose requires to take the complex conjugates of the entries of the respective vector. In particular, the inner product is defined as

$$\langle \boldsymbol{x}, \boldsymbol{y} \rangle = \sum_{i=1}^{n} \bar{x}_i y_i \ .$$

Perhaps, the most useful inequality between norms and inner products is the Cauchy-Schwarz inequality[5]

$$|\langle \boldsymbol{x}, \boldsymbol{y} \rangle| \leq \|\boldsymbol{x}\|_2 \ \|\boldsymbol{y}\|_2 \ ,$$

which holds true for any two complex vectors $\boldsymbol{x}$ and $\boldsymbol{y}$. More precisely, we have the following theorem:

Theorem 9.3 (Cauchy-Schwarz inequality). *Any vectors $\boldsymbol{x}, \boldsymbol{y} \in \mathbb{C}^n$ satisfy the inequality*

$$|\langle \boldsymbol{x}, \boldsymbol{y} \rangle| \leq \|\boldsymbol{x}\|_2 \ \|\boldsymbol{y}\|_2 \ .$$

Proof. Obviously, for $\boldsymbol{x} = \boldsymbol{0}$ or $\boldsymbol{y} = \boldsymbol{0}$ the Cauchy-Schwarz inequality holds as an equality $(0 = 0)$. For non-zero vectors $\boldsymbol{x}$ and $\boldsymbol{y}$ we see that for all $\theta \in \mathbb{R}$ we have

$$0 \leq \sum_{i=1}^{n} (\theta |x_i| + |y_i|)^2 = \theta^2 \sum_{i=1}^{n} |x_i|^2 + 2\theta \sum_{i=1}^{n} |x_i||y_i| + \sum_{i=1}^{n} |y_i|^2 \ . \tag{9.3}$$

The right hand side of this inequality is a quadratic polynomial for θ. The coefficient of the quadratic term θ^2 is positive since

$$\sum_{i=1}^{n} |x_i|^2 > 0 \ .$$

Since the quadratic polynomial in (9.3) is always positive, its discriminant D must be $D \leq 0$. Using this observation, we have

$$0 \geq D = 4 \left(\sum_{i=1}^{n} |x_i||y_i| \right)^2 - 4 \left(\sum_{i=1}^{n} |x_i|^2 \right) \left(\sum_{i=1}^{n} |y_i|^2 \right) ,$$

which gives

$$\left(\sum_{i=1}^{n} |\bar{x}_i y_i| \right)^2 \leq \left(\sum_{i=1}^{n} |x_i|^2 \right) \left(\sum_{i=1}^{n} |y_i|^2 \right) .$$

Taking square root on both sides of this inequality and using the triangle inequality $|x+y| \leq |x| + |y|$ we have

$$\left| \sum_{i=1}^{n} \bar{x}_i y_i \right| \leq \left(\sum_{i=1}^{n} |x_i|^2 \right)^{1/2} \left(\sum_{i=1}^{n} |y_i|^2 \right)^{1/2} ,$$

which is the Cauchy-Schwarz inequality $|\langle \boldsymbol{x}, \boldsymbol{y} \rangle| \leq \|\boldsymbol{x}\|_2 \|\boldsymbol{y}\|_2$. □

[5] Also called Cauchy-Schwarz-Bunyakovsky inequality named after Baron Augustin-Louis Cauchy (1789–1857), Karl Hermann Amandus Schwarz (1843–1921) and Viktor Yakovlevich Bunyakovsky (1804–1889).

We mentioned the triangle inequality for the absolute value of the sum of two numbers $|x+y| \leq |x|+|y|$. The exact same inequality holds for vectors and norms and it is the second property in the definition of a norm. It is left to the reader to verify that all the previously mentioned norms satisfy the norm properties 1–3. We only show here the triangle inequality for the Euclidean norm

$$\|\boldsymbol{x}+\boldsymbol{y}\|_2 \leq \|\boldsymbol{x}\|_2 + \|\boldsymbol{y}\|_2 \ .$$

To see why this is true, we write the norm of the sum as an inner product

$$\|\boldsymbol{x}+\boldsymbol{y}\|_2^2 = \langle \boldsymbol{x}+\boldsymbol{y}, \boldsymbol{x}+\boldsymbol{y}\rangle \ ,$$

and analyze the inner product as

$$\begin{aligned}
\|\boldsymbol{x}+\boldsymbol{y}\|_2^2 &= \langle \boldsymbol{x}+\boldsymbol{y}, \boldsymbol{x}+\boldsymbol{y}\rangle \\
&= \langle \boldsymbol{x}, \boldsymbol{x}\rangle + \langle \boldsymbol{x}, \boldsymbol{y}\rangle + \langle \boldsymbol{y}, \boldsymbol{x}\rangle + \langle \boldsymbol{y}, \boldsymbol{y}\rangle \\
&= \|\boldsymbol{x}\|_2^2 + 2\langle \boldsymbol{x}, \boldsymbol{y}\rangle + \|\boldsymbol{y}\|_2^2 \\
&\leq \|\boldsymbol{x}\|_2^2 + 2\|\boldsymbol{x}\|_2\|\boldsymbol{y}\|_2 + \|\boldsymbol{y}\|_2^2 \\
&= (\|\boldsymbol{x}\|_2 + \|\boldsymbol{y}\|_2)^2 \ .
\end{aligned}$$

Thus,

$$\|\boldsymbol{x}+\boldsymbol{y}\|_2^2 \leq (\|\boldsymbol{x}\|_2 + \|\boldsymbol{y}\|_2)^2 \ .$$

After taking the square root in the last inequality we obtain the triangle inequality $\|\boldsymbol{x}+\boldsymbol{y}\|_2 \leq \|\boldsymbol{x}\|_2 + \|\boldsymbol{y}\|_2$.

In analogy to vector norms we define matrix norms $\|\cdot\| : \mathbb{R}^{n,n} \to \mathbb{R}$ as a mapping that satisfies the following properties

- $\|\boldsymbol{A}\| \geq 0$, $\boldsymbol{A} \in \mathbb{R}^{n,n}$, $(\|\boldsymbol{A}\| = 0 \Leftrightarrow \boldsymbol{A} = 0)$
- $\|\boldsymbol{A}+\boldsymbol{B}\| \leq \|\boldsymbol{A}\| + \|\boldsymbol{B}\|$, $\boldsymbol{A}, \boldsymbol{B} \in \mathbb{R}^{n,n}$
- $\|c\boldsymbol{A}\| = |c|\|\boldsymbol{A}\|$ for $c \in \mathbb{R}$, $\boldsymbol{A} \in \mathbb{R}^{n,n}$
- $\|\boldsymbol{A}\boldsymbol{B}\| \leq \|\boldsymbol{A}\|\,\|\boldsymbol{B}\|$, $\boldsymbol{A}, \boldsymbol{B} \in \mathbb{R}^{n,n}$

and again, matrix norms determine the size of a matrix.

Note that the triangle inequality of matrix norms is inherited from vector norms. It is not hard to see that

$$\big|\|\boldsymbol{A}\| - \|\boldsymbol{B}\|\big| \leq \|\boldsymbol{A}-\boldsymbol{B}\| \ \leq \|\boldsymbol{A}\| + \|\boldsymbol{B}\| \ .$$

For example, the right inequality follows directly from the triangle inequality of $\|\boldsymbol{A}\| = \|\boldsymbol{A}-\boldsymbol{B}+\boldsymbol{B}\| \leq \|\boldsymbol{A}-\boldsymbol{B}\| + \|\boldsymbol{B}\|$. While the left, again as a consequence of triangle inequality, can be derived from $\|\boldsymbol{A}-\boldsymbol{B}\| = \|\boldsymbol{A}+(-\boldsymbol{B})\| \leq \|\boldsymbol{A}\| + \|-\boldsymbol{B}\| = \|\boldsymbol{A}\| + \|\boldsymbol{B}\|$.

The most commonly used matrix norms are those induced by vector norms. Such norms are called *natural norms* and are defined as

$$\|\boldsymbol{A}\| = \max_{\boldsymbol{x}\neq\boldsymbol{0}} \frac{\|\boldsymbol{A}\boldsymbol{x}\|}{\|\boldsymbol{x}\|} = \max\left\{\frac{\|\boldsymbol{A}\boldsymbol{x}\|}{\|\boldsymbol{x}\|},\ \boldsymbol{x}\neq\boldsymbol{0}\right\} \ .$$

Since $\|\boldsymbol{A}\|$ is the maximum of all quantities $\|\boldsymbol{A}\boldsymbol{x}\|/\|\boldsymbol{x}\|$ for all $\boldsymbol{x}\neq\boldsymbol{0}$, it is implied that

$$\|\boldsymbol{A}\| \geq \frac{\|\boldsymbol{A}\boldsymbol{x}\|}{\|\boldsymbol{x}\|} \quad \text{for all } \boldsymbol{x}\neq\boldsymbol{0} \ .$$

Thus, for $\boldsymbol{x} \neq \mathbf{0}$ and any square matrix $\boldsymbol{A}$ we have

$$\|\boldsymbol{A}\boldsymbol{x}\| \leq \|\boldsymbol{A}\|\|\boldsymbol{x}\| .$$

It is noted that the definition of matrix norm includes complex vectors $\boldsymbol{x} \in \mathbb{C}^n$. Also if $\boldsymbol{x} = \mathbf{0}$ the last inequality holds as equality. The most important matrix norms are the following:

$$\|\boldsymbol{A}\|_1 = \max_{j=1,\ldots,n} \sum_{i=1}^{n} |a_{ij}|, \quad (\ell_1\text{-norm})$$

$$\|\boldsymbol{A}\|_2 = \sqrt{\rho(\boldsymbol{A}^T\boldsymbol{A})}, \quad (\ell_2\text{-norm})$$

$$\|\boldsymbol{A}\|_\infty = \max_{i=1,\ldots,n} \sum_{j=1}^{n} |a_{ij}|, \quad (\ell_\infty\text{-norm})$$

$$\|\boldsymbol{A}\|_F = \sqrt{\sum_{j=1}^{n}\sum_{i=1}^{n} |a_{ij}|^2}, \quad (\text{Frobenius norm})$$

The norm $\|\boldsymbol{A}\|_F$ is called the Frobenius norm of $\boldsymbol{A}$ and is named after the German mathematician Ferdinand Georg Frobenius (1849–1917). In the definition of the norm $\|\boldsymbol{A}\|_2$, the $\rho(\boldsymbol{A})$ denotes the spectral radius of $\boldsymbol{A}$, which is the maximum in magnitude eigenvalue of $\boldsymbol{A}$

$$\rho(\boldsymbol{A}) = \max\{|\lambda_1|, |\lambda_2|, \ldots, |\lambda_k|\} ,$$

with $\det(\boldsymbol{A} - \lambda_i \boldsymbol{I}) = 0$.

9.3.2 Theoretical properties of matrix norms

Here we present a few theoretical properties of matrix norms. Their proofs require some advanced knowledge of linear algebra, and for the sake of completeness are included here.

In general, any two matrix or vector norms $\|\cdot\|_1$ and $\|\cdot\|_2$ are equivalent, in the sense that there are constact $C_1, C_2 > 0$ such that $C_1\|\cdot\|_2 \leq \|\cdot\|_1 \leq C_2\|\cdot\|_2$. Also all natural norms are related to the spectral radius. More precisely there is no way a matrix norm can become smaller than the spectral radius of $\boldsymbol{A}$:

Theorem 9.4. *For any natural matrix norm* $\|\cdot\|$ *and* $n \times n$ *complex matrix* $\boldsymbol{A}$ *we have* $\rho(\boldsymbol{A}) \leq \|\boldsymbol{A}\|$.

Proof. Let $\lambda \in \mathbb{C}$ be an eigenvalue of the matrix $\boldsymbol{A}$, and $\boldsymbol{x} \neq \mathbf{0}$ the corresponding eigenvector. This means that $\boldsymbol{A}\boldsymbol{x} = \lambda\boldsymbol{x}$. Taking the norm of both sides of the last relation we get

$$\|\boldsymbol{A}\boldsymbol{x}\| = \|\lambda\boldsymbol{x}\| .$$

Since $\|\cdot\|$ is assumed to be natural, we have $\|\boldsymbol{A}\boldsymbol{x}\| \leq \|\boldsymbol{A}\|\|\boldsymbol{x}\|$, and thus

$$\|\lambda\boldsymbol{x}\| = \|\boldsymbol{A}\boldsymbol{x}\| \leq \|\boldsymbol{A}\|\|\boldsymbol{x}\| .$$

But $\|\lambda\boldsymbol{x}\| = |\lambda|\|\boldsymbol{x}\|$ from the norm properties, and so

$$|\lambda|\|\boldsymbol{x}\| \leq \|\boldsymbol{A}\|\|\boldsymbol{x}\| .$$

Since $\|\boldsymbol{x}\| > 0$ we simplify the last inequality into $|\lambda| \leq \|\boldsymbol{A}\|$ and since λ was arbitrary, this holds for the maximum absolute value of eigenvalues $\rho(\boldsymbol{A}) \leq \|\boldsymbol{A}\|$. □

The inverse inequality is not true, but what one can prove is that the spectral radius of a matrix can be an estimate of its norm:

Theorem 9.5. *For any $n \times n$ (complex) matrix $\boldsymbol{A}$ and for all $\epsilon > 0$, there is a natural norm $\|\cdot\|$ such that $\|\boldsymbol{A}\| \leq \rho(\boldsymbol{A}) + \epsilon$.*

Proof. The proof requires the Jordan normal form[6] of matrix $\boldsymbol{A}$. So let $\boldsymbol{A} = \boldsymbol{PJP}^{-1}$, with $\boldsymbol{J} = \text{diag}(\boldsymbol{J}_1, \boldsymbol{J}_2, \ldots, \boldsymbol{J}_p)$ the Jordan normal form of $\boldsymbol{A}$. We define the matrix

$$\tilde{\boldsymbol{J}} = \boldsymbol{D}^{-1}\boldsymbol{J}\boldsymbol{D} ,$$

where $\boldsymbol{D} = \text{diag}(1, \epsilon, \epsilon^2, \ldots, \epsilon^{n-1})$. The matrix $\tilde{\boldsymbol{J}}$ has the exact same structure as $\boldsymbol{J}$ with blocks

$$\tilde{\boldsymbol{J}}_i = \begin{pmatrix} \lambda_i & \epsilon & 0 & \cdots & 0 & 0 \\ 0 & \lambda_i & \epsilon & \cdots & 0 & 0 \\ \vdots & \vdots & \vdots & \ddots & \vdots & \vdots \\ 0 & 0 & 0 & \cdots & \lambda_i & \epsilon \\ 0 & 0 & 0 & \cdots & 0 & \lambda_i \end{pmatrix} ,$$

for all the i blocks. Thus, $\|\tilde{\boldsymbol{J}}\|_\infty \leq |\lambda_i| + \epsilon$, and thus $\|\tilde{\boldsymbol{J}}\|_\infty \leq \rho(\tilde{\boldsymbol{J}}) + \epsilon$. From the definitions of $\boldsymbol{J}$ and $\tilde{\boldsymbol{J}}$ we have

$$\tilde{\boldsymbol{J}} = \boldsymbol{D}^{-1}\boldsymbol{J}\boldsymbol{D} = \boldsymbol{D}^{-1}\boldsymbol{P}^{-1}\boldsymbol{A}\boldsymbol{P}\boldsymbol{D} = (\boldsymbol{P}\boldsymbol{D})^{-1}\boldsymbol{A}(\boldsymbol{P}\boldsymbol{D}) = \boldsymbol{T}^{-1}\boldsymbol{A}\boldsymbol{T} ,$$

where $\boldsymbol{T} = \boldsymbol{P}\boldsymbol{D}$. Since $\boldsymbol{D} = \text{diag}(1, \epsilon, \ldots, \epsilon^{n-1})$, the columns of $\boldsymbol{T}$ are the columns of $\boldsymbol{P}$ multiplied with $1, \epsilon, \ldots, \epsilon^{n-1}$.

Let $\|\boldsymbol{A}\|$ be the matrix norm of $\boldsymbol{A}$ defined using the vector norm $\|\boldsymbol{x}\| = \|\boldsymbol{T}^{-1}\boldsymbol{x}\|_\infty$. Then

$$\begin{aligned} \|\boldsymbol{A}\| &= \max_{\boldsymbol{x}\in\mathbb{C}^n, \|\boldsymbol{x}\|=1} \|\boldsymbol{A}\boldsymbol{x}\| \\ &= \max_{\boldsymbol{x}\in\mathbb{C}^n, \|\boldsymbol{T}^{-1}\boldsymbol{x}\|_\infty=1} \|\boldsymbol{T}^{-1}\boldsymbol{A}\boldsymbol{x}\|_\infty \\ &= \max_{\boldsymbol{T}^{-1}\boldsymbol{x}\in\mathbb{C}^n, \|\boldsymbol{T}^{-1}\boldsymbol{x}\|_\infty=1} \|\boldsymbol{T}^{-1}\boldsymbol{A}\boldsymbol{T}(\boldsymbol{T}^{-1}\boldsymbol{x})\|_\infty \\ &= \max_{\boldsymbol{T}^{-1}\boldsymbol{x}\in\mathbb{C}^n, \|\boldsymbol{T}^{-1}\boldsymbol{x}\|_\infty=1} \|\tilde{\boldsymbol{J}}(\boldsymbol{T}^{-1}\boldsymbol{x})\|_\infty \\ &= \|\tilde{\boldsymbol{J}}\|_\infty \\ &\leq \rho(\boldsymbol{A}) + \epsilon , \end{aligned}$$

which completes the proof. □

Finally, we present a theorem of John von Neumann (1903–1957), which we will find useful in estimates and proofs later in this book.

Theorem 9.6 (Neumann). *If for an $n \times n$ complex matrix $\boldsymbol{A}$ and for any natural norm is $\|\boldsymbol{A}\| < 1$, then:*

(i) The matrix $\boldsymbol{I} - \boldsymbol{A}$ is invertible, which means that the inverse $(\boldsymbol{I} - \boldsymbol{A})^{-1}$ exists, and

(ii) $\frac{1}{1+\|\boldsymbol{A}\|} \leq \|(I - \boldsymbol{A})^{-1}\| \leq \frac{1}{1-\|\boldsymbol{A}\|}$.

Proof. We prove the first part of the theorem using a contradiction. If $\boldsymbol{I} - \boldsymbol{A}$ wasn't invertible, then the linear system $(\boldsymbol{I} - \boldsymbol{A})\boldsymbol{x} = \boldsymbol{0}$ would have a solution $\boldsymbol{x} \neq \boldsymbol{0}$. For that $\boldsymbol{x}$ we would also have $\boldsymbol{x} = \boldsymbol{A}\boldsymbol{x}$. Taking norms on both sides of $\boldsymbol{x} = \boldsymbol{A}\boldsymbol{x}$ we have $\|\boldsymbol{x}\| = \|\boldsymbol{A}\boldsymbol{x}\| \leq \|\boldsymbol{A}\|\|\boldsymbol{x}\|$. Since $\boldsymbol{x} \neq \boldsymbol{0}$ we have that $\|\boldsymbol{x}\| > \boldsymbol{0}$, and thus we have $\|\boldsymbol{A}\| \geq 1$, which is a contradiction since we assumed that $\|\boldsymbol{A}\| < 1$. Hence, the matrix $\boldsymbol{I} - \boldsymbol{A}$ is invertible.

[6] A discussion of the Jordan normal form can be found in Chapter 12 and in the Appendix.

In order to prove the second part of the theorem we make use of part (i). Since the matrix $\boldsymbol{I}-\boldsymbol{A}$ is invertible we have

$$\boldsymbol{I}=(\boldsymbol{I}-\boldsymbol{A})^{-1}(\boldsymbol{I}-\boldsymbol{A})=(\boldsymbol{I}-\boldsymbol{A})^{-1}-(\boldsymbol{I}-\boldsymbol{A})^{-1}\boldsymbol{A}\ .$$

Taking norms on both sides of the previous relation we have

$$1=\|\boldsymbol{I}\|=\|(\boldsymbol{I}-\boldsymbol{A})^{-1}-(\boldsymbol{I}-\boldsymbol{A})^{-1}\boldsymbol{A}\|\ . \tag{9.4}$$

On the other hand we have that

$$\begin{aligned}\|(\boldsymbol{I}-\boldsymbol{A})^{-1}\|-\|(\boldsymbol{I}-\boldsymbol{A})^{-1}\|\|\boldsymbol{A}\| &\leq \|(\boldsymbol{I}-\boldsymbol{A})^{-1}-(\boldsymbol{I}-\boldsymbol{A})^{-1}\boldsymbol{A}\|\\ &\leq \|(\boldsymbol{I}-\boldsymbol{A})^{-1}\|+\|(\boldsymbol{I}-\boldsymbol{A})^{-1}\|\|\boldsymbol{A}\|\ ,\end{aligned}$$

or better

$$\|(\boldsymbol{I}-\boldsymbol{A})^{-1}\|(1-\|\boldsymbol{A}\|)\leq\|(\boldsymbol{I}-\boldsymbol{A})^{-1}-(\boldsymbol{I}-\boldsymbol{A})^{-1}\boldsymbol{A}\|\leq\|(\boldsymbol{I}-\boldsymbol{A})^{-1}\|(1+\|\boldsymbol{A}\|)\ .$$

Hence,

$$1\leq\|(\boldsymbol{I}-\boldsymbol{A})^{-1}\|(1+\|\boldsymbol{A}\|)\ ,$$

which gives

$$\frac{1}{1+\|\boldsymbol{A}\|}\leq\|(\boldsymbol{I}-\boldsymbol{A})^{-1}\|\ . \tag{9.5}$$

From (9.4) we also have

$$\begin{aligned}1 &\geq \|(\boldsymbol{I}-\boldsymbol{A})^{-1}\|-\|(\boldsymbol{I}-\boldsymbol{A})^{-1}\boldsymbol{A}\|\\ &\geq \|(\boldsymbol{I}-\boldsymbol{A})^{-1}\|-\|(\boldsymbol{I}-\boldsymbol{A})^{-1}\|\|\boldsymbol{A}\|\\ &\geq \|(\boldsymbol{I}-\boldsymbol{A})^{-1}\|(1-\|\boldsymbol{A}\|)\ ,\end{aligned}$$

which gives (since $\|\boldsymbol{A}\|<1$)

$$\frac{1}{1-\|\boldsymbol{A}\|}\leq\|(\boldsymbol{I}-\boldsymbol{A})^{-1}\|\ . \tag{9.6}$$

Combining (9.5) and (9.6) we have that

$$\frac{1}{1+\|\boldsymbol{A}\|}\leq\|(\boldsymbol{I}-\boldsymbol{A})^{-1}\|\leq\frac{1}{1-\|\boldsymbol{A}\|}\ ,$$

which completes the proof. □

After having a deep insight of matrix and vector norms, we continue with the condition number of a matrix. The condition number gives us an estimate of how close a matrix is to be singular by numerical means.

9.3.3 Condition number

For any natural norm, the condition number of $\boldsymbol{A}$ is the number

$$\mathrm{cond}(\boldsymbol{A})=\|\boldsymbol{A}\|\cdot\|\boldsymbol{A}^{-1}\|\ .$$

Suppose that we want to solve a system

$$\boldsymbol{A}\boldsymbol{x}=\boldsymbol{b}\ . \tag{9.7}$$

To see how this number can determine the sensitivity of the solution $\boldsymbol{x}$ on $\boldsymbol{A}$ we make the following three assumptions: (i) We store $\boldsymbol{A}$ in memory using exact arithmetic; (ii) We store $\boldsymbol{b}$ with finite precision as $\boldsymbol{b}^* = \boldsymbol{b} + \boldsymbol{\delta b}$ instead of $\boldsymbol{b}$, where $\boldsymbol{\delta b}$ is the error-vector; (iii) We execute the Gaussian elimination with exact arithmetic.

Because of the error in the storage of vector $\boldsymbol{b}$, we compute the solution $\boldsymbol{x}^* = \boldsymbol{x} + \boldsymbol{\delta x}$ instead of $\boldsymbol{x}$ by solving exactly the system

$$\boldsymbol{A}\boldsymbol{x}^* = \boldsymbol{b}^* . \tag{9.8}$$

Subtracting systems (9.7) and (9.8) we get the equation for the error $\boldsymbol{\delta x}$

$$\boldsymbol{A}\ \boldsymbol{\delta x} = \boldsymbol{\delta b} ,$$

which implies

$$\boldsymbol{\delta x} = \boldsymbol{A}^{-1}\ \boldsymbol{\delta b} .$$

Taking norms in the last equation we obtain

$$\|\boldsymbol{\delta x}\| \leq \|\boldsymbol{A}^{-1}\|\ \|\boldsymbol{\delta b}\| . \tag{9.9}$$

Since $\boldsymbol{b} = \boldsymbol{A}\boldsymbol{x}$, taking norms on both sides and using the definition of matrix norm we obtain $\|\boldsymbol{b}\| \leq \|\boldsymbol{A}\|\|\boldsymbol{x}\|$, which is equivalent to

$$\|\boldsymbol{x}\| \geq \frac{\|\boldsymbol{b}\|}{\|\boldsymbol{A}\|} . \tag{9.10}$$

Dividing (9.9) by (9.10) we obtain the formula for the relative error

$$\frac{\|\boldsymbol{\delta x}\|}{\|\boldsymbol{x}\|} \leq \|\boldsymbol{A}\| \cdot \|\boldsymbol{A}^{-1}\| \frac{\|\boldsymbol{\delta b}\|}{\|\boldsymbol{b}\|} .$$

Equivalently, we write the formula of the relative error as

$$\frac{\|\boldsymbol{\delta x}\|}{\|\boldsymbol{x}\|} \leq \mathrm{cond}(\boldsymbol{A}) \frac{\|\boldsymbol{\delta b}\|}{\|\boldsymbol{b}\|} ,$$

where $\mathrm{cond}(\boldsymbol{A}) = \|\boldsymbol{A}\| \cdot \|\boldsymbol{A}^{-1}\|$ is the *condition number* of $\boldsymbol{A}$.

Since the relative error $\|\boldsymbol{\delta x}\|/\|\boldsymbol{x}\|$ is proportional to the condition number, we conclude that matrices with large condition number might lead to a solution with large relative error. In such cases we definitely need to use Gaussian elimination with pivoting.

Since $\boldsymbol{I} = \boldsymbol{A} \cdot \boldsymbol{A}^{-1}$, by taking any natural matrix norm we have

$$1 = \|\boldsymbol{I}\| = \|\boldsymbol{A}\boldsymbol{A}^{-1}\| \leq \|\boldsymbol{A}\|\|\boldsymbol{A}^{-1}\| ,$$

which yields

$$\mathrm{cond}(\boldsymbol{A}) \geq 1 .$$

This means that the best matrix for Gaussian elimination is the one with $\mathrm{cond}(\boldsymbol{A}) = 1$. The condition number of the identity matrix $\boldsymbol{I}$ is $\mathrm{cond}(\boldsymbol{I}) = 1$. Some times we write $\mathrm{cond}_p(\boldsymbol{A}) = \|\boldsymbol{A}\|_p\|\boldsymbol{A}^{-1}\|_p$.

The condition number can be computed numerically using NumPy and the function `cond` of the module `linalg`. The general call of the function `cond` is

```
cond(A,p)
```

where `A` is the matrix we search for its condition number and `p` corresponds to the ℓ_p-norm. By default `p=None` for computations with the ℓ_2-norm. Other values can be `'fro'` for the Frobenious, `inf` for the infinity, `1` for the ℓ_1-norm and `2` for the ℓ_2 norm. NumPy defines also the norms `-inf`, `-1` and `-2`, which we will not use here but the interested reader can look them up in the NumPy online reference for more information. The function `cond` returns the condition of matrix `A`. Here we consider a very simple example demonstration of the use of the function `cond`.

```
from numpy.linalg import cond
A = np.array([[1, 0, -1], [0, 1, 0], [1, 0, 1]])
cond(A)
cond(A, np.inf)
cond(A, 1)
cond(A, 2)
```

```
1.4142135623730951
2.0
2.0
1.4142135623730951
```

The specific condition number is not large. An *ill-conditioned* matrix usually has condition number greater than 100. An example of ill-conditioned matrices are the so-called Hilbert matrices with entries

$$a_{ij} = \frac{1}{i+j-1} .$$

Trying for example the 3×3 matrix

$$\boldsymbol{A} = \begin{pmatrix} 1 & 1/2 & 1/3 \\ 1/2 & 1/3 & 1/4 \\ 1/3 & 1/4 & 1/5 \end{pmatrix} ,$$

we compute $\mathrm{cond}_2(\boldsymbol{A}) \approx 524$. Such matrices occur in least square problems (polynomial regression) of Section 6.5.

9.4 Other Matrix Computations

The Gaussian elimination and *LU* factorization are very useful algorithms, not only for the solution of linear systems. We can employ these methods to perform other difficult computations such as the computation of inverse or determinant of a matrix. The computation of a determinant using the definition is extremely expensive. So expensive that it is almost impossible to compute the determinant of an 100×100 matrix into reasonable timeframes. This fact makes methods such as Cramer's rule useless for the solution of large linear systems.

9.4.1 Computation of inverse matrix

If A^{-1} is the inverse of an $n \times n$ matrix A, then $A A^{-1} = A^{-1} A = I$. We denote by x_i the i-th column of A^{-1} and by b_i the i-th column of the identity matrix I, i.e. $b_i = (0, 0, \ldots, 1, \ldots, 0)^T$. Observe that

$$A x_i = b_i \ .$$

Therefore, by solving the system $Ax_i = b_i$ for x_i we compute the i-th column of A^{-1}. This means that in order to compute all the columns of A^{-1} we need to solve n linear systems. In practice, we compute the inverse matrix of A applying Gaussian elimination to the augmented matrix

$$(A|I) \ ,$$

and backward substitution n times; or we compute the LU factorization once and apply Forward and Backward Substitution n times.

For example, in order to find the inverse of the matrix

$$A = \begin{pmatrix} 1 & 2 & -1 \\ 2 & 1 & 0 \\ -1 & 1 & 2 \end{pmatrix} ,$$

we consider the augmented matrix

$$A = \left(\begin{array}{rrr|rrr} 1 & 2 & -1 & 1 & 0 & 0 \\ 2 & 1 & 0 & 0 & 1 & 0 \\ -1 & 1 & 2 & 0 & 0 & 1 \end{array}\right) .$$

First, performing $(R_2 - 2R_1) \to (R_2)$ and $(R_3 + R_1) \to (R_3)$ we get

$$\left(\begin{array}{rrr|rrr} 1 & 2 & -1 & 1 & 0 & 0 \\ 0 & -3 & 2 & -2 & 1 & 0 \\ 0 & 3 & 1 & 1 & 0 & 1 \end{array}\right) ,$$

which after the operation $(R_3 + R_2) \to (R_3)$ gives the reduced form

$$\left(\begin{array}{rrr|rrr} 1 & 2 & -1 & 1 & 0 & 0 \\ 0 & -3 & 2 & -2 & 1 & 0 \\ 0 & 0 & 3 & -1 & 1 & 1 \end{array}\right) .$$

Backward substitution is performed on each of the three augmented matrices

$$\left(\begin{array}{rrr|r} 1 & 2 & -1 & 1 \\ 0 & -3 & 2 & -2 \\ 0 & 0 & 3 & -1 \end{array}\right) , \quad \left(\begin{array}{rrr|r} 1 & 2 & -1 & 0 \\ 0 & -3 & 2 & 1 \\ 0 & 0 & 3 & 1 \end{array}\right) , \quad \left(\begin{array}{rrr|r} 1 & 2 & -1 & 0 \\ 0 & -3 & 2 & 0 \\ 0 & 0 & 3 & 1 \end{array}\right) ,$$

to obtain the solutions

$$b_1 = \begin{pmatrix} -2/9 \\ 4/9 \\ -1/3 \end{pmatrix} , \quad b_2 = \begin{pmatrix} 5/9 \\ -1/9 \\ 1/3 \end{pmatrix} , \quad b_3 = \begin{pmatrix} -1/9 \\ 2/9 \\ 1/3 \end{pmatrix} .$$

The matrix A^{-1} is

$$A^{-1} = B = \begin{pmatrix} -\frac{2}{9} & \frac{5}{9} & -\frac{1}{9} \\ \frac{4}{9} & -\frac{1}{9} & \frac{2}{9} \\ -\frac{1}{3} & \frac{1}{3} & \frac{1}{3} \end{pmatrix} .$$

The numerical computation of the inverse of a matrix in NumPy can be done using the function `inv` of the module `linalg`. For example

```
import numpy.linalg as npl
A = np.array([[1., 2.], [3., 4.]])
ainv = npl.inv(A)
print(Ainv)
print(np.dot(A, Ainv))
```

```
[[-2.   1. ]
 [ 1.5 -0.5]]
[[1.00000000e+00 1.11022302e-16]
 [0.00000000e+00 1.00000000e+00]]
```

9.4.2 Computation of determinants

The computation of the determinant of an $n \times n$ matrix $\boldsymbol{A}$ using the methodology with the cofactors and the minors (see Section 2.2.10) requires $O(n!)$ floating point operations. Such a computation is perhaps impossible for large values of n. We bypass this obstacle with the help of LU factorization.

Assume that the LU factorization of the matrix $\boldsymbol{A}$ is given in the form $\boldsymbol{PA} = \boldsymbol{LU}$. Then

$$\det(\boldsymbol{A}) = \det(\boldsymbol{P}^{-1})\det(\boldsymbol{L})\det(\boldsymbol{U}) \ .$$

Any permutation matrix $\boldsymbol{P}$ factors as a product of S row-interchanging elementary permutation matrices $\boldsymbol{P}_i$, each having determinant -1, and thus $\det(\boldsymbol{P}^{-1}) = (-1)^S$. The determinants $\det(\boldsymbol{L}) = \prod_{i=1}^{n} l_{ii}$ and $\det(\boldsymbol{U}) = \prod_{i=1}^{n} u_{ii}$ since $\boldsymbol{L}$ and $\boldsymbol{U}$ are lower triangular and upper triangular, respectively. For the proofs of all these properties, we refer to basic linear algebra books such as [105]. Therefore,

$$\det(\boldsymbol{A}) = (-1)^S \left(\prod_{i=1}^{n} l_{ii}\right)\left(\prod_{i=1}^{n} u_{ii}\right) ,$$

where S is the number of row interchanges. Because the matrix $\boldsymbol{L}$ has only 1 in its main diagonal

$$\prod_{i=1}^{n} l_{ii} = 1 \ .$$

We conclude that the determinant $\det(\boldsymbol{A})$ can be computed with the product

$$\det(\boldsymbol{A}) = (-1)^S \left(\prod_{i=1}^{n} u_{ii}\right) \ .$$

In Section 9.2.2 we saw that the LU factorization of the matrix

$$A = \begin{pmatrix} 3 & 17 & 10 \\ 2 & 4 & -2 \\ 6 & 18 & -12 \end{pmatrix} ,$$

is the following

$$\begin{pmatrix} 0 & 0 & 1 \\ 1 & 0 & 0 \\ 0 & 1 & 0 \end{pmatrix}\begin{pmatrix} 3 & 17 & 10 \\ 2 & 4 & -2 \\ 6 & 18 & -12 \end{pmatrix} = \begin{pmatrix} 1 & 0 & 0 \\ 1/2 & 1 & 0 \\ 1/3 & -1/4 & 1 \end{pmatrix}\begin{pmatrix} 6 & 18 & -12 \\ 0 & 8 & 16 \\ 0 & 0 & 6 \end{pmatrix} .$$

In this case, $S = 2$ and thus, the determinant of the matrix A is $\det(A) = 6 \times 8 \times 6 = 288$. In Python we can compute the determinant of a matrix using the function `det` of the module `numpy.linalg`.

9.5 Symmetric Matrices

It happens very often in applications that the matrix $\boldsymbol{A}$ is symmetric. If $\boldsymbol{A}$ is symmetric, then we can store half of $\boldsymbol{A}$, i.e. only its lower triangular part, and save computer memory. If in addition $\boldsymbol{A}$ is positive definite, then we could make the LU factorization faster and stable. In particular, if $\boldsymbol{A}$ is symmetric and positive definite, then there is a factorization of the form $\boldsymbol{H}\boldsymbol{H}^T$. This factorization is called Cholesky decomposition named after André-Louis Cholesky (1875–1918). Recall that a matrix $\boldsymbol{A}$ is positive definite if

$$\boldsymbol{x}^T\boldsymbol{A}\boldsymbol{x} > 0 ,$$

for all n-dimensional vectors $\boldsymbol{x} \neq \boldsymbol{0}$.

As we saw in Section 2.3.2, the process of verifying the positive definiteness of $\boldsymbol{A}$ can be very difficult. There are a few, helpful facts in the verification process. First of all, a symmetric matrix $\boldsymbol{A}$ is positive definite if and only if the Gaussian elimination without row interchanges can be performed on the linear system $\boldsymbol{A}\boldsymbol{x} = \boldsymbol{b}$ with all pivot elements positive. The computations involving a positive definite matrix $\boldsymbol{A}$ are stable with respect to the growth factor of round-off errors. Moreover, any symmetric matrix $\boldsymbol{A}$ is positive definite if and only if $\boldsymbol{A}$ can be written in the form $\boldsymbol{H}\boldsymbol{H}^T$, where $\boldsymbol{H}$ is lower triangular with non-zero diagonal entries. But first let's see the result of the LU decomposition of symmetric matrices.

9.5.1 $\boldsymbol{LDL^T}$ factorization

The LDL^T factorization of a matrix $\boldsymbol{A}$ can be carried out if $\boldsymbol{A}$ is symmetric and has an ordinary LU factorization $\boldsymbol{A} = \boldsymbol{L}\boldsymbol{U}$, with $\boldsymbol{L}$ unit lower triangular, and is the decomposition of $\boldsymbol{A}$ into a product $\boldsymbol{L}\boldsymbol{D}\boldsymbol{L}^T$ where $\boldsymbol{D}$ is diagonal[7]. First we extract the main diagonal of $\boldsymbol{U}$ and we write $\boldsymbol{U} = \boldsymbol{D}\boldsymbol{U}_1$ where $\boldsymbol{U}_1$ is unit upper triangular. As a result, we write the LU decomposition of $\boldsymbol{A}$ as $\boldsymbol{A} = \boldsymbol{L}\boldsymbol{D}\boldsymbol{U}_1$. This is known as Crout decomposition named after Prescott Durand Crout (1907–1984). Due to the symmetry of $\boldsymbol{A}$, we have that $\boldsymbol{U}_1 = \boldsymbol{L}^T$ which makes the LDL^T factorization a special case of Crout decomposition. To see this, we start with

$$\boldsymbol{L}\boldsymbol{U} = \boldsymbol{A} = \boldsymbol{A}^T = (\boldsymbol{L}\boldsymbol{U})^T = \boldsymbol{U}^T\boldsymbol{L}^T .$$

Since $\boldsymbol{L}$ is unit lower triangular, it is invertible, and we can write $\boldsymbol{U} = \boldsymbol{L}^{-1}\boldsymbol{U}^T\boldsymbol{L}^T$. Then $\boldsymbol{U}(\boldsymbol{L}^T)^{-1} = \boldsymbol{L}^{-1}\boldsymbol{U}^T$. However, since the product of lower triangular matrices is lower triangular and the product of upper triangular matrices is upper triangular, the right side of this equation is lower triangular and the left side is upper triangular (see Section 2.2.8). Thus, both sides must be equal to a diagonal matrix, say, $\boldsymbol{D}$. The equation $\boldsymbol{U}(\boldsymbol{L}^T)^{-1} = \boldsymbol{D}$ implies

$$\boldsymbol{U} = \boldsymbol{D}\boldsymbol{L}^T ,$$

and thus

$$\boldsymbol{A} = \boldsymbol{L}\boldsymbol{U} = \boldsymbol{L}\boldsymbol{D}\boldsymbol{L}^T .$$

For example, let

$$\boldsymbol{A} = \begin{pmatrix} 4 & 3 & 2 & 1 \\ 3 & 3 & 2 & 1 \\ 2 & 2 & 2 & 1 \\ 1 & 1 & 1 & 1 \end{pmatrix} .$$

[7]If $\boldsymbol{A}$ is complex Hermitian, then the particular factorization is written as $\boldsymbol{A} = \boldsymbol{L}\boldsymbol{D}\boldsymbol{L}^H$

The LU factorization of $\boldsymbol{A}$ will be

$$\boldsymbol{A} = \begin{pmatrix} 1 & 0 & 0 & 0 \\ \frac{3}{4} & 1 & 0 & 0 \\ \frac{1}{2} & \frac{2}{3} & 1 & 0 \\ \frac{1}{4} & \frac{1}{3} & \frac{1}{2} & 1 \end{pmatrix} \begin{pmatrix} 4 & 3 & 2 & 1 \\ 0 & \frac{3}{4} & \frac{1}{2} & \frac{1}{4} \\ 0 & 0 & \frac{2}{3} & \frac{1}{3} \\ 0 & 0 & 0 & \frac{1}{2} \end{pmatrix} = \boldsymbol{LU} .$$

Extracting the diagonal entries from $\boldsymbol{U}$ and place them into a diagonal matrix $\boldsymbol{D}$ yields

$$\boldsymbol{U} = \begin{pmatrix} 4 & 0 & 0 & 0 \\ 0 & \frac{3}{4} & 0 & 0 \\ 0 & 0 & \frac{2}{3} & 0 \\ 0 & 0 & 0 & \frac{1}{2} \end{pmatrix} \begin{pmatrix} 1 & \frac{3}{4} & \frac{1}{2} & \frac{1}{4} \\ 0 & 1 & \frac{2}{3} & \frac{1}{3} \\ 0 & 0 & 1 & \frac{1}{2} \\ 0 & 0 & 0 & 1 \end{pmatrix} = \boldsymbol{DL}^T .$$

Clearly we have $\boldsymbol{A} = \boldsymbol{LDL}^T$.

9.5.2 Cholesky factorization

If $\boldsymbol{A}$ is symmetric and positive definite, then we can write $\boldsymbol{A}$ in the form

$$\boldsymbol{A} = \boldsymbol{HH}^T ,$$

where $\boldsymbol{H}$ is lower triangular matrix (without necessarily ones in the main diagonal). This is the Cholesky factorization of symmetric and positive definite matrix.

We can acquire the Cholesky factorization from the LDL^T factorization of $\boldsymbol{A}$. First, write $\boldsymbol{A} = \boldsymbol{LDL}^T$. The diagonal matrix $\boldsymbol{D}$ can be written as a product of diagonal matrices $\boldsymbol{D} = \boldsymbol{D}^{1/2}\boldsymbol{D}^{1/2}$ where

$$\boldsymbol{D}^{1/2} = \operatorname{diag}(\sqrt{d_{11}}, \sqrt{d_{22}}, \ldots, \sqrt{d_{nn}}) .$$

Then, we have

$$\boldsymbol{A} = \boldsymbol{LD}^{1/2}\boldsymbol{D}^{1/2}\boldsymbol{L}^T = \boldsymbol{LD}^{1/2}(\boldsymbol{LD}^{1/2})^T .$$

By setting $\boldsymbol{H} = \boldsymbol{LD}^{1/2}$ we obtain the Cholesky factorization $\boldsymbol{A} = \boldsymbol{HH}^T$. The positivity of the diagonal entries of $\boldsymbol{D}$ is guaranteed by the positivity of $\boldsymbol{A}$ (see Section 2.3.2).

For a 3×3 matrix $\boldsymbol{A}$ we have

$$\begin{pmatrix} a_{11} & a_{12} & a_{13} \\ a_{21} & a_{22} & a_{23} \\ a_{31} & a_{32} & a_{33} \end{pmatrix} = \begin{pmatrix} h_{11} & 0 & 0 \\ h_{21} & h_{22} & 0 \\ h_{31} & h_{32} & h_{33} \end{pmatrix} \begin{pmatrix} h_{11} & h_{21} & h_{31} \\ 0 & h_{22} & h_{32} \\ 0 & 0 & h_{33} \end{pmatrix} .$$

After performing the multiplication on the right-hand side we get

$$\begin{pmatrix} a_{11} & a_{12} & a_{13} \\ a_{21} & a_{22} & a_{23} \\ a_{31} & a_{32} & a_{33} \end{pmatrix} = \begin{pmatrix} h_{11}^2 & h_{11}h_{21} & h_{11}h_{31} \\ h_{11}h_{21} & h_{21}^2 + h_{22}^2 & h_{21}h_{31} + h_{22}h_{32} \\ h_{11}h_{31} & h_{21}h_{31} + h_{22}h_{32} & h_{31}^2 + h_{32}^2 + h_{33}^2 \end{pmatrix} .$$

Using matrix equality, and solving for h_{11}, h_{21}, h_{31} using the first row, for h_{22}, h_{32} using the second and for h_{33} using the third we obtain

$$\begin{aligned} h_{11} &= \sqrt{a_{11}} \\ h_{21} &= a_{21}/h_{11} \\ h_{31} &= a_{31}/h_{11} \\ h_{22} &= \sqrt{a_{32} - h_{21}^2} \end{aligned}$$

$$h_{32} = (a_{32} - h_{21}h_{31})/h_{22}$$

and finally

$$h_{33} = \sqrt{a_{33} - h_{31}^2 - h_{32}^2}$$

Following the same procedure for an $n \times n$ matrix we obtain a reduction algorithm: If

$$\boldsymbol{H} = \begin{pmatrix} h_{11} & \boldsymbol{0} \\ \hat{\boldsymbol{h}} & \hat{\boldsymbol{H}} \end{pmatrix}, \quad \boldsymbol{A} = \begin{pmatrix} a_{11} & \hat{\boldsymbol{a}} \\ \hat{\boldsymbol{a}}^T & \hat{\boldsymbol{A}} \end{pmatrix},$$

then writing $\boldsymbol{H}\boldsymbol{H}^T = \boldsymbol{A}$ leads to

$$h_{11} = \sqrt{a_{11}}\,,$$
$$\hat{\boldsymbol{H}}\hat{\boldsymbol{H}}^T = \hat{\boldsymbol{A}} - \hat{\boldsymbol{h}}\hat{\boldsymbol{h}}^T\,.$$

The last equation implies that for $i = 1, 2, \ldots, n$

$$h_{ii} = \sqrt{a_{ii} - \sum_{k=1}^{i-1} h_{ik}^2} \quad \text{and} \quad h_{ji} = \left(a_{ji} - \sum_{k=1}^{i-1} h_{ik}h_{jk}\right)/h_{jj} \quad \text{for } j = i+1, \ldots, n\,.$$

This procedure is summarized in Algorithm 32.

Algorithm 32 Cholesky factorization of symmetric and positive definite matrix $\boldsymbol{A}$

for $i = 1 : n$ **do**
  $\boldsymbol{H}(i,i) = \sqrt{\boldsymbol{A}(i,i) - \sum_{k=1}^{i-1} \boldsymbol{H}^2(i,k)}$
  for $j = i+1 : n$ **do**
    $\boldsymbol{H}(j,i) = \left(\boldsymbol{A}(j,i) - \sum_{k=1}^{i-1} \boldsymbol{H}(j,k)\boldsymbol{H}(i,k)\right)/\boldsymbol{H}(i,i)$
  end for
end for

For example, let

$$\boldsymbol{A} = \begin{pmatrix} 4 & -1 & 1 \\ -1 & 4.25 & 2.75 \\ 1 & 2.75 & 3.5 \end{pmatrix}$$

Following the Algorithm 32 or alternatively solving the equations $\boldsymbol{A} = \boldsymbol{H}\boldsymbol{H}^T$ we obtain

$$h_{11} = \sqrt{a_{11}} = \sqrt{4} = 2\,,$$
$$h_{21} = a_{21}/h_{11} = -1/2 = -0.5\,,$$
$$h_{31} = a_{31}/h_{11} = 1/2 = 0.5\,,$$
$$h_{22} = \sqrt{a_{22} - h_{21}^2} = \sqrt{4.25 - 0.25} = \sqrt{4} = 2\,,$$
$$h_{32} = (a_{32} - h_{21}h_{31})/h_{22} = (2.75 + 0.25)/2 = 1.5\,,$$
$$h_{33} = \sqrt{a_{33} - h_{31}^2 - h_{32}^2} = \sqrt{3.5 - 0.25 - 2.25} = 1\,.$$

Thus, $\boldsymbol{A}$ has the Cholesky factorization

$$\boldsymbol{A} = \boldsymbol{H}\boldsymbol{H}^T = \begin{pmatrix} 2 & 0 & 0 \\ -0.5 & 2 & 0 \\ 0.5 & 1.5 & 1 \end{pmatrix} \begin{pmatrix} 2 & -0.5 & 0.5 \\ 0 & 2 & 1.5 \\ 0 & 0 & 1 \end{pmatrix}.$$

Solving linear systems using Cholesky factorization

After having a symmetric and positive definite matrix $\boldsymbol{A}$ factorized in $\boldsymbol{H}\boldsymbol{H}^T$, we can employ forward and backward substitution to solve any linear system of the form $\boldsymbol{A}\boldsymbol{x} = \boldsymbol{b}$. We first solve the system $\boldsymbol{H}\boldsymbol{y} = \boldsymbol{b}$ and then the system $\boldsymbol{H}^T\boldsymbol{x} = \boldsymbol{b}$. This is exactly what we did in the *LU* factorization, with the only difference that instead of $\boldsymbol{U}$ we will have here $\boldsymbol{H}^T$, and in the forward substitution we need to divide by the diagonal entries h_{ii}.

Here we present an implementation of Cholesky factorization algorithm with forward and backward substitution.

```
def Cholesky(A):
    n = len(A)
    for i in range(n):
        try:
            A[i,i]=np.sqrt(A[i,i] - np.dot(A[i,0:i],A[i,0:i]))
        except ValueError:
            error.err('Matrix is not positive definite')
        for j in range(i+1,n):
            A[j,i] = (A[j,i]-np.dot(A[j,0:i],A[i,0:i]))/A[i,i]
    for k in range(1,n):
        A[0:k,k]=0.0
    return A
def solveCholesky(H,b):
    n = len(b)
    # Solve Hy=b
    for k in range(n):
        b[k] = (b[k] - np.dot(H[k,0:k],b[0:k]))/H[k,k]
    # Solve H^T x =y
    for k in range(n-1,-1,-1):
        b[k] = (b[k] - np.dot(H[k+1:n,k],b[k+1:n]))/H[k,k]
    return b
```

Trying our code with the matrix

$$\boldsymbol{A} = \begin{pmatrix} 4 & -1 & 1 \\ -1 & 4.25 & 2.75 \\ 1 & 2.75 & 3.5 \end{pmatrix},$$

and to the linear system $\boldsymbol{A}\boldsymbol{x} = \boldsymbol{b}$ with $\boldsymbol{b} = (4, 6, 7.25)^T$ we get the right answer, as expected.

```
A = np.array([[4.0, -1.0, 1.0],
              [-1.0, 4.25, 2.75],
              [1.0, 2.75, 3.5]])
H = Cholesky(A)
print(H)
b = np.array([4.0, 6.0, 7.25])
x = solveCholesky(H,b)
print(x)
```

```
[[ 2.   0.   0. ]
 [-0.5  2.   0. ]
 [ 0.5  1.5  1. ]]
[1. 1. 1.]
```

We close with the remark that the Cholesky factorization algorithm is always stable. It also requires $O(n^3/3)$ floating point operations. Taking into consideration the stability of the forward and backward substitution algorithms, Cholesky factorization leads to a safe method for the numerical solution of linear systems with symmetric and positive definite matrices.

> ☞ Cholesky factorization $\boldsymbol{A} = \boldsymbol{H}\boldsymbol{H}^T$ works for symmetric and positive definite matrices. The failure of this method is a proof that a matrix $\boldsymbol{A}$ is not positive definite, and can be used as a test of positive definiteness. To check if a matrix is symmetric is usually an easy task, for example, by checking whether $\boldsymbol{A} = \boldsymbol{A}^T$ or not.

9.6 The Module `scipy.linalg` Again

We have already discussed several functions for solving linear systems using the module `scipy.linalg` in Chapter 2. Here we discuss the implementations of the LU, LDL^T and Cholesky factorizations in SciPy.

9.6.1 *LU* factorization

The easiest way to obtain the LU factorization of a matrix $\boldsymbol{A}$ is by using its SciPy implementation, which is the function `scipy.linalg.lu`. The general call of the function `lu` is

```
lu(a, permute_l, overwrite_a, check_finite)
```

where here

`a`: Is the matrix $\boldsymbol{A}$ and is the only argument required to obtain its LU factorization

`permute_l`: Is a boolean variable in order to perform the multiplication $\boldsymbol{PL}$. (Optional with default value `False`)

`overwrite_a`: Is a boolean variable in order to overwrite $\boldsymbol{A}$ with the matrices $\boldsymbol{L}$ and $\boldsymbol{U}$. (Optional with default value `False`)

`check_finite`: Is a boolean variable in order to check that the input matrix contains only finite numbers. (Optional with default value `True`)

The output of the function `lu` is the three matrix $\boldsymbol{P}$, $\boldsymbol{L}$ and $\boldsymbol{U}$ as it is shown in the following example:

```
import scipy.linalg as spl
A = np.array([[ 3., 17., 10.],
              [ 2., 4.,-2.],
              [ 6.,18.,-12.]])
(P,L,U) = spl.lu(A)
print(P)
print(L)
print(U)
```

```
[[0. 1. 0.]
 [0. 0. 1.]
 [1. 0. 0.]]
[[ 1.          0.          0.        ]
 [ 0.5         1.          0.        ]
 [ 0.33333333 -0.25        1.        ]]
[[  6.  18. -12.]
 [  0.   8.  16.]
 [  0.   0.   6.]]
```

A similar function of the module `scipy.linalg` is the function `lu_factor`. This function's general call is:

```
lu_factor(a, overwrite_a, check_finite)
```

with arguments as described in `lu` function reference. The difference here is that the function returns the *LU* factorization in one matrix `lu` like our function `LUpivot` and the permutation matrix `perm`. The advantage of using the `lu_factor` instead of the function `lu` is that one can use its output to solve a linear system $Ax = b$ using the `scipy.linalg` function `lu_solve`. The general call of the `lu_solve` function is

```
lu_solve(lu_and_piv, b, trans, overwrite_b, check_finite)
```

with input arguments

`lu_and_piv`: The tuple `(lu, piv)` received in the output of `lu_factor`.

`b`: The right-hand side $\boldsymbol{b}$ of the linear system $\boldsymbol{Ax} = \boldsymbol{b}$.

`trans`: An integer in the set {`0`,`1`,`2`}. For `0` the function solves the system $\boldsymbol{Ax} = \boldsymbol{b}$, for `1` the system $\boldsymbol{A}^T\boldsymbol{x} = \boldsymbol{b}$ and for `2` the system $\boldsymbol{A}^H\boldsymbol{x} = \boldsymbol{b}$. (Optional with default value `0`)

`overwrite_b`: Is a boolean variable in order to overwrite $\boldsymbol{b}$ with the solution $\boldsymbol{x}$. (Optional with default value `False`)

`check_finite`: Is a boolean variable in order to check that the input matrix contains only finite numbers. (Optional with default value `True`)

The function `lu_solve` returns only the solution `x`.

As an example, we consider again in the following code the system $\boldsymbol{Ax} = \boldsymbol{b}$ with

$$\boldsymbol{A} = \begin{pmatrix} 3 & 17 & 10 \\ 2 & 4 & -2 \\ 6 & 18 & -12 \end{pmatrix} \quad \text{and} \quad \boldsymbol{b} = \begin{pmatrix} 30 \\ 4 \\ 12 \end{pmatrix} .$$

```python
A = np.array([[ 3., 17., 10.],
              [ 2., 4.,-2.],
              [ 6.,18.,-12.]])
b = np.array([30., 4., 12.])
(LU,P) = spl.lu_factor(A)
x = spl.lu_solve((LU,P),b)
print(x)
```

```
[1. 1. 1.]
```

9.6.2 Cholesky factorization

There are two implementations of the Cholesky factorization in Python. One is in module `numpy.linalg` and the other in `scipy.linalg`. Both functions have the same name, which is `cholesky` and require as input a symmetric and positive definite matrix $\boldsymbol{A}$, while they return the lower triangular matrix $\boldsymbol{H}$. The SciPy version accepts some other optional arguments and returns either the lower triangular matrix $\boldsymbol{H}$ or the upper triangular matrix $\boldsymbol{H}^T$. Here we present the usage of the NumPy version of Cholesky factorization for the same, simple example we discussed in our implementation.

Having the Cholesky factorization $\boldsymbol{A} = \boldsymbol{H}\boldsymbol{H}^T$, it is easy to solve any system of the form $\boldsymbol{A}\boldsymbol{x} = \boldsymbol{b}$. Since $\boldsymbol{H}\boldsymbol{H}^T\boldsymbol{x} = \boldsymbol{b}$ we set $\boldsymbol{y} = \boldsymbol{H}^T\boldsymbol{x}$, so we solve for $\boldsymbol{y}$ the system $\boldsymbol{H}\boldsymbol{y} = \boldsymbol{b}$ first, and then the system $\boldsymbol{H}^T\boldsymbol{x} = \boldsymbol{y}$ for $\boldsymbol{x}$. In order to solve the triangular systems $\boldsymbol{H}\boldsymbol{y} = \boldsymbol{b}$ and $\boldsymbol{H}^T\boldsymbol{x} = \boldsymbol{y}$ we can use the function `solve_triangular` of `scipy.linalg` (see Section 2.3.3). Utilizing this function for the lower triangular matrix $\boldsymbol{H}$ we need to use the option `lower=True`. In this way we can solve the system of the previous example using Python with the following commands:

```
A = np.array([[4.0, -1.0, 1.0],
              [-1.0, 4.25, 2.75],
              [1.0, 2.75, 3.5]])
b = np.array([4.0, 6.0, 7.25])
H = spl.cholesky(A)
print(H)
# We use the option lower=True since H is lower triangular matrix
y = spl.solve_triangular(H, b, lower=True)
# We use the option trans=1 to solve the system H^Tx=y
x = spl.solve_triangular(H, y, trans=1, lower=True)
print(x)
```

```
[[ 2.   0.   0. ]
 [-0.5  2.   0. ]
 [ 0.5  1.5  1. ]]
[1. 1. 1.]
```

In SciPy one can also find an implementation of Cholesky algorithm for banded matrices. This algorithm is discussed in the Exercises with some detail. Moreover, there is the function `cho_factor` which can be used in combination with the function `cho_solve` to solve linear systems. One can obtain first the cholesky factorization using the command `c=cho_factor(A)` and then obtain the solution by writing `x=cho_solve(c,b)`. The tupple `c` contains a triangular matrix and a boolean variable that specifies whether the output matrix is lower or upper triangular.

Similarly, the LDL^T factorization can also be obtained using pivoting techniques. This is implemented in the module `scipy.linalg` in the function `ldl`. This can take as input a symmetric matrix `A` to return the arrays `lu`, `d` and `perm`, where `lu` is the (possibly) permuted upper $\boldsymbol{L}^T$ or lower triangular $\boldsymbol{L}$ matrix, `d` is the diagonal matrix $\boldsymbol{D}$ and `perm` is the permutation matrix used during the factorization algorithm. We continue with a different class of numerical methods for the numerical solution of large and sparse systems of linear equations known to as iterative methods.

9.7 Iterative Methods

In previous sections we studied direct methods for dense linear systems. Direct methods find the exact solution when exact arithmetic is being used. In this section we discuss iterative methods for the numerical solution of sparse linear systems. These methods are used to compute an approximation to the unknown solution in a finite number of iterations. Iterative methods must converge to the exact solution in the limit of infinite iterations. The philosophy of these methods is exactly the same as the methods we studied for the solution of nonlinear equations. Like in Newton's methods, we start with an initial guess of the solution and the method estimates an approximation in a few iterations. Iterative methods perform better for large and sparse matrices. Direct methods are more often preferable for small and dense matrices. Also, direct methods can be our choice when it is hard to achieve convergence with iterative methods due to matrix properties.

But let's see first how we can derive iterative methods for linear systems. Perhaps the easiest way is to start with a system $\boldsymbol{A}\boldsymbol{x}=\boldsymbol{b}$ and write it as a fixed point equation in the form

$$\boldsymbol{x}=\boldsymbol{x}-\boldsymbol{A}\boldsymbol{x}+\boldsymbol{b}=(\boldsymbol{I}-\boldsymbol{A})\boldsymbol{x}+\boldsymbol{b}\ .$$

The iteration $\boldsymbol{x}^{(k+1)}=(\boldsymbol{I}-\boldsymbol{A})\boldsymbol{x}^{(k)}+\boldsymbol{b}$ along with an initial guess $\boldsymbol{x}^{(0)}$ define a sequence of approximations $\boldsymbol{x}^{(k)}$, $k=0,1,2,\ldots$. The superscript index in the notation $\boldsymbol{x}^{(k)}$ indicates the iteration, while the i-th entry of the vector $\boldsymbol{x}^{(k)}$ is denoted by $x_i^{(k)}$. This iterative method, perhaps the simplest of all, works well only in a few cases, unfortunately. For this reason, other methods have been derived based on a more sophisticated methodology.

9.7.1 Derivation of iterative methods

Let

$$\boldsymbol{A}\boldsymbol{x}=\boldsymbol{b}\ , \tag{9.11}$$

be a linear system of equations with large and sparse matrix $\boldsymbol{A}$. In order to derive an iterative method, and with the experience we have from fixed point methods, we will rewrite system (9.11) in the form $\boldsymbol{x}=\boldsymbol{G}(\boldsymbol{x})$. Then the implied iterative method will $\boldsymbol{x}^{(k+1)}=\boldsymbol{G}(\boldsymbol{x}^{(k)})$, with $\boldsymbol{x}^{(0)}$ an initial guess. To construct $\boldsymbol{G}$, we first split $\boldsymbol{A}$ into

$$\boldsymbol{A}=\boldsymbol{M}-\boldsymbol{N}\ , \tag{9.12}$$

where $\boldsymbol{M}$ and $\boldsymbol{N}$ can be anything leading to an efficient method. (Our first example can be seen as the method with $\boldsymbol{M}=\boldsymbol{I}$ and $\boldsymbol{N}=-\boldsymbol{I}-\boldsymbol{A}$). We will discuss examples of $\boldsymbol{M}$ and $\boldsymbol{N}$ later in this chapter but for the moment we consider general $\boldsymbol{M}$ and $\boldsymbol{N}$, with invertible $\boldsymbol{M}$. After splitting $\boldsymbol{A}=\boldsymbol{M}-\boldsymbol{N}$ we write the system $\boldsymbol{A}\boldsymbol{x}=\boldsymbol{b}$ as

$$(\boldsymbol{M}-\boldsymbol{N})\boldsymbol{x}=\boldsymbol{b} \quad \text{and then} \quad \boldsymbol{M}\boldsymbol{x}=\boldsymbol{N}\boldsymbol{x}+\boldsymbol{b}\ .$$

With the assumption that $\boldsymbol{M}$ is invertible, we write the last equation as

$$\boldsymbol{x}=\boldsymbol{T}\boldsymbol{x}+\boldsymbol{c}\ ,$$

where

$$\boldsymbol{T}=\boldsymbol{M}^{-1}\boldsymbol{N} \quad \text{and} \quad \boldsymbol{c}=\boldsymbol{M}^{-1}\boldsymbol{b}\ .$$

In other words, we transformed the equation $\boldsymbol{A}\boldsymbol{x}=\boldsymbol{b}$ into the fixed point equation $\boldsymbol{x}=\boldsymbol{T}\boldsymbol{x}+\boldsymbol{c}$. We define the fixed point iteration in the same manner as in Chapter 5 to be

$$\boldsymbol{x}^{(k+1)}=\boldsymbol{T}\boldsymbol{x}^{(k)}+\boldsymbol{c}\ ,$$

or more generally

$$\boldsymbol{M}\boldsymbol{x}^{(k+1)} = \boldsymbol{N}\boldsymbol{x}^{(k)} + \boldsymbol{b} \ . \tag{9.13}$$

The reason for keeping the equation in the form (9.13) is because, sometimes, it is easier to extract the equations from the matrix form without using the inverse $\boldsymbol{M}^{-1}$.

☞ Note that the index of the iteration k is a superscript $\boldsymbol{x}^{(k)}$ instead of subscript $\boldsymbol{x}_k$. This is because $\boldsymbol{x}$ is now a vector.

9.7.2 Classical iterative methods – Jacobi and Gauss-Seidel methods

Three classical iterative methods are the Jacobi, Gauss-Seidel and Successive-Over-Relaxation (SOR) methods. The last one is treated in the Exercises. The classical iterative methods are based on the splitting

$$\boldsymbol{A} = \boldsymbol{D} - \boldsymbol{L} - \boldsymbol{U} \ ,$$

where $\boldsymbol{D}$ is diagonal, $\boldsymbol{L}$ lower triangular and $\boldsymbol{U}$ upper triangular matrices such that

$$\boldsymbol{D} = \begin{pmatrix} a_{11} & 0 & \cdots & 0 \\ 0 & a_{22} & \ddots & 0 \\ \vdots & \ddots & \ddots & 0 \\ 0 & \cdots & 0 & a_{nn} \end{pmatrix}, \quad \boldsymbol{L} = \begin{pmatrix} 0 & 0\cdots & \cdots & 0 \\ -a_{21} & \ddots & & \vdots \\ \vdots & \ddots & \ddots & \vdots \\ -a_{n1} & \cdots & -a_{n,n-1} & 0 \end{pmatrix},$$

and

$$\boldsymbol{U} = \begin{pmatrix} 0 & -a_{12} & \cdots & -a_{1n} \\ \vdots & \ddots & \ddots & \vdots \\ \vdots & \ddots & \ddots & -a_{n-1,n} \\ 0 & \cdots & 0 & 0 \end{pmatrix} .$$

The matrices $\boldsymbol{L}$ and $\boldsymbol{U}$ have nothing in common with the LU factorization. They are usually denoted by the same letters because they are lower and upper triangular matrices.

The Jacobi method

The Jacobi method is the simplest iterative method. We choose for the splitting of $\boldsymbol{A} = \boldsymbol{M} - \boldsymbol{N}$ the matrices

$$\boldsymbol{M} = \boldsymbol{D} \quad \text{and} \quad \boldsymbol{N} = \boldsymbol{L} + \boldsymbol{U} \ .$$

Then the fixed point method $\boldsymbol{M}\boldsymbol{x}^{(k+1)} = \boldsymbol{N}\boldsymbol{x}^{(k)} + \boldsymbol{b}$ is written as

$$\boldsymbol{D}\boldsymbol{x}^{(k+1)} = (\boldsymbol{L} + \boldsymbol{U})\boldsymbol{x}^{(k)} + \boldsymbol{b} \ .$$

The choice of the matrix $\boldsymbol{M} = \boldsymbol{D}$ is not by chance of course. Since our method relies on the inversion of $\boldsymbol{M}$, we prefer $\boldsymbol{M}$ to be easily invertible, and there is no matrix more easily invertible than a diagonal matrix. After inverting the diagonal matrix $\boldsymbol{D}$, we have the Jacobi method written in vector form

$$\boldsymbol{x}^{(k+1)} = \boldsymbol{D}^{-1}(\boldsymbol{L} + \boldsymbol{U})\boldsymbol{x}^{(k)} + \boldsymbol{D}^{-1}\boldsymbol{b}, \quad k = 0, 1, 2, \dots \ . \tag{9.14}$$

Writing each equation of the Jacobi iteration (9.14) explicitly, we have for each entry of $\boldsymbol{x}^{(k+1)}$ that

$$x_i^{(k+1)} = \left(b_i - \sum_{\substack{j=1 \\ j\neq i}}^{n} a_{ij} x_j^{(k)} \right) \Big/ a_{ii}, \quad i = 1, \ldots, n \; . \tag{9.15}$$

From (9.15) we immediately deduce that we need $a_{ii} \neq 0$, for $i = 1, 2, \ldots, n$. If $\boldsymbol{A}$ is nonsingular but one of the values a_{ii} is 0, then reordering of the equations can be performed so that no a_{ii} is zero.

We start iterating using an initial guess $\boldsymbol{x}^{(0)}$. This can be any arbitrarily chosen vector. The question is when we should stop iterating, as we cannot wait to reach the exact solution in the limit $k \to \infty$. In the case of iterative methods for scalar equations, we could terminate the iterations when the difference of two successive approximations $|x_{k+1} - x_k|$ become less than a prescribed tolerance. In the case of systems where $\boldsymbol{x}^{(k)}$ are vectors we need to use norms in order to compute the difference of two successive approximations. So we can record the quantity $\|\boldsymbol{x}^{(k+1)} - \boldsymbol{x}^{(k)}\|$ for some vector-norm $\|\cdot\|$ or even better use the normalized error

$$\frac{\|\boldsymbol{x}^{(k)} - \boldsymbol{x}^{(k-1)}\|}{\|\boldsymbol{x}^{(k)}\|} ,$$

and compare with a prescribed tolerance.

The speed of convergence cannot be determined *a priori*, but we can increase it by rearranging the equations of the system. Rearranging the equations of a system can be described by a transformation of matrices $\boldsymbol{M}$, $\boldsymbol{N}$ and vector $\boldsymbol{b}$. We can increase the speed of the Jacobi method by rearranging the matrix $\boldsymbol{M} = \boldsymbol{D}$ so that the diagonal entries a_{ii} are large. The technique of modifying the matrix $\boldsymbol{M}$ in order to improve the speed of convergence is called *preconditioning*.

The implementation of the Jacobi method in Python can be straightforward using either (9.14) or (9.15). In order to compute each entry $x_i^{(k+1)}$ from (9.15) we only need the previous iteration $x_i^{(k)}$ for the same index i. If we had a high-performance computer with multiple processors, then we could compute each entry in parallel. Perhaps, the Jacobi method is the simplest example of a numerical method that can be parallelized trivially.

Here we present an implementation of the Jacobi method using the matrix-vector form (9.14).

```
def jacobi(A, b, x, tol = 1.e-5, maxit = 100):
    d = np.copy(np.diag(A))
    np.fill_diagonal(A,0.0)
    err = 1.0
    iters = 0
    while (err > tol and iters < maxit):
        iters += 1
        xnew = (b - np.dot(A,x)) / d
        err = npl.norm(xnew-x,np.inf)
        x = np.copy(xnew)
    print('iterations required for convergence:', iters)
    return x
```

The function `jacobi` takes as input the matrix `A`, vector `b` and initial guess `x` along with the tolerance `tol` and the maximum number of iterations `maxit` that we allow to happen. In the end, the function returns the approximation `x` and prints the number of iterations required by the Jacobi method to meet the stopping criteria. In order to compute the norm of the difference of two successive approximations, we use the module `numpy.linalg` and the function `norm`. Here we use the $\|\cdot\|_\infty$ but one can use any norm due to their equivalence.

We test the code for a system $\boldsymbol{A}\boldsymbol{x} = \boldsymbol{b}$ with matrix

$$\boldsymbol{A} = \begin{pmatrix} 2 & -1 & 0 \\ -1 & 3 & -1 \\ 0 & -1 & 2 \end{pmatrix} ,$$

and vector $\boldsymbol{b} = (1, 8, -5)^T$, which has exact solution $\boldsymbol{x} = (2, 3, -1)^T$. As initial guess we take the vector $\boldsymbol{x}^{(0)} = (0, 0, 0)^T$.

```
A = np.array([[2.0, -1.0,  0.0],
              [-1.0, 3.0, -1.0],
              [0.0, -1.0,  2.0]])
b = np.array([1.0, 8.0, -5.0])
x = np.zeros(3)
x = jacobi(A, b, x, 1.e-5, 100)
print(x)
```

```
iterations required for convergence: 24
[ 1.99999906  2.99999435 -1.00000094]
```

We observe that the solution is correct within the tolerance 10^{-5} and the solution has at least 5 correct digits.

The Gauss-Seidel method

We continue with the second classical iterative method, the Gauss-Seidel method. The Gauss-Seidel method uses the same splitting for the matrix $\boldsymbol{A} = \boldsymbol{M} - \boldsymbol{N} = \boldsymbol{D} - \boldsymbol{L} - \boldsymbol{U}$ as the Jacobi method, but instead of choosing $\boldsymbol{M}$ to be diagonal we choose $\boldsymbol{M}$ to be the lower triangular matrix $\boldsymbol{M} = \boldsymbol{D} - \boldsymbol{L}$. Therefore, $\boldsymbol{N}$ is the upper triangular matrix $\boldsymbol{N} = \boldsymbol{U}$. The fixed point equation then becomes

$$(\boldsymbol{D} - \boldsymbol{L})\boldsymbol{x} = \boldsymbol{U}\boldsymbol{x} + \boldsymbol{b} ,$$

and the Gauss-Seidel iteration is defined as

$$(\boldsymbol{D} - \boldsymbol{L})\boldsymbol{x}^{(k+1)} = \boldsymbol{U}\boldsymbol{x}^{(k)} + \boldsymbol{b}, \quad k = 0, 1, 2, \dots . \tag{9.16}$$

In order to solve for the next iteration $\boldsymbol{x}^{(k+1)}$ given the approximation $\boldsymbol{x}^{(k)}$ we either need to invert a lower triangular matrix or perform the forward substitution. After the inversion of the matrix $\boldsymbol{D} - \boldsymbol{L}$, the $k+1$ iteration can be written in the form

$$\boldsymbol{x}^{(k+1)} = (\boldsymbol{D} - \boldsymbol{L})^{-1}\boldsymbol{U}\boldsymbol{x}^{(k)} + (\boldsymbol{D} - \boldsymbol{L})^{-1}\boldsymbol{b} . \tag{9.17}$$

Performing forward substitution to the system (9.16) we obtain

$$x_i^{(k+1)} = \left(b_i - \sum_{j=1}^{i-1} a_{ij} x_j^{(k+1)} - \sum_{j=i+1}^{n} a_{ij} x_j^{(k)} \right) \Big/ a_{ii}. \quad i = 1, \dots, n . \tag{9.18}$$

Gauss-Seidel and Jacobi methods have many similarities, and big differences as well. The Gauss-Seidel method requires that $a_{ii} \neq 0$, for each $i = 1, 2, \dots, n$. Also, the speed of propagation depends on the magnitude of these values. In Gauss-Seidel method all the most recent available coordinates contribute in acquiring the next coordinate in the vector $\boldsymbol{x}^{(k+1)}$, while in Jacobi method only the coordinates of the previous steps are used. So in order to compute $x_i^{(k+1)}$ we need to use all the previously computed values $x_j^{(k+1)}$ for $j = 1, 2, \ldots, i-1$. Commenting finally on the speed of convergence of the Gauss-Seidel method we note that it is expected in general to converge faster than the Jacobi method.

The implementation of the Gauss-Seidel method in matrix-vector format is straightforward. In order to extract from $\boldsymbol{A}$ the matrix $\boldsymbol{M} = \boldsymbol{D} - \boldsymbol{L}$ we use the NumPy function `tril`, which extracts the lower triangular part of a matrix including its main diagonal. The matrix $\boldsymbol{M}$ is inverted and using (9.17) we compute the new approximation from the previous one.

```
def gauss_seidel(A, b, x, tol = 1.e-5, maxit = 100):
    n = len(b)
    err = 1.0
    iters = 0
    # Initialize the solution with the initial guess
    xnew = np.zeros_like(x)
    # Extract the lower triangular part of A
    M = np.tril(A)
    # Construct the upper triangular part of A
    U = A - M
    while (err > tol and iters < maxit):
        iters += 1
        # Compute the new approximation
        xnew = np.dot(npl.inv(M), b - np.dot(U, x))
        # Estimate convergence
        err = npl.norm(xnew-x,np.inf)
        x = np.copy(xnew)
    print('iterations required for convergence:', iters)
    return x
```

Testing the code for the same system as we did in the case of the Jacobi method we compute the same solution.

```
A = np.array([[2.0, -1.0,  0.0],
              [-1.0, 3.0, -1.0],
              [0.0, -1.0,  2.0]])
b = np.array([1.0, 8.0, -5.0])
x = np.zeros(3)
x = gauss_seidel(A, b, x, 1.e-5, 100)
print(x)
```

```
iterations required for convergence: 11
[ 1.99999577  2.99999718 -1.00000141]
```

What we observe though is that the Gauss-Seidel method converges in less iterations than the Jacobi method. Gauss-Seidel required only 11 iterations, which compared with the 24 iterations of the Jacobi method is a significant improvement. It is important to note that the inversion of the matrix $\boldsymbol{M}$ is a drawback of the previous implementation since inverting matrices is a slow procedure. An alternative, and perhaps more efficient implementation of the Gauss-Seidel method, can be based on formula (9.18). Here, the two sums in (9.18) are stored in the variables s1 and s2 and under the assumption $a_{ii} \neq 0$ the new iteration is computed via (9.18).

```
def gauss_seidel(A, b, x, tol = 1.e-5, maxit = 100):
    n = len(b)
    err = 1.0
    iters = 0
    xnew = np.zeros_like(x)
    while (err > tol and iters < maxit):
        iters += 1
        for i in range(n):
            s1 = np.dot(A[i, :i], xnew[:i])
            s2 = np.dot(A[i, i + 1:], x[i + 1:])
            xnew[i] = (b[i] - s1 - s2) / A[i, i]
        err = npl.norm(xnew-x,np.inf)
        x = np.copy(xnew)
    print('iterations required for convergence', iters)
    return x
```

The results obtained using the last function are identically the same as before. It is noted that NumPy arrays are mutable. For this reason, it is required to copy the parts of an array to a different one in order to extract the upper and lower triangular parts of $\boldsymbol{A}$.

The convergence of these two classical methods is the subject of the next section.

9.7.3 Convergence of iterative methods

So far, we derived iterative methods of the form

$$\boldsymbol{M}\boldsymbol{x}^{(k+1)} = \boldsymbol{N}\boldsymbol{x}^{(k)} + \boldsymbol{b}, \quad k = 0, 1, 2, \dots ,$$

that converge to the solution of the system

$$\boldsymbol{M}\boldsymbol{x} = \boldsymbol{N}\boldsymbol{x} + \boldsymbol{b} .$$

Subtracting these two equations we get the relationship

$$\boldsymbol{M}(\boldsymbol{x}^{(k+1)} - \boldsymbol{x}) = \boldsymbol{N}(\boldsymbol{x}^{(k)} - \boldsymbol{x}) ,$$

or better

$$\boldsymbol{x}^{(k+1)} - \boldsymbol{x} = \boldsymbol{G}(\boldsymbol{x}^{(k)} - \boldsymbol{x}) ,$$

where

$$\boldsymbol{G} = \boldsymbol{M}^{-1}\boldsymbol{N} .$$

The matrix $\boldsymbol{G}$ is called the iteration matrix.

If $\|\boldsymbol{G}\| < 1$ for any natural norm $\|\cdot\|$, then the error converges to 0. In order to see this consider the error

$$\boldsymbol{x}^{(k)} - \boldsymbol{x} = \boldsymbol{G}(\boldsymbol{x}^{(k-1)} - \boldsymbol{x}) = \boldsymbol{G}^2(\boldsymbol{x}^{(k-2)} - \boldsymbol{x}) = \cdots = \boldsymbol{G}^k(\boldsymbol{x}^{(0)} - \boldsymbol{x}) .$$

This yields $\|\boldsymbol{x}^{(k)} - \boldsymbol{x}\| \leq \|\boldsymbol{G}^k\| \, \|\boldsymbol{x}^{(0)} - \boldsymbol{x}\|$. The convergence of the error $\|\boldsymbol{x}^{(k)} - \boldsymbol{x}\| \to 0$ as $k \to \infty$ is equivalent with the fact that $\|\boldsymbol{G}^k\| \to 0$ as $k \to 0$ for any natural norm. This can happen if $\|\boldsymbol{G}\| < 1$ since then $\|\boldsymbol{G}^k\| \leq \|\boldsymbol{G}\|^k \to 0$. This is also equivalent to $\rho(\boldsymbol{G}) < 1$

where $\rho(\boldsymbol{G})$ is the spectral radius of $\boldsymbol{G}$. To see this assume that $\rho(G) < 1$. Recall that from Theorem 9.5 we have $\|\boldsymbol{G}\| \leq \rho(\boldsymbol{G}) + \epsilon$ for all $\epsilon > 0$. If we choose $\epsilon = (1 - \rho(\boldsymbol{G}))/2$, then,

$$\|\boldsymbol{G}\| \leq \rho(\boldsymbol{G}) + \epsilon = \rho(\boldsymbol{G})/2 + 1/2 < 1 ,$$

since $\rho(\boldsymbol{G}) < 1$. Inversely, if $\|\boldsymbol{G}\| < 1$, then the inequality $\rho(\boldsymbol{G}) \leq \|\boldsymbol{G}\|$ of Theorem 9.4 implies that $\rho(\boldsymbol{G}) < 1$.

In order for the Jacobi and Gauss-Seidel methods to converge, it can be shown that $\boldsymbol{A}$ must be *strictly row diagonally dominant matrix.* A matrix $\boldsymbol{A}$ is strictly row diagonally dominant if the values of the diagonal entries are greater than the sum of the off-diagonal entries in the same row. This means

$$|a_{ii}| > \sum_{\substack{j=1\\ j\neq i}}^{n} |a_{ij}|, \quad 1 \leq i \leq n . \tag{9.19}$$

For example, the matrix that we used to test our code

$$\boldsymbol{A} = \begin{pmatrix} 2 & -1 & 0 \\ -1 & 3 & -1 \\ 0 & -1 & 2 \end{pmatrix} ,$$

is strictly diagonally dominant, and therefore both classical iterative methods converge.

In what follows, we assume that $\boldsymbol{A}$ is strictly row diagonally dominant, and we prove the convergence of the Jacobi and Gauss-Seidel iteration. If the inequality in (9.19) is not strict, then $\boldsymbol{A}$ is just called row diagonally dominant.

Convergence of the Jacobi iteration

Denoting the iteration matrix for the Jacobi method by $\boldsymbol{G}_J = \boldsymbol{D}^{-1}(\boldsymbol{L} + \boldsymbol{U})$, we observe that

$$\boldsymbol{G}_J = \begin{pmatrix} 0 & -a_{12}/a_{11} & \cdots & -a_{1n}/a_{11} \\ -a_{21}/a_{22} & 0 & \cdots & -a_{2n}/a_{22} \\ \vdots & \vdots & \ddots & \vdots \\ -a_{n1}/a_{nn} & -a_{n2}/a_{nn} & \cdots & 0 \end{pmatrix} .$$

However, we have

$$\|\boldsymbol{G}_J\|_\infty = \max_i \sum_{\substack{j=1\\ j\neq i}}^{n} \left| -\frac{a_{ij}}{a_{ii}} \right| = \max_i \frac{\sum_{j=1,\ j\neq i}^{n} |a_{ij}|}{|a_{ii}|} < 1 ,$$

where the last inequality follows from the fact that $\boldsymbol{A}$ is strictly diagonally dominant matrix and the inequality (9.19). Therefore, the Jacobi iteration is convergent.

Convergence of the Gauss-Seidel iteration

Denote the Gauss-Seidel iteration matrix by $\boldsymbol{G}_{GS} = (\boldsymbol{D} - \boldsymbol{L})^{-1}\boldsymbol{U}$. Let λ be an eigenvalue of $\boldsymbol{G}_{GS}$ with corresponding eigenvector $\boldsymbol{x}$ such that $\boldsymbol{G}_{GS}\boldsymbol{x} = \lambda\boldsymbol{x}$. Substituting $\boldsymbol{G}_{GS}$ in the last equation we obtain $(\boldsymbol{D} - \boldsymbol{L})^{-1}\boldsymbol{U}\boldsymbol{x} = \lambda\boldsymbol{x}$ which implies that

$$\boldsymbol{U}\boldsymbol{x} = (\boldsymbol{D} - \boldsymbol{L})\lambda\boldsymbol{x} ,$$

and thus

$$-\sum_{j=i+1}^{n} a_{ij}x_j = \lambda \sum_{j=1}^{i} a_{ij}x_j, \quad i = 1, \ldots, n ,$$

or more conveniently

$$\lambda a_{ii}x_i = -\lambda\sum_{j=1}^{i-1} a_{ij}x_j - \sum_{j=i+1}^{n} a_{ij}x_j, \quad i = 1,\ldots,n\ .$$

If x_k is the entry of $\boldsymbol{x}$ with the largest magnitude, which is equal to 1, then taking absolute values in the previous relation we have

$$|\lambda||a_{kk}| \le |\lambda|\sum_{j=1}^{k-1} |a_{kj}|\cdot|x_j| + \sum_{j=k+1}^{n} |a_{kj}|\cdot|x_j|\ .$$

Solving for $|\lambda|$ we obtain

$$|\lambda| \le \frac{\sum_{j=k+1}^{n}|a_{kj}|}{|a_{kk}| - \sum_{j=1}^{k-1}|a_{kj}|}\ .$$

Since $\boldsymbol{A}$ is strictly row diagonally dominant we have

$$|a_{kk}| > \sum_{\substack{j=1\\ j\neq i}}^{n} |a_{kj}|\ ,$$

which can be written equivalently

$$|a_{kk}| - \sum_{j=1}^{k-1}|a_{kj}| > \sum_{j=k+1}^{n}|a_{kj}|\ .$$

We conclude that $|\lambda| < 1$ and since λ was an arbitrary eigenvalue of $\boldsymbol{G}_{GS}$ we have that the spectral radius $\rho(\boldsymbol{G}_{GS}) < 1$. Therefore, after the discussion of Section 9.7.3 we conclude that the Gauss-Seidel iteration converges.

☞ The classical iterative methods converge when the matrix $\boldsymbol{A}$ is also strictly column diagonally dominant. They also converge when $\boldsymbol{A}$ is symmetric and positive definite. For more general matrices we usually use other more sophisticated methods.

9.7.4 Sparse matrices in Python and the module `scipy.sparse`

Often in applications the resulting matrices are sparse. A matrix is sparse if it has most of its entries zero. An example of a sparse matrix is the tridiagonal matrix

$$\boldsymbol{A} = \begin{pmatrix} 2 & -1 & 0 & \cdots & 0\\ -1 & 2 & -1 & \cdots & 0\\ 0 & \ddots & \ddots & \ddots & 0\\ 0 & \cdots & -1 & 2 & -1\\ 0 & \cdots & 0 & -1 & 2 \end{pmatrix}\ .$$

Large and sparse matrices can be stored much easier than dense matrices. This can be done by storing only the non-zero entries in special data structures. We discussed the case of band matrices in Section 2.3.1. Here we continue the discussion for more general sparse matrices. SciPy provides tools to handle sparse matrices with the module `sparse`. There are seven different types of sparse matrix storage schemes available in Python, which are listed below:

`bsr_matrix`: Block Sparse Row matrix

`coo_matrix`: Coordinate format matrix

`csc_matrix`: Compressed sparse column matrix

`csr_matrix`: Compressed sparse row matrix

`dia_matrix`: Sparse matrix with diagonal storage

`dok_matrix`: Dictionary of keys based sparse matrix

`lil_matrix`: Row-based linked list sparse matrix

We will not analyze all these formats here, but we suggest the interested reader to search online for more information.

Here is an example of sparse storage using the `coo_matrix` format. The non-zero entries $a_{ij} \neq 0$ are stored in a one-dimensional array, while the corresponding indices i, j are stored in two different arrays. Consider the tridiagonal matrix

$$\boldsymbol{S} = \begin{pmatrix} 2 & -1 & 0 & 0 \\ -1 & 2 & -1 & 0 \\ 0 & -1 & 2 & -1 \\ 0 & 0 & -1 & 2 \end{pmatrix} .$$

We can store this matrix in sparse format using the following commands:

```
from scipy import sparse
A = np.array([[ 2.0,-1.0, 0.0, 0.0],
              [-1.0, 2.0,-1.0, 0.0],
              [ 0.0,-1.0, 2.0,-1.0],
              [ 0.0, 0.0,-1.0, 2.0]])
S = sparse.coo_matrix(S)
print(S)
```

```
  (0, 0) 2.0
  (0, 1) -1.0
  (1, 0) -1.0
  (1, 1) 2.0
  (1, 2) -1.0
  (2, 1) -1.0
  (2, 2) 2.0
  (2, 3) -1.0
  (3, 2) -1.0
  (3, 3) 2.0
```

Python stores in `S` only the non-zero entries and their indices. In order to extract the non-zero entries and their indices in separate arrays we can use the function `find()`.

```
(I,J,V)=sparse.find(S)
print(I,J,V)
```

```
[0 1 0 1 2 1 2 3 2 3]
[0 0 1 1 1 2 2 2 3 3]
[ 2. -1. -1.  2. -1. -1.  2. -1. -1.  2.]
```

The coordinate representation of sparse matrices is easy and convenient. We can refer to each non-zero entry a_{ij} by writing `V[I[i-1],J[j-1]]`.

Compressed Sparse Row format

A more sophisticated and very popular sparse storage format that also leads to efficient implementations of sparse matrix computations is the Compressed Sparse Row (CSR) format, [77]. In CSR format, all the non-zero entries of an $n \times m$ matrix A are stored row-wise in an array `A.data` while the indices are not stored as coordinates but in an implicit way:

First we define the array `A.indptr` with length equal to $n+1$. Its entries serve as counters to non-zero entries of `A`. Specifically, starting from `0`, the `i` entry of this array is equal to the number of the non-zero entries encountered up to the `i-1` row of `A`. Consequently, the difference `A.indptr[i+1]-A.indptr[i]` is equal to the number of non-zero entries in the row `i`. For example, consider the matrix

$$\boldsymbol{A} = \begin{pmatrix} 10 & 20 & 0 & 0 & 0 & 0 \\ 0 & 30 & 0 & 40 & 0 & 0 \\ 0 & 0 & 50 & 60 & 70 & 0 \\ 0 & 0 & 0 & 0 & 0 & 80 \end{pmatrix}$$

with 8 non-zero entries. The CSR format is the following

```
A.data = [10 20 30 40 50 60 70 80]
A.indptr = [0 2 4 7 8]
```

Moreover, we define the array `A.indices` of size equal to the size of `A.data`. The entries of `A.indices` specify the columns where the corresponding entries of `A.data` are located. The column indices of row `i` of `A` are stored in `A.indices[A.indptr[i]:A.indptr[i+1]]`. Their corresponding values are kept in `A.data[A.indptr[i]:A.indptr[i+1]]`.

```
A.indices = [0 1 1 3 2 3 4 5]
```

This means for example that the first row (`i=0`) has

`A.indptr[1]-A.indptr[0]=2-0=2`

nonzero entries and these are located at the columns

`A.indices[0]=0` and `A.indices[1]=1`

and these are

`A.data[0]=10` and `A.data[1]=20`.

The second row (`i=1`) has

`A.indptr[2]-A.indptr[1]=4-2=2`

non-zero entries located at the columns

`A.indices[2]=1` and `A.indices[3]=3`

and these are

`A.data[2]=30` and `A.data[3]=40`.

The third row (`i=2`) has

`A.indptr[3]-A.indptr[2]=7-4=3`

non-zero entries located at the columns

`A.indices[4]=2`, `A.indices[5]=3` and `A.indices[6]=4`

and these are

`A.data[4]=50`, `A.data[5]=60` and `A.data[6]=70`.

Finally, the fourth row (`i=3`) has

`A.indptr[4]-A.indptr[3]=8-7=1`

non-zero entries located at the column

`A.indices[7]=5`

and this is `A.data[7]=80`.

In Python we can get the CSR format easily using the `sparse.csr_matrix` function as it is demonstrated in the following example.

```
import numpy as np

A = np.array([[10, 20,  0,  0,  0,  0],
              [ 0, 30,  0, 40,  0,  0],
              [ 0,  0, 50, 60, 70,  0],
              [ 0,  0,  0,  0,  0, 80]])
S = sparse.csr_matrix(A)
print(S.data[:])
print(S.indptr[:])
print(S.indices[:])
```

```
[10 20 30 40 50 60 70 80]
[0 2 4 7 8]
[0 1 1 3 2 3 4 5]
```

It is noted that in addition to the CSR format, there is also the Compressed Sparse Column (CSC) format, which does exactly the same job with the CSR but for the columns of the sparse matrix. It happens that Python implementations work faster with the CSC format than the CSR.

9.7.5 Sparse Gauss-Seidel implementation

We remind that the Gauss-Seidel method for the approximation of the solution of the system $\boldsymbol{Ax} = \boldsymbol{b}$ is defined by the iteration $\boldsymbol{x}^{(n+1)} = \boldsymbol{G}\boldsymbol{x}^{(n)} + \boldsymbol{c}$ where $\boldsymbol{G} = (\boldsymbol{D}-\boldsymbol{L})^{-1}\boldsymbol{U}$ and $\boldsymbol{c} = (\boldsymbol{D}-\boldsymbol{L})^{-1}\boldsymbol{b}$.

We first present a naive implementation of the method for sparse matrices, where we just store the matrix $\boldsymbol{A}$ into the CSR sparse format in Python. We consider the matrix $\boldsymbol{A} = \text{tridiag}(-1, 2, -1)$, which can be written analytically

$$\boldsymbol{A} = \begin{pmatrix} 2 & -1 & 0 & 0 \\ -1 & 2 & -1 & 0 \\ 0 & -1 & 2 & -1 \\ 0 & 0 & -1 & 2 \end{pmatrix},$$

and the vector $\boldsymbol{b} = (1, 0, 0, 1)$. Then the system $\boldsymbol{Ax} = \boldsymbol{b}$ has the unique solution $\boldsymbol{x} = (1, 1, 1, 1)$. We get the CSR format of $\boldsymbol{A}$ as before.

```
A = np.array([[ 2.0,-1.0, 0.0, 0.0],
              [-1.0, 2.0,-1.0, 0.0],
              [ 0.0,-1.0, 2.0,-1.0],
              [ 0.0, 0.0,-1.0, 2.0]])
S = sparse.csr_matrix(A)
x = np.ones(4)
b = S.dot(x)
```

In this code we create the right-hand side using the command `b = S.dot(x)` which performs sparse matrix vector multiplication. This multiplication takes into account the sparsity of $\boldsymbol{A}$ and is faster than the full matrix multiplication. In the following code, we present an implementation of the Gauss-Seidel method for sparse matrices.

```
def sp_gauss_seidel(S, b, x, tol = 1.e-5, maxit = 100):
    n = len(b)
    xnew = np.zeros_like(x) # create new vector
    D = sparse.spdiags(S, 0, n, n, format = 'csc')
    L = sparse.tril(S, 0, format = 'csc')
    U = sparse.triu(S, 1, format = 'csc')
    G = -(sparse.linalg.inv(L)).dot(U)
    c = (sparse.linalg.inv(L)).dot(b)
    iters = 0
    err = 1.0
    while (err > tol and iters < maxit):
        iters += 1
        xnew = G*x + c
        err = npl.norm(xnew-x, np.inf)
        x = xnew
    print('iterations required for convergence:', iters)
    return x
```

In the previous implementation we formed the iterations using the default sparse matrix multiplication. For example, we compute the matrix $\boldsymbol{G}$ just by multiplying the inverse of the sparse matrix $\boldsymbol{D}-\boldsymbol{L}$ which we store in the array L with the sparse matrix $\boldsymbol{U}$, which is stored in the array U, by using the command `G = -(sparse.linalg.inv(L)).dot(U)`. Similarly, we compute the vector $\boldsymbol{c}$ with the command `c = (sparse.linalg.inv(L)).dot(b)`. Basically, this is the main ingredient for efficient implementation of iterative methods: The multiplication of two matrices and the matrix vector multiplication. Finally, we test our code taking as initial guess the vector $\boldsymbol{x}^{(0)} = (0, 0, 0, 0)^T$.

```
x = np.zeros(len(b))
x, niters = sp_gauss_seidel(S, b, x, tol = 1.e-5, maxit = 100)
print(x)
```

```
iterations required for convergence: 27
[0.99998763 0.9999838  0.9999869  0.99999345]
```

Avoiding inverse matrix

A sophisticated implementation of Gauss-Seidel method should avoid the computation of the inverse of $\boldsymbol{D}-\boldsymbol{L}$, and make use of the indices of the non-zero entries only. In order to implement the Gauss-Seidel method efficiently, we consider writing the method in vector format: Let $\boldsymbol{A}_i$ be the i-th row of $\boldsymbol{A}$. The two sums in the formula

$$x_i^{(k+1)} = \left(b_i - \sum_{j=1}^{i-1} a_{ij}x_j^{(k+1)} - \sum_{j=i+1}^{n} a_{ij}x_j^{(k)}\right) / a_{ii} ,$$

can be represented as the regular vector product of $\boldsymbol{A}_i$ with the vector

$$\boldsymbol{z}^{(k)} = (x_1^{(k+1)}, \dots, x_{i-1}^{(k+1)}, 0, x_{i+1}^{(k)}, \dots, x_n^{(k)})^T .$$

Subsequently, the Gauss-Seidel iteration takes the form

$$x_i^{(k+1)} = \left(b_i - \boldsymbol{A}_i \boldsymbol{z}^{(k)}\right) / a_{ii} .$$

The implementation of this technique can be the following:

```
def sp_gauss_seidel(S, b, x, tol = 1.e-5, maxit = 100):
    n = len(b)
    err = 1.0
    iters = 0
    D = S.diagonal()
    xnew = np.zeros_like(x)
    while (err > tol and iters < maxit):
        iters += 1
        for i in range(n):
            rowstart = S.indptr[i]
            rowend = S.indptr[i+1]
            z = np.copy(x)
            z[:i] = xnew[:i]
            z[i]=0.0
            s = np.dot(S.data[rowstart:rowend], z[S.indices[rowstart:rowend]])
            xnew[i] = (b[i] - s) / D[i]
        err = np.linalg.norm(xnew-x,np.inf)
        x = np.copy(xnew)
    print('iterations required for convergence:', iters)
    return x
```

Testing the code using the same matrices as before we get

```
A = np.array([[ 2.0,-1.0, 0.0, 0.0],
              [-1.0, 2.0,-1.0, 0.0],
              [ 0.0,-1.0, 2.0,-1.0],
              [ 0.0, 0.0,-1.0, 2.0]])
S = sparse.coo_matrix(A)
x = np.ones(4)
S = S.tocsr()
b = S.dot(x)
x = np.zeros(len(b))
x = sp_gauss_seidel(S, b, x, tol = 1.e-5, maxit = 100)
print(x)
```

```
iterations required for convergence: 27
[0.99998763 0.9999838  0.9999869  0.99999345]
```

We observe that the more sophisticated implementation gives exactly the same results as before. The difference is based on the speed only, and the reason for presenting it here is for pedagogical reasons as Python offers ready-to-use sparse solvers as we shall see later.

9.7.6 Sparse linear systems and the module `scipy.sparce`

SciPy offers an efficient sparse matrix multiplication. In order to construct a sparse linear system with exact solution $\boldsymbol{x} = (1, 1, 1, 1)^T$ we consider first transforming the matrix $\boldsymbol{A}$ into CSC or CSR format. Then we can solve the system of equations $\boldsymbol{S}\boldsymbol{x} = \boldsymbol{b}$ given that $\boldsymbol{S}$ is in sparse format using the function `spsolve` of `scipy.sparse.linalg`. The simplest use of the function `spsolve` requires the sparse matrix $\boldsymbol{S}$ and the right-hand side vector $\boldsymbol{b}$, while the general call is

```
spsolve(A, b, permc_spec, use_umfpack)
```

where

A: Is an array or sparse matrix that will be converted into CSC or CSR format. The algorithm behind spsolve is faster when matrix $\boldsymbol{A}$ is given in CSC format.

b: The right-hand side $\boldsymbol{b}$ of the linear system $\boldsymbol{Ax} = \boldsymbol{b}$.

permc_spec: Is a string variable that specifies how to permute the columns of the coefficient matrix for sparsity preservation. It can take the values NATURAL for natural ordering; MMD_ATA for minimum degree ordering on $\boldsymbol{A}^T\boldsymbol{A}$; MMD_AT_PLUS_A for minimum degree ordering on the structure of $\boldsymbol{A}^T + \boldsymbol{A}$; and COLAMD, which is an approximate minimum degree column ordering [45] (Optional with default value COLAMD)

use_umfpack: A boolean variable determining whether to use the umfpack for the solution of the system. Umfpack is a widely used library for the numerical solution of linear systems written in C programming language. (Optional with default value True)

The function returns just the solution vector x. In the following example, we demonstrate the simplest use of the function spsolve.

```
from scipy import sparse
from scipy.sparse import linalg
import numpy as np

A = np.array([[ 2.0,-1.0, 0.0, 0.0],
              [-1.0, 2.0,-1.0, 0.0],
              [ 0.0,-1.0, 2.0,-1.0],
              [ 0.0, 0.0,-1.0, 2.0]])
S= sparse.csc_matrix(A)
x = np.ones(4)
b = S.dot(x)
x = linalg.spsolve(S,b)
print(x)
```

```
[1. 1. 1. 1.]
```

Sometimes, it is useful to know where the non-zero entries are located in a sparse matrix $\boldsymbol{A}$ and if they follow a certain pattern. We can visualize the non-zero structure of a matrix using the MatPlotLib function spy of the module matplotlib.pyplot, and inspect the pattern of a matrix. For example, we can see the tridiagonal structure of the previous matrix with the following commands:

```
plt.spy(S)
plt.show()
```

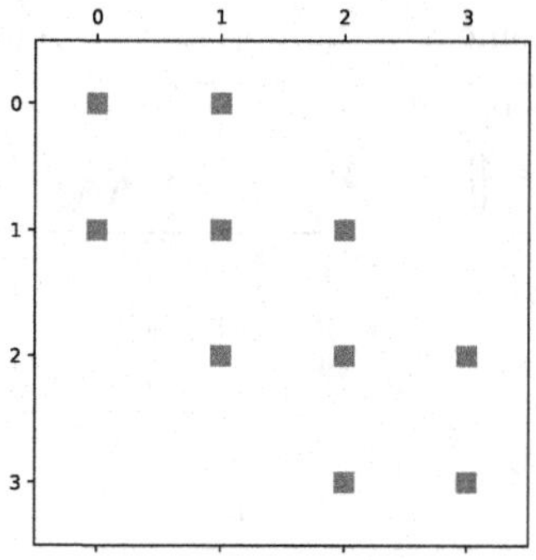

We proceed now with a more precise definition of sparsity. Let N be the total number of entries of a matrix $\boldsymbol{A}$. If N_z is the number of zero entries, then we define the *sparsity index* of $\boldsymbol{A}$ to be the number

$$I_s = \frac{N_z}{N} \ .$$

Obviously, $I_s \in [0,1]$. A dense matrix will have sparsity index closer to 1, while the smaller the number I_s is, the sparser the matrix $\boldsymbol{A}$ is. The sparsest matrix will be the zero matrix with $I_s = 0$.

In Python, we can find the number of non-zero entries of an array `A` using the function `count_non-zero(A)` of NumPy. Similarly, the SciPy method `S.count_non-zero()` returns the number of the non-zero entries of a sparse matrix in CSR format. The last command is equivalent to `count_non-zero(S.toarray())` command. In general we can return to a dense format (or NumPy array) either by using the method `toarray()` or the method `todense()`. The first command returns an array while the former returns a matrix. For example, we can find the non-zero entries of the previous matrix `S` as follows.

```
A = np.array([[ 2.0,-1.0, 0.0, 0.0],
              [-1.0, 2.0,-1.0, 0.0],
              [ 0.0,-1.0, 2.0,-1.0],
              [ 0.0, 0.0,-1.0, 2.0]])
S = sparse.csr_matrix(A)
print(S.count_non-zero())
print(np.count_non-zero(S.toarray()))
```

The result is in both cases 10 as it is expected. Since the sparsity index requires the number of zero entries N_z, we can compute the sparsity index of $\boldsymbol{S}$ using the formula

```
Is = 1.0-S.count_non-zero()/np.size(S.toarray())
```

which returns 0.375.

9.7.7 Application in electric circuits

Here we study simple electric circuits consisting of closed connections of batteries, resistors and wires. The presence of a battery in a circuit generates a flow of electrons current. The current flows out of the positive terminal of the battery and flows back into the negative terminal. In the battery representation (see Table 9.2) the positive terminal is the longer vertical bar.

TABLE 9.2
Basic elements of an electrical circuit.

battery	resistor
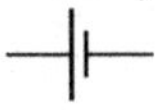	

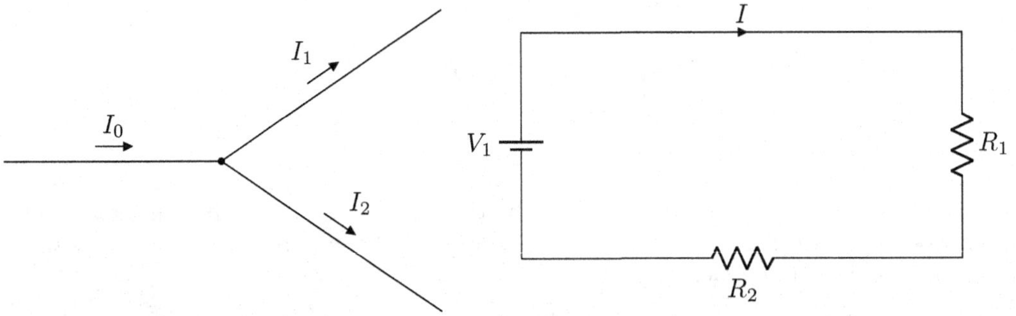

(a) First law: $I_1 - I_2 - I_3 = 0$

(b) Second law: $V_1 - V_2 - V_3 = 0$

FIGURE 9.2
Kirchhoff's laws.

Other devices that are connected to the batteries, such as motors, lightbulbs and heating coils, generate an opposing force to the flow. For this reason, we call them resistances. The resistance is a measure of that opposing force.

An electric current is denoted by I and is measured in Amperes (A), a resistance by R and is measured in Ohms (Ω) and an electrical potential difference (voltage) by V and is measured in Volts (V). These three quantities are related with Ohm's law which determines the amount of power need to make the current flow through a resistor. This fundamental law of electricity states that the voltage drop V across a resistor R is $V = I \cdot R$.

In addition to Ohm's law, for any electric circuit we have two more conservation laws known as Kirchhoff's laws. The first law, which is called current law, states that the sum of the currents flowing into a node is equal to the sum of the currents flowing out of the same node. An example of the current law is presented in Figure 9.2 where a current I_1 is split into I_2 and I_3 at the particular node with $I_1 = I_2 + I_3$.

The second law states that the sum of the voltage drops around any circuit is equal to the total voltage around the circuit. This means that the directed sum of the voltages (potential differences) around a closed loop should be zero. In Figure 9.2, the voltage of the battery V_1 should be equal to the sum of the voltages of the two resistors $V_2 + V_3$. Due to Ohm's law $V_2 = I \cdot R_1$ and $V_3 = I \cdot R_2$. Thus, we have that $V_1 = I \cdot (R_1 + R_2)$.

To demonstrate the application of linear systems to electric circuits we compute all the electric currents running through the circuit shown in Figure 9.3. This particular circuit consists of three loops (indicated in the figure with an arc) and four junctions (indicated in the figure with a dot).

Using Kirchhoff's first law at the three junctions E_1, E_2, E_3, we obtain the following three equations:

$$
\begin{array}{llllll}
E_1: & I_0 & -I_1 & -I_4 & = & 0\,, \\
E_2: & I_1 & -I_2 & -I_3 & = & 0\,, \\
E_3: & I_3 & +I_4 & -I_5 & = & 0\,.
\end{array}
$$

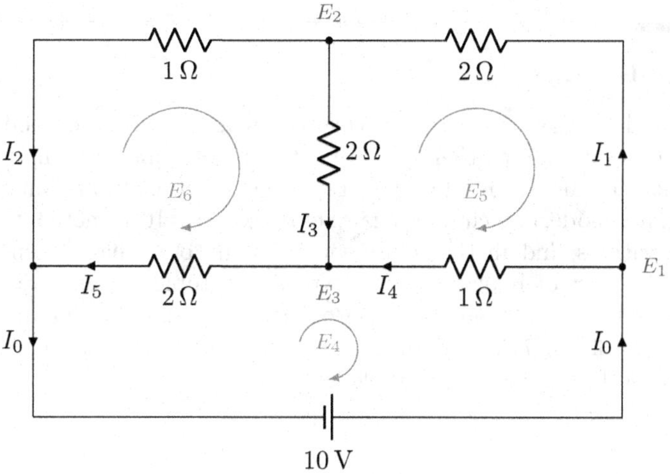

FIGURE 9.3
An electric circuit.

Applying Kirchhoff's voltage law at the three closed loops E_4, E_5, E_6, we obtain the following three equations:

$$\begin{array}{rrrrrcl} E_4: & & & I_4 & +2I_5 & = & 10\,, \\ E_5: & 2I_1 & +2I_3 & -I_4 & & = & 0\,, \\ E_6: & I_2 & -2I_3 & -2I_5 & & = & 0\,. \end{array}$$

For the 6 unknowns I_i, $i = 0, 1, \dots, 5$ we obtained 6 equations. Our choice of six equations form the system $\boldsymbol{Ax} = \boldsymbol{b}$ with

$$\boldsymbol{A} = \begin{pmatrix} 1 & -1 & 0 & 0 & -1 & 0 \\ 0 & 1 & -1 & -1 & 0 & 0 \\ 0 & 0 & 0 & 1 & 1 & -1 \\ 0 & 0 & 0 & 0 & 1 & 2 \\ 0 & 2 & 0 & 2 & -1 & 0 \\ 0 & 0 & 1 & -2 & 0 & -2 \end{pmatrix}, \quad \boldsymbol{x} = \begin{pmatrix} I_0 \\ I_1 \\ I_2 \\ I_3 \\ I_4 \\ I_5 \end{pmatrix} \quad \text{and} \quad \boldsymbol{b} = \begin{pmatrix} 0 \\ 0 \\ 0 \\ 10 \\ 0 \\ 0 \end{pmatrix},$$

where $\boldsymbol{A}$ is sparse with more than half of its entries zero. With the help of the function `spsolve` we compute the solution $\boldsymbol{x} = (7, 3, 4, -1, 4, 3)$. Similarly, we could use the function `solve` of `numpy.linalg` to solve the system $\boldsymbol{Ax} = \boldsymbol{b}$ and obtain the same result.

```
A = np.array([[ 1.0,-1.0, 0.0, 0.0,-1.0, 0.0],
              [ 0.0, 1.0,-1.0,-1.0, 0.0, 0.0],
              [ 0.0, 0.0, 0.0, 1.0, 1.0,-1.0],
              [ 0.0, 0.0, 0.0, 0.0, 1.0, 2.0],
              [ 0.0, 2.0, 0.0, 2.0,-1.0, 0.0],
              [ 0.0, 0.0, 1.0,-2.0, 0.0,-2.0]])
b = np.array([0, 0, 0, 10, 0, 0])
S = sparse.csc_matrix(A)
I = linalg.spsolve(S,b)
print('I =', I)
```

```
I = [ 7.  3.  4. -1.  4.  3.]
```

9.8 Further Reading

Methods of numerical linear algebra are covered in all classical textbooks of numerical analysis and computational mathematics due to their significance in natural sciences. Specialized book with algorithms focusing more on direct methods but with all the details, intuition and modern techniques is the book [48]. It is noted that the style of some of the algorithms and methods presented in our book have been influenced substantially by [48]. A list of books covering extensively topics of numerical linear algebra includes [135, 29, 45, 24, 8, 32, 69, 37, 35]. Books with focus on iterative methods include [112, 137, 142, 75, 53, 96, 77]. There are also important books on numerical algorithms designed for high-performance computing [47].

Chapter Highlights

- There are two basic groups of numerical methods aiming to solve systems of linear equations, namely direct and iterative methods. Direct methods are usually more efficient for solving systems represented by dense matrices, contrary to iterative methods which are ideal for large and sparse matrices.
- The most general direct method is the Gaussian elimination, which is applied in matrix form using row operations.
- Gaussian elimination leads to the LU factorization of a matrix $\boldsymbol{A}$.
- The LU factorization of $\boldsymbol{A}$ is a product of a lower triangular matrix $\boldsymbol{L}$ with ones in the main diagonal and an upper triangular matrix $\boldsymbol{U}$ such that $\boldsymbol{A} = \boldsymbol{LU}$.
- Having the $\boldsymbol{A} = \boldsymbol{LU}$ factorization we can solve the system $\boldsymbol{Ax} = \boldsymbol{b}$ by considering the system $\boldsymbol{LUx} = \boldsymbol{b}$. We first solve the system $\boldsymbol{Ly} = \boldsymbol{b}$ for $\boldsymbol{y}$ (forward substitution) and then we solve for $\boldsymbol{x}$ the system $\boldsymbol{Ux} = \boldsymbol{y}$ (backward substitution).
- Floating point arithmetic can lead to failure of Gaussian elimination and thus to the LU factorization. A remedy for this problem is the interchange of rows (or columns, or both) using pivoting techniques. The interchange in rows of $\boldsymbol{A}$ can be represented by a permutation matrix $\boldsymbol{P}$ and the LU factorization can be written as $\boldsymbol{PA} = \boldsymbol{LU}$.
- The matrices $\boldsymbol{L}$ and $\boldsymbol{U}$ in the $\boldsymbol{PA} = \boldsymbol{LU}$ factorization with pivoting are not the same as in the $\boldsymbol{A} = \boldsymbol{LU}$ factorization without pivoting.
- Solving a system $\boldsymbol{Ax} = \boldsymbol{b}$ with LU factorization and pivoting leads to the solution of the equivalent system $\boldsymbol{PAx} = \boldsymbol{Pb}$. Then we solve the system $\boldsymbol{LUx} = \boldsymbol{Pb}$ as before.
- The sensitivity of a matrix $\boldsymbol{A}$ to small perturbations can be determined with the help of the condition number defined as $\mathrm{cond}(\boldsymbol{A}) = \|\boldsymbol{A}^{-1}\| \, \|\boldsymbol{A}\|$.
- The LU factorization of a matrix $\boldsymbol{A}$ can be used to compute efficaciously its inverse $\boldsymbol{A}^{-1}$ matrix and its determinant $\det(\boldsymbol{A})$.
- The LU factorization of a symmetric matrix leads to the LDL^T factorization, where $\boldsymbol{U} = \boldsymbol{DL}^T$.
- Symmetric and positive definite matrices can be written in the form $\boldsymbol{A} = \boldsymbol{HH}^T$ where $\boldsymbol{H}$ is a lower triangular matrix. This factorization of a matrix $\boldsymbol{A}$ is known as Cholesky factorization.
- Cholesky factorization can be used as a test for positive definiteness of matrices.
- Two classical iterative methods for the solution of large, sparse systems of equations are the Jacobi and Gauss-Seidel methods with similar properties and implementation. Gauss-Seidel converges faster than Jacobi but Jacobi can be implemented in parallel computers in a trivial way.

Exercises

1. Consider the matrices

$$A = \begin{pmatrix} 2 & -1 & 0 \\ -1 & 2 & -1 \\ 0 & -1 & 2 \end{pmatrix}, \quad B = \begin{pmatrix} 4 & -1 & -1 & 0 \\ -1 & 4 & 0 & -1 \\ -1 & 0 & 4 & -1 \\ 0 & -1 & -1 & 4 \end{pmatrix}, \quad C = \begin{pmatrix} 1 & 0 & 2 & 1 \\ 0 & 4 & 8 & 10 \\ 2 & 8 & 29 & 22 \\ 1 & 10 & 22 & 42 \end{pmatrix}.$$

   (a) Use Gaussian elimination to solve the systems

$$\boldsymbol{A}\boldsymbol{x} = \boldsymbol{a}, \quad \boldsymbol{B}\boldsymbol{x} = \boldsymbol{b}, \quad \boldsymbol{C}\boldsymbol{x} = \boldsymbol{c}\ ,$$

   where

$$\boldsymbol{a} = (2, 1, 0)^T, \quad \boldsymbol{b} = (6, 1, 1, 1)^T, \quad \boldsymbol{c} = (4, 22, 61, 75)^T\ .$$

   (b) Determine the LU factorization of $\boldsymbol{A}, \boldsymbol{B}, \boldsymbol{C}$.

   (c) Use the LU factorization to solve the systems $\boldsymbol{A}\boldsymbol{x} = \boldsymbol{a}$, $\boldsymbol{B}\boldsymbol{x} = \boldsymbol{b}$ and $\boldsymbol{C}\boldsymbol{x} = \boldsymbol{c}$.

2. Consider the matrix

$$A = \begin{pmatrix} 0 & 1 & 2 \\ 2 & -2 & 1 \\ 5 & 3 & 1 \end{pmatrix}.$$

   (a) Compute the LU factorization of $\boldsymbol{A}$ using partial pivoting.

   (b) Compute the determinant $\det(\boldsymbol{A})$ using the fact that $\boldsymbol{PA} = \boldsymbol{LU}$.

   (c) Compute the inverse of $\boldsymbol{A}$ using the $\boldsymbol{PA} = \boldsymbol{LU}$ factorization of $\boldsymbol{A}$.

3. Consider the matrix

$$A = \begin{pmatrix} 1 & 0 & 2 & 1 \\ 0 & 4 & 8 & 10 \\ 2 & 8 & 29 & 22 \\ 1 & 10 & 22 & 42 \end{pmatrix}.$$

   (a) Compute the LDL^T factorization of $\boldsymbol{A}$.

   (b) Prove that $\boldsymbol{A}$ is positive definite matrix.

   (c) Compute the Cholesky factorization of $\boldsymbol{A}$.

4. Consider a real $n \times n$ symmetric and positive definite matrix $\boldsymbol{A}$.

   (a) Prove that the matrix $\alpha\boldsymbol{A} + \beta\boldsymbol{z}\boldsymbol{z}^T$ is symmetric and positive definite for all $\alpha > 0$, $\beta \geq 0$ and $\boldsymbol{z} \in \mathbb{R}^n$.

   (b) Using the previous result, prove that the matrix

$$B = \begin{pmatrix} \alpha+\beta & \beta & \cdots & \beta \\ \beta & \alpha+\beta & \cdots & \beta \\ \vdots & \vdots & \ddots & \vdots \\ \beta & \beta & \cdots & \alpha+\beta \end{pmatrix},$$

   is symmetric and positive definite.

(c) Prove that the matrix

$$C = \begin{pmatrix} 3 & 1 & 1 \\ 1 & 3 & 1 \\ 1 & 1 & 3 \end{pmatrix} ,$$

is symmetric and positive definite.

(d) Compute the Cholesky factorization of $\boldsymbol{C}$.

5. Let $\|\cdot\|$ be a norm in $\mathbb{R}^n$ and $\boldsymbol{A}$ invertible $n \times n$ matrix. Prove the following:

(a) If $\boldsymbol{Ax} = \boldsymbol{b}$, $(\boldsymbol{A} + \delta\boldsymbol{A})(\boldsymbol{x} + \delta\boldsymbol{x}) = \boldsymbol{b}$ and $\|\boldsymbol{A}^{-1}\|\, \|\delta\boldsymbol{A}\| < 1$, then $\boldsymbol{A} + \delta\boldsymbol{A}$ is invertible and

$$\frac{\|\delta\boldsymbol{x}\|}{\|\boldsymbol{x}\|} \le \frac{\mathrm{cond}(\boldsymbol{A})}{1 - \|\boldsymbol{A}^{-1}\|\, \|\delta\boldsymbol{A}\|} \frac{\|\delta\boldsymbol{A}\|}{\|\boldsymbol{A}\|} .$$

(b) If $\boldsymbol{Ax} = \boldsymbol{b}$, $(\boldsymbol{A} + \delta\boldsymbol{A})(\boldsymbol{x} + \delta\boldsymbol{x}) = \boldsymbol{b} + \delta\boldsymbol{b}$ and $\|\boldsymbol{A}^{-1}\|\, \|\delta\boldsymbol{A}\| < 1$, then $\boldsymbol{A} + \delta\boldsymbol{A}$ is invertible and

$$\frac{\|\delta\boldsymbol{x}\|}{\|\boldsymbol{x}\|} \le \frac{\mathrm{cond}(\boldsymbol{A})}{1 - \|\boldsymbol{A}^{-1}\|\, \|\delta\boldsymbol{A}\|} \left\{ \frac{\|\delta\boldsymbol{A}\|}{\|\boldsymbol{A}\|} + \frac{\|\delta\boldsymbol{b}\|}{\|\boldsymbol{b}\|} \right\} .$$

6. Let $n = 1000$. Construct in Python the $n \times n$ banded matrix $\boldsymbol{A}$

$$\boldsymbol{A} = \mathrm{pentadiag}(1, -4, 6, -4, 1) .$$

The specific matrix is a symmetric, banded, matrix consisting of 5 diagonals. The main diagonal consists of 6 only while the other four diagonals have -4 and 1, respectively. This matrix has lower and upper bandwidth $p = 2$, which means that it has two non-zero diagonals above and two below the main diagonals. The total number of non-zero diagonals is $2 \cdot p + 1 = 5$.

Taking advantage of the properties of this matrix we can extend the Cholesky factorization algorithm for banded matrices. Given a symmetric, positive definite matrix $\boldsymbol{A} \in \mathbb{R}^{n,n}$ with bandwidth p, the following Algorithm 33 computes a lower triangular matrix $\boldsymbol{H}$ with lower bandwidth p such that $\boldsymbol{A} = \boldsymbol{H}\boldsymbol{H}^T$. For all $i \ge j$, $\boldsymbol{H}(i,j)$ overwrites $\boldsymbol{A}(i,j)$.

To solve a system $\boldsymbol{Ax} = \boldsymbol{b}$ with banded, symmetric and positive definite matrix $\boldsymbol{A}$ we modify the forward and backward substitution algorithms appropriately to Algorithms 34 and 35.

Algorithm 33 Band Cholesky

for $j = 1 : n$ **do**
  for $k = \max(1, j - p) : j - 1$ **do**
    $\lambda = \min(k + p, n)$
    $\boldsymbol{A}(j : \lambda, j) = \boldsymbol{A}(j : \lambda, j) - \boldsymbol{A}(j, k) \cdot \boldsymbol{A}(j : \lambda, k)$
  end for
  $\lambda = \min(j + p, n)$
  $\boldsymbol{A}(j : \lambda, j) = \boldsymbol{A}(j : \lambda, j) / \sqrt{\boldsymbol{A}(j, j)}$
end for

(a) Compute the number of non-zero entries of $\boldsymbol{A}$

(b) Prove that $\boldsymbol{A}$ is positive definite.

(c) Use the function `cholesky` implementing Algorithm 32 to find the Cholesky factorization of $\boldsymbol{A}$.

Algorithm 34 Forward Substitution

for $j = 1 : n$ **do**
  $\boldsymbol{b}(j) = \boldsymbol{b}(j)/\boldsymbol{A}(j, j)$
  $\boldsymbol{b}(j+1 : \min(j+p, n)) = \boldsymbol{b}(j+1 : \min(j+p, n)) - \boldsymbol{A}(j+1 : \min(j+p, n), j) \cdot \boldsymbol{b}(j)$
end for

Algorithm 35 Backward Substitution

for $j = n : -1 : 1$ **do**
  $\boldsymbol{b}(j) = \boldsymbol{b}(j)/\boldsymbol{A}(j, j)$
  $\boldsymbol{b}(\max(1, j-p) : j-1) = \boldsymbol{b}(\max(1, j-p) : j-1) - \boldsymbol{A}(j, \max(1, j-p) : j-1)^T \cdot \boldsymbol{b}(j)$
end for

(d) Write a function `bandcholesky` implementing the banded Cholesky factorization given by Algorithm 33. Use your function to factorize the matrix $\boldsymbol{A}$.

(e) Compare your functions with Python function `numpy.linalg.cholesky`.

(f) Write functions `solve` and `bandsolve` to solve the linear system $\boldsymbol{A}\boldsymbol{x} = \boldsymbol{b}$ with

$$b_i = \sum_{j=1}^{n} a_{ij} \ .$$

To verify the accuracy of the results, compute the norm of the error $\|\tilde{\boldsymbol{x}} - \boldsymbol{x}\|_2$ where $\tilde{\boldsymbol{x}}$ is the numerical solution and $\boldsymbol{x}$ the exact solution with $x_i = 1$ for all $i = 1, 2, \ldots, n$.

(g) Improve your code by using symmetric band storage. Specifically, taking advantage of the symmetry and the banded structure of $\boldsymbol{A}$, store the main diagonal and all the left diagonals in the rows of a new $p \times n$ (dense) matrix $\boldsymbol{A}_b$ using the formula:

$$\boldsymbol{A}_b(1 + i - j, j) = \boldsymbol{A}(i, j), \qquad \text{for} \quad j \le i \le \min(n, j + p).$$

For example, when $n = 5$ and $p = 2$ the matrix $\boldsymbol{A}$ is being stored in the 3×5 matrix $\boldsymbol{A}_b$:

$$\boldsymbol{A} = \begin{pmatrix} a_{11} & a_{12} & a_{13} & & \\ a_{21} & a_{22} & a_{23} & a_{24} & \\ a_{31} & a_{32} & a_{33} & a_{34} & a_{35} \\ & a_{42} & a_{43} & a_{44} & a_{45} \\ & & a_{53} & a_{54} & a_{55} \end{pmatrix}, \qquad \boldsymbol{A}_b = \begin{pmatrix} a_{11} & a_{22} & a_{33} & a_{44} & a_{55} \\ a_{21} & a_{32} & a_{43} & a_{54} & * \\ a_{31} & a_{42} & a_{53} & * & * \end{pmatrix}.$$

For more information about the specific band-storage format see Section 2.3.1.

(h) Explore the options of the function `cholesky_banded` of the module `scipy.linalg` and use it to get the Cholesky factorization of the particular matrix $\boldsymbol{A}$. Compare with the previous results.

(i) Compare the efficiency of all these methods. Among other indicators present the CPU time required for each implementation. For more information on how to measure the performance of your code see Section 2.2.7.

7. Consider the linear system $\boldsymbol{A}\boldsymbol{x} = \boldsymbol{b}$ with

$$\boldsymbol{A} = \begin{pmatrix} 2 & -1 & 0 \\ -1 & 2 & -1 \\ 0 & -1 & 2 \end{pmatrix} \quad \text{and} \quad \boldsymbol{b} = \begin{pmatrix} 3 \\ -5 \\ 5 \end{pmatrix}.$$

(a) Show that the sequences of approximations generated by the Jacobi and Gauss-Seidel methods converge to the exact solution.

(b) Use the functions `jacobi` and `gauss_seidel` of Section 9.7.2 to solve the linear system $\boldsymbol{Ax} = \boldsymbol{b}$ and compare the results. As initial guess for the solution of the system consider the vector $\boldsymbol{x}^{(0)} = (1, 1, 1)^T$.

(c) Use the function `sp_gauss_seidel` of Section 9.7.5 to solve the linear system $\boldsymbol{Ax} = \boldsymbol{b}$ using the same initial guess of the solution as before.

(d) Write a new function `sp_jacobi` implementing the Jacobi method for matrices stored is CSR format and solve the linear system $\boldsymbol{Ax} = \boldsymbol{b}$ using the same initial guess of the solution as before. Compare the results with the previous methods.

8. Let $n = 1000$. Consider the linear system $\boldsymbol{Ax} = \boldsymbol{b}$ with

$$b_i = \sum_{j=1}^{n} a_{ij} ,$$

and the $n \times n$ matrix $\boldsymbol{A} = \text{pentadiag}(1, -4, 6, -4, 1)$.

(a) Show that $\boldsymbol{A}$ is row diagonally dominant matrix.

(b) Store matrix $\boldsymbol{A}$ is dense format and use the naive Jacobi and Gauss-Seidel methods to solve the linear system $\boldsymbol{Ax} = \boldsymbol{b}$.

(c) Store matrix $\boldsymbol{A}$ in CSR format.

(d) Solve the linear system using the function `spsolve` of module `scipy.linalg`.

(e) Compare the results in terms of efficiency and performance. For this reason, use the norm of your preference.

9. (Successive Over-Relaxation) In order to improve the Jacobi and Gauss-Seidel methods for the numerical solution of a system $\boldsymbol{Ax} = \boldsymbol{b}$ we consider an extrapolation (relaxation) technique. Let $\omega \in \mathbb{R}$. Then the system $\boldsymbol{Ax} = \boldsymbol{b}$ is written in the following way:

$$\begin{aligned} \boldsymbol{Ax} &= \boldsymbol{b} \\ \omega\boldsymbol{Ax} &= \omega\boldsymbol{b} \\ \omega(\boldsymbol{D} - \boldsymbol{L} - \boldsymbol{U})\boldsymbol{x} &= \omega\boldsymbol{b} \\ (\boldsymbol{D} - \boldsymbol{D} + \omega\boldsymbol{D} - \omega\boldsymbol{L} - \omega\boldsymbol{U})\boldsymbol{x} &= \omega\boldsymbol{b} \\ (\boldsymbol{D} - (1-\omega)\boldsymbol{D} - \omega\boldsymbol{L} - \omega\boldsymbol{U})\boldsymbol{x} &= \omega\boldsymbol{b} \end{aligned}$$

We define the iteration

$$(\boldsymbol{D} - \omega\boldsymbol{L})\boldsymbol{x}^{(k+1)} = [(1-\omega)\boldsymbol{D} + \omega\boldsymbol{U}]\boldsymbol{x}^{(k)} + \omega\boldsymbol{b} .$$

The last method is written:

$$\boldsymbol{x}^{(k+1)} = \mathcal{L}_\omega \boldsymbol{x}^{(k)} + \boldsymbol{c}_\omega, \quad k = 0, 1, 2, \ldots,$$

where $\mathcal{L}_\omega = (\boldsymbol{D} - \omega\boldsymbol{L})^{-1}[(1-\omega)\boldsymbol{D} + \omega\boldsymbol{U}]$ and $\boldsymbol{c}_\omega = \omega(\boldsymbol{D} - \omega\boldsymbol{L})^{-1}$. This method is known as the Successive Over-Relaxation (SOR) method and it is known to converge to general matrices as long as $\omega \in (0, 2)$. Evaluating the SOR method yields that each entry of $\boldsymbol{x}^{(k+1)}$ is given by the formula

$$x_i^{(k+1)} = (1-\omega)x_i^{(k)} + \omega\left(b_i - \sum_{j=1}^{i-1} a_{ij}x_j^{(k+1)} - \sum_{j=i+1}^{n} a_{ij}x_j^{(k)}\right)/a_{ii}, \quad i = 1, \ldots, n .$$

(a) Write a Python function implementing efficiently the SOR method for general $\omega \in (0, 2)$. Then solve the linear system with the same matrix $\boldsymbol{A}$ as before for various values of ω. Try to implement the SOR method for sparse matrices using the `csr` format.

(b) Consider the optimal value ω_0 to be the value of ω for which the SOR method converges with the least iterations. Find an approximation of the optimal value of the relaxation parameter $\omega_0 \in (0, 2)$ by solving the system $\boldsymbol{Ax} = \boldsymbol{b}$ for the values $\omega_i = 0.1, 0.2, \ldots, 1.9$.

(c) Compare the convergence of the three methods (Jacobi, Gauss-Seidel and SOR with the optimal ω).

10. (Sherman-Morrison-Woodbury algorithm for banded matrices) Consider an $n \times n$ matrix $\boldsymbol{A}$ with entries $a_{ij} \neq 0$ only when

- $i \leq n_k$ and $j \geq n - n_k + 1$, or
- $i \geq n - n_k + 1$ and $j \leq n_k$, or
- $|i - j| \leq n_a$

where $2n_a + 1 + 2n_k < n$ and $n_k \leq n_a$. Let $\bar{\boldsymbol{A}}$ be an $n \times n$ invertible matrix and $\boldsymbol{u}, \boldsymbol{v} \in \mathbb{R}^n$ vectors such that $\bar{\boldsymbol{A}} = \boldsymbol{A} + \boldsymbol{uv}^T$ and $\boldsymbol{v}^T \bar{\boldsymbol{A}}^{-1} \boldsymbol{u} \neq 1$.

(a) Show that $\boldsymbol{A}$ is invertible and

$$\boldsymbol{A}^{-1} = \bar{\boldsymbol{A}}^{-1} + \alpha(\bar{\boldsymbol{A}}^{-1}\boldsymbol{u})(\boldsymbol{v}^T\bar{\boldsymbol{A}}^{-1}) ,$$

where

$$\alpha = \frac{1}{1 - \boldsymbol{v}^T\bar{\boldsymbol{A}}^{-1}\boldsymbol{u}} .$$

This is the Sherman-Morrison-Woodbury formula which provides a fast and memory efficient way to compute the inverse of $\boldsymbol{A} = \bar{\boldsymbol{A}} - \boldsymbol{uv}^T$.

(b) Let $\boldsymbol{b} \in \mathbb{R}^n$. Show that the linear system $\bar{\boldsymbol{A}}\bar{\boldsymbol{x}} = \boldsymbol{b}$ can be solved using the Sherman-Morrison-Woodbury algorithm [73]

Algorithm 36 Sherman-Morrison-Woodbury algorithm

Solve the system $\bar{\boldsymbol{A}}\bar{\boldsymbol{x}} = \boldsymbol{b}$
Solve the system $\bar{\boldsymbol{A}}\boldsymbol{y} = \boldsymbol{u}$
Solve the system $\bar{\boldsymbol{A}}^T\boldsymbol{z} = \boldsymbol{v}$
Compute $\alpha = 1/(1 - \boldsymbol{v}^T\boldsymbol{y})$
Compute $\beta = \boldsymbol{z}^T\boldsymbol{b}$
Compute the solution $\boldsymbol{x} = \bar{\boldsymbol{x}} + \alpha\beta\boldsymbol{y}$

(c) Show that the complexity of Algorithm 36 is $O(n^2)$.

(d) Write a function with input a matrix $\boldsymbol{A}$, the parameters n_a and n_k and the vector $\boldsymbol{b}$. This function should generate appropriate vectors $\boldsymbol{u}$, $\boldsymbol{v}$ and matrix $\boldsymbol{A}$. Then the function will compute approximate solution $\tilde{\boldsymbol{x}}$ of the system $\boldsymbol{Ax} = \boldsymbol{b}$ implementing Algorithm 36. The function should also estimate the accuracy of the solution by computing appropriate norm of the residual vector $\boldsymbol{r} = \boldsymbol{Ax} - \boldsymbol{b}$.

(e) Test your code for the system of Exercise 10 of Chapter 2.

(f) Compute the LU factorization of $\boldsymbol{A}$ and solve the system $\boldsymbol{Ax} = \boldsymbol{b}$ using forward and backward substitution. Compare the speed of the two methods.

11. Consider the $n \times n$ tridiagonal matrix

$$\boldsymbol{A} = \begin{pmatrix} a_1 & b_1 & & & 0 \\ c_2 & a_2 & b_2 & & \\ & \ddots & \ddots & \ddots & \\ & & c_{n-1} & a_{n-1} & b_{n-1} \\ 0 & & & c_n & a_n \end{pmatrix},$$

where $a_i, b_i, c_i \in \mathbb{R}$ for all i.

(a) Assume that $\boldsymbol{A}$ is invertible and has the LU-factorization $\boldsymbol{A} = \boldsymbol{LU}$ with

$$\boldsymbol{L} = \begin{pmatrix} d_1 & & & & 0 \\ c_2 & d_2 & & & \\ & \ddots & \ddots & & \\ & & c_{n-1} & d_{n-1} & \\ 0 & & & c_n & d_n \end{pmatrix}, \quad \boldsymbol{U} = \begin{pmatrix} 1 & e_1 & & & 0 \\ & 1 & e_2 & & \\ & & \ddots & \ddots & \\ & & & 1 & e_{n-1} \\ 0 & & & & 1 \end{pmatrix}.$$

Verify that $d_1 = a_1$, $b_1 = d_1 \cdot e_1$, and in general $d_k = a_k - c_k \cdot e_{k-1}$ and $b_k = d_k \cdot e_k$.

(b) Using the previous observations devise the Algorithm 37 for the computation of the LU-factorization of the tridiagonal matrix $\boldsymbol{A}$.

Algorithm 37 LU-factorization for tridiagonal matrices

Given the three diagonals $\boldsymbol{a}$, $\boldsymbol{b}$, $\boldsymbol{c}$ of $\boldsymbol{A}$
The algorithm returns the diagonals $\boldsymbol{d}$ and $\boldsymbol{e}$
Set $d_1 = a_1$
Set $e_1 = b_1/d_1$
for $k = 2 : n-1$ **do**
  $d_k = a_k - c_k \cdot e_{k-1}$
  $e_k = b_k/d_k$
end for
$d_n = a_n - c_n \cdot e_{n-1}$

(c) Prove that if $|a_1| > |b_1|$, $|a_n| > |c_n|$ and $|a_k| \geq |b_k| + |c_k|$, for $k = 2, \ldots, n-1$ then the matrix $\boldsymbol{A}$ is invertible. [Hint: Use mathematical induction to show that $d_k \neq 0$ for all $k = 1, 2, \ldots, n$, and that $|e_k| < 1$ for $k = 1, 2, \ldots, n-1$]

(d) Adapt the forward and backward substitution Algorithms 30 and 28 for the solution of the tridiagonal system $\boldsymbol{Ax} = \boldsymbol{f}$. [Hint: See Algorithms 38 and 39]

Algorithm 38 Forward substitution for tridiagonal matrices

Given the two diagonals $\boldsymbol{c}$ and $\boldsymbol{d}$ of the matrix $\boldsymbol{L}$ and the right-hand side
the algorithm solves the system $\boldsymbol{Ly} = \boldsymbol{f}$
Set $y_1 = f_1/d_1$
for $k = 2 : n$ **do**
  $y_k = (f_k - c_k \cdot y_{k-1})/d_k$
end for

Algorithm 39 Backward substitution for tridiagonal matrices

Given the diagonal $\boldsymbol{e}$ of the matrix $\boldsymbol{U}$ and the right-hand side $\boldsymbol{y}$
the algorithm solves the system $\boldsymbol{U}\boldsymbol{x} = \boldsymbol{y}$
Set $x_n = y_n$
for $k = n - 1 : 1 : -1$ **do**
    $x_k = y_k - e_k \cdot x_{k-1}$
end for

(e) Implement the new algorithms in Python where the input will be the three diagonals of $\boldsymbol{A}$ let's say $\boldsymbol{a}$, $\boldsymbol{b}$ and $\boldsymbol{c}$.

(f) Test your code for the system with $\boldsymbol{A} = \text{tridiag}(-1, 3, -1)$ and exact solution $x_i = 1$ for $i = 1, 2, \ldots, n$.

12. Consider the electric circuit depicted in Figure 9.4.

(a) Analyze the particular circuit using Kirchhoff's laws, and form a linear system with 8 equations and 8 unknowns.

(b) Solve the linear system using dense matrix techniques. Verify that your solution satisfies all the equations derived from Kirchhoff's laws.

(c) Solve the linear system using sparse matrix techniques and absolute error tolerance 10^{-6}.

(d) Compare the speed of the two numerical methods and say which one is faster.

(e) Compute the difference of the two solutions and describe their accuracy.

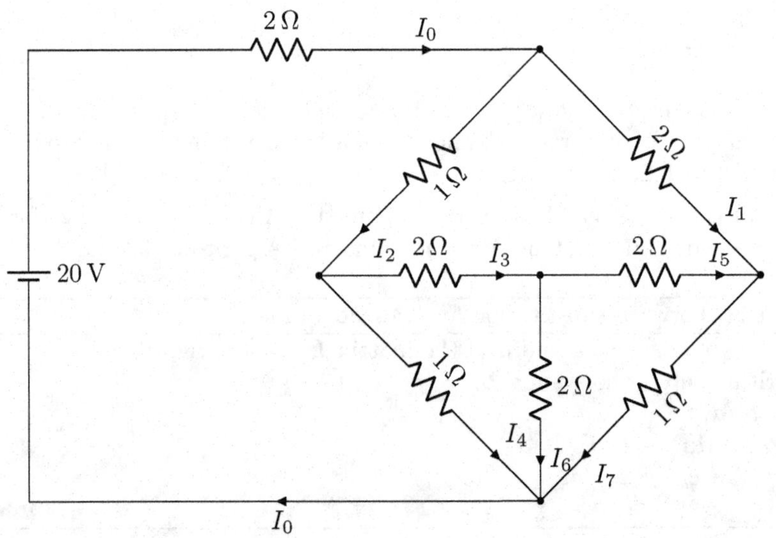

FIGURE 9.4
An electric circuit.

Part III

Advanced Topics

10

Best Approximations

In this chapter, we discuss rigorously the notion of best approximation, and in particular best approximation of vectors, matrices and functions. This is a more advanced topic compared to previous topics as it introduces some abstract notions. On the other hand, the discussion is just as important as the previous chapters. We develop the Singular Value Decomposition (SVD) of a matrix $\boldsymbol{A}$ and compute its best approximation. The SVD is used extensively in statistics, data science and machine learning, and also in engineering. We also present the Gram-Schmidt orthonormalization process and the QR factorization, which are necessary tools for the development of the SVD. In QR factorization, Q is an orthogonal matrix and R is an upper triangular. We close this chapter with an application in image compression, where the compressed image is basically its best approximation. First, we review the properties of vector spaces.

10.1 Vector Spaces, Norms and Approximations

In Chapter 6 we introduced the notion of approximation of functions by polynomials via interpolation. To measure the distance between a function and its interpolating polynomial we introduced the notion of norm. Later in Chapter 9 we introduced norms for vectors and matrices and we measured the accuracy of approximations. Norms can measure distances between exact and numerical approximations, but what is the best approximation of a solution and how can I compute it? In this section we answer to this question by studying best approximations in abstract vector spaces.

10.1.1 Vector spaces

For the moment we will take the objects to be in a specific linear, real vector space. A set X of elements $\boldsymbol{u}, \boldsymbol{v}, \boldsymbol{w}, \ldots$ is called *vector space* (over $\mathbb{R}$ or $\mathbb{C}$) with the operations of addition and scalar multiplication if:

(a) For every pair of elements $\boldsymbol{u}, \boldsymbol{v} \in X$, their sum $\boldsymbol{u} + \boldsymbol{v}$ is an element of X

(b) For every number λ and every $\boldsymbol{u} \in X$ the product $\lambda \boldsymbol{u}$ is an element of X

(c) For all $\boldsymbol{u}, \boldsymbol{v}, \boldsymbol{w} \in X$ and for numbers λ, μ, the operations of addition and scalar multiplication have the following properties:

  (i) $\boldsymbol{u} + \boldsymbol{v} = \boldsymbol{v} + \boldsymbol{u}$

  (ii) $(\boldsymbol{u} + \boldsymbol{v}) + \boldsymbol{w} = \boldsymbol{u} + (\boldsymbol{v} + \boldsymbol{w})$

  (iii) $\boldsymbol{u} + \boldsymbol{0} = \boldsymbol{u}$

  (iv) $1 \cdot \boldsymbol{u} = \boldsymbol{u} \cdot 1 = \boldsymbol{u}$

DOI: 10.1201/9781003287292-10

(v) $\lambda(\mu \boldsymbol{u}) = (\lambda\mu)\boldsymbol{u}$

(vi) $(\lambda + \mu)\boldsymbol{u} = \lambda \boldsymbol{u} + \mu \boldsymbol{u}$

(vii) $\lambda(\boldsymbol{u} + \boldsymbol{v}) = \lambda \boldsymbol{u} + \lambda \boldsymbol{v}$.

Vector spaces (sometimes called linear spaces) consist of objects that we call vectors or elements. These elements can be anything from functions to arrays. For example $\boldsymbol{u}$ and $\boldsymbol{v}$ can be any two continuous functions in the space $X = C[a, b]$. We know that the linear combination of two continuous functions is continuous. On the other hand $\boldsymbol{u}$ and $\boldsymbol{v}$ can be real n-dimensional vectors, or even $m \times n$ matrices. If X is a vector space, then any subset $Y \subset X$ that is also a vector space is called subspace of X.

We discuss the notion of linear dependence and independence. The elements $\boldsymbol{u}_1, \boldsymbol{u}_2, \dots, \boldsymbol{u}_n$ of a vector space X are called *linearly dependent* if there are constants $\lambda_1, \lambda_2, \dots, \lambda_n$ not all zero such that

$$\lambda_1 \boldsymbol{u}_1 + \lambda_2 \boldsymbol{u}_2 + \cdots + \lambda_n \boldsymbol{u}_n = 0 \ .$$

On the other hand, their linear combination is zero only when $\lambda_i = 0$ for all $i = 1, \dots, n$, in which case the n vectors are called *linearly independent*. This means that for linearly independent vectors $\boldsymbol{u}_1, \boldsymbol{u}_2, \dots, \boldsymbol{u}_n$, we have that

$$\lambda_1 \boldsymbol{u}_1 + \lambda_2 \boldsymbol{u}_2 + \cdots + \lambda_n \boldsymbol{u}_n = 0 \ ,$$

only when $\lambda_1 = \lambda_2 = \cdots = \lambda_n = 0$.

Importance of linear independence

Consider n linearly independent column vectors $\boldsymbol{u}_1, \boldsymbol{u}_2, \dots, \boldsymbol{u}_n$ of $\mathbb{R}^n$ and an $n \times n$ matrix $\boldsymbol{A}$ with columns the vectors $\boldsymbol{u}_i$. Expressing matrix multiplication as $\boldsymbol{A}\boldsymbol{x} = \boldsymbol{u}_1 x_1 + \boldsymbol{u}_2 x_2 + \cdots + \boldsymbol{u}_n x_n$ implies that the system $\boldsymbol{A}\boldsymbol{x} = \boldsymbol{0}$ has only the solution $\boldsymbol{x} = \boldsymbol{0}$, which means that $\boldsymbol{A}$ is invertible. If the columns of an $n \times n$ matrix $\boldsymbol{A}$ are not linearly independent, then $\boldsymbol{A}$ will not be invertible since the system $\boldsymbol{A}\boldsymbol{x} = \boldsymbol{0}$ will have in addition to the solution $\boldsymbol{x} = \boldsymbol{0}$ another solution[1] $\boldsymbol{x} \neq \boldsymbol{0}$.

> ☞ Matrices with linear independent columns or rows are invertible.

Consider an $m \times n$ matrix $\boldsymbol{A}$ with r linearly independent columns. The number of linearly independent columns of $\boldsymbol{A}$ is called the *rank* of $\boldsymbol{A}$ and is denoted as $\mathrm{rank}(\boldsymbol{A})$. It happens that the number of linear independent columns and rows of a matrix will be always the same. Moreover, r cannot exceed the minimum dimension of $\boldsymbol{A}$, i.e. if $m < n$, then $r \leq m$, while if $n < m$, then $r \leq n$. A matrix $\boldsymbol{A}$ is called *full-rank* if $\mathrm{rank}(\boldsymbol{A}) = \min\{m, n\}$.

10.1.2 Inner products and norms

The next step towards understanding the best approximation of elements of vector spaces is to define inner products and introduce some geometry. Inner product is a map or bilinear form $\langle \cdot, \cdot \rangle : X \times X \to \mathbb{R}$ and is a real number. For example, if $\boldsymbol{u} = (u_1, u_2)^T$ and $\boldsymbol{v} = (v_1, v_2)^T$ are two vectors of $\mathbb{R}^2$, then we define the inner product of $\boldsymbol{u}$ and $\boldsymbol{v}$ as

$$\langle \boldsymbol{u}, \boldsymbol{v} \rangle = \boldsymbol{u} \cdot \boldsymbol{v} = \boldsymbol{u}^T \boldsymbol{v} = u_1 v_1 + u_2 v_2 \ .$$

[1]The system $\boldsymbol{A}\boldsymbol{x} = \boldsymbol{0}$ is characterized by the non-uniqueness of its solutions.

The inner product can also be defined as a map onto $\mathbb{C}$ but for the purposes of this book we focus on real vector spaces with real inner products.

The inner product must satisfy the following properties for all $\boldsymbol{u}, \boldsymbol{v}, \boldsymbol{w} \in X$:

(i) $\langle \boldsymbol{u}, \boldsymbol{u} \rangle > 0$ for all $\boldsymbol{u} \neq \boldsymbol{0}$

(ii) $\langle \boldsymbol{u}, \boldsymbol{v} \rangle = \langle \boldsymbol{v}, \boldsymbol{u} \rangle$

(iii) $\langle a\boldsymbol{u}, \boldsymbol{v} \rangle = \langle \boldsymbol{u}, a\boldsymbol{v} \rangle = a\langle \boldsymbol{u}, \boldsymbol{v} \rangle$ for any $a \in \mathbb{R}$

(iv) $\langle \boldsymbol{u} + \boldsymbol{v}, \boldsymbol{w} \rangle = \langle \boldsymbol{u}, \boldsymbol{w} \rangle + \langle \boldsymbol{v}, \boldsymbol{w} \rangle$.

If we work with complex inner product instead of real, then the property (ii) can be replaced by $\langle \boldsymbol{u}, \boldsymbol{v} \rangle = \overline{\langle \boldsymbol{v}, \boldsymbol{u} \rangle}$, while in (iii) we can take $a \in \mathbb{C}$.

If $\boldsymbol{u} = f$ and $\boldsymbol{v} = g$ functions of the vector space $C[a,b]$, we can define an inner product of $C[a,b]$ to be the integral

$$\langle f, g \rangle = \int_a^b f(x)\, g(x)\, dx \ , \tag{10.1}$$

for any $f, g \in C[a,b]$. This is one of the most commonly used inner products between functions. We saw this before when we discussed the definition of norms of functions in Chapter 6.

A vector space equipped with inner product is called inner product space. We will denote the general inner product space by X. An inner product enhances a vector space with geometry and things can be visualized. This is mainly because the inner product leads to the notion of orthogonality. It doesn't really matter what the objects in the space are, we can always think of them as points in general spaces. Now that we have in our arsenal inner products, we define again the norm $\|\cdot\| : X \to [0, +\infty)$ as

$$\|\boldsymbol{u}\| = \sqrt{\langle \boldsymbol{u}, \boldsymbol{u} \rangle}, \qquad \text{for any } \boldsymbol{u} \in X \ .$$

The norm of $\boldsymbol{u}$ determines the magnitude of $\boldsymbol{u}$. The magnitude of $\boldsymbol{u}$ is practically its distance from the origin $\boldsymbol{0}$. So if we want to study the distance between two objects $\boldsymbol{u}$ and $\boldsymbol{v}$, then it suffices to study the norm $\|\boldsymbol{u} - \boldsymbol{v}\|$. This will give us the magnitude of the error between $\boldsymbol{u}$ and $\boldsymbol{v}$.

If $\boldsymbol{u} = (u_1, u_2)^T$ is a real vector, then $\|\boldsymbol{u}\| = \sqrt{u_1^2 + u_2^2}$, which is exactly the well-known modulus of vectors. Let's discuss the norm of a function $\boldsymbol{u} = f$. Since $\|f\| = \sqrt{\langle f, f \rangle}$, using the inner product (10.1) we have

$$\|f\| = \sqrt{\int_a^b f^2(x)\, dx} \ .$$

This is the norm $\|\cdot\|_2$ of Chapter 6.

In any case, we need to measure the accuracy of our approximations, so we need to define the best approximation in a formal way. Let's say we have our general vector space X and its subspace Y as shown in Figure 10.1. We will define the best approximation of a solution $\boldsymbol{u} \in X$ to be the closest to $\boldsymbol{u}$ element of Y. So the best approximation will always depend on the choice of the subspace Y.

For practical reasons think of X as a very complicated space with many elements (dimensions), while Y is a low-dimensional space. One such example is the infinite dimensional space $X = C(a,b)$ of continuous functions and the two-dimensional space of linear polynomials $Y = \mathbb{P}_1$. Our object $\boldsymbol{u}$ lives in X and we search for its best approximation $\boldsymbol{v} \in Y$. Obviously, we don't want to find the best approximation in a totally irrelevant space, and

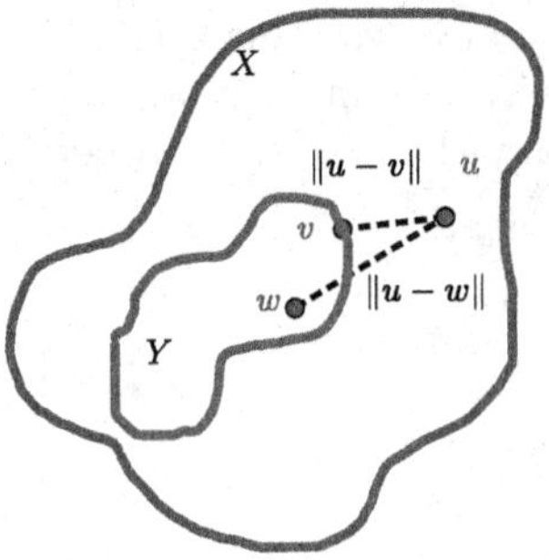

FIGURE 10.1
The space X and its subspace Y.

that is why we choose Y to be a subspace of X. On the other hand, we could have chosen a more complicated subspace but we need to be very careful so as everything make sense. Figure 10.1 shows a sketch of a space X and its subspace Y. We repeat that the best approximation of $\boldsymbol{u}$ will be the element $\boldsymbol{v} \in Y$, which is the closest to $\boldsymbol{u}$.

Before we proceed with a definition of the best approximation, we revise the notion of orthogonality in analogy to what we know in geometry. Two objects $\boldsymbol{u}$, $\boldsymbol{v}$ of a vector space X are orthogonal if and only if $\langle \boldsymbol{u}, \boldsymbol{v}\rangle = 0$.

$$\boldsymbol{u} \perp \boldsymbol{v} \iff \langle \boldsymbol{u}, \boldsymbol{v}\rangle = 0$$

The Pythagoras theorem states that for any two orthogonal elements $\boldsymbol{u}$ and $\boldsymbol{v}$ we have

$$\|\boldsymbol{u} + \boldsymbol{v}\|^2 = \|\boldsymbol{u}\|^2 + \|\boldsymbol{v}\|^2 . \tag{10.2}$$

This is true because

$$\begin{aligned} \|\boldsymbol{u} + \boldsymbol{v}\|^2 &= \langle \boldsymbol{u} + \boldsymbol{v}, \boldsymbol{u} + \boldsymbol{v}\rangle \\ &= \langle \boldsymbol{u}, \boldsymbol{u}\rangle + 2\langle \boldsymbol{u}, \boldsymbol{v}\rangle + \langle \boldsymbol{v}, \boldsymbol{v}\rangle \\ &= \|\boldsymbol{u}\|^2 + 2\langle \boldsymbol{u}, \boldsymbol{v}\rangle + \|\boldsymbol{v}\|^2 , \end{aligned}$$

and since $\langle \boldsymbol{u}, \boldsymbol{v}\rangle = 0$, then we have the result. Note that the relationship $\|\boldsymbol{u} + \boldsymbol{v}\|^2 = \|\boldsymbol{u}\|^2 + 2\langle \boldsymbol{u}, \boldsymbol{v}\rangle + \|\boldsymbol{v}\|^2$ generalizes the identity $(a+b)^2 = a^2 + 2ab + b^2$ in the case of vectors.

10.1.3 Best approximations

The best approximation of an object is very similar to an orthogonal projection. The formal definition of the best approximation is the following:

Definition 10.1. *Let $\boldsymbol{u} \in X$ and $Y \subset X$. The best approximation of $\boldsymbol{u}$ in Y is the element $\boldsymbol{v} \in Y$ such that*

$$\|\boldsymbol{u} - \boldsymbol{v}\| \leq \|\boldsymbol{u} - \boldsymbol{w}\| ,$$

for all $\boldsymbol{w} \in Y$.

With this definition of the best approximation we can explore the geometry of an inner product space. We know from geometry that the shortest distance between a point and a line in a space is the vertical distance. In Figure 10.2, the space $X = \mathbb{R}^2$ and Y is a straight line. The best approximation of $\boldsymbol{u} \in X$ in $Y \subset X$ is the element $\boldsymbol{v} \in Y$ such as $\|\boldsymbol{u} - \boldsymbol{v}\| \leq \|\boldsymbol{u} - \boldsymbol{w}\|$ for all $\boldsymbol{w} \in Y$.

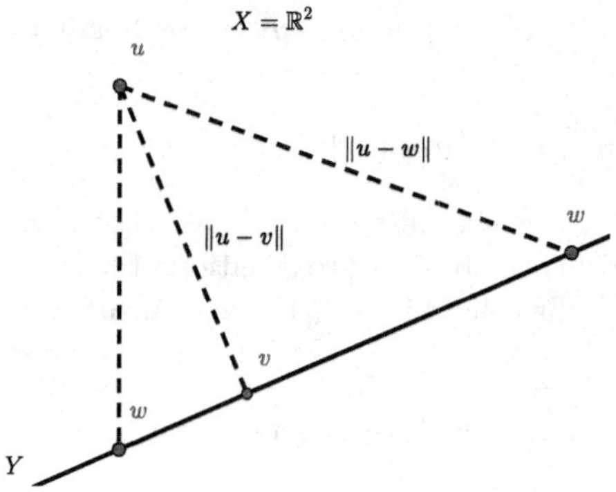

FIGURE 10.2
Practical representation of best approximation and the orthogonal projection.

The element $\boldsymbol{v}$ is practically the orthogonal projection of $\boldsymbol{u}$ onto Y. In Figure 10.2, the vector $\boldsymbol{u} - \boldsymbol{v}$ connecting $\boldsymbol{u}$ and $\boldsymbol{v}$ is orthogonal to the line connecting $\mathbf{0}$ and $\boldsymbol{w}$ (all vector spaces contain $\mathbf{0}$). Since orthogonal vectors have zero inner product, we have that

$$\langle \boldsymbol{u} - \boldsymbol{v}, \boldsymbol{w} \rangle = 0 \ .$$

We therefore conclude that for $\boldsymbol{u}$ in an inner product space X, its best approximation $\boldsymbol{v}$ in the subspace Y is the orthogonal projection defined as

$$\langle \boldsymbol{u}, \boldsymbol{w} \rangle = \langle \boldsymbol{v}, \boldsymbol{w} \rangle \quad \text{for all} \quad \boldsymbol{w} \in Y \ .$$

The fact that $\boldsymbol{v}$ is the closest to $\boldsymbol{u}$ element of Y can be expressed by the inequality

$$\|\boldsymbol{u} - \boldsymbol{v}\| \leq \|\boldsymbol{u} - \boldsymbol{w}\| \quad \text{for all} \quad \boldsymbol{w} \in Y \ .$$

This is described formally with the following theorem:

Theorem 10.2. *Let Y be a subspace of an inner product space X and $\boldsymbol{u} \in X$. A vector $\boldsymbol{v} \in Y$ is the best approximation of $\boldsymbol{u}$ if and only if*

$$\langle \boldsymbol{v}, \boldsymbol{w} \rangle = \langle \boldsymbol{u}, \boldsymbol{w} \rangle, \quad \textit{for all } \boldsymbol{w} \in Y \ . \tag{10.3}$$

Proof. If (10.3) is true, then it suffices to prove that

$$\|\boldsymbol{u} - \boldsymbol{v}\| \leq \|\boldsymbol{u} - \boldsymbol{w}\|, \quad \text{for all } \boldsymbol{w} \in Y \ .$$

From (10.3) we have

$$\langle \boldsymbol{u} - \boldsymbol{v}, \boldsymbol{w} \rangle = 0, \quad \text{for all } \boldsymbol{w} \in Y \ ,$$

and since $\boldsymbol{v} \in Y$ we can take $\boldsymbol{w} = \boldsymbol{v}$ and thus

$$\langle \boldsymbol{u} - \boldsymbol{v}, \boldsymbol{v} \rangle = 0 \ .$$

Subtracting the last two relations we obtain

$$\langle \boldsymbol{u} - \boldsymbol{v}, \boldsymbol{v} - \boldsymbol{w} \rangle = 0 \ .$$

This means that the vectors $\boldsymbol{u}-\boldsymbol{v}$ and $\boldsymbol{v}-\boldsymbol{w}$ are orthogonal, and thus from Pythagoras' theorem we have

$$\|\boldsymbol{u}-\boldsymbol{w}\|^2 = \|(\boldsymbol{u}-\boldsymbol{v})+(\boldsymbol{v}-\boldsymbol{w})\|^2 = \|\boldsymbol{u}-\boldsymbol{v}\|^2 + \|\boldsymbol{v}-\boldsymbol{w}\|^2 \geq \|\boldsymbol{u}-\boldsymbol{v}\|^2 ,$$

since $\|\boldsymbol{v}-\boldsymbol{w}\|^2 \geq 0$. Thus, $\boldsymbol{v}$ is the best approximation of $\boldsymbol{u}$ in Y.

On the other hand, if $\boldsymbol{v}$ is the best approximation of $\boldsymbol{u}$ in Y, we can prove (10.3) using a contradiction. We assume that there is $\boldsymbol{w} \in Y$, $\boldsymbol{w} \neq \boldsymbol{0}$, such that $\langle \boldsymbol{u}-\boldsymbol{v}, \boldsymbol{w}\rangle \neq 0$. We define the function $\phi : \mathbb{R} \to \mathbb{R}$ with

$$\phi(x) = \|\boldsymbol{u}-(\boldsymbol{v}+x\boldsymbol{w})\|^2 .$$

For this function $\phi(x)$ with $x \in \mathbb{R}$ we have that

$$\phi(x) = \langle \boldsymbol{u}-\boldsymbol{v}-x\boldsymbol{w}, \boldsymbol{u}-\boldsymbol{v}-x\boldsymbol{w}\rangle = \|\boldsymbol{u}-\boldsymbol{v}\|^2 - 2\langle \boldsymbol{u}-\boldsymbol{v}, \boldsymbol{w}\rangle x + x^2\|\boldsymbol{w}\|^2 ,$$

which is a quadratic polynomial in x. Its minimum can be achieved at the point $x = \langle \boldsymbol{u}-\boldsymbol{v}, \boldsymbol{w}\rangle/\|\boldsymbol{w}\|$, and thus

$$\min_{x\in\mathbb{R}} \phi(x) = \phi\left(\frac{\langle \boldsymbol{u}-\boldsymbol{v}, \boldsymbol{w}\rangle}{\|\boldsymbol{w}\|^2}\right) < \phi(0) = \|\boldsymbol{u}-\boldsymbol{v}\|^2 ,$$

which is a contradiction since $\boldsymbol{v}$ is the best approximation. Thus, $\langle \boldsymbol{u}-\boldsymbol{v}, \boldsymbol{w}\rangle = 0$. □

All the above apply to vectors, matrices and functions by using appropriate inner products and norms. The inner product is not always necessary to define the best approximation, as the norm suffices by itself. The importance of the inner product is the geometry induced by it, so we can visualize the orthogonality of the projections.

10.2 Linear Least Squares

In Chapter 6 we studied the problem of least squares where in order to approximate a set of data (x_i, y_i), $i = 0, \ldots, N$ by a function $f(x)$ that depends on the parameters a_i, $i = 0, \ldots, m$, we used the parameters a_i to minimize the error

$$S(a_0, \ldots, a_m) = \sum_{i=0}^{N} |y_i - f(x_i)|^2 .$$

When we chose the function $f(x)$ to be linear, and in particular of the form $f(x) = a_0x + a_1$, the problem was reduced in finding the coefficients a_0 and a_1 such that the error

$$S(a_0, a_1) = \sum_{i=0}^{N} |y_i - a_0x_i - a_1|^2 ,$$

becomes minimum.

Considering the notation for the matrix

$$\boldsymbol{A} = \begin{pmatrix} x_0 & 1 \\ x_1 & 1 \\ \vdots & \\ x_N & 1 \end{pmatrix} ,$$

and the vectors

$$\boldsymbol{x} = \begin{pmatrix} a_0 \\ a_1 \end{pmatrix} \quad \text{and} \quad \boldsymbol{b} = \begin{pmatrix} y_0 \\ y_1 \\ \vdots \\ y_N \end{pmatrix} ,$$

we can write the linear least squares problem in the equivalent form

$$S(a_0, a_1) = \sum_{i=0}^{N} |b_i - a_0 x_i - a_1|^2 = \|\boldsymbol{b} - \boldsymbol{A}\boldsymbol{x}\|_2^2 .$$

This problem is a special case of the more general linear least squares problem: If $\boldsymbol{A}$ is an $m \times n$ matrix and $\boldsymbol{b}$ an m-dimensional vector, then the linear least squares solution $\boldsymbol{x}^*$ is the n-dimensional vector that minimizes the error function $\|\boldsymbol{b} - \boldsymbol{A}\boldsymbol{x}\|_2^2$. In this sense, the linear least squares approximation is a best approximation. Note that in linear least squares problem $m > n$.

The least squares problem can also be understood as a special case of the general optimization problem of finding $\boldsymbol{x}^*$ that minimizes a function $\boldsymbol{F}(\boldsymbol{x})$. We usually call the solution $\boldsymbol{x}^*$ *optimal solution* or *minimizer* and we write $\boldsymbol{x}^* = \text{argmin}\boldsymbol{F}(\boldsymbol{x})$. The optimal solution of the linear least squares problem is the $\boldsymbol{x}^* = \text{argmin}\boldsymbol{F}(\boldsymbol{x})$ where $\boldsymbol{F}(\boldsymbol{x}) = \|\boldsymbol{b} - \boldsymbol{A}\boldsymbol{x}\|_2^2$.

10.2.1 Theoretical properties of linear least squares

The linear least squares problem always has a solution, which is also unique if the kernel of the $m \times n$ matrix $\boldsymbol{A}$ with $m \geq n$ consists of the n-dimensional zero vector only, that is $\ker(\boldsymbol{A}) = \{\boldsymbol{0}\}$. Recall that the kernel of an $m \times n$ matrix $\boldsymbol{A}$ is the set of solutions of the equation $\boldsymbol{A}\boldsymbol{x} = \boldsymbol{0}$

$$\ker(\boldsymbol{A}) = \{\boldsymbol{x} \in \mathbb{R}^n : \boldsymbol{A}\boldsymbol{x} = \boldsymbol{0}\} .$$

Proposition 10.3. *Let $\boldsymbol{A}$ be any $m \times n$ matrix. There is always an optimal solution to the linear least squares problem $\boldsymbol{x}^* = \text{argmin}\|\boldsymbol{b} - \boldsymbol{A}\boldsymbol{x}\|_2^2$. The optimal solution is also unique if and only if the kernel of $\boldsymbol{A}$ is $\ker(\boldsymbol{A}) = \{\boldsymbol{0}\}$.*

Proof. We denote by $\mathcal{R}(\boldsymbol{A}) = \text{range}(\boldsymbol{A}) = \{\boldsymbol{A}\boldsymbol{x} \in \mathbb{R}^m : \ \boldsymbol{x} \in \mathbb{R}^n\}$ the range of $\boldsymbol{A}$. It is easy to check that if $\boldsymbol{y} \in \mathcal{R}(\boldsymbol{A})$, then $\boldsymbol{y}$ is an m-dimensional vector. Let also

$$\mathcal{R}^{\perp}(\boldsymbol{A}) = \{\boldsymbol{z} \in \mathbb{R}^m : \ \langle \boldsymbol{z}, \boldsymbol{x} \rangle = 0 \ \text{ for all } \boldsymbol{x} \in \mathcal{R}(\boldsymbol{A})\} .$$

The space $\mathcal{R}^{\perp}(\boldsymbol{A})$ is called the orthogonal complement of $\mathcal{R}(\boldsymbol{A})$, and consists of m-dimensional vectors that are perpendicular to every element of $\mathcal{R}(\boldsymbol{A})$. We know from linear algebra that every vector $\boldsymbol{b} \in \mathbb{R}^m$ can be analyzed into two components $\boldsymbol{b}_1 \in \mathcal{R}(\boldsymbol{A})$ and $\boldsymbol{b}_2 \in R^{\perp}(\boldsymbol{A})$ such that $\boldsymbol{b} = \boldsymbol{b}_1 + \boldsymbol{b}_2$. Therefore, for any $\boldsymbol{y} \in \mathbb{R}^m$, we have

$$\boldsymbol{b} - \boldsymbol{A}\boldsymbol{y} = (\boldsymbol{b}_1 - \boldsymbol{A}\boldsymbol{y}) + \boldsymbol{b}_2 .$$

Since $\boldsymbol{b}_1 \in \mathcal{R}(\boldsymbol{A})$ and also $\boldsymbol{A}\boldsymbol{y} \in \mathcal{R}(\boldsymbol{A})$, then $\boldsymbol{b}_1 - \boldsymbol{A}\boldsymbol{y} \in \mathcal{R}(\boldsymbol{A})$, and thus $\boldsymbol{b}_1 - \boldsymbol{A}\boldsymbol{y} \perp \boldsymbol{b}_2$, or else $\langle \boldsymbol{b}_1 - \boldsymbol{A}\boldsymbol{y}, \boldsymbol{b}_2 \rangle = 0$.

We now consider the function $\boldsymbol{F}(\boldsymbol{x}) = \|\boldsymbol{b} - \boldsymbol{A}\boldsymbol{x}\|_2^2$. Using the previous analysis of the vector $\boldsymbol{b} = \boldsymbol{b}_1 + \boldsymbol{b}_2$ and the Pythagoras' Theorem we have

$$\boldsymbol{F}(\boldsymbol{x}) = \|\boldsymbol{b} - \boldsymbol{A}\boldsymbol{x}\|_2^2 = \|\boldsymbol{b}_1 - \boldsymbol{A}\boldsymbol{x}\|_2^2 + \|\boldsymbol{b}_2\|_2^2 .$$

Thus, the optimal solution of the problem $\text{argmin}\|\boldsymbol{b}_1 - \boldsymbol{A}\boldsymbol{x}\|_2^2$ will be the optimal solution of the problem $\text{argmin}\boldsymbol{F}(\boldsymbol{x})$ as well. Since $\boldsymbol{b}_1 \in \mathcal{R}(\boldsymbol{A})$, we know that there exists an $\boldsymbol{x}^*$ such that $\boldsymbol{b}_1 = \boldsymbol{A}\boldsymbol{x}^*$. This implies that $\|\boldsymbol{b}_1 - \boldsymbol{A}\boldsymbol{x}^*\|_2^2 = 0$. On the other hand $\|\boldsymbol{b}_1 - \boldsymbol{A}\boldsymbol{x}\|_2^2 \geq 0 = \|\boldsymbol{b}_1 - \boldsymbol{A}\boldsymbol{x}^*\|_2^2$ for all $\boldsymbol{x} \in \mathbb{R}^n$. Therefore, $\boldsymbol{x}^* = \text{argmin}\|\boldsymbol{b}_1 - \boldsymbol{A}\boldsymbol{x}\|_2^2$. This proves the existence of the optimal solution to the linear least squares problem.

In order to prove uniqueness, we need to observe that the linear system $\boldsymbol{A}\boldsymbol{x} = \boldsymbol{b}_1$ has a unique solution if and only if the homogenous system $\boldsymbol{A}\boldsymbol{x} = \boldsymbol{0}$ has the unique solution $\boldsymbol{x} = \boldsymbol{0}$, which is equivalent to the fact that $\ker(\boldsymbol{A}) = \{\boldsymbol{0}\}$. □

10.2.2 Solution of linear least squares problem

In this section we derive the solution of the linear least squares problem in closed form. Let $\boldsymbol{x}^* \in \mathbb{R}^m$ be the optimal solution of the linear least squares problem

$$\boldsymbol{x}^* = \text{argmin}\|\boldsymbol{b} - \boldsymbol{A}\boldsymbol{x}\|_2^2 .$$

If $\boldsymbol{r} = \boldsymbol{b} - \boldsymbol{A}\boldsymbol{x}$ is the residual, then from the previous section we have that

$$\boldsymbol{r} = \boldsymbol{b}_1 - \boldsymbol{A}\boldsymbol{x}^* + \boldsymbol{b}_2 = \boldsymbol{b}_2 ,$$

where $\boldsymbol{b}_1 + \boldsymbol{b}_2 = \boldsymbol{b}$, $\boldsymbol{b}_1 \in \mathcal{R}(\boldsymbol{A})$ and $\boldsymbol{b}_2 \in R^{\perp}(\boldsymbol{A})$. Thus, we have that $\boldsymbol{r} = \boldsymbol{b}_2$, and since $\boldsymbol{b}_2 \in \mathcal{R}^{\perp}(\boldsymbol{A})$, we have that

$$\boldsymbol{A}^T\boldsymbol{r} = \boldsymbol{A}^T\boldsymbol{b} = \boldsymbol{0} .$$

The last relationship is very interesting since it guarantees that the residual $\boldsymbol{r}$ is also orthogonal to the columns of $\boldsymbol{A}$. Anyway, we have

$$\boldsymbol{0} = \boldsymbol{A}^T\boldsymbol{r} = \boldsymbol{A}^T(\boldsymbol{b} - \boldsymbol{A}\boldsymbol{x}^*) = \boldsymbol{A}^T\boldsymbol{b} - \boldsymbol{A}^T\boldsymbol{A}\boldsymbol{x}^* .$$

Thus, we have

$$\boldsymbol{A}^T\boldsymbol{A}\boldsymbol{x}^* = \boldsymbol{A}^T\boldsymbol{b} ,$$

even if $\boldsymbol{b} - \boldsymbol{A}\boldsymbol{x}^* \neq 0$.

☞ The solution of the linear least squares problem $\boldsymbol{x}^* = \text{argmin}\|\boldsymbol{b} - \boldsymbol{A}\boldsymbol{x}\|_2^2$ is the solution of the linear system $\boldsymbol{A}^T\boldsymbol{A}\boldsymbol{x} = \boldsymbol{A}^T\boldsymbol{b}$.

The matrix $\boldsymbol{A}^T\boldsymbol{A}$ is an $n \times n$ symmetric matrix since $(\boldsymbol{A}^T\boldsymbol{A})^T = \boldsymbol{A}^T(\boldsymbol{A}^T)^T = \boldsymbol{A}^T\boldsymbol{A}$. If we also assume that $\ker(\boldsymbol{A}) = \{\boldsymbol{0}\}$, then the matrix $\boldsymbol{A}^T\boldsymbol{A}$ is positive definite and thus invertible. Therefore, the optimal solution of the linear least squares problem is given by the formula

$$\boldsymbol{x}^* = (\boldsymbol{A}^T\boldsymbol{A})^{-1}\boldsymbol{A}^T\boldsymbol{b} . \tag{10.4}$$

The matrix

$$\boldsymbol{A}^+ = (\boldsymbol{A}^T\boldsymbol{A})^{-1}\boldsymbol{A}^T ,$$

is called the *pseudoinverse* or *Moore-Penrose* inverse and is of significant importance in statistics, data science and other modern fields of science and engineering. The matrix $\boldsymbol{A}^+$ is also called *generalized inverse* since in the special case of invertible $n \times n$ matrix $\boldsymbol{A}$ it coincides with the classical inverse matrix $\boldsymbol{A}^{-1}$. This can be verified easily by taking

$$\boldsymbol{A}^+ = (\boldsymbol{A}^T\boldsymbol{A})^{-1}\boldsymbol{A}^T = \boldsymbol{A}^{-1}(\boldsymbol{A}^T)^{-1}\boldsymbol{A}^T = \boldsymbol{A}^{-1} .$$

The pseudoinverse matrix $\boldsymbol{A}^+$ can help to define a condition number for the linear least squares problem. The condition number of any $m \times n$ matrix $\boldsymbol{A}$ is defined as $\text{cond}(\boldsymbol{A}) = \|\boldsymbol{A}\|\|\boldsymbol{A}^+\|$ generalizing the condition number of invertible $n \times n$ matrices.

The module `numpy.linalg` provides the function `pinv` for efficient computation of the pseudoinverse $\boldsymbol{A}^+$. The input of the function `pinv` is a matrix $\boldsymbol{A}$ and optionally a tolerant (`rcond` with default value `1.e-15`). This tolerance is related with the singular values of $\boldsymbol{A}$ (the square roots of eigenvalues of $\boldsymbol{A}^T\boldsymbol{A}$). Optionally also accepts a boolean variable `hermitian` (with default value `False`) specifying if the matrix $\boldsymbol{A}$ is Hermitian or not.

☞ The pseudoinverse matrix $\boldsymbol{A}^+$ of any $m \times n$ matrix $\boldsymbol{A}$ generalizes the classical inverse of invertible matrices. If $\boldsymbol{A}^T\boldsymbol{A}$ is invertible, then the pseudoinverse is given explicitly as $\boldsymbol{A}^+ = (\boldsymbol{A}^T\boldsymbol{A})^{-1}\boldsymbol{A}^T$ and provides the solution of the linear least squares problem $\boldsymbol{x}^* = \text{argmin}\|\boldsymbol{b} - \boldsymbol{A}\boldsymbol{x}\|_2^2 = \boldsymbol{A}^+\boldsymbol{b}$.

In terms of best approximations, when we solve the linear least squares problem, we compute a solution (the optimal solution) such that the residual $\boldsymbol{r}$ is orthogonal to the range $\mathcal{R}(\boldsymbol{A})$. This means that the optimal solution $\boldsymbol{x}^*$ minimizes the distance of $\boldsymbol{A}\boldsymbol{x}^*$ from $\boldsymbol{b}$ in the space $\mathcal{R}(\boldsymbol{A})$. To see this consider an $\boldsymbol{x} \in \mathcal{R}(\boldsymbol{A})$, and take the inner product

$$\langle \boldsymbol{A}\boldsymbol{x}, \boldsymbol{b} - \boldsymbol{A}\boldsymbol{x}^* \rangle = (\boldsymbol{A}\boldsymbol{x})^T \cdot (\boldsymbol{b} - \boldsymbol{A}\boldsymbol{x}^*) = 0 \ .$$

The solution of linear least squares problem is a best approximation.

10.2.3 Linear least squares problem in Python

As we have already noticed, if $\ker(\boldsymbol{A}) = \{\boldsymbol{0}\}$ then the optimal solution $\boldsymbol{x}^*$ of the linear least squares problem satisfies the linear system

$$\boldsymbol{A}^T\boldsymbol{A}\boldsymbol{x} = \boldsymbol{A}^T\boldsymbol{b} \ . \tag{10.5}$$

This set of equations is called the *normal equations.* Since, in this case, the matrix $\boldsymbol{A}^T\boldsymbol{A}$ is symmetric and positive definite, one can solve the linear system (10.5) using the Cholesky factorization as we saw in Section 9.5. Although, this procedure can be efficient, the Cholesky factorization of $\boldsymbol{A}^T\boldsymbol{A}$ can be inaccurate. The reason for this inaccuracy is that if $\boldsymbol{A}$ is not well-conditioned, in the sense that its condition number $\text{cond}(\boldsymbol{A})$ is large, then the condition number $\text{cond}(\boldsymbol{A}^T\boldsymbol{A}) = (\text{cond}(\boldsymbol{A}))^2$ can be even larger. Therefore, we conclude that solving the linear least squares problem by solving the system of the normal equations is not a good idea. We will see later that in order to solve efficiently linear least squares problems we can use other factorization methods, such as the singular value decomposition.

In Python, one can use the function `lstsq` of the module `numpy.linalg` to solve linear least squares problems. The general syntax of this function is

```
lstsq(a,b,rcond)
```

where

- `a`: Is an $m \times n$ matrix $\boldsymbol{A}$
- `b`: Is the right hand side $\boldsymbol{b}$
- `rcond`: Is a float that serves as a tolerance in the numerical computations of the solution. When it is not set, then warnings are printed on the screen. (Optional with default value `rcond='warn'`)

The function lstsq returns a tuple (x, residuals, rank, s) where

x: Is the optimal solution $\boldsymbol{x}^*$

residuals: Is the quantity $\|\boldsymbol{b}-\boldsymbol{A}\boldsymbol{x}^*\|^2$ for each column of b

rank: Is the rank of $\boldsymbol{A}$

s: Is a vector with the singular values of $\boldsymbol{A}$. These are the eigenvalues of the matrix $\boldsymbol{A}^T\boldsymbol{A}$.

For the moment we will try to find the argmin$\|\boldsymbol{b}-\boldsymbol{A}\boldsymbol{x}\|^2$ with

$$\boldsymbol{A}=\begin{pmatrix}1 & 1\\ -1 & 1\\ 1 & 1\end{pmatrix}, \text{ and } \boldsymbol{b}=\begin{pmatrix}2\\1\\3\end{pmatrix} .$$

This can be done easily with the following code.

```
import numpy.linalg as npl
A = np.array([[1.0 , 1.0],
              [-1.0, 1.0],
              [1.0, 1.0]])
b = np.array([2.0],
             [1.0],
             [3.0]])
(x, residual, rank,s) = npl.lstsq(A,b)
print('x= '); print(x)
print('residual= '); print(residual)
print('rank= '); print(rank)
print('singular values= '); print(s)
```

```
x=
[[0.75]
 [1.75]]
residual=
[0.5]
rank=
2
singular values=
[2.         1.41421356]
```

Similarly, there is a function in the module scipy.linalg with the same name lstsq and some extra capabilities.

Solving the system of the normal equations $\boldsymbol{A}^T\boldsymbol{A}\boldsymbol{x}=\boldsymbol{A}^T\boldsymbol{b}$ can be done by computing first the pseudoinverse $\boldsymbol{A}^+=(\boldsymbol{A}^T\boldsymbol{A})^{-1}\boldsymbol{A}^T$ and then the product $\boldsymbol{x}=\boldsymbol{A}^+\boldsymbol{b}$. In the previous case we can compute the optimal solution $\boldsymbol{x}=(3/4,7/4)^T$ analytically. We verify this result with Python once more. We first compute the matrix $(\boldsymbol{A}^T\boldsymbol{A})^{-1}$ and we store it in the variable tmp. We compute the pseudoinverse $(\boldsymbol{A}^T\boldsymbol{A})^{-1}\boldsymbol{A}^T$ and we store it in the array Amp. Finally, we compute the solution $\boldsymbol{x}=\boldsymbol{A}^+\boldsymbol{b}$ and we verify the result.

```
1 # Compute the Moore-Penrose inverse
2 tmp = npl.inv(np.dot(np.transpose(A),A))
3 Amp= np.dot(tmp,np.transpose(A))
4 x = np.dot(Amp,b)
5 print('x= '); print(x)
6 print('x= '); print(npl.pinv(A)@b)
```

```
x=
[[0.75]
 [1.75]]
x=
[[0.75]
 [1.75]]
```

In the last line of the previous code (line 6) we verify the result using the function pinv of numpy.linalg, while we left for the reader to test the scipy.linalg version of the function pinv.

The last implementation is not useful for large matrices since the computation of the inverse of a full and large matrix is not an efficient operation. The next sections are devoted to tools and methods that are required for the development of the singular value decomposition. They are also very important developments by themselves because of their numerous applications. We present first the Gram-Schmidt orthonormalization.

10.3 Gram-Schmidt Orthonormalization

Let X be a (finite) n-dimensional vector space with basis $\{\boldsymbol{e}_1, \boldsymbol{e}_2, \ldots, \boldsymbol{e}_n\}$. For example, think of the space $\mathbb{R}^n$ of n-dimensional real vectors[2]. Any vector $\boldsymbol{x} \in X$ can be expressed as a linear combination of these basis functions

$$\boldsymbol{x} = c_1\boldsymbol{e}_1 + c_2\boldsymbol{e}_2 + \cdots + c_n\boldsymbol{e}_n \ . \tag{10.6}$$

If we take the inner product of (10.6) with the basis vector $\boldsymbol{e}_i$ for $i = 1, 2, \ldots, n$ we form the following equations

$$\begin{aligned}
\langle \boldsymbol{e}_1, \boldsymbol{e}_1\rangle c_1 + \langle \boldsymbol{e}_2, \boldsymbol{e}_1\rangle c_2 + \cdots + \langle \boldsymbol{e}_n, \boldsymbol{e}_1\rangle c_n &= \langle \boldsymbol{x}, \boldsymbol{e}_1\rangle \ , \\
\langle \boldsymbol{e}_1, \boldsymbol{e}_2\rangle c_1 + \langle \boldsymbol{e}_2, \boldsymbol{e}_2\rangle c_2 + \cdots + \langle \boldsymbol{e}_n, \boldsymbol{e}_2\rangle c_n &= \langle \boldsymbol{x}, \boldsymbol{e}_2\rangle \ , \\
&\ \vdots \\
\langle \boldsymbol{e}_1, \boldsymbol{e}_n\rangle c_1 + \langle \boldsymbol{e}_2, \boldsymbol{e}_n\rangle c_2 + \cdots + \langle \boldsymbol{e}_n, \boldsymbol{e}_n\rangle c_n &= \langle \boldsymbol{x}, \boldsymbol{e}_n\rangle \ .
\end{aligned} \tag{10.7}$$

These are the normal equations we discussed before. Assume in addition that the basis vectors are orthogonal but not necessarily normal. This means that $\langle \boldsymbol{e}_i, \boldsymbol{e}_j\rangle = 0$ if $i \neq j$, while $\|\boldsymbol{e}_i\|^2 = \langle \boldsymbol{e}_i, \boldsymbol{e}_i\rangle$ is not necessarily equal to 1. Since $\langle \boldsymbol{e}_i, \boldsymbol{e}_j\rangle = 0$ for $i \neq j$, then the set

[2] The dimension of the space X need not be n.

of the normal equations is reduced to

$$
\begin{aligned}
\|e_1\|^2 c_1 &= \langle \boldsymbol{x}, e_1\rangle \ , \\
\|e_2\|^2 c_2 &= \langle \boldsymbol{x}, e_2\rangle \ , \\
&\ \vdots \\
\|e_n\|^2 c_n &= \langle \boldsymbol{x}, e_n\rangle \ .
\end{aligned}
\tag{10.8}
$$

Solving for the coefficients c_i yields

$$c_i = \frac{\langle \boldsymbol{x}, e_i\rangle}{\|e_i\|^2}, \quad i = 1, 2, \ldots, n \ .$$

Thus, the vector $\boldsymbol{x}$ can be expressed as

$$\boldsymbol{x} = \frac{\langle \boldsymbol{x}, e_1\rangle}{\|e_1\|^2} e_1 + \frac{\langle \boldsymbol{x}, e_2\rangle}{\|e_2\|^2} e_2 + \cdots + \frac{\langle \boldsymbol{x}, e_n\rangle}{\|e_n\|^2} e_n \ . \tag{10.9}$$

This is an analysis of the vector $\boldsymbol{x}$ into coordinates along the directions of the basis elements $\boldsymbol{e}_i$. The component $\frac{\langle \boldsymbol{x}, e_i\rangle}{\|e_i\|^2} \boldsymbol{e}_i$ is the projection of the vector $\boldsymbol{x}$ onto $\boldsymbol{e}_i$ and is written usually as

$$\mathrm{proj}_{e_i} \boldsymbol{x} = \frac{\langle \boldsymbol{x}, e_i\rangle}{\|e_i\|^2} e_i \ .$$

From this procedure, it is clear that in order to decompose a vector $\boldsymbol{x}$ into its coordinates we need to know the basis vectors $\boldsymbol{e}_i$, which might not be trivial in a general case. Orthonormal vectors are orthogonal and at the same time their norm equals 1.

The Gram-Schmidt orthonormalization is a method to produce a set of orthonormal vectors from an arbitrary set of linearly independent vectors. The resulting vectors can serve as basis of the vector space, but as we shall see in the next section, the Gram-Schmidt orthonormalization leads to a matrix factorization different from the *LU* decomposition. The basic Gram-Schmidt orthonormalization algorithm is based on the construction of projections onto vertical directions to generate orthogonal vectors.

Let $\boldsymbol{x}_1, \boldsymbol{x}_2, \ldots, \boldsymbol{x}_n$ be linearly independent vectors. Then we define the vectors $e'_1, e'_2, \ldots, e'_n$ such that

$$
\begin{aligned}
e'_1 &= \boldsymbol{x}_1 \ , \\
e'_2 &= \boldsymbol{x}_2 - \frac{\langle \boldsymbol{x}_2, e'_1\rangle}{\langle e'_1, e'_1\rangle} e'_1 \ , \\
e'_3 &= \boldsymbol{x}_3 - \frac{\langle \boldsymbol{x}_3, e'_1\rangle}{\langle e'_1, e'_1\rangle} e'_1 - \frac{\langle \boldsymbol{x}_3, e'_2\rangle}{\langle e'_2, e'_2\rangle} e'_2 \ , \\
&\ \vdots \\
e'_n &= \boldsymbol{x}_n - \frac{\langle \boldsymbol{x}_n, e'_1\rangle}{\langle e'_1, e'_1\rangle} e'_1 - \cdots - \frac{\langle \boldsymbol{x}_n, e'_{n-1}\rangle}{\langle e'_{n-1}, e'_{n-1}\rangle} e_{n-1} \ ,
\end{aligned}
\tag{10.10}
$$

which are the same with

$$
\begin{aligned}
e'_1 &= \boldsymbol{x}_1 \ , \\
e'_2 &= \boldsymbol{x}_2 - \mathrm{proj}_{e'_1} \boldsymbol{x}_2 \ , \\
e'_3 &= \boldsymbol{x}_3 - \mathrm{proj}_{e'_1} \boldsymbol{x}_3 - \mathrm{proj}_{e'_2} \boldsymbol{x}_3 \ ,
\end{aligned}
$$

$$\vdots$$

$$e'_n = x_n - \sum_{j=1}^{n-1} \text{proj}_{e'_j} x_n \ .$$

It is easy to check that $\langle e'_1, e'_2 \rangle = 0$, and also, generally, that $\langle e'_i, e'_j \rangle = 0$ for $j = 1, 2, \dots, i-1$, which implies that the vectors $e'_1, e'_2, \dots, e'_n$ are orthogonal. In order to normalize them, we divide by their norms. We define the orthonormal vectors

$$e_i = \frac{e'_i}{\|e'_i\|}, \quad i = 1, 2, \dots, n \ .$$

The Gram-Schmidt orthonormalization is summarized in Algorithm 40. In this algorithm

Algorithm 40 Naive Gram-Schmidt orthonormalization procedure

```
for i = 1 : n do
    E'(:, i) = X(:, i)
    for j = 1 : i − 1 do
        E'(:, i) = E'(:, i) − (E(:, j)^T X(:, i))E(:, j)
    end for
    E(:, i) = X(:, i)/||X(: .i)||_2
end for
```

we store the linearly independent vectors $x_i \in \mathbb{R}^m$, $i = 1, \dots, n$ into the columns of an $m \times n$ matrix X. We compute the vectors e'_j as columns of matrix E' and the orthonormal vectors e_i are stored into the columns of matrix E.

10.3.1 Numerical implementation

Direct implementation of the classical Gram-Schmidt process is not recommended because of rounding errors that can result in non-orthogonal vectors. As a more stable alternative the following modified Gram-Schmidt process is recommended. In this process the computation of the vector e_i happens in $i-1$ steps, for $i = 1, \dots, n$:

$$\begin{aligned}
e_i^{(1)} &= x_i - \text{proj}_{e_1} x_i \ , \\
e_i^{(2)} &= e_i^{(1)} - \text{proj}_{e_2} e_i^{(1)} \ , \\
&\vdots \\
e_i^{(i-2)} &= e_i^{(i-3)} - \text{proj}_{e_{i-2}} e_i^{(i-3)} \ , \\
e_i^{(i-1)} &= e_i^{(i-2)} - \text{proj}_{e_{i-1}} e_i^{(i-2)} \ , \\
e_i &= e_i^{(i-1)} / \|e_i^{(i-1)}\| \ ,
\end{aligned}$$

Note that the normalization follows immediately after the computation of the orthogonal vector. This method contains smaller rounding errors and leads to a more stable algorithm. But again this might not be an ideal algorithm for the generation of orthonormal vectors. This modified Gram-Schmidt method is summarized in Algorithm 41, where the vectors x_i are stored in the columns of an $m \times n$ matrix X and the orthonormal set of vectors is returned in the columns of an $m \times n$ matrix E .

The implementation of the modified Gram-Schmidt algorithm is presented in the next function `gram_schmidt`:

Algorithm 41 Modified Gram-Schmidt orthonormalization procedure

$\boldsymbol{E}(:,0) = \boldsymbol{X}(:,0)/\|\boldsymbol{X}(:,0)\|_2^2$
for $i = 2, n$ **do**
  $\boldsymbol{E}(:,i) = \boldsymbol{X}(:,i)$
  for $j = 1 : i-1$ **do**
    $\boldsymbol{E}(:,i) = \boldsymbol{E}(:,i) - \left(\boldsymbol{E}(:,i)^T\boldsymbol{E}(:,j)\right)\boldsymbol{E}(:,j)/\|\boldsymbol{E}(:,j)\|^2$
  end for
  $\boldsymbol{E}(:,i) = \boldsymbol{E}(:,i)/\|\boldsymbol{E}(:,i)\|_2$
end for

```
def gram_schmidt(X):
    E = np.zeros(np.shape(X))
    (m,n)=np.shape(X)
    E[:,0] = X[:,0]/np.sqrt(np.inner(X[:,0],X[:,0]))
    for i in range(1,n):
        E[:,i] = X[:,i]
        for j in range(0,i):
            proj=np.inner(E[:,i],E[:,j])/np.inner(E[:,j],E[:,j])*E[:,j]
            E[:,i] = E[:,i]-proj
        E[:,i] = E[:,i]/np.sqrt(np.inner(E[:,i],E[:,i]))
    return E
```

Now, we test our code with the vectors

$$\boldsymbol{x}_1 = \begin{pmatrix} 1 \\ -1 \\ -1 \\ 1 \end{pmatrix}, \quad \boldsymbol{x}_2 = \begin{pmatrix} 2 \\ 1 \\ 0 \\ 1 \end{pmatrix} \quad \text{and} \quad \boldsymbol{x}_3 = \begin{pmatrix} 2 \\ 2 \\ 1 \\ 2 \end{pmatrix}.$$

In order to check the orthonormality of the columns of the resulting matrix $\boldsymbol{E}$ we compute the product $\boldsymbol{E}^T\boldsymbol{E}$ which must be the identity matrix.

```
import numpy as np
X = np.array([[ 1, 2, 2],
              [-1, 1, 2],
              [-1, 0, 1],
              [ 1, 1, 2]])
E = gram_schmidt(X)
print('E='); print(E)
print('I='); print(np.dot(np.transpose(E),E))
```

```
E=
[[ 0.5         0.67082039 -0.40824829]
 [-0.5         0.67082039  0.        ]
 [-0.5         0.2236068   0.40824829]
 [ 0.5         0.2236068   0.81649658]]
I=
[[1. 0. 0.]
 [0. 1. 0.]
 [0. 0. 1.]]
```

In practice, the Gram-Schmidt orthonormalization algorithm is not used at all. Instead one can use the QR factorization, which for pedagogical reasons we introduce as a consequence of the Gram-Schmidt method, and then we use it to generate orthonormal vectors.

10.4 *QR* Factorization

Suppose that we have n linearly independent vectors $\boldsymbol{a}_1, \boldsymbol{a}_2, \dots, \boldsymbol{a}_n \in \mathbb{R}^m$ forming the n columns of an $m \times n$ matrix $\boldsymbol{A}$ with $m \geq n$. Applying the Gram-Schmidt orthonormalization procedure to the n column-vectors $\boldsymbol{a}_i$ we construct n orthonormal vectors $\boldsymbol{e}_i$, for which

$$\begin{aligned} \boldsymbol{a}_1 &= r_{11}\boldsymbol{e}_1 , \\ \boldsymbol{a}_2 &= r_{12}\boldsymbol{e}_1 + r_{22}\boldsymbol{e}_2 , \\ &\vdots \\ \boldsymbol{a}_n &= r_{1n}\boldsymbol{e}_1 + r_{2n}\boldsymbol{e}_2 + \cdots + r_{nn}\boldsymbol{e}_n , \end{aligned} \tag{10.11}$$

with $r_{ij} = \langle \boldsymbol{e}_i, \boldsymbol{a}_j \rangle$ being the coefficients obtained from (10.10). Writing (10.11) in matrix-vector form, we obtain the famous QR factorization of $\boldsymbol{A}$

$$\boldsymbol{A} = \begin{pmatrix} \boldsymbol{a}_1 & \boldsymbol{a}_2 & \cdots & \boldsymbol{a}_n \end{pmatrix} = \begin{pmatrix} \boldsymbol{e}_1 & \boldsymbol{e}_2 & \cdots & \boldsymbol{e}_n \end{pmatrix} \cdot \begin{pmatrix} r_{11} & r_{12} & \cdots & r_{1n} \\ 0 & r_{22} & \cdots & r_{2n} \\ \vdots & \vdots & \ddots & \vdots \\ 0 & 0 & \cdots & r_{nn} \end{pmatrix} = \boldsymbol{QR} .$$

The matrix $\boldsymbol{Q}$ is $m \times n$ matrix with orthonormal columns ($\boldsymbol{Q}^{-1} = \boldsymbol{Q}^T$) the vectors from the Gram-Schmidt orthonormalization procedure, while the matrix $\boldsymbol{R}$ is an $n \times n$ upper-triangular matrix with diagonal entries $r_{ii} \neq 0$ for all $i = 1, 2, \dots, n$. If one of those was zero, let's say $r_{jj} = 0$, then the column $\boldsymbol{a}_j$ would have been written as a linear combination of the vectors $\boldsymbol{a}_1, \dots, \boldsymbol{a}_{j-1}$. This contradicts to the assumption that all $\boldsymbol{a}_i$ are linearly independent. Therefore, $\boldsymbol{R}$ is invertible, and $\boldsymbol{Q} = \boldsymbol{A}\boldsymbol{R}^{-1}$. Hence, $\text{rank}(\boldsymbol{Q}) = \text{rank}(\boldsymbol{A})$.

☞ An $m \times n$, full-rank, complex matrix $\boldsymbol{A}$ with $m \geq n$ has a QR factorization with $r_{ii} \neq 0$ for all $i = 1, 2, \dots, n$.

The QR decomposition of a matrix $\boldsymbol{A}$, contrary to the LU decomposition, is a stable algorithm and the result is not susceptible to rounding errors. Thus, QR decomposition can be helpful in various problems. Here we discuss a few applications.

10.4.1 Linear least squares using *QR* factorization

We have seen that the solution of the linear least squares problem can be found using the pseudoinverse. The knowledge of the QR decomposition of an $m \times n$ matrix $\boldsymbol{A}$ can lead to an efficient computation of its pseudoinverse $\boldsymbol{A}^+$. Assume that we have the QR decomposition of $\boldsymbol{A} = \boldsymbol{QR}$ of an $m \times n$ matrix $\boldsymbol{A}$. Using the properties of the orthogonal matrix $\boldsymbol{Q}$ we have

$$\begin{aligned}
A^+ &= (A^T A)^{-1} A^T \\
&= ((QR)^T (QR))^{-1} (QR)^T \\
&= (R^T Q^T Q R)^{-1} R^T Q^T \\
&= (R^T R)^{-1} R^T Q^T \quad (QQ^T = I) \\
&= R^{-1} R^{-T} R^T Q^T \quad (R \text{ is nonsingular}) \\
&= R^{-1} Q^T
\end{aligned}$$

Since the pseudoinverse of a nonsingular, square matrix coincides with its inverse A^{-1}, we can use the QR decomposition to compute the inverse of a square matrix A using the same formula $A^{-1} = R^{-1}Q^T$.

10.4.2 Computation of range

Recall that range of an $m \times n$ matrix A is the set range$(A) = \{Ax : x \in \mathbb{R}^n\}$. If A has linearly independent columns and $A = QR$, then range(A) = range(Q). In order to see this, consider a vector $y \in$ range(A). This means that there is an $x \in \mathbb{R}^n$ such that $y = Ax$. This is equivalent to

$$y = QRx = Qz ,$$

with $z = Rx \in \mathbb{R}^n$, which is equivalent to $y \in$ range(Q).

10.4.3 Householder matrices and QR factorization

Although the modified Gram-Schmidt orthonormalization appears to be very stable in obtaining the QR factorization of a matrix A, in practice we obtain the QR factorization using an even more stable algorithm, which is based on series of reflections. For any vector u, a representation of the reflection about the hyper-plane perpendicular to u is given by the *Householder matrix*

$$H = I - \frac{2uu^T}{\|u\|^2} .$$

For the sake of convenience we often set $v = u/\|u\|$ and we write $H = I - 2vv^T$. Such matrices have been introduced by A. S. Householder (1904–1993) in [68]. Any Householder matrix H is symmetric since

$$H^T = \left(I - 2vv^T\right)^T = I^T - 2(vv^T)^T = I - 2(v^T)^T v^T = H .$$

Moreover, H is orthonormal with

$$H^{-1}H = H^T H = H^2 = I .$$

To see this take

$$H^T H = \left(I - 2vv^T\right)^2 = I - 2 \cdot 2vv^T + 4vv^T vv^T = I - 4vv^T + 4v\|v\|^2 v^T = I ,$$

because $\|v\| = \|u/\|u\|\| = \|u\|/\|u\| = 1$.

Consider a matrix A with columns $a_1, a_2, \dots, a_n$. If a is a column vector, we define the Householder matrix H_1 with $u = a + r_1$ where

$$r_1 = \text{sign}(a_1)\|a\| e_1 = \begin{pmatrix} \text{sign}(a_1)\|a\| \\ 0 \\ \vdots \\ 0 \end{pmatrix} .$$

We observe that $\|\boldsymbol{r}_1\| = \|\boldsymbol{a}\|$ and subsequently

$$\begin{aligned}
\boldsymbol{H}_1\boldsymbol{a} &= \left(\boldsymbol{I} - \frac{2\boldsymbol{u}\boldsymbol{u}^T}{\|\boldsymbol{u}\|^2}\right)\boldsymbol{a} \\
&= \boldsymbol{a} - (\boldsymbol{a}+\boldsymbol{r}_1)\frac{2(\boldsymbol{a}+\boldsymbol{r}_1)^T\boldsymbol{a}}{(\boldsymbol{a}+\boldsymbol{r}_1)^T(\boldsymbol{a}+\boldsymbol{r}_1)} \\
&= \boldsymbol{a} - (\boldsymbol{a}+\boldsymbol{r}_1)\frac{2\|\boldsymbol{a}\|^2 + 2\boldsymbol{r}_1^T\boldsymbol{a}}{\|\boldsymbol{a}\|^2 + \boldsymbol{a}^T\boldsymbol{r}_1 + \boldsymbol{r}_1^T\boldsymbol{a} + \|\boldsymbol{r}_1\|^2} \\
&= \boldsymbol{a} - (\boldsymbol{a}+\boldsymbol{r}_1)\frac{2\|\boldsymbol{a}\|^2 + 2\boldsymbol{r}_1^T\boldsymbol{a}}{2\|\boldsymbol{a}\|^2 + 2\boldsymbol{r}_1^T\boldsymbol{a}} \\
&= \boldsymbol{a} - (\boldsymbol{a}+\boldsymbol{r}_1) \\
&= -\boldsymbol{r}_1 .
\end{aligned}$$

The matrix $\boldsymbol{H}_1$ practically is used to transform the first column of $\boldsymbol{A}$ into the first column of $\boldsymbol{R}$. Subsequently, we consider the Householder matrix $\boldsymbol{H}_2$ with $\boldsymbol{u} = \boldsymbol{a} + \boldsymbol{r}_2$ with

$$\boldsymbol{r}_2 = \text{sign}(a_2)\|\boldsymbol{a}\|\boldsymbol{e}_2 = \begin{pmatrix} 0 \\ \text{sign}(a_2)\|\boldsymbol{a}\| \\ 0 \\ \vdots \\ 0 \end{pmatrix},$$

and $\boldsymbol{a}$ the second column of the matrix $\boldsymbol{H}_1\boldsymbol{A}$ with 0 above the principal diagonal. Then the first row and first column of matrix $\boldsymbol{H}_2\boldsymbol{H}_1\boldsymbol{A}$ remain unchanged since the first entry of $\boldsymbol{u}$ is 0. Repeating the same process for every column $\boldsymbol{a}_i$ we obtain the product $\boldsymbol{H}_{n-1}\cdots\boldsymbol{H}_2\boldsymbol{H}_1\boldsymbol{A} = \boldsymbol{R}$, which after inversion of the orthogonal matrices $\boldsymbol{H}_i$ gives us

$$\boldsymbol{A} = (\boldsymbol{H}_1\boldsymbol{H}_2\cdots\boldsymbol{H}_{n-1})\boldsymbol{R} = \boldsymbol{Q}\boldsymbol{R} ,$$

since $\boldsymbol{H}_i^{-1} = \boldsymbol{H}_i^T = \boldsymbol{H}_i$. The matrices $\boldsymbol{H}_i$ are also known as Householder reflections or reflectors. This is the $\boldsymbol{QR}$ factorization of a matrix $\boldsymbol{A}$ obtained without the use of the Gram-Schmidt process.

The matrix $\boldsymbol{H} = \boldsymbol{I} - \frac{2\boldsymbol{u}\boldsymbol{u}^T}{\|\boldsymbol{u}\|^2}$ is called *reflector* because the product $\boldsymbol{H}\boldsymbol{x}$ is a reflection of a vector $\boldsymbol{x}$ through the perpendicular to $\boldsymbol{u}$ line $\{\boldsymbol{z} : \boldsymbol{u}^T\boldsymbol{z} = 0\}$. In practice, we never compute the matrix $\boldsymbol{H}$ but only the result of its application to another matrix. This is made clear in the Householder QR algorithm as it is presented in Algorithm 42. In order to decrease the complexity of the Householder algorithm, instead of working with the full matrix $\boldsymbol{A}$, we work with sub-matrices at each step.

Algorithm 42 Householder QR

```
for k = 1, 2, ..., n do
    Define a = A(k : m, k)
    Compute w = a + sign(a(1))||a||e_1 and u = w/||w||
    Compute A(k : m, k : n) = A(k : m, k : n) − 2u(u^T A(k : m, k : n))
end for
The algorithm overwrites A with R
```

This algorithm reduces $\boldsymbol{A}$ to upper-triangular matrix $\boldsymbol{R}$ but it does not explicitly computes the matrix $\boldsymbol{Q}$. This is because the computation of the matrix $\boldsymbol{Q}$ requires many operations. Moreover, in applications we usually need the products $\boldsymbol{Q}^T\boldsymbol{b}$ and $\boldsymbol{Q}\boldsymbol{x}$. These products can be computed efficiently using the knowledge of the vectors $\boldsymbol{u}$, [135].

Although this procedure can lead to a more stable algorithm to implement, again, it is not the most efficient and stable way to obtain the QR factorization. Other commonly used methods for the QR factorization can be found in books that specialize in numerical linear algebra.

10.4.4 Python implementation and orthonormalization

The QR factorization of an $m \times n$ matrix $\boldsymbol{A}$ can be computed using the function `qr` of `numpy.linalg`. The general call of this function is

```
q, r, (h,tau) = qr(a, mode)
```

The function `qr` accepts as input a matrix $\boldsymbol{A}$ in the variable `a` and returns in the output arrays `q` and `r` the matrices $\boldsymbol{Q}$, $\boldsymbol{R}$, respectively, and also the tuple `(h,tau)` where `h` is an array with the Householder reflectors. The array `tau` contains scaling factors for the Householder reflectors. All the output parameters are optional, so we can call the function `qr` using any of the output arguments we need. In addition to the array `a`, the function `qr` accepts as input the optional argument `mode`, which takes one of the following string values: `'reduced'`, the default value that asks the code to return `q` and `r` with dimensions $m \times k$ and $k \times n$; `'complete'`, that asks the code to return `q` and `r` with dimensions $m \times m$ and $m \times n$; `'r'` that asks the code to return only the array `r` with dimensions $k \times n$; and `'raw'`, that asks the code to return only the tuple (h,tau) with dimensions $n \times k$ and `(k,)`.

Similarly, there is a SciPy implementation of the QR factorization, namely the function `qr` of `scipy.linalg`. The SciPy implementation can be also called easily as `qr(a)` since all its rest arguments are optional.

For example, consider the matrix

$$\boldsymbol{A} = \begin{pmatrix} 2 & -2 & 18 \\ 2 & 1 & 0 \\ 1 & 2 & 0 \end{pmatrix} .$$

We will use the Householder reflections method to compute the QR decomposition analytically first, and then we will proceed with its numerical computation. Initially, we take the first column of $\boldsymbol{A}$, which is $\boldsymbol{a} = (2,2,1)^T$ with

$$\|\boldsymbol{a}\| = \sqrt{2^2 + 2^2 + 1^2} = 3 .$$

The vector $\boldsymbol{r}_1$ is $\boldsymbol{r}_1 = (3,0,0)^T$ and the vector $\boldsymbol{u} = \boldsymbol{a} + \boldsymbol{r}_1 = (5,2,1)^T$ with $\|\boldsymbol{u}\| = \sqrt{30}$. Thus,

$$\boldsymbol{H}_1 = \boldsymbol{I} - \frac{2\boldsymbol{u}\boldsymbol{u}^T}{\|\boldsymbol{u}\|^2} = \begin{pmatrix} 1 & 0 & 0 \\ 0 & 1 & 0 \\ 0 & 0 & 1 \end{pmatrix} - \frac{2}{30} \begin{pmatrix} 25 & 10 & 15 \\ 10 & 4 & 2 \\ 5 & 2 & 1 \end{pmatrix} = \frac{1}{3} \begin{pmatrix} -2 & -2 & -1 \\ -2 & 11/5 & -2/5 \\ -1 & -2/5 & 14/5 \end{pmatrix} .$$

So, we have

$$\boldsymbol{H}_1\boldsymbol{A} = \begin{pmatrix} -3 & 0 & -12 \\ 0 & 9/5 & -12 \\ 0 & 12/5 & -6 \end{pmatrix} .$$

Observe that the first column of $\boldsymbol{H}_1\boldsymbol{A}$ is the column vector $-\boldsymbol{r}_1$.

We proceed now with $\boldsymbol{a} = (0, 9/5, 12/5)^T$, the second column of the matrix $\boldsymbol{H}_1\boldsymbol{A}$, where $\|\boldsymbol{a}\| = 3$. Here, $\boldsymbol{r}_2 = (0,3,0)^T$ and $\boldsymbol{u} = \boldsymbol{a} + \boldsymbol{r}_2 = (0, 24/5, 12/5)^T$ with $\|\boldsymbol{u}\| = 12/\sqrt{5}$. Then,

$$\boldsymbol{H}_2 = \boldsymbol{I} - \frac{2\boldsymbol{u}\boldsymbol{u}^T}{\|\boldsymbol{u}\|^2} = \begin{pmatrix} 1 & 0 & 0 \\ 0 & 1 & 0 \\ 0 & 0 & 1 \end{pmatrix} - \frac{2}{10} \begin{pmatrix} 0 & 0 & 0 \\ 0 & 8 & 4 \\ 0 & 4 & 2 \end{pmatrix} = \frac{1}{3} \begin{pmatrix} 1 & 0 & 0 \\ 0 & -3/5 & -4/5 \\ 0 & -4/5 & 3/5 \end{pmatrix} .$$

So, we have

$$H_2H_1A = \begin{pmatrix} -3 & 0 & -12 \\ 0 & -3 & 12 \\ 0 & 0 & 6 \end{pmatrix} .$$

Thus,

$$R = \begin{pmatrix} -3 & 0 & -12 \\ 0 & -3 & 12 \\ 0 & 0 & 6 \end{pmatrix} \quad \text{and} \quad Q = H_1H_2 = \frac{1}{3}\begin{pmatrix} -2 & 2 & 1 \\ -2 & -1 & -2 \\ -1 & -2 & 2 \end{pmatrix} .$$

We can check the validity of the result by computing the product QR and verifying that Q is orthogonal. Next, we compute a QR factorization of A using Python and the following simple script:

```
import numpy as np
from numpy.linalg import qr
A = np.array([[2.0, -2.0, 18.0],
              [2.0,  1.0,  0.0],
              [1.0,  2.0,  0.0]])
Q, R = qr( A )
print('R= '); print(R)
print('Q= '); print(Q)
```

```
R=
[[-3.0000000e+00  4.4408921e-16 -1.2000000e+01]
 [ 0.0000000e+00 -3.0000000e+00  1.2000000e+01]
 [ 0.0000000e+00  0.0000000e+00  6.0000000e+00]]
Q=
[[-0.66666667  0.66666667  0.33333333]
 [-0.66666667 -0.33333333 -0.66666667]
 [-0.33333333 -0.66666667  0.66666667]]
```

We observe that the numerical results verify our manual computations. We refer the interested reader to the references for more sophisticated algorithms related to the computation of the QR factorization.

10.5 Singular Value Decomposition

A very important decomposition for both Statistics and Machine Learning is the Singular Value Decomposition (SVD). The SVD has a wide spectrum of applications in many fields of science, even in cryptography and in image processing. At the end of this chapter, we present an application of SVD in image processing. A digital image can be represented by an $m \times n$ matrix with entries values that represent the intensity of the colors of the corresponding pixels. Processing the matrix with the colors of the pixels will result in a modified image. As you understood very well, we will deal with general rectangular matrices rather than square matrices. The singular value decomposition of an $m \times n$ matrix A is a product of three matrices $U\Sigma V^T$, where Σ is $m \times n$ diagonal matrix, U and V are $m \times m$ and $n \times n$ orthogonal matrices, respectively. Recall, that because U and V are orthogonal matrices we have that $U^TU = I$ and $V^TV = I$, and that U and V have orthonormal rows and columns.

10.5.1 Derivation of *SVD*

The singular value decomposition is based on computations of eigenvalues and eigenvectors. Here, $\boldsymbol{A}$ is $m \times n$, so we cannot compute any eigenvalues unless $m = n$. Although it might seem to have our hands tight, there are various miracles happening with the *SVD*, if the actual *SVD* is not a miracle. We start by exploring the matrix $\boldsymbol{A}^T\boldsymbol{A}$, which is $n \times n$ and symmetric since $(\boldsymbol{A}^T\boldsymbol{A})^T = \boldsymbol{A}^T(\boldsymbol{A}^T)^T = \boldsymbol{A}^T\boldsymbol{A}$. A direct consequence of this, and the fact that $\boldsymbol{A}^T\boldsymbol{A}$ is symmetric and square, is that $\boldsymbol{A}^T\boldsymbol{A}$ has real and positive eigenvalues. To see this, consider an eigenvalue λ of $\boldsymbol{A}^T\boldsymbol{A}$ and the corresponding eigenvector $\boldsymbol{x} \neq \boldsymbol{0}$ with real entries. Since $\boldsymbol{A}^T\boldsymbol{A}\boldsymbol{x} = \lambda\boldsymbol{x}$ we have

$$\begin{aligned} 0 \leq \|\boldsymbol{A}\boldsymbol{x}\|^2 &= \boldsymbol{A}\boldsymbol{x}\cdot\boldsymbol{A}\boldsymbol{x} = (\boldsymbol{A}\boldsymbol{x})^T(\boldsymbol{A}\boldsymbol{x}) = \boldsymbol{x}^T\boldsymbol{A}^T\boldsymbol{A}\boldsymbol{x} \\ &= \boldsymbol{x}^T(\boldsymbol{A}^T\boldsymbol{A}\boldsymbol{x}) = \boldsymbol{x}^T(\lambda\boldsymbol{x}) = \lambda\boldsymbol{x}^T\boldsymbol{x} \\ &= \lambda\|\boldsymbol{x}\|^2 , \end{aligned}$$

from which we obtain that $\lambda\|\boldsymbol{x}\|^2 \geq 0$, and thus $\lambda \geq 0$ since $\|\boldsymbol{x}\| > 0$.

The eigenvalues of $\boldsymbol{A}^T\boldsymbol{A}$ are of great importance. The square roots of these eigenvalues are called *singular values* of $\boldsymbol{A}$ and are denoted by $\sigma_1, \sigma_2, \ldots, \sigma_n$. In other words, if λ_i is the i-th eigenvalue of $\boldsymbol{A}^T\boldsymbol{A}$ with $\boldsymbol{x}_i$ the corresponding (normalized) eigenvector, then $\sigma_i = \sqrt{\lambda_i} = \|\boldsymbol{A}\boldsymbol{x}_i\|$. It is also very important to consider them ordered in the sense that $\sigma_1 \geq \sigma_2 \geq \cdots \geq \sigma_n \geq 0$.

Assume that the nonzero singular values are the $\sigma_1, \sigma_2, \ldots, \sigma_r$ and all the rest $\sigma_{r+1} = \cdots = \sigma_n = 0$. Since we want to produce a factorization $\boldsymbol{A} = \boldsymbol{U}\boldsymbol{\Sigma}\boldsymbol{V}^T$, we consider first an $m \times n$ matrix $\boldsymbol{\Sigma}$ with diagonal values the singular values of $\boldsymbol{A}$

$$\boldsymbol{\Sigma} = \begin{pmatrix} \boldsymbol{D} & \boldsymbol{0} \\ \boldsymbol{0} & \boldsymbol{0} \end{pmatrix} \quad \text{and} \quad \boldsymbol{D} = \begin{pmatrix} \sigma_1 & 0 & \cdots & 0 \\ 0 & \sigma_2 & \cdots & 0 \\ \vdots & & \ddots & \vdots \\ 0 & 0 & \cdots & \sigma_r \end{pmatrix} .$$

Since we want the matrix $\boldsymbol{V}$ of the *SVD* to be orthogonal, we take $\boldsymbol{V} = (\boldsymbol{v}_1, \boldsymbol{v}_2, \ldots, \boldsymbol{v}_n)$ where $\boldsymbol{v}_i$ are the orthonormal vectors produced by the Gram-Schmidt orthonormalization of the eigenvectors $\boldsymbol{x}_i$ of $\boldsymbol{A}^T\boldsymbol{A}$. At this stage, we have chosen $\boldsymbol{\Sigma}$ and $\boldsymbol{V}$ in the analysis $\boldsymbol{A} = \boldsymbol{U}\boldsymbol{\Sigma}\boldsymbol{V}^T$ and it remains to find matrix $\boldsymbol{U}$. From the equation $\boldsymbol{A} = \boldsymbol{U}\boldsymbol{\Sigma}\boldsymbol{V}^T$, since $\boldsymbol{V}^T\boldsymbol{V} = \boldsymbol{I}$ (or else since $\boldsymbol{V}^{-1} = \boldsymbol{V}^T$) we can solve easily for $\boldsymbol{U}\boldsymbol{\Sigma}$ and obtain

$$\boldsymbol{U}\boldsymbol{\Sigma} = \boldsymbol{A}\boldsymbol{V} .$$

Moreover, we observe that the columns of $\boldsymbol{A}\boldsymbol{V}$ consist of the vectors $\boldsymbol{A}\boldsymbol{v}_i$, and thus

$$\boldsymbol{A}\boldsymbol{V} = (\boldsymbol{A}\boldsymbol{v}_1, \boldsymbol{A}\boldsymbol{v}_2, \ldots, \boldsymbol{A}\boldsymbol{v}_n) .$$

The columns $\boldsymbol{A}\boldsymbol{v}_i$, for $i = 1, \ldots, n$ are orthogonal. To see this take $i \neq j$, and then

$$(\boldsymbol{A}\boldsymbol{v}_i)\cdot(\boldsymbol{A}\boldsymbol{v}_j) = (\boldsymbol{A}\boldsymbol{v}_i)^T(\boldsymbol{A}\boldsymbol{v}_j) = \boldsymbol{v}_i^T\boldsymbol{A}^T\boldsymbol{A}\boldsymbol{v}_j = \boldsymbol{v}_i^T\lambda_j\boldsymbol{v}_j = \lambda_j\boldsymbol{v}_i\cdot\boldsymbol{v}_j = \boldsymbol{0} .$$

If $\boldsymbol{\Sigma}$ was invertible, then we would have defined $\boldsymbol{U} = \boldsymbol{A}\boldsymbol{V}\boldsymbol{\Sigma}^{-1}$. Although this is not the case, it gives us the idea to define the first r columns of $\boldsymbol{U}$ as $\boldsymbol{u}_i = \boldsymbol{A}\boldsymbol{v}_i/\sigma_i$ for $i = 1, 2, \ldots, r$. This could have been adequate if we could extend this set of r orthogonal column-vectors to a set of n orthogonal column-vectors. This again can be done by considering the rest $\boldsymbol{u}_i$, $i = r+1, \ldots, m$ to be the result of the Gram-Schmidt orthonormalization procedure of the set $\{\boldsymbol{u}_1, \boldsymbol{u}_2, \ldots, \boldsymbol{u}_r, \boldsymbol{e}_{r+1}, \ldots, \boldsymbol{e}_m\}$ where $\boldsymbol{e}_{r+1}, \ldots, \boldsymbol{e}_m$ are the standard basis vectors of $\mathbb{R}^m$.

With this construction of $\boldsymbol{\Sigma}$, $\boldsymbol{U}$ and $\boldsymbol{V}$ we have that $\boldsymbol{A} = \boldsymbol{U}\boldsymbol{\Sigma}\boldsymbol{V}^T$, which can be written in the form

$$\boldsymbol{A} = \begin{pmatrix} \boldsymbol{u}_1 & \boldsymbol{u}_2 & \cdots & \boldsymbol{u}_r & | & \boldsymbol{u}_{r+1} & \cdots & \boldsymbol{u}_m \end{pmatrix} \left(\begin{array}{cccc|c} \sigma_1 & 0 & \cdots & 0 & \\ 0 & \sigma_2 & \cdots & 0 & \\ \vdots & & \ddots & \vdots & \boldsymbol{O} \\ 0 & 0 & \cdots & \sigma_r & \\ \hline & & \boldsymbol{O} & & \boldsymbol{O} \end{array}\right) \begin{pmatrix} \boldsymbol{v}_1^T \\ \boldsymbol{v}_2^T \\ \vdots \\ \boldsymbol{v}_r^T \\ \hline \boldsymbol{v}_{r+1}^T \\ \vdots \\ \boldsymbol{v}_n^T \end{pmatrix} .$$

Remark 10.4. *The singular values σ_i are the square roots of the eigenvalues λ_i of the matrix $\boldsymbol{A}^T\boldsymbol{A}$. If $\boldsymbol{A}$ is symmetric matrix with eigenvalues κ_i then $\sigma_i = |\kappa_i|$. This is because for symmetric matrix $\boldsymbol{A}$ the product $\boldsymbol{A}^T\boldsymbol{A} = \boldsymbol{A}^2$ with eigenvalues $\lambda_i = \kappa_i^2$.*

And still we haven't explored all the magic of this method.

☞ The Singular Value Decomposition of an $m \times n$ matrix $\boldsymbol{A}$ is the factorization $\boldsymbol{A} = \boldsymbol{U}\boldsymbol{\Sigma}\boldsymbol{V}^T$ where $\boldsymbol{\Sigma}$ is a diagonal matrix with values in the main diagonal the r eigenvalues of the matrix $\boldsymbol{A}^T\boldsymbol{A}$. $\boldsymbol{V}$ is $n\times n$ orthogonal matrix with columns the Gram-Schmidt orthonormalized eigenvectors of $\boldsymbol{A}^T\boldsymbol{A}$, and $\boldsymbol{U}$ is $m \times m$ orthogonal matrix with columns the result of the Gram-Schmidt orthonormalization of the vectors $\boldsymbol{u}_i = A\boldsymbol{v}_i/\sigma_i$ for $i = 1, 2, \ldots, r$ and $\boldsymbol{u}_i = \boldsymbol{e}_i$ for $i = r + 1, \ldots, m$. The columns of U are called right singular vectors and the columns of V left singular vectors.

For example, consider the 2×3 matrix

$$\boldsymbol{A} = \begin{pmatrix} 1 & 2 & 3 \\ 3 & 2 & 1 \end{pmatrix} ,$$

where $m = 2$ and $n = 3$. First, we form the matrix

$$\boldsymbol{A}^T\boldsymbol{A} = \begin{pmatrix} 10 & 8 & 6 \\ 8 & 8 & 8 \\ 6 & 8 & 10 \end{pmatrix} ,$$

which has eigenvalues

$$\lambda_1 = 24, \quad \lambda_2 = 4, \quad \text{and} \quad \lambda_2 = 0 ,$$

and (normilized) eigenvectors

$$\boldsymbol{v}_1 = \frac{1}{\sqrt{3}} \begin{pmatrix} -1 \\ -1 \\ -1 \end{pmatrix}, \quad \boldsymbol{v}_2 = \frac{1}{\sqrt{2}} \begin{pmatrix} 1 \\ 0 \\ -1 \end{pmatrix}, \quad \text{and} \quad \boldsymbol{v}_3 = \frac{1}{\sqrt{6}} \begin{pmatrix} 1 \\ -2 \\ 1 \end{pmatrix} .$$

Therefore, the singular values of $\boldsymbol{A}$ are $\sigma_i = \sqrt{\lambda_i}$ for $i = 1, 2$. More precisely,

$$\sigma_1 = 2\sqrt{6}, \quad \text{and} \quad \sigma_2 = 2 .$$

Here $r = 2$, which coincides with the rank of $\boldsymbol{A}$ since $\sigma_2 \neq 0$ and $\sigma_3 = 0$. Hence, the 2×3 matrix $\boldsymbol{\Sigma}$ is written as

$$\boldsymbol{\Sigma} = \begin{pmatrix} 2\sqrt{6} & 0 & 0 \\ 0 & 2 & 0 \end{pmatrix} .$$

The 3×3 orthogonal matrix $\boldsymbol{V}$ is defined by the orthonormalized eigenvectors $\boldsymbol{v}_i$ and can be written as

$$\boldsymbol{V} = \frac{1}{\sqrt{6}} \begin{pmatrix} -\sqrt{2} & \sqrt{3} & 1 \\ -\sqrt{2} & 0 & -2 \\ -\sqrt{2} & -\sqrt{3} & 1 \end{pmatrix} .$$

Finally, in order to compute the 2×2 matrix $\boldsymbol{U}$ we consider the vectors

$$\boldsymbol{u}_1 = \boldsymbol{A}\boldsymbol{v}_1/\sigma_1 = \frac{1}{\sqrt{2}}\begin{pmatrix} -1 \\ -1 \end{pmatrix} \quad \text{and} \quad \boldsymbol{u}_2 = \boldsymbol{A}\boldsymbol{v}_2/\sigma_2 = \frac{1}{\sqrt{2}}\begin{pmatrix} -1 \\ 1 \end{pmatrix} .$$

These two vectors are adequate to form the matrix

$$\boldsymbol{U} = \frac{1}{\sqrt{2}} \begin{pmatrix} -1 & -1 \\ -1 & 1 \end{pmatrix} .$$

It is easy now to verify that $\boldsymbol{A} = \boldsymbol{U}\boldsymbol{\Sigma}\boldsymbol{V}^T$.

10.5.2 Theoretical considerations

Taking the *SVD* of $\boldsymbol{A}$ we have

$$\begin{aligned}
\boldsymbol{A} &= \boldsymbol{U}\boldsymbol{\Sigma}\boldsymbol{V}^T \\
&= \left(\boldsymbol{u}_1 \quad \boldsymbol{u}_2 \quad \cdots \quad \boldsymbol{u}_r \;\middle|\; \boldsymbol{u}_{r+1} \quad \cdots \quad \boldsymbol{u}_m \right)
\left(\begin{array}{cccc|c} \sigma_1 & 0 & \cdots & 0 & \\ 0 & \sigma_2 & \cdots & 0 & \boldsymbol{O} \\ \vdots & & \ddots & \vdots & \\ 0 & 0 & \cdots & \sigma_r & \\ \hline & \boldsymbol{O} & & & \boldsymbol{O} \end{array}\right)
\left(\begin{array}{c} \boldsymbol{v}_1^T \\ \boldsymbol{v}_2^T \\ \vdots \\ \boldsymbol{v}_r^T \\ \hline \boldsymbol{v}_{r+1}^T \\ \vdots \\ \boldsymbol{v}_n^T \end{array}\right) \\
&= \left(\boldsymbol{u}_1 \quad \boldsymbol{u}_2 \quad \cdots \quad \boldsymbol{u}_r \right)
\begin{pmatrix} \sigma_1 & 0 & \cdots & 0 \\ 0 & \sigma_2 & \cdots & 0 \\ \vdots & & \ddots & \vdots \\ 0 & 0 & \cdots & \sigma_r \end{pmatrix}
\begin{pmatrix} \boldsymbol{v}_1^T \\ \boldsymbol{v}_2^T \\ \vdots \\ \boldsymbol{v}_r^T \end{pmatrix} \\
&= \left(\sigma_1\boldsymbol{u}_1 \quad \sigma_2\boldsymbol{u}_2 \quad \cdots \quad \sigma_r\boldsymbol{u}_r \right)
\begin{pmatrix} \boldsymbol{v}_1^T \\ \boldsymbol{v}_2^T \\ \vdots \\ \boldsymbol{v}_r^T \end{pmatrix} \\
&= \sigma_1\boldsymbol{u}_1\boldsymbol{v}_1^T + \sigma^2\boldsymbol{u}_2\boldsymbol{v}_2^T + \cdots + \sigma^r\boldsymbol{u}_r\boldsymbol{v}_r^T \ .
\end{aligned}$$

In the previous computations we wrote $\boldsymbol{A}$ as a sum of r matrices $\boldsymbol{A} = \boldsymbol{B}_1 + \boldsymbol{B}_2 + \cdots + \boldsymbol{B}_r$ where $\boldsymbol{B}_i = \sigma_i\boldsymbol{u}_i\boldsymbol{v}_i^T$. Something which is intrinsic is that

$$\|\boldsymbol{B}_i\| = \sigma_i\|\boldsymbol{u}_i\boldsymbol{v}_i^T\| = \sigma_i \ ,$$

and thus $\|\boldsymbol{B}_1\| \geq \|\boldsymbol{B}_2\| \geq \cdots \geq \|\boldsymbol{B}_r\|$. In this way we have obtained a hierarchical analysis of $\boldsymbol{A}$ in matrices where the first of them is the most important. Now, you might be able to guess what is the biggest miracle of the singular values of $\boldsymbol{A}$. This is that the *SVD* of $\boldsymbol{A}$ contains all the important information of the matrix $\boldsymbol{A}$ in a hierarchical form, in the sense that the largest singular value contains most of the information of the matrix, the second largest singular value something less and so on. So if we take the first $k < r$ terms in the sum of $\boldsymbol{B}_i$'s, we construct an approximate matrix of $\boldsymbol{A}$. Specifically, the matrix

$$\boldsymbol{A}_k = \boldsymbol{U}\boldsymbol{\Sigma}_k\boldsymbol{V}^T = \sum_{i=1}^{k} \boldsymbol{B}_i ,$$

where $\boldsymbol{\Sigma}_k$ is the $m \times n$ diagonal matrix with entries on the principal diagonal only the first k singular values $\sigma_1, \ldots, \sigma_k$. This is of great importance not only because it is an approximation of $\boldsymbol{A}$ but because it is the best approximation of rank k. Specifically, we have the following theorem:

Theorem 10.5. *The matrix*

$$\boldsymbol{A}_k = \sum_{i=1}^{k} \boldsymbol{B}_i = \sum_{i=1}^{k} \sigma_i \boldsymbol{u}_i \boldsymbol{v}_i^T ,$$

is the best rank-k approximation of matrix $\boldsymbol{A}$ *by minimizing the* ℓ_2*-norm of the error*

$$\|\boldsymbol{A} - \boldsymbol{A}_k\|_2 = \min_{\text{rank}(\boldsymbol{B}) \leq k} \|\boldsymbol{A} - \boldsymbol{B}\|_2 . \tag{10.12}$$

Moreover,

$$\|\boldsymbol{A} - \boldsymbol{A}_k\|_2 = \sigma_{k+1} .$$

Proof. First we observe that $\text{rank}(\boldsymbol{A}_k) = k$. This is because

$$\text{rank}(\boldsymbol{A}_k) = \text{rank}(\boldsymbol{U}\boldsymbol{\Sigma}_k\boldsymbol{V}^T) = \text{rank}(\boldsymbol{\Sigma}_k) = k .$$

For the minimization problem, we consider a matrix $\boldsymbol{B}$ with $\text{rank}(\boldsymbol{B}) = k$. Then the kernel (null space) $\ker(\boldsymbol{B})$ is of dimension $n - k$. Take also the space X produced by the vectors $\{\boldsymbol{v}_1, \ldots, \boldsymbol{v}_{k+1}\}$. Since both $\ker(\boldsymbol{B})$ and X are subspaces of $\mathbb{R}^n$, and the sum of their dimensions is greater than n we conclude that they have nonempty intersection. If $\boldsymbol{z}$ is a vector in their intersection with $\|\boldsymbol{z}\|_2 = 1$ such that $\boldsymbol{B}\boldsymbol{z} = \boldsymbol{0}$, then $\boldsymbol{z} = c_1\boldsymbol{v}_1 + c_2\boldsymbol{v}_2 + \cdots + c_{k+1}\boldsymbol{v}_{k+1}$. Since $\|\boldsymbol{z}\|_2 = 1$ and $\boldsymbol{v}_i$'s are orthonormal, we have $|c_1|^2 + |c_2|^2 + \cdots + |c_{k+1}|^2 = 1$. So,

$$(\boldsymbol{A} - \boldsymbol{B})\boldsymbol{z} = \boldsymbol{A}\boldsymbol{z} = \boldsymbol{A}\sum_{i=1}^{k+1} c_i \boldsymbol{v}_i = \sum_{i=1}^{k+1} c_i \boldsymbol{A}\boldsymbol{v}_i = \sum_{i=1}^{k+1} c_i \sigma_i \boldsymbol{u}_i .$$

Because $\boldsymbol{u}_i$'s are orthogonal, we have

$$\|(\boldsymbol{A} - \boldsymbol{B})\boldsymbol{z}\|_2^2 = \sum_{i=1}^{k+1} |\sigma_i c_i|^2 \geq \sigma_{k+1}^2 \sum_{i=1}^{k+1} |c_i|^2 = \sigma_{k+1}^2 .$$

Hence,

$$\|\boldsymbol{A} - \boldsymbol{B}\|_2 \geq \frac{\|(\boldsymbol{A} - \boldsymbol{B})\boldsymbol{z}\|_2}{\|\boldsymbol{z}\|_2} \geq \sigma_{k+1} .$$

To show that $\sigma_{k+1} = \|\boldsymbol{A} - \boldsymbol{A}_k\|_2$ we take the difference

$$\|\boldsymbol{A} - \boldsymbol{A}_k\|_2 = \|\boldsymbol{U}(\boldsymbol{\Sigma} - \boldsymbol{\Sigma}_k)\boldsymbol{V}^T\|_2 = \|\boldsymbol{\Sigma} - \boldsymbol{\Sigma}_k\|_2 = \sigma_{k+1} ,$$

and we have that

$$\|\boldsymbol{A} - \boldsymbol{B}\|_2 \geq \|\boldsymbol{A} - \boldsymbol{A}_k\|_2 ,$$

for any $m \times n$ matrix $\boldsymbol{B}$ of rank k. □

This best approximation property of the *SVD* perhaps is the most applicable. It is also known as *Principal Component Analysis.* This lies on the fact that from all the ordered singular values $\sigma_1 \geq \sigma_2 \geq \cdots \geq \sigma_r$ only some of them are important and the rest can be discarded. Applications areas are numerous and include machine learning where we can extract information from a matrix $\boldsymbol{A}$ without using data, and also in dimension reduction problems where, obviously, by ignoring negligible singular values, the resulting problem is much simpler and smaller compared to the original one. For an example in image processing see Section 10.5.6.

On the other hand, the *SVD* has some other very useful properties. For example, it helps in the computations with norms.

Theorem 10.6. *For any $m \times n$ matrix $\boldsymbol{A}$ with singular values $\sigma_1 \geq \sigma_2 \geq \cdots \geq \sigma_r$. Then,*

(i) $\|\boldsymbol{A}\|_2 = \sigma_1$

(ii) $\|\boldsymbol{A}\|_F = \sqrt{\sigma_1^2 + \sigma_2^2 + \cdots + \sigma_r^2}$

Proof. In order to prove (i) we write

$$\begin{aligned}\|\boldsymbol{A}\|_2 &= \|\boldsymbol{U\Sigma V}^T\|_2 = \sqrt{\sigma(\boldsymbol{U\Sigma V}^T(\boldsymbol{U\Sigma V}^T)^T)} = \sqrt{\rho(\boldsymbol{U\Sigma V}^T\boldsymbol{V}(\boldsymbol{U\Sigma})^T)} \\ &= \sqrt{\rho(((\boldsymbol{U\Sigma})^T\boldsymbol{U\Sigma})^T)} = \sqrt{\rho((\boldsymbol{\Sigma}^T\boldsymbol{U}^T\boldsymbol{U\Sigma})^T)} = \sqrt{\rho(\boldsymbol{\Sigma\Sigma}^T)} \\ &= \|\boldsymbol{\Sigma}\|_2 = \max_i \sigma_i = \sigma_1 \ .\end{aligned}$$

For the property (ii), similarly, we have

$$\|\boldsymbol{A}\|_F = \|\boldsymbol{U\Sigma V}^T\|_F = \|\boldsymbol{\Sigma}\|_F = \sqrt{\sigma_1^2 + \sigma_2^2 + \cdots + \sigma_r^2} \ .$$

□

If $\boldsymbol{A}$ is square $n \times n$ and invertible (rank$(\boldsymbol{A}) = n$), then we have some even more interesting properties.

Theorem 10.7. *For any non-singular $n \times n$ matrix $\boldsymbol{A}$ we have that*

$$\|\boldsymbol{A}^{-1}\|_2 = \frac{1}{\sigma_n} \quad \textit{and thus} \quad \operatorname{cond}(\boldsymbol{A}) = \frac{\sigma_1}{\sigma_n} \ .$$

Proof. Since, $\boldsymbol{A} = \boldsymbol{U\Sigma V}^T$ we have that $\boldsymbol{A}^{-1} = \boldsymbol{V\Sigma}^{-1}\boldsymbol{U}^T$, and thus $\|\boldsymbol{A}^{-1}\|_2 = 1/\sigma_1$ according to the previous theorem. Moreover,

$$\operatorname{cond}(\boldsymbol{A}) = \|\boldsymbol{A}\|_2 \cdot \|\boldsymbol{A}^{-1}\|_2 = \sigma_n/\sigma_1 \ ,$$

since $\|\boldsymbol{A}\|_2 = \sigma_n$ and $\|\boldsymbol{A}^{-1}\|_2 = 1/\sigma_1$. □

10.5.3 Pseudoinverse and *SVD*

The pseudoinverse matrix $\boldsymbol{A}^+$ can be computed in a trivial way using the singular value decomposition of $\boldsymbol{A}$. Recall that the pseudoinverse of an $m \times n$ matrix $\boldsymbol{A}$ is the $m \times n$ matrix $\boldsymbol{A}^+ = (\boldsymbol{A}^T\boldsymbol{A})^{-1}\boldsymbol{A}^T$. Since $\boldsymbol{A}^T = (\boldsymbol{U\Sigma V}^T)^T = \boldsymbol{V\Sigma U}^T$, we have that

$$\begin{aligned}\boldsymbol{A}^+ &= (\boldsymbol{A}^T\boldsymbol{A})^{-1}\boldsymbol{A}^T = (\boldsymbol{V\Sigma U}^T\boldsymbol{U\Sigma V}^T)^{-1}(\boldsymbol{V\Sigma U}^T) \\ &= \boldsymbol{V}(\boldsymbol{\Sigma}^T\boldsymbol{\Sigma})^{-1}\boldsymbol{V}^T\boldsymbol{V\Sigma U}^T = \boldsymbol{V}(\boldsymbol{\Sigma}^T\boldsymbol{\Sigma})^{-1}\boldsymbol{\Sigma}^T\boldsymbol{U}^T \ ,\end{aligned}$$

and thus we write

$$A^+ = V\Sigma^+ U^T , \tag{10.13}$$

where Σ^+ is the $n \times m$ matrix

$$\Sigma^+ = \left(\begin{array}{cccc|c} 1/\sigma_1 & 0 & \cdots & 0 & \\ 0 & 1/\sigma_2 & \cdots & 0 & \\ \vdots & \vdots & \ddots & \vdots & O \\ 0 & 0 & \cdots & 1/\sigma_r & \\ \hline & & O & & O \end{array}\right) .$$

10.5.4 Linear least squares problem and *SVD*

A very important application of the singular value decomposition is the solution of the linear least squares problem. We try to minimize the error-norm $\|b - Ax\|_2^2$ for an $m \times n$ matrix A, given that the singular value decomposition of $A = U\Sigma V^T$, with singular values $\sigma_1 \geq \sigma_2 \geq \cdots \geq \sigma_r > 0$. Note that because U is orthonormal, then

$$\|Ux\|_2 = \sqrt{(Ux)^T(Ux)} = \sqrt{x^T U^T U x} = \sqrt{x^T x} = \|x\|_2 ,$$

for any vector x. Using the orthogonality of U and V, we have

$$\begin{aligned} \|Ax - b\|_2^2 &= \|U^T(Ax - b)\|_2^2 = \|\Sigma V^T x - U^T b\|_2^2 \\ &= \sum_{i=1}^{r} (\sigma_i z_i - u_i^T b)^2 + \sum_{i=r+1}^{m} (u_i^T b)^2 , \end{aligned}$$

where z_i are the entries of the vector $z = V^T x$. Thus, the minimum $\min_x \|Ax - b\|_2^2$ can be achieved when $\sigma_i z_i - u_i^T b = 0$ for $i = 1, 2, \ldots, r$, and for arbitrary values z_i for $i = r + 1, \ldots, n$. So, the solution of the linear least squares problem is $x^* = Vz$, where $z \in \mathbb{R}^n$ is the vector with entries

$$\begin{aligned} z_i &= \frac{u_i^T b}{\sigma_i}, \quad \text{for } i = 1, \ldots, r , \\ z_i &= 0, \quad \text{for } i = r + 1, \ldots, n . \end{aligned}$$

Since $z = V^T x^*$, the minimum norm solution is

$$x^* = Vz = \sum_{i=1}^{r} \frac{u_i^T b}{\sigma_i} v_i .$$

Moreover, we have

$$\min_x \|Ax - b\|_2^2 = \sum_{i=r+1}^{m} (u_i^T b)^2 .$$

Given the pseudoinverse $A^+ = V\Sigma^+ U^T$ we can also compute the solution of the linear least squares problem using the formula (10.4) as $x^* = A^+ b$.

10.5.5 Singular value decomposition in Python

We can compute the Singular Value Decomposition of any matrix A in Python just by using the function `svd` of the module `numpy.linalg`. For example, typing `U, S, V = svd(`

A) we get the three matrices $\boldsymbol{U}$, $\boldsymbol{\Sigma}$ and $\boldsymbol{V}^T$ of the *SVD*. The matrix S contains only the nonzero diagonal entries of $\boldsymbol{\Sigma}$. The general call of the function `svd` is

```
svd(a, full_matrices, compute_uv,hermitian)
```

where

`a`: Is an $m \times n$ array

`full_matrices`: Is a boolean variable, which is `True` if the output arrays `u` and `vh` have shapes `(m,m)` and `(n,n)`, respectively. Otherwise, the shapes are `(m,k)` and `(k,n)`, respectively, where `k=min(m,n)`. (Optional with default value `True`)

`compute_uv`: Is a boolean variable that specifies whether to compute the matrices `u` and `vh` in addition to `s`. (Optional with default value `True`)

`hermitian`: A boolean variable determining whether the matrix $\boldsymbol{A}$ is Hermitian (symmetric if real-valued). (Optional with default value `True`)

The function `svd` returns the arrays `u`, `s` and `vh` with the matrices $\boldsymbol{U}$, $\boldsymbol{\Sigma}$ and $\boldsymbol{V}^T$ respectively.

For example, let us consider the matrix

$$\boldsymbol{A} = \begin{pmatrix} 1 & 2 & 0 \\ 2 & 1 & 2 \\ 0 & 2 & 1 \end{pmatrix} ,$$

which is symmetric. Because $\boldsymbol{A}$ is symmetric, its singular values coincide with the absolute values of the eigenvalues of $\boldsymbol{A}$. Let's see if this is true using the following code

```
A = np.array([[1.0, 2.0, 0.0],
              [2.0, 1.0, 2.0],
              [0.0, 2.0, 1.0]])
U, S, V = npl.svd( A )
print(S)
print(npl.eigvals(A))
```

```
[3.82842712 1.82842712 1.        ]
[ 3.82842712  1.         -1.82842712]
```

Observe that S contains the absolute values of the eigenvalues of $\boldsymbol{A}$, ordered from the one with maximum magnitude to the minimum. This is because given an eigenvalue λ of $\boldsymbol{A}$, then the eigenvalues of $\boldsymbol{A}^T\boldsymbol{A}$ satisfy

$$\boldsymbol{A}^T\boldsymbol{A}\boldsymbol{x} = \boldsymbol{A}^T\lambda\boldsymbol{x} = \lambda\boldsymbol{A}\boldsymbol{x} = \lambda^2\boldsymbol{x} ,$$

and thus are the eigenvalues of $\boldsymbol{A}$ squared. If we want to verify the result of the `svd` we need to transform the diagonal matrix S into a rectangular matrix. In the previous example we can verify the result $\boldsymbol{A} = \boldsymbol{U}\boldsymbol{\Sigma}\boldsymbol{V}^T$ by typing

```
# Convert matrix S into rectangular matrix
Sigma = np.zeros((A.shape[0], A.shape[1]))
Sigma[:A.shape[1], :A.shape[1]] = np.diag(S)
print(U.dot(Sigma.dot(V)))
```

```
[[ 1.00000000e+00  2.00000000e+00 -6.66133815e-16]
 [ 2.00000000e+00  1.00000000e+00  2.00000000e+00]
 [-4.99600361e-16  2.00000000e+00  1.00000000e+00]]
```

10.5.6 Application in image compression

Here, we demonstrate how to use the *SVD* in image compression. We have seen in Section 6.6.6 how to manipulate digital images using arrays. Any digital color image is an arranged set of pixels. Each pixel has each own color, which can be represented by its values of the intensity of red, green and blue colors (RGB). These can be a tuple of the form (R, G, B) where R stands for Red, G for Green and B for Blue. The digital colors can take values from 0 to 255 and usually are integer values. The value 0 indicates the complete absence of a specific color. For example, the blue color in the RGB scale will be $(0, 0, 255)$, indicating that there is no red and green since their values are zero and there is only blue with value 255.

In Python we can read an image using the command `imread` of the module `imageio`. Let's try using a photo with the name `Newton.jpg`. The `imread` function in the following code reads the digital image stored in the file `Newton.jpg` and stores it into the rectangular array `photo`. Printing the shape of the array `photo`, we get that this array is $559 \times 407 \times 3$, which means that our image consists of 559×407 pixels, with RGB values stored in three slices of the array `photo`. We can print the photo on the screen with the command `imshow` of MatPlotLib.

```
import imageio
photo = imageio.imread("Newton.jpg") # Read the image into the array photo
print(photo.shape)
plt.imshow(photo) # Plot the image on the screen
plt.show()
```

In order to do some image processing we store each slice of the array `photo` in a different matrix, for example storing the values of the red intensity in the matrix `Red`, the values of the green intensity in `Green` and the respective values for the blue in the matrix `Blue`. In this way we will have three 559×407 matrices. If we want to explore the intensity of the colors, then we will need to put the three matrices back in three dimensional arrays and plot them using the command `imshow`.

```
# Plot the different matrices using imshow
f, axs = plt.subplots(2,2,figsize=(15,15))
# Separate the three basic colors
Red[:,:,0] = photo[:,:,0]; Green[:,:,1] = photo[:,:,1];
Blue[:,:,2] = photo[:,:,2]
plt.subplot(2,2,1); plt.imshow(Red); plt.subplot(2,2,2); plt.imshow(Green)
plt.subplot(2,2,3); plt.imshow(Blue); plt.subplot(2,2,4); plt.imshow(photo)
plt.show()
```

In this figure we observe that the first three matrices contain only red, green and blue colors of different intensity, while the last image, which is the original one contains all the information of the picture.

Now we perform the *SVD* to these three matrices, and approximate them by keeping only k singular values. Specifically, we approximate the arrays `Red`, `Green` and `Blue`, using the property $\boldsymbol{A} \approx \boldsymbol{A}_k = \boldsymbol{U}\boldsymbol{\Sigma}_k\boldsymbol{V}^T$. Then we combine the three approximate matrices into one (putting them in a three-dimensional array), and we plot the approximate matrix. We will try several values of k, including small and large values, to see how good the approximation is. Observe first by executing the following code that the rank of the original matrix is 407. So we choose to compute approximations of the original matrix of rank $5, 10, 20, 40, 100$ and 400. We observe that there are practically no significant differences between the approximation up to rank 40. The image becomes blurred as we use smaller rank matrices (less singular values) but even with the tiny matrix of rank 5 we can still recognize the image.

```
U_r,S_r,V_r = svd(Red); U_g,S_g,V_g = svd(Green); U_b,S_b,V_b = svd(Blue)
s = len(S_r)
sequence = [5, 10, 20, 40, 100, 400]
f, axs = plt.subplots(2,3,figsize=(15,15))
j=0
for k in sequence:
    U_r_c = U_r[:,0:k]; V_r_c = V_r[0:k,:]; S_r_c = diag(S_r[0:k])
    U_g_c = U_g[:,0:k]; V_g_c = V_g[0:k,:]; S_g_c = diag(S_g[0:k])
    U_b_c = U_b[:,0:k]; V_b_c = V_b[0:k,:]; S_b_c = diag(S_b[0:k])
    comp_img_r = dot(U_r_c, dot(S_r_c,V_r_c))
    comp_img_g = dot(U_g_c, dot(S_g_c,V_g_c))
    comp_img_b = dot(U_b_c, dot(S_b_c,V_b_c))
    comp_img = zeros((row, col, 3))
    comp_img[:,:,0] = comp_img_r
    comp_img[:,:,1] = comp_img_g
    comp_img[:,:,2] = comp_img_b
    comp_img[comp_img < 0] = 0; comp_img[comp_img > 1] = 1
```

```
    j=j+1
    plt.subplot(2,3,j); plt.title('Rank %d'%(k))
    plt.imshow(comp_img)
plt.show()
```

Compared to the original image we can see that the compressed images of the *SVD* are indeed of very good quality. Moreover, instead of 407 singular values and the respective singular eigenvectors even 20 are enough for a good quality image. This can be very useful especially when we deal with huge amount of data. We can approximate a huge matrix by using its *SVD* and then analyze its properties using approximations and still get safe results. This is why the *SVD* is considered one of the pillars of machine learning.

10.6 Further Reading

The material covered in this chapter such as the *QR* factorization and the *SVD* decomposition are usually included in all classical numerical analysis and linear algebra textbooks. Many details can be found in more specialized books such as [48, 135, 24, 29, 12]. The Gram-Schmidt orthonormalization is also described in most linear algebra books [128, 129].

Chapter Highlights

- The best approximation $\boldsymbol{v} \in Y \subset X$ of an element $\boldsymbol{u} \in X$ is a solution that minimizes the error norm of the form $\|\boldsymbol{u} - \boldsymbol{w}\|$ for all $\boldsymbol{w} \in Y$.

- The linear least squares problem is the problem of finding a solution $\boldsymbol{x}^*$ that minimizes the error $\|\boldsymbol{b} - \boldsymbol{A}\boldsymbol{x}\|_2^2$ with $\boldsymbol{A} \in \mathbb{R}^{m,n}$ $(m \geq n)$ and $\boldsymbol{b} \in \mathbb{R}^n$. This is written as $\boldsymbol{x}^* = \text{argmin}\|\boldsymbol{b} - \boldsymbol{A}\boldsymbol{x}\|_2^2$.

- We can compute the solution of the linear least squares problem by solving the linear system $\boldsymbol{A}^T\boldsymbol{A}\boldsymbol{x} = \boldsymbol{A}^T\boldsymbol{b}$.

- The pseudoinverse of an $m \times n$ matrix $\boldsymbol{A}$ is a matrix $\boldsymbol{A}^+$ such that $\boldsymbol{A}^+\boldsymbol{A} = \boldsymbol{I}$. If $\boldsymbol{A}^T\boldsymbol{A}$ is invertible, then the pseudoinverse is given explicitly by the formula $\boldsymbol{A}^+ = (\boldsymbol{A}^T\boldsymbol{A})^{-1}\boldsymbol{A}^T$.

- The pseudoinverse leads to the solution $\boldsymbol{x}^* = \boldsymbol{A}^+\boldsymbol{b}$ of linear least squares problem after multiplying both sides of the equation $\boldsymbol{A}\boldsymbol{x} = \boldsymbol{b}$ with $\boldsymbol{A}^+$.

- The Gram-Schmidt orthonormalization is an algorithm that generates a set of orthonormal vectors from a set of linearly independent vectors. The orthonormal set of vectors generated by the Gram-Schmidt orthonormalization algorithm can serve as a basis for a linear space.

- Any $m \times n$ matrix $\boldsymbol{A}$ (with $m \geq n$) can be factorized in the form $\boldsymbol{A} = \boldsymbol{Q}\boldsymbol{R}$ where $\boldsymbol{Q}$ is $m \times m$ matrix with orthonormal columns and $\boldsymbol{R}$ is an $m \times n$ upper triangular matrix with nonzero diagonal entries.

- In practice we use the QR decomposition of an $m \times n$ matrix $\boldsymbol{A}$ instead of the Gram-Schmidt orthonormalization due to the complexity and numerical instability of its algorithm. Applications of the QR factorization includes the computation of the pseudoinverse matrix, the linear least squares problem and the range of a matrix.

- Householder matrix is a matrix of the form $\boldsymbol{H} = \boldsymbol{I} - 2\boldsymbol{u}\boldsymbol{u}^T/\|\boldsymbol{u}\|^2$, for any vector $\boldsymbol{u}$. Householder matrices are symmetric and orthogonal in the sense $\boldsymbol{H}^{-1} = \boldsymbol{H}^T = \boldsymbol{H}$. Householder matrices can be used for the computation of the QR decomposition.

- The Singular Value Decomposition of an $m \times n$ matrix $\boldsymbol{A}$ is the factorization $\boldsymbol{A} = \boldsymbol{U}\boldsymbol{\Sigma}\boldsymbol{V}^T$ where $\boldsymbol{U}$ is $m \times m$ and $\boldsymbol{V}$ is $n \times n$ orthogonal matrices, while $\boldsymbol{\Sigma}$ is an $m \times n$ diagonal matrix with the r eigenvalues of the matrix $\boldsymbol{A}^T\boldsymbol{A}$ in the main diagonal. The eigenvalues of $\boldsymbol{A}^T\boldsymbol{A}$ are called singular values of $\boldsymbol{A}$.

- The best rank k approximation matrix of $\boldsymbol{A} = \boldsymbol{U}\boldsymbol{\Sigma}\boldsymbol{V}^T$ is the matrix $\boldsymbol{A}_k = \boldsymbol{U}\boldsymbol{\Sigma}_k\boldsymbol{V}^T$ where $\boldsymbol{\Sigma}_k$ contains only the first k singular values of $\boldsymbol{A}$.

- Applications of SVD include the solution of the linear least squares problem, the computation of the pseudoinverse matrix $\boldsymbol{A}^+$, the condition number and various matrix norms as well as applications in other fields of science such as image processing and machine learning.

Exercises

1. Consider the space $\mathbb{P}_k$ of polynomials of degree k, with the inner product

$$\langle f, g\rangle = \int_0^1 f(x)g(x)\, dx\ .$$

   (a) Show that the polynomials $p_0(t) = 1$, $p_1(t) = t$, $p_2(t) = t^2$ and $p_3(t) = t^3$ are linearly independent in $\mathbb{P}_3$, and that any polynomial of degree 3 can be written as a linear combination of these functions.

   (b) Starting with the set $\{p_0, p_1, p_2\}$ construct using the Gram-Schmidt orthonormalization process a set of orthonormal polynomials $\{u_0, u_1, u_2\} = \{1, t-1/2. t^2-t+1/6\}$.

   (c) Find a quadratic polynomial $p(t) \in \mathbb{P}_2$, which is the closest to $p_3(t) = t^3$. [Hint: $p(t) = 3/2t^2 - 3/5t + 1/20$.

2. Find the best approximation of the function $f(x) = e^x$ in the space $\mathbb{P}_3[-1,1]$ of cubic polynomials in $[-1,1]$ with the norm

$$\|v\|_2 = \sqrt{\int_{-1}^1 |v(x)|^2\, dx}, \quad v \in C[-1,1]\ .$$

3. Consider the matrix

$$\boldsymbol{A} = \begin{pmatrix} 1 & 0 & 1 \\ 1 & 0 & -1 \\ 0 & 1 & 1 \\ 0 & 1 & -1 \end{pmatrix} \qquad \text{and the vector} \qquad \boldsymbol{b} = \begin{pmatrix} 1 \\ 1 \\ 1 \\ 1 \end{pmatrix}\ .$$

   (a) Solve the linear least squares problem $\operatorname{argmin}\|\boldsymbol{b} - \boldsymbol{A}\boldsymbol{x}\|_2^2$ using the corresponding normal equations. In order to solve the normal equations employ appropriate numerical method of your choice such as Cholesky decomposition.

   (b) Solve the same problem using the QR decomposition of $\boldsymbol{A}$.

   (c) Find the value $\min\|\boldsymbol{b} - \boldsymbol{A}\boldsymbol{x}\|_2^2$.

4. Let $\boldsymbol{A}$ be an $m \times n$ matrix and $\|\cdot\|$ the Euclidean vector norm. Prove that

   (a) $\|\boldsymbol{A}\| = \max\{|\langle \boldsymbol{A}\boldsymbol{x}, \boldsymbol{y}\rangle : \boldsymbol{x} \in \mathbb{R}^n,\ \boldsymbol{y} \in \mathbb{R}^m,\ \|\boldsymbol{x}\| = \|\boldsymbol{y}\| = 1\}$,

   (b) $\|\boldsymbol{A}\| = \|\boldsymbol{A}^T\|$,

   (c) $\|\boldsymbol{A}^T\boldsymbol{A}\| = \|\boldsymbol{A}\|^2$.

5. Find the singular value decomposition SVD of the following matrices:

   (a) $\boldsymbol{A}_1 = \begin{pmatrix} -2 & 0 \\ 0 & -1 \end{pmatrix}$,

   (b) $\boldsymbol{A}_2 = \begin{pmatrix} 1 & 2 \\ 2 & 1 \end{pmatrix}$,

   (c) $\boldsymbol{A}_3 = \begin{pmatrix} 1 & 1 & 0 & 1 \\ 0 & 0 & 0 & 1 \\ 1 & 1 & 0 & 0 \end{pmatrix}$.

6. Consider the matrix

$$A = \begin{pmatrix} 1 & 1 & 1 \\ 1 & 1 & 0 \\ -1 & 0 & 1 \\ 0 & 0 & 1 \end{pmatrix} .$$

(a) Use the Python function `qr` to compute the QR decomposition of $\boldsymbol{A}$.

(b) Use the QR decomposition to compute the solution of the linear least squares problems $\boldsymbol{Ax} = \boldsymbol{b}$ with $\boldsymbol{b} = (1, 2, 3, 4)^T$.

(c) Compute the SVD of $\boldsymbol{A}$ using the Python function `svd`.

7. Prove that the pseudoinverse $\boldsymbol{A}^+$ of an $m \times n$ matrix $\boldsymbol{A}$ is the unique matrix satisfying:

(a) $\boldsymbol{AA}^+\boldsymbol{A} = \boldsymbol{A}$,

(b) $\boldsymbol{A}^+\boldsymbol{AA}^+ = \boldsymbol{A}^+$,

(c) $(\boldsymbol{AA}^+)^T = \boldsymbol{AA}^+$,

(d) $(\boldsymbol{A}^+\boldsymbol{A})^T = \boldsymbol{A}^+\boldsymbol{A}$.

8. Prove that for all $\boldsymbol{A} \in \mathbb{R}^{m\times n}$,

(a) $(\boldsymbol{A}^T)^+ = (\boldsymbol{A}^+)^T$,

(b) $(\boldsymbol{A}^+)^+ = \boldsymbol{A}$

(c) $(\boldsymbol{A}^T\boldsymbol{A})^+ = \boldsymbol{A}^+(\boldsymbol{A}^T)^+$,

(d) $(\boldsymbol{AA}^T)^+ = (\boldsymbol{A}^T)^+$,

(e) $\mathcal{R}(\boldsymbol{A}^+) = \mathcal{R}(\boldsymbol{A}^T) = \mathcal{R}(\boldsymbol{A}^+\boldsymbol{A}) = \mathcal{R}(\boldsymbol{A}^T\boldsymbol{A})$, where $\mathcal{R}(\boldsymbol{C}) = \{\boldsymbol{y} \in \mathbb{R}^m : \boldsymbol{Cx} = \boldsymbol{y}\}$ is the range of $\boldsymbol{C} \in \mathbb{R}^{m\times n}$,

(f) $\mathcal{N}(\boldsymbol{A}^+) = \mathcal{N}(\boldsymbol{AA}^+) = \mathcal{N}((\boldsymbol{AA}^T)^+) = \mathcal{N}(\boldsymbol{AA}^T) = \mathcal{N}(\boldsymbol{A}^T)$, where $\mathcal{N}(\boldsymbol{C}) = \{\boldsymbol{x} \in \mathbb{R}^n : \boldsymbol{Cx} = \boldsymbol{0}\}$ is the kernel (or null space) of $\boldsymbol{C} \in \mathbb{R}^{m\times n}$,

(g) If $\boldsymbol{A} \in \mathbb{R}^{n\times n}$ is normal such that $\boldsymbol{AA}^T = \boldsymbol{A}^T\boldsymbol{A}$ then $\boldsymbol{A}^k\boldsymbol{A}^+ = \boldsymbol{A}^+\boldsymbol{A}^k$ for all $k > 0$, and $(\boldsymbol{A}^k)^+ = (\boldsymbol{A}^+)^k$ for all $k > 0$.

9. Let $\boldsymbol{A} \in \mathbb{R}^{m\times n}$, $\boldsymbol{b} \in \mathbb{R}^n$, and suppose $\boldsymbol{AA}^+\boldsymbol{b} = \boldsymbol{b}$. Then for any vector $\boldsymbol{y} \in \mathbb{R}^n$ the vector $\boldsymbol{x} = \boldsymbol{A}^+\boldsymbol{b} + (I - \boldsymbol{A}^+\boldsymbol{A})\boldsymbol{y}$ is a solution to the system $\boldsymbol{Ax} = \boldsymbol{b}$.

10. For any $\boldsymbol{A} \in \mathbb{R}^{m\times n}$, prove that

(a) $\mathcal{R}(\boldsymbol{A}) = \mathcal{R}(\boldsymbol{AA}^T)$,

(b) $\mathcal{R}(\boldsymbol{A}^+) = \mathcal{R}(\boldsymbol{A}^T)$.

11. A planet P is moving in the space following an elliptic orbit. If (x, y) is the location of the planet at some time in the space, then the equation describing its orbit is given by the formula

$$c_1x^2 + c_2xy + c_3y^2 + c_4x + c_5y = 634.37 ,$$

for some parameters $c = (c_1, c_2, \ldots, c_5)^T$. Suppose that the following set of observed locations (x_k, y_k), $k = 1, \ldots, 11$ has been recorded

x	−0.14	−3.76	3.85	−3.31	−2.84	−1.97	0.42	2.15	2.06	1.53	0.31	−2.51
y	1.51	−2.32	−1.27	−0.18	0.33	0.95	1.50	−0.27	−0.72	−1.67	−2.69	−3.24

(a) Find an overdetermined system $\boldsymbol{Ac} = \boldsymbol{b}$ by using the recorded data for the coefficients c_i, $i = 1, \ldots, 5$.

(b) Solve the system $\boldsymbol{Ac} = \boldsymbol{b}$ using the function `lstsq` of `numpy.linalg`.

(c) Repeat the previous questions using 9, 10 and 11 recorded locations instead of 12 and plot the different orbits using the following script

```
import numpy as np
import matplotlib.pyplot as plt
xx = [-0.14, -3.76,-3.85,-3.31,-2.84,-1.97,
         0.42,2.15,2.06,1.53,0.31,-2.51]
yy = [1.51, -2.32,-1.27,-0.18,0.33,0.95,1.5,
         -0.27,-0.72,-1.67,-2.69,-3.24]
x = np.linspace(-5,5,1000)
y = np.linspace(-5,5,1000)
X,Y = np.meshgrid(x,y)
eqn = c[0]* X**2 + c[1]*X*Y + c[2]*Y**2 + c[3]*X + c[4]*Y-634.37
Z = 1
plt.plot(xx,yy,'o')
plt.contour(X,Y,eqn,[Z])
plt.xlim([-5,5])
plt.ylim([-5,5])
plt.show()
```

(d) In all cases compute the condition number of $\boldsymbol{A}$. What do you observe?

(e) The exact values of the coefficients are given by the vector

$$\boldsymbol{c} = (93.21, -75.32, 146.26, 94.78, 196, 29)^T \ .$$

What is the error between the linear least squares approximations and the exact values?

11

Unconstrained Optimization and Neural Networks

The main objective of optimization is the study of methods for computing best approximations with special focus on the minimization of functions. We have already studied the least squares problem, which is the problem of finding the vector $\boldsymbol{x}^* \in \mathbb{R}^n$ that minimizes the error function

$$E(\boldsymbol{x}) = \|\boldsymbol{b} - \boldsymbol{A}\boldsymbol{x}\|_2^2 ,$$

for appropriate matrix $\boldsymbol{A} \in \mathbb{R}^{m,n}$ and vector $\boldsymbol{b} \in \mathbb{R}^m$. To solve such problems we employed direct methods such as the QR decomposition. In this chapter, we will study numerical methods for the approximation of vectors $\boldsymbol{x}^* \in \mathbb{R}^n$ that minimize general functions $f(\boldsymbol{x})$. These methods are iterative methods and are used often in problems of data science and machine learning. For this reason, at the end of this chapter, we present an introduction to neural networks with some applications.

11.1 Gradient Methods for Unconstrained Optimization

We consider a real function $f : \mathbb{R}^n \to \mathbb{R}$ (often called *objective function*) with

$$f := f(x_1, x_2, \ldots, x_n) \equiv f(\boldsymbol{x}), \quad \boldsymbol{x} \in \mathbb{R}^n .$$

A point $\boldsymbol{x}^* \in \mathbb{R}^n$ is called *minimizer* of f if

$$f(\boldsymbol{x}^*) \leq f(\boldsymbol{x}) \quad \text{for all} \quad \boldsymbol{x} \in D \subset \mathbb{R}^n ,$$

where D is a part or the whole of the domain of f. If D is the entire domain of the function f, we call $\boldsymbol{x}^*$ *global minimizer*. If there is another minimum value in a different region of the domain of f, then $\boldsymbol{x}^*$ is called *local minimizer*. There is only one global minimizer, while there might be several local minimizers. The vector $\boldsymbol{x}^*$ that minimizes the function f is called *minimizer*, and we write

$$\boldsymbol{x}^* = \text{argmin} f(\boldsymbol{x}) .$$

Similar definitions can be used for the maxima of functions, but because finding maxima of a function f is the same with finding the minima of the function $-f$, i.e. $\max(f) = -\min(-f)$, we focus on finding minimizers only[1]. Moreover, we will only consider minimizers that are not on the boundary of the domain of f but rather an interior point of D where f is sufficiently smooth.

Convexity is also a factor that plays crucial role in the discussion below. For this reason, the branch of optimization of convex functions is called convex optimization f, [13]. A function f is called *convex* if

$$f(p\boldsymbol{x} + (1-p)\boldsymbol{y}) < pf(\boldsymbol{x}) + (1-p)f(\boldsymbol{y}) \quad \text{for} \quad p \in (0,1) \quad \text{and} \quad \boldsymbol{x}, \boldsymbol{y} \in D .$$

[1]Let $y = \max(f)$, then for all x we have $f(x) \geq y$, which implies $-f(x) \leq -y$ for all x. Therefore, $\min(-f) = -y$ and thus $y = -\min(-f)$.

DOI: 10.1201/9781003287292-11

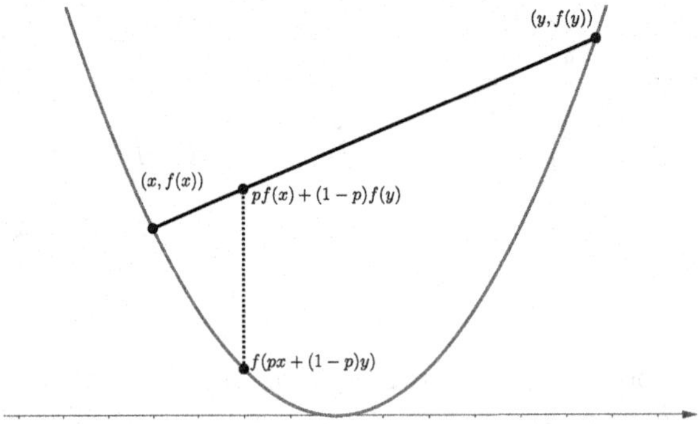

FIGURE 11.1
An example of convex function.

(a) Saddle point of $f(\boldsymbol{x}) = x_1^2 - x_2^2$ (b) Global minimum of $f(\boldsymbol{x}) = x_1^2 + x_2^2$

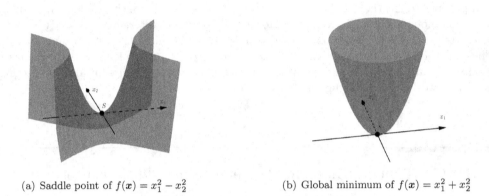

FIGURE 11.2
Multivariable functions and its extreme values.

Note that this definition requires $p\boldsymbol{x} + (1-p)\boldsymbol{y} \in D$ since D is the domain of f. The definition implies that the line segment $pf(\boldsymbol{x}) + (1-p)f(\boldsymbol{y})$ with $p \in (0,1)$ that connects the point $(\boldsymbol{x}, f(\boldsymbol{x}))$ with the point $(\boldsymbol{y}, f(\boldsymbol{y}))$ is above the graph of the function. Convexity ensures the existence of minima, and thus the solution to the optimization problem. An example of a convex function is the function $f(x) = x^2$ (see Figure 11.1). In general, we can imagine convex functions to have similar shape to that of the x^2 function.

Optimization of functions relies also on the gradient of a function f, which is defined as the vector with the partial derivatives of f

$$\nabla f(\boldsymbol{x}) = \begin{pmatrix} \frac{\partial f}{\partial x_1} & \frac{\partial f}{\partial x_2} & \cdots & \frac{\partial f}{\partial x_n} \end{pmatrix}^T .$$

Recall from multivariable calculus that the local minima of a smooth function f will be the *critical points* of f. If f is smooth, then the critical points coincide with the zeros of the gradient. More precisely, a horizontal tangent plane to f has slope

$$\nabla f(\boldsymbol{x}) = 0 .$$

Note that there might be critical points that are not minimizers of f. These are called *saddle points*. For example, the point $x_1 = 0$ is a saddle point of the function $f(\boldsymbol{x}) = x_1^3$. Figure 11.2 shows a case with a saddle in three dimensions.

In general, we assume that the objective function f is sufficiently smooth, in the sense that all the derivatives are well defined and continuous. The Taylor expansion of a smooth, multivariable function f in a neighborhood of a vector $\boldsymbol{x}$ can be written in the form

$$f(\boldsymbol{x}+\boldsymbol{u}) = f(\boldsymbol{x}) + \nabla f(\boldsymbol{x})^T \boldsymbol{u} + \frac{1}{2}\boldsymbol{u}^T \boldsymbol{H}(\boldsymbol{x})\boldsymbol{u} + O(\|\boldsymbol{u}\|^3) , \tag{11.1}$$

where $\boldsymbol{u}$ is a vector close enough to $\boldsymbol{x}$, and $\boldsymbol{H}$ is the Hessian matrix with entries

$$H_{ij}(\boldsymbol{x}) = \frac{\partial^2 f(\boldsymbol{x})}{\partial x_i \partial x_j} .$$

If ∇f and $\boldsymbol{H}$ are well defined, and if $\boldsymbol{x}^*$ is a minimizer of f, then $\nabla f(\boldsymbol{x}^*) = 0$, and $\boldsymbol{H}(\boldsymbol{x}^*)$ is positive semidefinite. Conversely, if $\nabla f(\boldsymbol{x}^*) = 0$ and $\boldsymbol{H}(\boldsymbol{x}^*)$ is positive definite, then $\boldsymbol{x}^*$ is a local minimizer. This generalizes the second derivative test of single variable calculus.

If we consider the first order Taylor expansion of the function f in (11.1), we have

$$f(\boldsymbol{x}+\boldsymbol{u}) = f(\boldsymbol{x}) + \nabla f(\boldsymbol{\xi})^T \boldsymbol{u} ,$$

for some $\boldsymbol{\xi} \in D$. Choosing a direction $\boldsymbol{u}$ such that $\nabla f(\boldsymbol{\xi})^T \boldsymbol{u} < 0$, then we observe that

$$f(\boldsymbol{x}+\boldsymbol{u}) < f(\boldsymbol{x}) .$$

This direction $\boldsymbol{u}$ is called *descent* direction since it leads to smaller values of f.

Denote $\langle \cdot, \cdot \rangle \equiv \langle \cdot, \cdot \rangle_2$ and $\|\cdot\| \equiv \|\cdot\|_2$ the Euclidean inner product and norm, respectively. From multivariable calculus we know that the gradient $\nabla f(\boldsymbol{x}_0)$ is orthogonal to the tangent line of the level set $f(\boldsymbol{x}) = f(\boldsymbol{x}_0)$, and behaves like a compass which shows the direction of most rapid increase of f. This means that if we are moving on the graph of $f(\boldsymbol{x})$ and we wish to move in the direction with the most rapid increase, we shall follow the direction of the vector $\nabla f(\boldsymbol{x}_0)$. Moreover, if we choose to move along the direction of a different vector $\boldsymbol{u}$, with $\|\boldsymbol{u}\| = 1$, then the rate of increase of f in the direction of $\boldsymbol{u}$ at the point $\boldsymbol{x}_0$ is the inner product $\langle \nabla f(\boldsymbol{x}_0), \boldsymbol{u} \rangle = \nabla f(\boldsymbol{x}_0)^T \boldsymbol{u}$. Recall that the inner product $\langle \nabla f(\boldsymbol{x}_0), \boldsymbol{u} \rangle$ is the directional derivative of f and is usually denoted by

$$\frac{\partial f(\boldsymbol{x})}{\partial \boldsymbol{u}} = \nabla f(\boldsymbol{x})^T \boldsymbol{u} .$$

In order to verify that moving towards the direction of $\nabla f(\boldsymbol{x}_0)$ we move towards higher points, we take the inner product between $\nabla f(\boldsymbol{x}_0)$ and a vector $\boldsymbol{u}$ with $\|\boldsymbol{u}\| = 1$. Using the Cauchy-Schwarz inequality (Theorem 9.3) we obtain

$$\langle \nabla f(\boldsymbol{x}_0), \boldsymbol{u} \rangle \leq \|\nabla f(\boldsymbol{x}_0)\| \cdot \|\boldsymbol{u}\| = \|\nabla f(\boldsymbol{x}_0)\| .$$

This means that the rate of increase has the upper bound $\|\nabla f(\boldsymbol{x}_0)\|$.

On the other hand, for a particular value of $\boldsymbol{u} = \boldsymbol{u}_0$ with

$$\boldsymbol{u}_0 = \frac{\nabla f(\boldsymbol{x}_0)}{\|\nabla f(\boldsymbol{x}_0)\|} ,$$

we have

$$\|\nabla f(\boldsymbol{x}_0)\| = \langle \nabla f(\boldsymbol{x}_0), \boldsymbol{u}_0 \rangle ,$$

which means that the maximum rate of increase can be achieved if we follow the direction of the vector $\boldsymbol{u}_0$.

If we take $\boldsymbol{u} \in \mathbb{R}^n$, $\boldsymbol{u} \neq \boldsymbol{0}$, fixed and scaled by $\lambda \in \mathbb{R}$, then the Taylor expansion of the function f around the vector $\boldsymbol{x}_0$ implies for the function $g(\lambda) = f(\boldsymbol{x}_0 + \lambda \boldsymbol{u})$ that

$$g(\lambda) = f(\boldsymbol{x}_0 + \lambda \boldsymbol{u}) = f(\boldsymbol{x}_0) + \lambda \langle \nabla f(\boldsymbol{x}_0), \boldsymbol{u} \rangle + O(\lambda^2) \ . \tag{11.2}$$

Obviously, $g(0) = f(\boldsymbol{x}_0)$ and $g'(0) = \langle \nabla f(\boldsymbol{x}_0), \boldsymbol{u} \rangle$ and thus, g is increasing if $\langle f(\boldsymbol{x}_0), \boldsymbol{u} \rangle > 0$ and is decreasing when $\langle f(\boldsymbol{x}_0), \boldsymbol{u} \rangle < 0$. Solving (11.2) for $\lambda \langle \nabla f(\boldsymbol{x}_0), \boldsymbol{u} \rangle$ we observe that for small enough $\lambda > 0$, the difference $f(\boldsymbol{x}_0 + \lambda \boldsymbol{u}) - f(\boldsymbol{x}_0)$ has the same sign as $\langle \nabla f(\boldsymbol{x}_0), \boldsymbol{u} \rangle$. So when $g(\lambda)$ is decreasing, then the decreasing rate is maximal if and only if $\langle \nabla f(\boldsymbol{x}_0), \boldsymbol{u} \rangle = \|\nabla f(\boldsymbol{x}_0)\| \|\boldsymbol{u}\|$. This means that the direction of $\boldsymbol{u} = -\nabla f(\boldsymbol{x}_0)$ is the one that leads to the fastest descent.

From the previous analysis we deduce that moving in the direction of $-\nabla f(\boldsymbol{x}_0)$ leads to a lower point such that

$$f(\boldsymbol{x}_0 - \lambda \nabla f(\boldsymbol{x}_0)) = f(\boldsymbol{x}_0) - \lambda \|\nabla f(\boldsymbol{x}_0)\|^2 + O(\lambda^2) < f(\boldsymbol{x}_0) + O(\lambda^2) \ .$$

This leads to the definition of an iterative method for which starting from an initial guess of the minimizer $\boldsymbol{x}^{(0)}$ we approximate the minimizer $\boldsymbol{x}^*$ of $f(\boldsymbol{x})$ by the iteration

$$\boldsymbol{x}^{(k+1)} = \boldsymbol{x}^{(k)} - \lambda \nabla f(\boldsymbol{x}^{(k)}) \quad k = 0, 1, 2, \dots \ . \tag{11.3}$$

In order to show that the sequence $\boldsymbol{x}^{(k)}$ converges to a minimizer, assume that $\boldsymbol{x}^{(k)} \to \boldsymbol{x}^*$ as $k \to \infty$ for some $\boldsymbol{x}^*$. Then the same holds true for the $\boldsymbol{x}^{(k+1)}$, and thus we have that $\nabla f(\boldsymbol{x}^{(k)}) \to \nabla f(\boldsymbol{x}^*) = 0$.

This is the main idea behind gradient methods. The parameter λ in the language of machine learning is called *learning rate*, and can take any arbitrary positive value. To achieve optimal speed of convergence we assign to λ different values $\lambda = \lambda_k$, at every iteration $k = 0, 1, 2, \dots$. This is the subject of the next few paragraphs.

11.1.1 Estimation of convergence

Let us first estimate the convergence of general gradient method through an example. In particular, consider the general gradient method of the form

$$\boldsymbol{x}^{(k+1)} = \boldsymbol{x}^{(k)} - \lambda \nabla f(\boldsymbol{x}^{(k)}) \ , \tag{11.4}$$

for the minimization of $f(\boldsymbol{x}) = \frac{1}{2}(2x_1^2 + x_2^2)$. In this case, the Hessian matrix is

$$\boldsymbol{H} = \begin{pmatrix} 2 & 0 \\ 0 & 1 \end{pmatrix} \ .$$

Thus, for every $\boldsymbol{y} \in \mathbb{R}^2$

$$\boldsymbol{y}^T \boldsymbol{H} \boldsymbol{y} = \begin{pmatrix} y_1 & y_2 \end{pmatrix} \begin{pmatrix} 2 & 0 \\ 0 & 1 \end{pmatrix} \begin{pmatrix} y_1 \\ y_2 \end{pmatrix} = 2y_1^2 + 1y_2^2 \leq 2\|\boldsymbol{y}\|^2 \ ,$$

and similarly that

$$\boldsymbol{y}^T \boldsymbol{H} \boldsymbol{y} \geq \|\boldsymbol{y}\|^2 \ .$$

Then from the Taylor expansion (11.1) we estimate

$$\begin{aligned} f(\boldsymbol{x}^{(k+1)}) &= f(\boldsymbol{x}^{(k)}) + \nabla f(\boldsymbol{x}^{(k)})^T (\boldsymbol{x}^{(k+1)} - \boldsymbol{x}^{(k)}) + \frac{1}{2} (\boldsymbol{x}^{(k+1)} - \boldsymbol{x}^{(k)})^T \boldsymbol{H} (\boldsymbol{x}^{(k+1)} - \boldsymbol{x}^{(k)}) \\ &\leq f(\boldsymbol{x}^{(k)}) + \nabla f(\boldsymbol{x}^{(k)})^T (\boldsymbol{x}^{(k+1)} - \boldsymbol{x}^{(k)}) + \|\boldsymbol{x}^{(k+1)} - \boldsymbol{x}^{(k)}\|^2 \ . \end{aligned}$$

Equation (11.4) implies $\boldsymbol{x}^{(k+1)} - \boldsymbol{x}^{(k)} = -\lambda \nabla f(\boldsymbol{x}^{(k)})$ and thus

$$f(\boldsymbol{x}^{(k+1)}) = f(\boldsymbol{x}^{(k)}) - \lambda \|\nabla f(\boldsymbol{x}^{(k)})\|^2 + \lambda^2 \|\nabla f(\boldsymbol{x}^{(k)})\|^2 .$$

The value λ_{opt} that minimizes the quadratic polynomial on the right-hand side of the above equation can be found easily to be 1/2. Thus,

$$f(\boldsymbol{x}^{(k+1)}) \leq f(\boldsymbol{x}^{(k)}) - \frac{1}{4}\|\nabla f(\boldsymbol{x}^{(k)})\|^2 . \tag{11.5}$$

Working in the same way we obtain

$$f(\boldsymbol{x}) \geq f(\boldsymbol{x}^{(k)}) + \nabla f(\boldsymbol{x}^{(k)})^T(\boldsymbol{x} - \boldsymbol{x}^{(k)}) + \frac{1}{2}\|\boldsymbol{x} - \boldsymbol{x}^{(k)}\|^2 . \tag{11.6}$$

We observe that the left side of (11.6) has minimum at $\boldsymbol{x} = \boldsymbol{x}^*$ while the right-hand side for $\boldsymbol{x} = \boldsymbol{x}^{(k)} - \nabla f(\boldsymbol{x}^{(k)})$. Thus, minimizing both sides with respect to $\boldsymbol{x}$ we obtain

$$f(\boldsymbol{x}^*) \geq f(\boldsymbol{x}^{(k)}) - \frac{1}{2}\|\nabla f(\boldsymbol{x}^{(k)})\|^2 ,$$

or equivalently

$$2f(\boldsymbol{x}^*) \geq f(x^*) + f(\boldsymbol{x}^{(k)}) - \frac{1}{2}\|\nabla f(\boldsymbol{x}^{(k)})\|^2 . \tag{11.7}$$

Multiplying (11.5) with 2 and subtracting (11.7) yields

$$|f(x^{(k+1)}) - f(x^*)| \leq \frac{1}{2}|f(x^{(k)}) - f(x^*)| .$$

This shows that a gradient method can converge at least linearly. Even if this is not optimal, it is optimistic. The previous description does not promise convergence in general and for any value of λ. In particular, there are cases where the method converges only for appropriate values of λ.

In general, the convergence of the gradient descent method depends on the eigenvalues of the Hessian matrix. Let the Hessian matrix $\boldsymbol{H}$ be positive definite. Then the eigenvalues of $\boldsymbol{H}$ are $\kappa_i > 0$, and f is strictly convex. Assume also that $0 < m \leq \kappa_i \leq M$ for all i. We will see that the convergence is in general linear, and its speed depends on the magnitude of the ratio m/M.

The Taylor expansion of $f(\boldsymbol{x}^{(k+1)})$ around $\boldsymbol{x}^{(k)}$ gives

$$f(\boldsymbol{x}^{(k+1)}) = f(\boldsymbol{x}^{(k)}) + \nabla f(\boldsymbol{x}^{(k)})^T(\boldsymbol{x}^{(k+1)} - \boldsymbol{x}^{(k)}) + \frac{1}{2}(\boldsymbol{x}^{(k+1)} - \boldsymbol{x}^{(k)})^T \boldsymbol{H}(\boldsymbol{\xi})(\boldsymbol{x}^{(k+1)} - \boldsymbol{x}^{(k)}) ,$$

for some $\boldsymbol{\xi}$ close to $\boldsymbol{x}^{(k+1)}$ and $\boldsymbol{x}^{(k)}$. Since $\boldsymbol{H}$ is positive definite, then from the Rayleigh inequality[2] we have

$$(\boldsymbol{x}^{(k+1)} - \boldsymbol{x}^{(k)})^T \boldsymbol{H}(\boldsymbol{\xi})(\boldsymbol{x}^{(k+1)} - \boldsymbol{x}^{(k)}) \leq M\|\boldsymbol{x}^{(k+1)} - \boldsymbol{x}^{(k)}\|^2 .$$

Thus, we have

$$\begin{aligned} f(\boldsymbol{x}^{(k+1)}) &\leq f(\boldsymbol{x}^{(k)}) + \nabla f(\boldsymbol{x}^{(k)})^T(\boldsymbol{x}^{(k+1)} - \boldsymbol{x}^{(k)}) + \frac{M}{2}\|\boldsymbol{x}^{(k+1)} - \boldsymbol{x}^{(k)}\|^2 \\ &= f(\boldsymbol{x}^{(k)}) - \lambda \nabla f(\boldsymbol{x}^{(k)})^T \nabla f(\boldsymbol{x}^{(k)}) + \frac{M}{2}\|\boldsymbol{x}^{(k+1)} - \boldsymbol{x}^{(k)}\|^2 , \end{aligned}$$

[2] The Rayleigh inequality states that for any $\boldsymbol{H} = \boldsymbol{H}^T$ we have $\kappa_{\min}\|\boldsymbol{x}\|^2 \leq \boldsymbol{x}^T \boldsymbol{H} \boldsymbol{x} \leq \kappa_{\max}\|\boldsymbol{x}\|^2$, where $\kappa_{\min}$ and κ_{max} are the maximum and minimum eigenvalues of $\boldsymbol{H}$.

which gives

$$f(\boldsymbol{x}^{(k+1)}) \leq f(\boldsymbol{x}^{(k)}) - \lambda \|\nabla f(\boldsymbol{x}^{(k)})\|^2 + \frac{M}{2}\lambda^2 \|\nabla f(\boldsymbol{x}^{(k)})\|^2 .$$

The right-hand side of the last inequality is a quadratic polynomial for the variable λ. By choosing $\lambda = 1/M$ we make the right-hand side to be the smallest possible, giving also

$$f(\boldsymbol{x}^{(k+1)}) \leq f(\boldsymbol{x}^{(k)}) + \frac{1}{2M}\|\nabla f(\boldsymbol{x}^{(k)})\|^2 . \tag{11.8}$$

Similarly, we have that

$$f(\boldsymbol{v}) \geq f(\boldsymbol{w}) + \nabla f(\boldsymbol{w})^T(\boldsymbol{v} - \boldsymbol{w}) + \frac{m}{2}\|\boldsymbol{v} - \boldsymbol{w}\|^2 .$$

Treating the right-hand side as a function of $\boldsymbol{v}$ with $\boldsymbol{w} = \boldsymbol{x}^{(k)}$, and minimizing both sides in terms of $\boldsymbol{v}$ we obtain $f(\boldsymbol{x}^*) \geq f(\boldsymbol{x}^{(k)}) - \frac{1}{2m}\|\nabla f(\boldsymbol{x}^{(k)})\|^2$, which implies

$$-\|\nabla f(\boldsymbol{x}^{(k)})\|^2 \leq -2m(f(\boldsymbol{x}^{(k)}) - f(\boldsymbol{x}^*)) . \tag{11.9}$$

The last inequality is known to as the Polyak-Łojasiewicz inequality, [74]. The estimation of the convergence rate follows from (11.8) after substitution of the gradient using (11.9) we obtain

$$f(\boldsymbol{x}^{(k+1)}) \leq f(\boldsymbol{x}^{(k)}) - \frac{m}{M}(f(\boldsymbol{x}^{(k)}) - f(\boldsymbol{x}^*)) .$$

Subtracting $f(\boldsymbol{x}^*)$ from both sides implies

$$f(\boldsymbol{x}^{(k+1)}) - f(\boldsymbol{x}^*) \leq f(\boldsymbol{x}^{(k)}) - f(\boldsymbol{x}^*) - \frac{m}{M}(f(\boldsymbol{x}^{(k)}) - f(\boldsymbol{x}^*)) ,$$

and factorizing the right-hand side yields

$$f(\boldsymbol{x}^{(k+1)}) - f(\boldsymbol{x}^*) \leq \left(1 - \frac{m}{M}\right)(f(\boldsymbol{x}^{(k)}) - f(\boldsymbol{x}^*)) . \tag{11.10}$$

Inequality (11.10) shows that the general gradient method converges linearly. Moreover, applying (11.10) recursively, we obtain

$$\begin{aligned} f(\boldsymbol{x}^{(k)}) - f(\boldsymbol{x}^*) &\leq \left(1 - \frac{m}{M}\right)(f(\boldsymbol{x}^{(k-1)}) - f(\boldsymbol{x}^*)) \\ &\leq \left(1 - \frac{m}{M}\right)^2 (f(\boldsymbol{x}^{(k-2)}) - f(\boldsymbol{x}^*)) \\ &\cdots \\ &\leq \left(1 - \frac{m}{M}\right)^k (f(\boldsymbol{x}^{(0)}) - f(\boldsymbol{x}^*)) . \end{aligned}$$

The constant $C = (1 - m/M)$ characterizes the speed of convergence: Since we will always have $0 < m \leq M$, we deduce that the smaller the quantity C is, the faster the error $f(\boldsymbol{x}^{(k)}) - f(\boldsymbol{x}^*)$ converges to 0.

☞ General gradient descent methods converge linearly.

11.1.2 Method of steepest descent

The main idea behind gradient methods is to construct approximations $\boldsymbol{x}^{(k+1)}$ from previous approximations $\boldsymbol{x}^{(k)}$ in the direction of the vector $\boldsymbol{u}^{(k)} = -\nabla f(\boldsymbol{x}^{(k)})$. The method of steepest descent is the gradient method for which the parameter $\lambda = \lambda_k$ in the general iteration formula

$$\boldsymbol{x}^{(k+1)} = \boldsymbol{x}^{(k)} - \lambda_k \nabla f(\boldsymbol{x}^{(k)}) \quad k = 1, 2, \ldots ,$$

is chosen to minimize the function $g(\lambda) = f(\boldsymbol{x}^{(k)} - \lambda \nabla f(\boldsymbol{x}^{(k)}))$ and is updated at every iteration. This value minimizes $f(\boldsymbol{x}^{(k+1)})$, which means that for $\lambda = \lambda_k$ we have $g'(\lambda_k) = 0$.

The derivative

$$g'(\lambda) = \frac{d}{d\lambda} g(\lambda) = \nabla f(\boldsymbol{x}^{(k)} - \lambda \nabla f(\boldsymbol{x}^{(k)}))^T (-\nabla f(\boldsymbol{x}^{(k)})) ,$$

implies for $\lambda = \lambda_k$ that

$$0 = g'(\lambda_k) = -\langle \nabla f(\boldsymbol{x}^{(k)} - \lambda_k \nabla f(\boldsymbol{x}^{(k)})), \nabla f(\boldsymbol{x}^{(k)}) \rangle . \tag{11.11}$$

Because $\boldsymbol{x}^{(k+1)} = \boldsymbol{x}^{(k)} - \lambda_k \nabla f(\boldsymbol{x}^{(k)})$ we have that

$$\langle \nabla f(\boldsymbol{x}^{(k+1)}), \nabla f(\boldsymbol{x}^{(k)}) \rangle = 0 .$$

Thus, the directions of the vectors $\nabla f(\boldsymbol{x}^{(k+1)})$ and $\nabla f(\boldsymbol{x}^{(k)})$ are orthogonal.

This observation leads to the following property of the method of steepest descent: Consider three iterations $\boldsymbol{x}^{(k)}$, $\boldsymbol{x}^{(k+1)}$ and $\boldsymbol{x}^{(k+2)}$, then

$$\langle \boldsymbol{x}^{(k+1)} - \boldsymbol{x}^{(k)}, \boldsymbol{x}^{(k+2)} - \boldsymbol{x}^{(k+1)} \rangle = \lambda_k \lambda_{k+1} \langle \nabla f(\boldsymbol{x}^{(k)}), \nabla f(\boldsymbol{x}^{(k+1)}) \rangle = 0 ,$$

or in other words, for every three iterations the directions we followed are orthogonal.

None of the previous observations show how to find the actual value of λ_k. To find the optimal value λ_k one can use methods of Chapter 5 for the approximation of the roots of $g'(\lambda_k) = 0$. If the minimizer $\boldsymbol{x}^*$ is a point in the interior of the domain of f, and if f is strictly convex smooth function with

$$f(p\boldsymbol{x} + (1-p)\boldsymbol{y}) < pf(\boldsymbol{x}) + (1-p)f(\boldsymbol{y}), \quad \text{for} \quad 0 < p < 1 ,$$

then the Hessian matrix $\boldsymbol{H}(\boldsymbol{x})$ is positive definite with entries

$$H_{ij}(\boldsymbol{x}) = \frac{\partial^2 f(\boldsymbol{x})}{\partial x_i \partial x_j}, \quad i, j = 1, 2, \ldots, n .$$

Moreover, the Hessian matrix is symmetric due to the properties of the second derivatives of f. Then the Taylor expansion of $\nabla f(\boldsymbol{x}^{(k)} - \lambda \nabla f(\boldsymbol{x}^{(k)}))$ around $\boldsymbol{x}^{(k)}$ is

$$\nabla f(\boldsymbol{x}^{(k+1)}) = \nabla f(\boldsymbol{x}^{(k)} - \lambda_k \nabla f(\boldsymbol{x}^{(k)})) = \nabla f(\boldsymbol{x}^{(k)}) - \lambda_k \boldsymbol{H}(\boldsymbol{x}^{(k)}) \nabla f(\boldsymbol{x}^{(k)}) + O(\lambda_k^2) .$$

Substituting this formula into (11.11), and discarding the high-order terms of $O(\lambda_k^2)$ we obtain

$$0 \approx \langle \nabla f(\boldsymbol{x}^{(k)}) - \lambda_k \boldsymbol{H}(\boldsymbol{x}^{(k)}) \nabla f(\boldsymbol{x}^{(k)}), \nabla f(\boldsymbol{x}^{(k)}) \rangle .$$

Solving for λ_k we obtain

$$\lambda_k \approx \frac{\|\nabla f(\boldsymbol{x}^{(k)})\|^2}{\langle \boldsymbol{H}(\boldsymbol{x}^{(k)}) \nabla f(\boldsymbol{x}^{(k)}), \nabla f(\boldsymbol{x}^{(k)}) \rangle} .$$

Note that this formula can be useful if the initial approximation $\boldsymbol{x}_k$ is close to the minimizer $\boldsymbol{x}^*$ and $\|\nabla f(\boldsymbol{x}^{(k)})\|^2 < \langle \boldsymbol{H}(\boldsymbol{x}^{(k)})\nabla f(\boldsymbol{x}^{(k)}), \nabla f(\boldsymbol{x}^{(k)})\rangle$.

An alternative computation of the parameter λ_k is the Barzilai-Borwein method, [9], where

$$\lambda_k = \frac{|(\boldsymbol{x}^{(k)} - \boldsymbol{x}^{(k-1)})^T(\nabla f(\boldsymbol{x}^{(k)}) - \nabla f(\boldsymbol{x}^{(k-1)}))|}{\|\nabla f(\boldsymbol{x}^{(k)}) - \nabla f(\boldsymbol{x}^{(k-1)})\|^2} .$$

A description of the general method of steepest descent can be found in Algorithm 43.

Algorithm 43 Method of steepest descent

Set a tolerance TOL for the accuracy
Initialize $\boldsymbol{x}^{(k)}$ in the domain of f and set $k = 0$
while $\|\nabla f(\boldsymbol{x}^{(k)})\| > TOL$ or $\|\boldsymbol{x}^{(k+1)} - \boldsymbol{x}^{(k)}\| > TOL$ **do**
  Find $\lambda_k = \text{argmin}_{\lambda>0} f(\boldsymbol{x}^{(k)} - \lambda\nabla f(\boldsymbol{x}^{(k)}))$
  Compute $\boldsymbol{x}^{(k+1)} = \boldsymbol{x}^{(k)} - \lambda_k \nabla f(\boldsymbol{x}^{(k)})$
  Increase the counter $k = k + 1$
end while
Return the approximation $\boldsymbol{x}^{(k)}$

This algorithm requires the minimization of the function $f(\boldsymbol{x}^{(k)} - \lambda\nabla f(\boldsymbol{x}^{(k)}))$ as a function of the single variable λ. An alternative of Chapter 5 methods to solve this simple minimization problem is a variant of a search method (like bisection method) known as the *golden section search algorithm*, which is presented in the exercises at the end of this chapter[3]. An implementation of this algorithm is the function `golden` of the module `scipy.optimize`. The specific function can be used to obtain an approximation to the value λ_k. The general call of function `golden` is the following:

```
golden(func, args, brac, tol, full_output, maxiter)
```

where

`func`: Can be the function $g(\lambda) = f(\boldsymbol{x} - \lambda\boldsymbol{r})$

`args`: A tuple containing other input arguments of the function `func`. (Optional with default value `func=()`)

`brack`: A tuple `(a,b,c)`, where `func(b)<func(a), func(c)` used in the process to find the minimum of f. (Optional with default value `brack=None`)

`tol`: The stopping criterion tolerance. (Optional with default value `tol=1.5e-8`)

`full_output`: Is a bolean variable. If it is `True` then the function returns optional output. (Optional with default value `full_output=0`)

`maxiter`: The maximum number of iterations allowed. (Optional with default value `maxiter=5000`)

The output is an approximation of the minimizer. With the help of the function `golden` we present an implementation of Algorithm 43 in the following block of code:

[3]Of course one can use any of the methods presented in Chapter 5.

```
import numpy as np
import numpy.linalg as npl
import scipy.optimize as spo

def steepestdescent(f,df,x0,tol=1.e-3,maxit=50):
    x = x0
    r = df(x0)
    iters = 0
    while ( np.abs(npl.norm(r))>tol and iters<maxit ):
        lambda_k = spo.golden(g,(x,r))
        x = x - lambda_k * r
        r = df(x)
        iters += 1
    return x
```

As an example, we consider the function $f(x_1, x_2) = x_1^2/2 + 5x_2^2/2$ which has its minimum value at the point $\boldsymbol{x}^* = (0, 0)$. As an initial guess we consider the value $\boldsymbol{x}^{(0)} = (2, 1)$. Using a tolerance `tol=1.e-8` we are able to achieve a very accurate solution within a few iterations.

```
def f(x):
    return 0.5*x[0]**2 + 2.5*x[1]**2
def df(x):
    return np.array([x[0], 5*x[1]])
def g(lambda_k,x,r):
    return f(x - lambda_k*r)
x0 = np.array([2.0,1.0])
x=steepestdescent(f, df, x0, tol = 1.e-8, maxit = 50)
print('x = ', x)
```

```
x =  [ 6.2403221e-09 -4.9922576e-10]
```

In order to observe the sequence of the approximations and the orthogonality between the subsequent vectors generated by the solution we record all the approximations for each step of the steepest descent method in the following code:

```
def steepestdescent(f,df,x0,tol=1.e-3,maxit=50):
    xk = x0
    x = [xk]
    r = df(x0)
    iters = 0
    while (np.abs(npl.norm(r))>tol and iters<maxit):
        lambda_k = sopt.golden(g,(xk,r))
        xk = xk - lambda_k * r
        r = df(xk)
        x.append(xk)
        iters += 1
    return x
x0 = np.array([2.0,1.0])
x = steepestdescent(f,df,x0,tol=1.e-3,maxit=50)
xmesh, ymesh = np.mgrid[-3:3:50j,-2:2:50j]
fmesh = f(np.array([xmesh, ymesh]))
```

```
plt.axis("equal")
plt.contour(xmesh, ymesh, fmesh, 20)
it_array = np.array(x)
plt.plot(it_array.T[0], it_array.T[1], "x-")
plt.show()
```

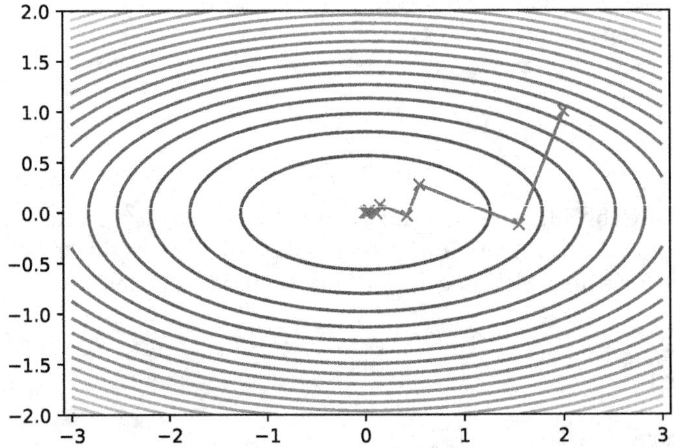

In this example we used a quite large value for tolerance since our intention was to demonstrate the orthogonality property of the relevant approximations and the convergence of the method.

11.1.3 Solving linear systems using optimization

Although the method of steepest descent can be applied to very general functions $f(\boldsymbol{x})$, it is of special interest the case where $f(\boldsymbol{x})$ is a quadratic functional and especially a functional of the form

$$f(\boldsymbol{x}) = \frac{1}{2}\langle \boldsymbol{A}\boldsymbol{x}, \boldsymbol{x}\rangle - \langle \boldsymbol{b}, \boldsymbol{x}\rangle, \quad \boldsymbol{x} \in \mathbb{R}^n \ , \tag{11.12}$$

where $\boldsymbol{A}$ is an $n \times n$ matrix and $\boldsymbol{b}$ an $n \times 1$ vector. This function is of special interest because its gradient is the residual of the linear system $\boldsymbol{A}\boldsymbol{x} = \boldsymbol{b}$. Specifically, we have

$$\nabla f(\boldsymbol{x}) = \boldsymbol{A}\boldsymbol{x} - \boldsymbol{b} \ . \tag{11.13}$$

If $\boldsymbol{x}^*$ is a minimizer of f, then $\nabla f(\boldsymbol{x}^*) = \boldsymbol{A}\boldsymbol{x}^* - \boldsymbol{b} = \boldsymbol{0}$, and thus it is a solution of the linear system $\boldsymbol{A}\boldsymbol{x} = \boldsymbol{b}$. Note that the inner product is the Euclidean inner product and the norm is the Euclidean norm

$$\langle \boldsymbol{x}, \boldsymbol{y}\rangle = \sum_{i=1}^{n} x_i y_i \quad \text{and} \quad \|\boldsymbol{x}\|_2 = \sqrt{\sum_{i=1}^{n} x_i^2} \ .$$

To prove (11.13), we compute the partial derivatives of f

$$\frac{\partial}{\partial x_k}f(\boldsymbol{x}) = \frac{\partial}{\partial x_k}\left\{\frac{1}{2}\langle \boldsymbol{A}\boldsymbol{x},\boldsymbol{x}\rangle - \langle \boldsymbol{b},\boldsymbol{x}\rangle\right\} = \frac{1}{2}\frac{\partial}{\partial x_k}\sum_{i=1}^{n}\left(x_i\sum_{j=1}^{n}a_{ij}x_j\right) - \frac{\partial}{\partial x_k}\sum_{i=1}^{n}b_i x_i$$
$$= \frac{1}{2}\left(\sum_{j=1}^{n}a_{kj}x_j + \sum_{i=1}^{n}x_i a_{ik}\right) - b_k = \sum_{i=1}^{n}a_{ki}x_i - b_k = (\boldsymbol{A}\boldsymbol{x}-\boldsymbol{b})_k\ .$$

If $\boldsymbol{A}$ is symmetric ($\boldsymbol{A}^T = \boldsymbol{A}$) and positive definite ($\langle \boldsymbol{A}\boldsymbol{x},\boldsymbol{x}\rangle > 0$ for all $\boldsymbol{x}\in\mathbb{R}^n$, $\boldsymbol{x}\neq\boldsymbol{0}$), then since $\nabla f(\boldsymbol{x}^*) = \boldsymbol{0}$, the minimizer is the solution $\boldsymbol{x}^* = \boldsymbol{A}^{-1}\boldsymbol{b}$ of the corresponding linear system. If $\boldsymbol{x} = \boldsymbol{x}^* + \boldsymbol{y}$, $\boldsymbol{y}\in\mathbb{R}^n$ is any other vector, then

$$\begin{aligned} f(\boldsymbol{x}) = f(\boldsymbol{x}^*+\boldsymbol{y}) &= \frac{1}{2}\langle \boldsymbol{A}(\boldsymbol{x}^*+\boldsymbol{y}),\boldsymbol{x}^*+\boldsymbol{y}\rangle - \langle \boldsymbol{b},\boldsymbol{x}^*+\boldsymbol{y}\rangle \\ &= \frac{1}{2}\langle \boldsymbol{A}\boldsymbol{x}^*,\boldsymbol{x}^*\rangle + \frac{1}{2}\langle \boldsymbol{A}\boldsymbol{x}^*,\boldsymbol{y}\rangle + \frac{1}{2}\langle \boldsymbol{A}\boldsymbol{y},\boldsymbol{x}^*\rangle + \frac{1}{2}\langle \boldsymbol{A}\boldsymbol{y},\boldsymbol{y}\rangle - \langle \boldsymbol{b},\boldsymbol{x}^*\rangle - \langle \boldsymbol{b},\boldsymbol{y}\rangle \\ &= f(\boldsymbol{x}^*) + \langle \boldsymbol{A}\boldsymbol{x}^*-\boldsymbol{b},\boldsymbol{y}\rangle + \frac{1}{2}\langle \boldsymbol{A}\boldsymbol{y},\boldsymbol{y}\rangle \\ &> f(\boldsymbol{x}^*)\ . \end{aligned}$$

Due to the simplicity of the function f, we can explicitly compute the values of the parameter $\lambda > 0$ to achieve optimal descent rates. Suppose that an initial guess $\boldsymbol{x}^{(0)}$ is given. We define the residual of the k-th iteration to be

$$\boldsymbol{r}^{(k)} = \boldsymbol{b} - \boldsymbol{A}\boldsymbol{x}^{(k)} = -\nabla f(\boldsymbol{x}^{(k)})\ . \tag{11.14}$$

If $\boldsymbol{r}^{(k)} = \boldsymbol{0}$, then the vector $\boldsymbol{x}^{(k)}$ is the solution of the system $\boldsymbol{A}\boldsymbol{x} = \boldsymbol{b}$. Otherwise, we define the iteration

$$\boldsymbol{x}^{(k+1)} = \boldsymbol{x}^{(k)} + \lambda\boldsymbol{r}^{(k)}\ , \tag{11.15}$$

for $\lambda > 0$ to be determined so as to minimize the function

$$g(\lambda) = f(\boldsymbol{x}^{(k)} + \lambda\boldsymbol{r}^{(k)})\ .$$

Using the linearity of the inner product we can analyze the function $g(\lambda)$ such as

$$\begin{aligned} f(\boldsymbol{x}^{(k)}+\lambda\boldsymbol{r}^{(k)}) &= \frac{1}{2}\langle \boldsymbol{A}(\boldsymbol{x}^{(k)}+\lambda\boldsymbol{r}^{(k)}),\boldsymbol{x}^{(k)}+\lambda\boldsymbol{r}^{(k)}\rangle + \langle \boldsymbol{b},\boldsymbol{x}^{(k)}+\lambda\boldsymbol{r}^{(k)}\rangle \\ &= \frac{1}{2}\langle \boldsymbol{A}\boldsymbol{x}^{(k)},\boldsymbol{x}^{(k)}\rangle + \frac{1}{2}\lambda\langle \boldsymbol{A}\boldsymbol{x}^{(k)},\boldsymbol{r}^{(k)}\rangle + \frac{1}{2}\lambda\langle \boldsymbol{A}\boldsymbol{r}^{(k)},\boldsymbol{x}^{(k)}\rangle + \frac{1}{2}\lambda^2\langle \boldsymbol{A}\boldsymbol{r}^{(k)},\boldsymbol{r}^{(k)}\rangle \\ &\quad - \langle \boldsymbol{b},\boldsymbol{x}^{(k)}\rangle - \lambda\langle \boldsymbol{b},\boldsymbol{r}^{(k)}\rangle \\ &= f(\boldsymbol{x}^{(k)}) + \frac{1}{2}\lambda^2\langle \boldsymbol{A}\boldsymbol{r}^{(k)},\boldsymbol{r}^{(k)}\rangle + \lambda\langle \boldsymbol{A}\boldsymbol{x}^{(k)},\boldsymbol{r}^{(k)}\rangle - \lambda\langle \boldsymbol{b},\boldsymbol{r}^{(k)}\rangle \\ &= f(\boldsymbol{x}^{(k)}) + \frac{1}{2}\lambda^2\langle \boldsymbol{A}\boldsymbol{r}^{(k)},\boldsymbol{r}^{(k)}\rangle - \lambda\langle \boldsymbol{r}^{(k)},\boldsymbol{r}^{(k)}\rangle\ . \end{aligned}$$

The last equation is a quadratic polynomial of λ. Since $\langle \boldsymbol{A}\boldsymbol{r}^{(k)},\boldsymbol{r}^{(k)}\rangle > 0$ due to the positive definiteness of $\boldsymbol{A}$, and $\boldsymbol{r}^{(k)}\neq\boldsymbol{0}$, the quadratic polynomial achieves its minimum value for

$$\lambda = \lambda_k = \frac{\langle \boldsymbol{r}^{(k)},\boldsymbol{r}^{(k)}\rangle}{\langle \boldsymbol{A}\boldsymbol{r}^{(k)},\boldsymbol{r}^{(k)}\rangle}\ . \tag{11.16}$$

Observe that for the specific value of λ we have

$$f(\boldsymbol{x}^{(k)} + \lambda_k \boldsymbol{r}^{(k)}) = f(\boldsymbol{x}^{(k)}) - \frac{\langle \boldsymbol{r}^{(k)}, \boldsymbol{r}^{(k)} \rangle^2}{2\langle \boldsymbol{A}\boldsymbol{r}^{(k)}, \boldsymbol{r}^{(k)} \rangle} < f(\boldsymbol{x}^{(k)}) .$$

The algorithm for the method of steepest descent in the case of quadratic functions is presented in Algorithm 44. This algorithm computes the parameter λ_k using (11.16), the residual (11.14), and then updates the approximation of the minimizer via (11.15). We consider here the zero vector as an initial guess for the minimizer, but any other vector can be used, and then of course, the initialization of the residual should be updated with the appropriate residual as defined by (11.14).

Algorithm 44 Steepest descent method for linear equations

Set a tolerance TOL for the accuracy
Initialize $\boldsymbol{x}^{(0)} = \boldsymbol{0}$ and $\boldsymbol{r}^{(0)} = \boldsymbol{b}$
Initialize $k = 0$
while $\|\boldsymbol{r}^{(k)})\| > TOL$ or $\|\boldsymbol{x}^{(k+1)} - \boldsymbol{x}^{(k)}\| > TOL$ **do**
  Find $\lambda_k = \dfrac{\langle \boldsymbol{r}^{(k)}, \boldsymbol{r}^{(k)} \rangle}{\langle \boldsymbol{A}\boldsymbol{r}^{(k)}, \boldsymbol{r}^{(k)} \rangle}$
  Compute $\boldsymbol{x}^{(k+1)} = \boldsymbol{x}^{(k)} - \lambda_k \boldsymbol{r}^{(k)}$
  Compute $\boldsymbol{r}^{(k+1)} = \boldsymbol{b} - \boldsymbol{A}\boldsymbol{x}^{(k+1)}$
  Increase the counter $k = k + 1$
end while
Return the approximation $\boldsymbol{x}^{(k)}$

11.1.4 Theoretical considerations

The way we devised the algorithm of the method of steepest descent leads to the conclusion that the residual $\boldsymbol{r}^{(k)} = -\nabla f(\boldsymbol{x}^{(k)})$ is in practice the direction in which we move in order to approach the minimizer. It is interesting to see that, as in the case of a general function $f(\boldsymbol{x})$, these directions are orthogonal, in the sense $\langle \boldsymbol{r}^{(k)}, \boldsymbol{r}^{(k-1)} \rangle = 0$. More precisely, we have

$$\begin{aligned}
\langle \boldsymbol{r}^{(k)}, \boldsymbol{r}^{(k-1)} \rangle &= \langle \boldsymbol{r}^{(k-1)} - \frac{\langle \boldsymbol{r}^{(k-1)}, \boldsymbol{r}^{(k-1)} \rangle}{\langle \boldsymbol{A}\boldsymbol{r}^{(k-1)}, \boldsymbol{r}^{(k-1)} \rangle} \boldsymbol{A}\boldsymbol{r}^{(k-1)}, \boldsymbol{r}^{(k-1)} \rangle \\
&= \langle \boldsymbol{r}^{(k-1)}, \boldsymbol{r}^{(k-1)} \rangle - \frac{\langle \boldsymbol{r}^{(k-1)}, \boldsymbol{r}^{(k-1)} \rangle}{\langle \boldsymbol{A}\boldsymbol{r}^{(k-1)}, \boldsymbol{r}^{(k-1)} \rangle} \langle \boldsymbol{A}\boldsymbol{r}^{(k-1)}, \boldsymbol{r}^{(k-1)} \rangle \\
&= 0 .
\end{aligned}$$

Since the iteration of the method of steepest descent is defined as $\boldsymbol{x}^{(k+1)} = \boldsymbol{x}^{(k)} + \lambda_k \boldsymbol{r}^{(k)}$ we can conclude that the vectors connecting three successive iterations are orthogonal, or in other words

$$\langle \boldsymbol{x}^{(k+1)} - \boldsymbol{x}^{(k)}, \boldsymbol{x}^{(k)} - \boldsymbol{x}^{(k-1)} \rangle = \langle \lambda_k \boldsymbol{r}^{(k)}, \lambda_{k-1} \boldsymbol{r}^{(k-1)} \rangle = 0 .$$

Proof of convergence

Showing the convergence of the method of steepest descent is not straightforward, and the information included in this section can appear to be very technical but it is included for the sake of completeness. We first assume that $\boldsymbol{A}$ is symmetric and positive definite, and we define the *weighted* norm

$$\|\boldsymbol{x}\|_{\boldsymbol{A}} = \sqrt{\langle \boldsymbol{A}\boldsymbol{x}, \boldsymbol{x} \rangle} .$$

In order to see that this inner product forms a vector norm indeed, consider the Cholesky decomposition of $\boldsymbol{A} = \boldsymbol{H}\boldsymbol{H}^T$, denote $\boldsymbol{H}^T = \boldsymbol{A}^{1/2}$ and write

$$\sqrt{\langle \boldsymbol{A}\boldsymbol{x}, \boldsymbol{x}\rangle} = \sqrt{\langle \boldsymbol{H}^T\boldsymbol{x}, \boldsymbol{H}^T\boldsymbol{x}\rangle} = \sqrt{\langle \boldsymbol{A}^{1/2}\boldsymbol{x}, \boldsymbol{A}^{1/2}\boldsymbol{x}\rangle} = \|\boldsymbol{A}^{1/2}\boldsymbol{x}\| \ .$$

Then we have the following theorem:

Theorem 11.1. *If* $\kappa = \rho_{\max}/\rho_{\min}$, *where* $\rho_{\max}$ *and* $\rho_{\min}$ *the maximum and minimum eigenvalues of* $\boldsymbol{A}$, *respectively, then if* $\mathbf{e}^{(k)} = \boldsymbol{x}^{(k)} - \boldsymbol{x}$, *is the error between the* k*-th approximation* $\boldsymbol{x}^{(k)}$ *of the method of steepest descent and the solution* $\boldsymbol{x}$ *of the linear system* $\boldsymbol{A}\boldsymbol{x} = \boldsymbol{b}$, *then*

$$\|\mathbf{e}^{(k)}\|_{\boldsymbol{A}} \leq \left(\frac{\kappa - 1}{\kappa + 1}\right)^k \|\mathbf{e}^{(0)}\|_{\boldsymbol{A}}, \quad k = 1, 2, \ldots \ . \tag{11.17}$$

Proof. We first note that

$$\boldsymbol{A}\mathbf{e}^{(k+1)} = \boldsymbol{A}(\boldsymbol{x}^{(k+1)} - \boldsymbol{x}) = \boldsymbol{A}\boldsymbol{x}^{(k+1)} - \boldsymbol{b} = -\boldsymbol{r}^{(k+1)} \ .$$

Thus,

$$\langle \boldsymbol{A}\mathbf{e}^{(k+1)}, \boldsymbol{r}^{(k)}\rangle = -\langle \boldsymbol{r}^{(k+1)}, \boldsymbol{r}^{(k)}\rangle = 0 \ .$$

Furthermore, we observe that

$$\mathbf{e}^{(k+1)} = \boldsymbol{x}^{(k+1)} - \boldsymbol{x} = \boldsymbol{x}^{(k)} + \lambda_k \boldsymbol{r}^{(k)} - \boldsymbol{x} = \mathbf{e}^{(k)} + \lambda_k \boldsymbol{r}^{(k-1)}.$$

Hence, if $\lambda \in \mathbb{R}$ we have

$$\begin{aligned}
\|\mathbf{e}^{(k+1)}\|_{\boldsymbol{A}}^2 &= \langle \boldsymbol{A}\mathbf{e}^{(k+1)}, \mathbf{e}^{(k+1)}\rangle \\
&= \langle \boldsymbol{A}\mathbf{e}^{(k+1)}, \mathbf{e}^{(k)} + \lambda_k \boldsymbol{r}^{(k)}\rangle \\
&= \langle \boldsymbol{A}\mathbf{e}^{(k+1)}, \mathbf{e}^{(k)}\rangle + \lambda_k \langle \boldsymbol{A}\mathbf{e}^{(k+1)}, \boldsymbol{r}^{(k)}\rangle \\
&= \langle \boldsymbol{A}\mathbf{e}^{(k+1)}, \mathbf{e}^{(k)}\rangle + \lambda \langle \boldsymbol{A}\mathbf{e}^{(k+1)}, \boldsymbol{r}^{(k)}\rangle \\
&= \langle \boldsymbol{A}\mathbf{e}^{(k+1)}, \mathbf{e}^{(k)} + \lambda \boldsymbol{r}^{(k)}\rangle \ ,
\end{aligned}$$

where we replaced λ_k with general $\lambda \in \mathbb{R}$, since $\langle \boldsymbol{A}\mathbf{e}^{(k+1)}, \boldsymbol{r}^{(k)}\rangle = 0$.

Using the Cauchy-Schwarz inequality (Theorem 9.3) we have

$$\begin{aligned}
\|\mathbf{e}^{(k+1)}\|_{\boldsymbol{A}}^2 = \langle \boldsymbol{A}\mathbf{e}^{(k+1)}, \mathbf{e}^{(k)} + \lambda \boldsymbol{r}^{(k)}\rangle &= \langle \boldsymbol{A}^{1/2}\mathbf{e}^{(k+1)}, \boldsymbol{A}^{1/2}(\mathbf{e}^{(k)} + \lambda \boldsymbol{r}^{(k)})\rangle \\
&\leq \|\mathbf{e}^{(k+1)}\|_{\boldsymbol{A}} \ \|\mathbf{e}^{(k)} + \lambda \boldsymbol{r}^{(k)}\|_{\boldsymbol{A}} \ ,
\end{aligned}$$

or equivalently

$$\|\mathbf{e}^{(k+1)}\|_{\boldsymbol{A}}^2 \leq \|\mathbf{e}^{(k)} + \lambda \boldsymbol{r}^{(k)}\|_{\boldsymbol{A}}^2 = \langle \boldsymbol{A}(\mathbf{e}^{(k)} + \lambda \boldsymbol{r}^{(k)}), \mathbf{e}^{(k)} + \lambda \boldsymbol{r}^{(k)}\rangle \ ,$$

for all $\lambda \in \mathbb{R}$. Since $\mathbf{e}^{(k)} + \lambda \boldsymbol{r}^{(k)} = (\boldsymbol{I} - \lambda \boldsymbol{A})\mathbf{e}^{(k)}$ and $\boldsymbol{A}(\mathbf{e}^{(k)} + \lambda \boldsymbol{r}^{(k)}) = (\boldsymbol{I} - \lambda \boldsymbol{A})\mathbf{e}^{(k)}$ we have

$$\|\mathbf{e}^{(k+1)}\|_{\boldsymbol{A}}^2 \leq \inf_{\lambda \in \mathbb{R}} \langle (\boldsymbol{I} - \lambda \boldsymbol{A})\boldsymbol{A}\mathbf{e}^{(k)}, (\boldsymbol{I} - \lambda \boldsymbol{A})\mathbf{e}^{(k)}\rangle, \quad k \geq 0 \ . \tag{11.18}$$

Here, we denote the eigenvalues of $\boldsymbol{A}$ by ρ, $i = 1, 2, \ldots$. Since $\boldsymbol{A}$ is symmetric and positive definite we have that

$$0 < \rho_{\min} = \rho_1 \leq \rho_2 \leq \cdots \leq \rho_n = \rho_{\max} \ .$$

Let z_i be the eigenvector of the eigenvalues ρ_i. For each $u \in \mathbb{R}^n$ we have that

$$u = \sum_{i=1}^{n} \langle u, z_i \rangle z_i \ .$$

This is true because the eigenvectors of the specific A can form a basis of $\mathbb{R}^n$. Furthermore, for any real polynomial $P(x)$, we have that

$$P(A)u = \sum_{i=1}^{n} P(\rho_i) \langle u, z_i \rangle z_i \ .$$

If

$$e^{(k)} = \sum_{i=1}^{n} \beta_i z_i \ ,$$

with $\beta_i = \langle e^{(k)}, z_i \rangle$, for all $i = 1, 2, \ldots$, then, for each $\lambda \in \mathbb{R}$, the inner product in the infimum of (11.18) can be estimated as follows

$$\begin{aligned}
\langle (I - \lambda A) A e^{(k)}, (I - \lambda A) e^{(k)} \rangle &= \langle (I - \lambda A) A \sum_{i=1}^{n} \beta_i z_i, (I - \lambda A) \sum_{i=1}^{n} \beta_i z_i \rangle \\
&= \langle \sum_{i=1}^{n} (1 - \lambda \rho_i) \rho_i \beta_i z_i, \sum_{i=1}^{n} (1 - \lambda \rho_i) \beta_i z_i \rangle \\
&= \sum_{i=1}^{n} (1 - \lambda \rho_i)^2 \rho_i \beta_i^2 \\
&\le (\max_i |1 - \lambda \rho_i|)^2 \sum_{i=1}^{n} \rho_i \beta_i^2 \\
&= (\max_i |1 - \lambda \rho_i|)^2 \langle \sum_{i=1}^{n} \rho_i \beta_i z_i, \sum_{i=1}^{n} \beta_i z_i \rangle \\
&= (\max_i |1 - \lambda \rho_i|)^2 \langle \sum_{i=1}^{n} \beta_i A z_i, \sum_{i=1}^{n} \beta_i z_i \rangle \\
&= (\max_i |1 - \lambda \rho_i|)^2 \langle A e^{(k)}, e^{(k)} \rangle \\
&= (\max_i |1 - \lambda \rho_i|)^2 \| e^{(k)} \|_A^2 \ .
\end{aligned}$$

Equation (11.18) then becomes

$$\| e^{(k+1)} \|_A^2 \le \inf_{\lambda \in \mathbb{R}} \left(\max_{\rho \in [\rho_{\min}, \rho_{\max}]} |1 - \lambda \rho| \right) \| e^{(k)} \|_A, \quad k \ge 0 \ . \tag{11.19}$$

In this min-max problem the optimal value of ρ can be found to be $2/(\rho_{\min} + \rho_{\max})$. Then we have that

$$\| e^{(k+1)} \|_A \le \frac{\rho_{\max} - \rho_{\min}}{\rho_{\max} + \rho_{\min}} \| e^{(k)} \|_A \ ,$$

which implies directly that

$$\| e^{(k)} \|_A \le \left(\frac{\kappa - 1}{\kappa + 1} \right)^k \| e^{(0)} \|_A \ ,$$

and the proof is complete. □

From the last relationship we can see easily that

$$\lim_{k\to\infty} \boldsymbol{e}^{(k)} = 0 ,$$

and thus the convergence can be achieved. For more details, we refer to specialized books on numerical linear algebra.

11.2 Conjugate Gradient Method

Thus far, we approximated the minimizer $\boldsymbol{x}^*$ of a function $f(\boldsymbol{x})$ following the direction $\boldsymbol{r}^{(k)} = -\nabla f(\boldsymbol{x}^{(k)})$, for $k = 0, 1, 2, \ldots$. If instead we start with an arbitrary direction $\boldsymbol{r}^{(k)} = \boldsymbol{p}^{(k)}$, and we try to minimize the value of the function $f(\boldsymbol{x}^{(k+1)})$ using the approximation $\boldsymbol{x}^{(k+1)} = \boldsymbol{x}^{(k)} + \lambda \boldsymbol{p}^{(k)}$ we obtain a new method. This new method is known as the method of *general directions.*

For simplicity, we consider the quadratic function

$$f(\boldsymbol{x}) = \frac{1}{2}\langle \boldsymbol{A}\boldsymbol{x}, \boldsymbol{x}\rangle - \langle \boldsymbol{b}, \boldsymbol{x}\rangle ,$$

where $\boldsymbol{A}$ being a positive definite matrix, and we estimate the value $f(\boldsymbol{x}^{(k+1)})$.

$$\begin{aligned}
f(\boldsymbol{x}^{(k+1)}) &= f(\boldsymbol{x}^{(k)} + \lambda \boldsymbol{p}^{(k)}) \\
&= \frac{1}{2}\langle \boldsymbol{A}(\boldsymbol{x}^{(k)} + \lambda \boldsymbol{p}^{(k)}), \boldsymbol{x}^{(k)} + \lambda \boldsymbol{p}^{(k)}\rangle - \langle \boldsymbol{b}, \boldsymbol{x}^{(k)} + \lambda \boldsymbol{p}^{(k)}\rangle \\
&= \frac{1}{2}\langle \boldsymbol{A}\boldsymbol{x}^{(k)}, \boldsymbol{x}^{(k)}\rangle + \lambda\langle \boldsymbol{A}\boldsymbol{x}^{(k)}, \boldsymbol{p}^{(k)} + \frac{1}{2}\lambda^2\langle \boldsymbol{A}\boldsymbol{p}^{(k)}, \boldsymbol{p}^{(k)}\rangle - \langle \boldsymbol{b}, \boldsymbol{x}^{(k)}\rangle - \lambda \boldsymbol{b}, \boldsymbol{p}^{(k+1)}\rangle \\
&= f(\boldsymbol{x}^{(k)}) + \frac{1}{2}\lambda^2\langle \boldsymbol{A}\boldsymbol{p}^{(k)}, \boldsymbol{p}^{(k)}\rangle - \lambda\langle \boldsymbol{p}^{(k)}, \boldsymbol{r}^{(k)}\rangle .
\end{aligned}$$

From the last equality we conclude that eliminating the positive inner product we achieve the minimization of $f(\boldsymbol{x}^{(k+1)})$. This can be achieved if $\langle \boldsymbol{p}^{(k)}, \boldsymbol{r}^{(k)}\rangle \neq 0$ by taking

$$\lambda = \lambda_k = \frac{\langle \boldsymbol{p}^{(k)}, \boldsymbol{r}^{(k)}\rangle}{\langle \boldsymbol{A}\boldsymbol{p}^{(k)}, \boldsymbol{p}^{(k)}\rangle} ,$$

leading to

$$f(\boldsymbol{x}^{(k+1)}) = f(\boldsymbol{x}^{(k)}) - \frac{\langle \boldsymbol{p}^{(k)}, \boldsymbol{r}^{(k)}\rangle^2}{2\langle \boldsymbol{A}\boldsymbol{p}^{(k)}, \boldsymbol{p}^{(k)}\rangle} < f(\boldsymbol{x}^{(k)}) .$$

In such a case, the sequence of the values $f(\boldsymbol{x}^{(k)})$ will be strictly decreasing and bounded from below by the value $f(\boldsymbol{A}^{-1}\boldsymbol{b})$, and thus will converge to the limit

$$\lim_{k\to\infty} f(\boldsymbol{x}^{(k)}) \geq f(\boldsymbol{A}^{-1}\boldsymbol{b}) .$$

This fact of course doesn't guarantee the convergence of the method to the solution of the linear system $\boldsymbol{x} = \boldsymbol{A}^{-1}\boldsymbol{b}$ since this would require

$$\lim_{k\to\infty} f(\boldsymbol{x}^{(k)}) = f(\boldsymbol{A}^{-1}\boldsymbol{b}) ,$$

which cannot be proved. The idea, though, of choosing other directions than those of the method of steepest descent leads to the very important method of *conjugate gradient* which was introduced by Magnus Hestenes (1906–1991) and Edward Stiefel (1909–1978), [60].

11.2.1 Conjugate gradient method and linear systems

In the previous analysis, where we used a general direction $\boldsymbol{p}^{(k)}$ instead of the directions of the steepest descent $\boldsymbol{r}^{(k)} = -\nabla f(\boldsymbol{x}^{(k)})$, we observed that each iteration $\boldsymbol{x}^{(k+1)}$ is a linear combination of the previous iteration $\boldsymbol{x}^{(k)}$ and the direction $\boldsymbol{p}^{(k)}$. This implies that

$$\begin{aligned}\boldsymbol{x}^{(k+1)} &= \boldsymbol{x}^{(k)} + \lambda_k \boldsymbol{p}^{(k)}\\ &= \boldsymbol{x}^{(k-1)} + \lambda_{k-1}\boldsymbol{p}^{(k-1)} + \lambda_k \boldsymbol{p}^{(k)}\\ &\vdots\\ &= \boldsymbol{x}^{(0)} + \sum_{i=0}^{k} \lambda_k \boldsymbol{p}^{(k)} .\end{aligned}$$

This is also true for the method of steepest descent, where

$$\boldsymbol{x}^{(k+1)} = \boldsymbol{x}^{(0)} + \sum_{i=0}^{k} \lambda_k \boldsymbol{r}^{(k)} .$$

Thus, in both methods the difference $\boldsymbol{x}^{(k+1)} - \boldsymbol{x}^{(0)}$ is a linear combination of the previously chosen directions. If all these directions were linearly independent, then they would form a linear space. If the initial guess lies in the same space, then the same will happen to the approximate solution $\boldsymbol{x}^{(k)}$ for any $k = 1, 2, \ldots$. We proceed with the generation of such linearly independent directions $\boldsymbol{p}^{(k)}$.

Let the $n \times n$ matrix $\boldsymbol{A}$ be symmetric and positive-definite ($\boldsymbol{x}^T \boldsymbol{A}\boldsymbol{x} > 0$ for all non-zero vectors $\boldsymbol{x} \in \mathbb{R}^n$). Then two vectors $\boldsymbol{u}, \boldsymbol{v} \in \mathbb{R}^n$, $\boldsymbol{u} \neq \boldsymbol{v}$ are called *conjugate* with respect to $\boldsymbol{A}$ (or $\boldsymbol{A}$-orthogonal) if

$$\boldsymbol{u}^T \boldsymbol{A}\boldsymbol{v} = 0 .$$

This product is simply the inner product $\langle \boldsymbol{A}\boldsymbol{v}, \boldsymbol{u}\rangle = \boldsymbol{u}^T\boldsymbol{A}\boldsymbol{v}$. In the sequel, we refer to such vectors as conjugate without mentioning their dependence on the matrix $\boldsymbol{A}$.

Suppose that we have $k+1$ conjugate vectors $\boldsymbol{p}^{(0)}, \boldsymbol{p}^{(1)}, \ldots, \boldsymbol{p}^{(k)}$. These vectors are linearly independent. To see this consider the linear combination

$$c_0\boldsymbol{p}^{(0)} + c_1\boldsymbol{p}^{(1)} + \cdots + c_k\boldsymbol{p}^{(k)} = \boldsymbol{0} ,$$

with $c_i \in \mathbb{R}$ for $i = 0, 1, \ldots, k$. Taking the inner product with $\boldsymbol{A}\boldsymbol{p}^{(j)}$ for all $j = 0, 1, \ldots, k$ we have

$$0 = \langle \boldsymbol{A}\boldsymbol{p}^{(j)}, \sum_{i=0}^{k} c_i\boldsymbol{p}^{(i)}\rangle = \sum_{i=0}^{k} c_i\langle \boldsymbol{A}\boldsymbol{p}^{(j)}, \boldsymbol{p}^{(i)}\rangle = c_j\langle \boldsymbol{A}\boldsymbol{p}^{(j)}, \boldsymbol{p}^{(j)}\rangle .$$

Since $\boldsymbol{A}$ is positive definite we have that $\langle \boldsymbol{A}\boldsymbol{p}^{(j)}, \boldsymbol{p}^{(j)}\rangle > 0$ for all $j = 0, 1, \ldots, k$. Thus, we have that $c_j = 0$ for all $j = 0, 1, \ldots, k$.

Given the initial guess $\boldsymbol{x}^{(0)}$, we define the first direction to be $\boldsymbol{p}^{(0)} = \boldsymbol{b} - \boldsymbol{A}\boldsymbol{x}^{(0)}$, which coincides with the direction of the steepest descent $-\nabla f(\boldsymbol{x}^{(0)})$ for the function $f(\boldsymbol{x}) = \frac{1}{2}\langle \boldsymbol{A}\boldsymbol{x}, \boldsymbol{x}\rangle + \langle \boldsymbol{b}, \boldsymbol{x}\rangle$. The rest of the vectors should be conjugate to the gradient and to each other.

Note that the residual

$$\boldsymbol{r}^{(k)} = \boldsymbol{b} - \boldsymbol{A}\boldsymbol{x}^{(k)} = \boldsymbol{b} - \boldsymbol{A}(\boldsymbol{x}^{(k-1)} + \lambda_{k-1}\boldsymbol{p}^{(k-1)}) = \boldsymbol{r}^{(k-1)} - \lambda_{k-1}\boldsymbol{A}\boldsymbol{p}^{(k-1)} .$$

Given the conjugate direction $\boldsymbol{p}^{(k)}$ it holds (can be proved using induction) that

$$\langle \boldsymbol{r}^{(k)}, \boldsymbol{p}^{(j)}\rangle = 0, \quad j = 0, 1, 2, \ldots k-1 .$$

What remains is to find the conjugate direction $\boldsymbol{p}^{(k)}$ given the conjugate direction $\boldsymbol{p}^{(k-1)}$. The new vector will be also conjugate to the others by construction. We choose the new direction $\boldsymbol{p}^{(k)}$ to be a linear combination of the residual $-\boldsymbol{r}^{(k)}$ and the previous conjugate direction $\boldsymbol{p}^{(k-1)}$. Specifically, we consider the new direction to be

$$\boldsymbol{p}^{(k)} = -\boldsymbol{r}^{(k)} + \beta_k \boldsymbol{p}^{(k-1)} .$$

We compute the values β_k as follows: We first take the inner product of $\boldsymbol{p}^{(k)}$ with $\boldsymbol{A}\boldsymbol{p}^{(k-1)}$ to get

$$\langle \boldsymbol{A}\boldsymbol{p}^{(k-1)}, \boldsymbol{p}^{(k)} \rangle = -\langle \boldsymbol{A}\boldsymbol{p}^{(k-1)}, \boldsymbol{r}^{(k)} \rangle + \beta_k \langle \boldsymbol{A}\boldsymbol{p}^{(k-1)}, \boldsymbol{p}^{(k-1)} \rangle . \tag{11.20}$$

Since $\langle \boldsymbol{A}\boldsymbol{p}^{(k-1)}, \boldsymbol{p}^{(k)} \rangle = 0$, solving equation (11.20) yields

$$\beta_k = \frac{\langle \boldsymbol{A}\boldsymbol{p}^{(k-1)}, \boldsymbol{r}^{(k)} \rangle}{\langle \boldsymbol{A}\boldsymbol{p}^{(k-1)}, \boldsymbol{p}^{(k-1)} \rangle} .$$

The last formula can be simplified using the facts that $\lambda_k \boldsymbol{A}\boldsymbol{p}^{(k)} = \boldsymbol{r}^{(k+1)} - \boldsymbol{r}^{(k)}$, $\langle \boldsymbol{r}^{(k)}, \boldsymbol{p}^{(j)} \rangle = 0$ for all $j < k$, and $\boldsymbol{p}^{(k+1)} = -\boldsymbol{r}^{(k+1)} + \beta_k \boldsymbol{p}^{(k)}$, into

$$\beta_k = \frac{\langle \boldsymbol{r}^{(k)}, \boldsymbol{r}^{(k)} \rangle}{\langle \boldsymbol{r}^{(k-1)}, \boldsymbol{r}^{(k-1)} \rangle} .$$

In this way less matrix-vector multiplications are required.

We summarize all the above into the Conjugate Gradient Algorithm 45.

Algorithm 45 Conjugate Gradient method for linear equations

Set a tolerance TOL for the accuracy
Initialize $\boldsymbol{x}^{(0)} = 0$, $\boldsymbol{r}^{(0)} = \boldsymbol{A}\boldsymbol{x}^{(0)} - \boldsymbol{b}$, $\boldsymbol{p}^{(0)} = -\boldsymbol{r}^{(0)}$
Initialize $k = 0$
while $\|\boldsymbol{r}^{(k)}\| > TOL$ or $\|\boldsymbol{x}^{(k+1)} - \boldsymbol{x}^{(k)}\| > TOL$ **do**
  Compute $\lambda_k = \frac{\langle \boldsymbol{r}^{(k)}, \boldsymbol{r}^{(k)} \rangle}{\langle \boldsymbol{p}^{(k)}, \boldsymbol{A}\boldsymbol{p}^{(k)} \rangle}$
  Compute $\boldsymbol{x}^{(k+1)} = \boldsymbol{x}^{(k)} - \lambda_k \boldsymbol{p}^{(k)}$
  Compute $\boldsymbol{r}^{(k+1)} = \boldsymbol{r}^{(k)} + \lambda_k \boldsymbol{A}\boldsymbol{p}^{(k)}$
  Compute $\beta_{k+1} = \frac{\langle \boldsymbol{r}^{(k+1)}, \boldsymbol{r}^{(k+1)} \rangle}{\langle \boldsymbol{r}^{(k)}, \boldsymbol{r}^{(k)} \rangle}$
  Compute $\boldsymbol{p}^{(k+1)} = -\boldsymbol{r}^{(k+1)} + \beta_{k+1} \boldsymbol{p}^{(k)}$
  Increase the counter $k = k + 1$
end while
Return the approximation $\boldsymbol{x}^{(k)}$

An implementation of the Conjugate Gradient Algorithm 45 can be the following

```
def conjgrad(A,b,x0,tol=1.e-3,maxit=50):
    x = x0.copy()
    r = np.dot(A,x0) - b
    p = - r.copy()
    iters = 0
    while (npl.norm(r)>tol and iters<maxit):
        # Compute the products
        Ap = np.dot(A,p)
        r_dot_r = np.sum(r*r)
        p_dot_Ap = np.sum(p*Ap)
```

```
        # Compute the parameters and the unknowns
        lam = r_dot_r / p_dot_Ap
        x = x + lam * p
        r = r + lam * Ap
        beta = np.sum(r*r)/r_dot_r
        p = - r + beta * p
        iters += 1
    return x
```

As we mentioned before, the Conjugate Gradient method converges if $\boldsymbol{A}$ is symmetric and positive definite. For this reason, in order to test our implementation we consider the linear system $\boldsymbol{Ax} = \boldsymbol{b}$ with

$$\boldsymbol{A} = \begin{pmatrix} 2 & -1 & 0 \\ -1 & 2 & -1 \\ 0 & -1 & 2 \end{pmatrix} \quad \text{and} \quad \boldsymbol{b} = \begin{pmatrix} 1 \\ 0 \\ 1 \end{pmatrix} ,$$

which has the exact solution $\boldsymbol{x} = (1, 1, 1)^T$. As initial condition we choose the vector $\boldsymbol{x}^{(0)} = (0, 0, 0)^T$.

```
A = np.array([[2.0, -1.0, 0.0],[-1.0, 2.0, -1.0],[0.0, -1.0, 2.0]])
b = np.array([[1],[0],[1]])
x0 = np.zeros((3,1))
x= conjgrad(A, b, x0)
print('x = '); print(x)
```

```
x =
[[1.]
 [1.]
 [1.]]
```

Printing the number of iterations `iters` variable we observe that for the 3×3 matrix of our example, it is required only 2 iterations to converge within the prescribed `tol`. Similarly, fast convergence can be achieved even if we try smaller values of `tol`. In principle, if $\boldsymbol{A}$ is $n \times n$ symmetric and positive definite matrix, then the Conjugate Gradient method converges in at most n iterations when exact arithmetic is being used.

11.2.2 Convergence and preconditioning

Approximations of the conjugate gradient method can be shown to converge for symmetric and positive definite $n \times n$ matrix $\boldsymbol{A}$. More precisely, if $\kappa = \text{cond}(\boldsymbol{A})$ is the condition number of $\boldsymbol{A}$ based on the ℓ_2-norm, then the error $\boldsymbol{e}^{(k)} = \boldsymbol{x}^{(k)} - \boldsymbol{x}$ is

$$\|\boldsymbol{e}^{(k)}\|_{\boldsymbol{A}} \leq 2 \left[\left(\frac{\sqrt{\kappa} - 1}{\sqrt{\kappa} + 1} \right)^k + \left(\frac{\sqrt{\kappa} + 1}{\sqrt{\kappa} - 1} \right)^k \right]^{-1} \|\boldsymbol{e}^{(0)}\|_{\boldsymbol{A}}, \quad k = 1, 2, \ldots, n - 1 .$$

Although convergence can be achieved, for large values of n, it happens that this can be a slow process. Moreover, the speed of convergence is an increasing function of the condition number $\text{cond}(\boldsymbol{A})$.

In order to improve the speed of convergence of iterative methods including the conjugate gradient method we usually employ a preconditioning technique. Preconditioning will improve the condition number of a matrix $\boldsymbol{A}$ or just change its structure. For this reason, we multiply the system $\boldsymbol{Ax}=\boldsymbol{b}$ with the inverse of an $n\times n$ matrix $\boldsymbol{M}$. The matrix $\boldsymbol{M}$ is called *preconditioner* and is chosen symmetric and positive definite. Instead of working with the system $\boldsymbol{Ax}=\boldsymbol{b}$ we work with the system

$$\boldsymbol{M}^{-1}\boldsymbol{Ax}=\boldsymbol{M}^{-1}\boldsymbol{b}\ , \tag{11.21}$$

with the hope that the condition number of the matrix $\boldsymbol{M}^{-1}\boldsymbol{A}$ is smaller compared to $\boldsymbol{A}$.

Since $\boldsymbol{M}$ is symmetric and positive definite, then we can find a matrix $\boldsymbol{C}$ such that $\boldsymbol{C}^2=\boldsymbol{M}$. Using the matrix $\boldsymbol{C}$ we can write (11.21) in the form

$$\boldsymbol{C}^{-1}\boldsymbol{AC}^{-1}\boldsymbol{Cx}=\boldsymbol{C}^{-1}\boldsymbol{b}\ ,$$

or

$$\tilde{\boldsymbol{A}}\tilde{\boldsymbol{x}}=\tilde{\boldsymbol{b}}\ , \tag{11.22}$$

with $\tilde{\boldsymbol{A}}=\boldsymbol{C}^{-1}\boldsymbol{AC}^{-1}$, $\tilde{\boldsymbol{x}}=\boldsymbol{Cx}$, and $\tilde{\boldsymbol{b}}=\boldsymbol{C}^{-1}\boldsymbol{b}$.

If $\sigma(\boldsymbol{A})$ denotes the spectrum of $\boldsymbol{A}$, i.e. the set of the eigenvalues of $\boldsymbol{A}$ in absolute value, then

$$\sigma(\tilde{\boldsymbol{A}})=\sigma(\boldsymbol{C}^{-1}\boldsymbol{AC}^{-1})=\sigma(\boldsymbol{C}^{-2}\boldsymbol{A})=\sigma(\boldsymbol{M}^{-1}\boldsymbol{A})\ .$$

An example of preconditioner is the Jacobi matrix

$$\boldsymbol{M}=\mathrm{diag}(a_{11},a_{22},\dots,a_{nn})\ ,$$

which can be inverted trivially. The resulting conjugate gradient method is known as the Jacobi-Conjugate gradient method.

The algorithm for the preconditioned Conjugate Gradient method is described in Algorithm 46. We leave the implementation of the preconditioned Conjugate Gradient method as an exercise for the reader.

Algorithm 46 Preconditioned Conjugate Gradient method for linear equations

---

Set a tolerance TOL for the accuracy
Initialize $\boldsymbol{x}^{(0)}=\boldsymbol{0}$, $\boldsymbol{r}^{(0)}=\boldsymbol{Ax}^{(0)}-\boldsymbol{b}$, $\boldsymbol{p}^{(0)}=-\boldsymbol{r}^{(0)}$
Solve the system $\boldsymbol{My}^{(0)}=\boldsymbol{r}^{(0)}$ for $\boldsymbol{y}^{(0)}$
Initialize $k=0$
while $\|\boldsymbol{r}^{(k)}\|>TOL$ or $\|\boldsymbol{x}^{(k+1)}-\boldsymbol{x}^{(k)}\|>TOL$ **do**
  Compute $\lambda_k=\frac{\langle \boldsymbol{r}^{(k)},\boldsymbol{y}^{(k)}\rangle}{\langle \boldsymbol{p}^{(k)},\boldsymbol{Ap}^{(k)}\rangle}$
  Compute $\boldsymbol{x}^{(k+1)}=\boldsymbol{x}^{(k)}-\lambda_k\boldsymbol{p}^{(k)}$
  Compute $\boldsymbol{r}^{(k+1)}=\boldsymbol{r}^{(k)}+\lambda_k\boldsymbol{Ap}^{(k)}$
  Solve the system $\boldsymbol{My}^{(k+1)}=\boldsymbol{r}^{(k+1)}$ for $\boldsymbol{y}^{(k+1)}$
  Compute $\beta_{k+1}=\frac{\langle \boldsymbol{r}^{(k+1)},\boldsymbol{y}^{(k+1)}\rangle}{\langle \boldsymbol{r}^{(k)},\boldsymbol{y}^{(k)}\rangle}$
  Compute $\boldsymbol{p}^{(k+1)}=-\boldsymbol{y}^{(k+1)}+\beta_{k+1}\boldsymbol{p}^{(k)}$
  Increase the counter $k=k+1$
end while
Return the approximation $\boldsymbol{x}^{(k)}$

11.2.3 Extensions to nonlinear systems

The conjugate gradient method has been extended for the minimization of general functions. The idea of its extension to nonlinear equations has been attributed to Fletcher and Reeves

[41]. Other extensions are by Polak and Ribière, Hestenes and Stiefel, and Dai and Yuan [104, 60, 28].

In the Fletcher-Reeves extension we consider the general function $f(\boldsymbol{x})$, then as before, we start with the direction $p^{(0)} = -\nabla f(\boldsymbol{x}^{(0)})$. The new approximation $\boldsymbol{x}^{(k+1)} = \boldsymbol{x}^{(k)} + \lambda_k \boldsymbol{p}^{(k)}$ is computed with

$$\lambda_k = \text{argmin}_{\lambda>0} f(x^{(k)} - \lambda p^{(k)}) \ .$$

Then the new direction $\boldsymbol{p}^{(k+1)}$ is chosen so as

$$\boldsymbol{p}^{(k+1)} = \boldsymbol{r}^{(k+1)} + \beta_{k+1}\boldsymbol{p}^{(k)} \ ,$$

with $\boldsymbol{r}^{(k+1)} = -\nabla f(\boldsymbol{x}^{(k+1)})$ and

$$\beta_{k+1} = \frac{\langle \boldsymbol{r}^{(k+1)}, \boldsymbol{r}^{(k+1)} \rangle}{\langle \boldsymbol{r}^{(k)}, \boldsymbol{r}^{(k)} \rangle} \ .$$

If the function $f(\boldsymbol{x})$ is a quadratic function, then the method reduces to the linear conjugate gradient method.

Algorithm 47 Conjugate gradient method for nonlinear equations

Set a tolerance TOL for the accuracy
Initialize $\boldsymbol{x}^{(0)} = \boldsymbol{0}$, $\boldsymbol{r}^{(0)} = \nabla f(\boldsymbol{x}^{(0)})$, $\boldsymbol{p}^{(0)} = -\boldsymbol{r}^{(0)}$
Initialize $k = 0$
while $\|\boldsymbol{r}^{(k)})\| > TOL$ or $\|\boldsymbol{x}^{(k+1)} - \boldsymbol{x}^{(k)}\| > TOL$ **do**
  Compute $\lambda_k = \text{argmin}_{\lambda>0} f(\boldsymbol{x}^{(k)} - \lambda \boldsymbol{p}^{(k)})$
  Compute $\boldsymbol{x}^{(k+1)} = \boldsymbol{x}^{(k)} - \lambda_k \boldsymbol{p}^{(k)}$
  Compute $\boldsymbol{r}^{(k+1)} = \boldsymbol{r}^{(k)} + \lambda_k \boldsymbol{A}\boldsymbol{p}^{(k)}$
  Compute $\beta_{k+1} = \frac{\langle \boldsymbol{r}^{(k+1)}, \boldsymbol{r}^{(k+1)} \rangle}{\langle \boldsymbol{r}^{(k)}, \boldsymbol{r}^{(k)} \rangle}$
  Compute $\boldsymbol{p}^{(k+1)} = -\boldsymbol{r}^{(k+1)} + \beta_{k+1}\boldsymbol{p}^{(k)}$
  Increase the counter $k = k + 1$
end while
Return the approximation $\boldsymbol{x}^{(k)}$

The Fletcher-Reeves algorithm is presented in Algorithm 47. The other methods are improvements of the Fletcher-Reeves algorithm and are based on different computations of the parameter β_k. We summarize the various choices of these values in Table 11.1.

TABLE 11.1
Variations of nonlinear conjugate gradient method

Fletcher-Reeves	$\beta_k = \dfrac{\langle \boldsymbol{r}^{(k)}, \boldsymbol{r}^{(k)} \rangle}{\langle \boldsymbol{r}^{(k-1)}, \boldsymbol{r}^{(k-1)} \rangle}$
Polak-Ribière	$\beta_k = \dfrac{\langle \boldsymbol{r}^{(k)}, \boldsymbol{r}^{(k)} - \boldsymbol{r}^{(k-1)} \rangle}{\langle \boldsymbol{r}^{(k-1)}, \boldsymbol{r}^{(k-1)} \rangle}$
Hestenes-Stiefel	$\beta_k = \dfrac{\langle \boldsymbol{r}^{(k)}, \boldsymbol{r}^{(k)} - \boldsymbol{r}^{(k-1)} \rangle}{\langle \boldsymbol{p}^{(k-1)}, \boldsymbol{r}^{(k)} - \boldsymbol{r}^{(k-1)} \rangle}$
Dai-Yuan	$\beta_k = \dfrac{\langle \boldsymbol{r}^{(k)}, \boldsymbol{r}^{(k)} \rangle}{\langle \boldsymbol{p}^{(k-1)}, \boldsymbol{r}^{(k)} - \boldsymbol{r}^{(k-1)} \rangle}$

The Polak-Ribière modification appears to lead to a more robust method compared to the original Fletcher-Reeves algorithm. Sometimes, the Fletcher-Reeves algorithm can fail to converge if the direction $\boldsymbol{p}_k$ does not have the descent direction property. We can overcome this problem by choosing in Algorithm 47

$$\boldsymbol{p}^{(k+1)} = \boldsymbol{r}^{(k)} + \beta_{k+1}\boldsymbol{p}^{(k)} .$$

If the choice of the parameter λ_k minimizes the function f, then the $\boldsymbol{p}^{(k)}$ is indeed a direction of descent. Otherwise, there are modifications that can help us overcome the specific problem, but we do not mention them here. Also, the Polak-Ribière modification can appear to solve convergence problems but a small modification of the Algorithm 47 can lead to more stable algorithm. For example, choosing the parameter

$$\beta_{k+1}^{+} = \max\{\beta_{k+1}, 0\} ,$$

we can ensure that the resulting directions are descent directions.

More precisely, to guarantee the convergence of such Conjugate Gradient methods, the parameter λ_k needs to satisfy the following *Wolfe conditions*,

$$f(\boldsymbol{x}^{(k)} + \lambda_k\boldsymbol{p}^{(k)}) \leq f(\boldsymbol{x}^{(k)}) + c_1\lambda_k\langle\nabla f(\boldsymbol{x}^{(k)}), \boldsymbol{p}^{(k)}\rangle ,$$
$$|\langle f(\boldsymbol{x}^{(k)} + \lambda_k\boldsymbol{p}^{(k)}), \boldsymbol{p}^{(k)}\rangle| \leq c_2|\langle\nabla f(\boldsymbol{x}^{(k)}), \boldsymbol{p}^{(k)}\rangle| ,$$

where $0 < c_1 < c_2 < 1/2$. One can verify that both Fletcher-Reeves and Polak-Ribière methods, with the small modification described above, satisfy the Wolfe conditions.

Python implementations

Conjugate Gradient implementations have been included in Python for both linear and nonlinear equations. There are all included in submodules of `scipy`, which we describe below. We start with the Python implementation of the Conjugate Gradient method for linear systems.

The Conjugate Gradient implementation for linear systems is the function `cg` of the module `scipy.sparse.linalg`. The method has been included in the specific module as it appears to be exceptionally popular for systems with sparse matrix $\boldsymbol{A}$. The specific function can also be used with preconditioner. Its general call is the following:

```
cg(A,b,x0,tol,maxiter,M,callback,atol)
```

where

`A`: Is the $n \times n$ matrix $\boldsymbol{A}$ of the linear system. This can be sparse or dense, and must be symmetric (or Hermitian if it is complex) and positive definite

`b`: Is the right hand side $\boldsymbol{b}$

`x0`: Is the initial guess of the solution. (Optional with default value `x0=None`)

`tol`: Used for the termination of the algorithm alongside `atol`. (Optional with default value `tol=1.e-5`)

`atol`: Used with the `tol` for the termination of the algorithm with default value `None`. The termination criterion is the following

$$\|r_k\| \leq \max\{\texttt{tol}\|b\|, \texttt{atol}\} .$$

(Optional with default value `atol=None`)

`maxiter`: Is the maximum number of iterations allowed. (Optional with default value `maxiter=None`)

`M`: It can be a sparse or dense matrix and is the preconditioner $\boldsymbol{M}$ described in Section 11.2.2. (Optional with default value `M=None`)

`callback`: Is a user-supplied function to call after each iteration. (Optional with default value is `callback=None`)

The function `cg` returns the approximation of the solution vector `x` and an integer variable `info` which can be either 0 indicating that the convergence was successful, positive, indicating that convergence was not achieved, or negative, indicating that other problems occurred.

Solving the linear system $\boldsymbol{A}\boldsymbol{x} = \boldsymbol{b}$ (see also Section 11.2) can be done in straightforward way as described in the following code:

```
from scipy.sparse.linalg import cg
A = np.array([[2.0, -1.0, 0.0],[-1.0, 2.0, -1.0],[0.0, -1.0, 2.0]])
b = np.array([[1],[0],[1]])
x0 = np.zeros((3,1))
print( cg(A,b,x0) )
```

```
(array([1., 1., 1.]), 0)
```

We observe that the solution is the correct one, and also the integer parameter `info` is 0 indicating the convergence of the Conjugate Gradient method without any problems.

Various extensions of the Conjugate Gradient method for the minimization of nonlinear functions can be found in the module `scipy.optimize` such as `fmin_cg`, `fmin_ncg`, `fmin_powel` and others. The `fmin_cg` function is based on the Polak-Ribière algorithm [104] and is suggested when f has a unique global minimum, and no other local minima or stationary points. It is also required that f is smooth in the sense that both f and its gradient are continuous. The function `fmin_ncg` is based on a Newton-Conjugate Gradient method and perhaps is more powerful in terms of convergence rate. All the minimization algorithms can be called through the function `minimize` of scipy.optimize, which is analyzed later in this chapter.

11.3 Newton's Method in Optimization

Finding a minimizer for a smooth function $f(\boldsymbol{x}) = 0$ is equivalent to finding a root of the system of equations $\nabla f(\boldsymbol{x}) = \boldsymbol{0}$. As we have already discussed in Section 5.5.6, Newton's method can be an effective method for the solution of such equations.

11.3.1 Newton's iteration

For a vector function $\boldsymbol{F}(\boldsymbol{x})$ with $\boldsymbol{x} \in \mathbb{R}^n$ the Newton iteration takes the form

$$\boldsymbol{x}^{(k+1)} = \boldsymbol{x}^{(k)} - [\boldsymbol{DF}(\boldsymbol{x}^{(k)})]^{-1}\boldsymbol{F}(\boldsymbol{x}^{(k)}) ,$$

where $\boldsymbol{DF}(\boldsymbol{x}^{(k)}) = \boldsymbol{J}$ is the Jacobian matrix with the first derivatives of $\boldsymbol{F}$. In optimization, we often seek for the roots of the vector function $\nabla f(\boldsymbol{x})$ where f is a real valued function. In this case, the Jacobian of the vector function $\nabla f(\boldsymbol{x})$ is the Hessian matrix with the second derivatives of the function $f(\boldsymbol{x})$. Consider for example the Taylor expansion of the function $\nabla f(\boldsymbol{x})$ around $\boldsymbol{x}^{(k+1)}$

$$\nabla f(\boldsymbol{x}^{(k+1)}) \approx \nabla f(\boldsymbol{x}^{(k)}) + \boldsymbol{H}(\boldsymbol{x}^{(k)})(\boldsymbol{x}^{(k+1)} - \boldsymbol{x}^{(k)}) ,$$

where $\boldsymbol{H}$ is the Hessian matrix of f with entries the second derivatives of the function f,

$$H_{ij} = \frac{\partial^2 f}{\partial x_i \partial x_j} .$$

Alternatively, one can think of the Hessian matrix $\boldsymbol{H}(\boldsymbol{x}^{(k)})$ as the Jacobian matrix of the $\nabla f(\boldsymbol{x}^{(k)})$. Taking $\nabla f(\boldsymbol{x}^{(k+1)}) \approx 0$ we obtain

$$\boldsymbol{0} \approx \nabla f(\boldsymbol{x}^{(k)}) + \boldsymbol{H}(\boldsymbol{x}^{(k)})(\boldsymbol{x}^{(k+1)} - \boldsymbol{x}^{(k)}) .$$

Given an initial guess $\boldsymbol{x}^{(0)}$ of the solution $\boldsymbol{x}^*$ of the equation $\nabla f(\boldsymbol{x}) = \boldsymbol{0}$, the Newton iteration for $k = 0, 1, \ldots$ takes the form

$$\boldsymbol{x}^{(k+1)} = \boldsymbol{x}^{(k)} - \boldsymbol{H}^{-1}(\boldsymbol{x}^{(k)})\nabla f(\boldsymbol{x}^{(k)}) , \tag{11.23}$$

Newton's iteration (11.23) can be seen as the gradient method of the form

$$\boldsymbol{x}^{(k+1)} = \boldsymbol{x}^{(k)} + \lambda_k \boldsymbol{p}^{(k)} ,$$

with direction $\boldsymbol{p}^{(k)} = -\boldsymbol{H}^{-1}(\boldsymbol{x}^{(k)})\nabla f(\boldsymbol{x}^{(k)})$ and $\lambda_k = 1$, which is quadratically convergent, in the sense

$$\|\boldsymbol{x}^{(k+1)} - \boldsymbol{x}^*\| \le C\|\boldsymbol{x}^{(k)} - \boldsymbol{x}^*\|^2 .$$

Sometimes the choice $\lambda_k \in (0, 1)$ can improve the convergence of Newton's iteration by satisfying the so-called Wolfe conditions. The inclusion of such small parameter λ_k can be seen as a generic technique of performing smaller step from $\boldsymbol{x}^{(k)}$ to obtain $\boldsymbol{x}^{(k+1)}$. In principle Newton's method can be convergent if the function $\nabla f(\boldsymbol{x})$ satisfies a Lipschitz condition

$$\|\nabla f(\boldsymbol{x}) - \nabla f(\boldsymbol{y})\| \le L\|\boldsymbol{x} - \boldsymbol{y}\| ,$$

for some constant $L > 0$.

11.3.2 Quasi-Newton methods

Quasi-Newton methods can be thought of as generalizations of the secant method, where instead of using the Hessian matrix, we consider approximations $\boldsymbol{B}^{(k)}$ of it. A general quasi-Newton iteration can be expressed as

$$\boldsymbol{x}^{(k+1)} = \boldsymbol{x}^{(k)} - [\boldsymbol{B}^{(k)}]^{-1}\nabla f(\boldsymbol{x}^{(k)}) .$$

A straightforward approximation of the Hessian matrix can be obtained by using the Taylor expansion of $\nabla f(\boldsymbol{x}^{(k)})$ around $\boldsymbol{x}^{(k-1)}$. This is

$$\nabla f(\boldsymbol{x}^{(k)}) = \nabla f(\boldsymbol{x}^{(k-1)}) + \boldsymbol{B}^{(k)}\boldsymbol{p}^{(k-1)} ,$$

where $\boldsymbol{p}^{(k-1)} = \boldsymbol{x}^{(k-1)} - \boldsymbol{x}^{(k)}$. Solving for $\boldsymbol{B}^{(k)}$ requires the solution of a linear system of course, but this technique can be useful when the computation of the Hessian matrix is not possible[4].

[4]Alternatively, one can use the complex-step approximation of the second derivative to bypass the barrier of the Hessian. For more information see the exercises of Chapter 5.

One of the most well-known quasi-Newton methods is Broyden's method, in which the use of Sherman-Morrison formula was suggested for the computation of the matrix $\boldsymbol{B}^{(k)}$ from the previous approximation $\boldsymbol{B}^{(k-1)}$. More precisely, Broyden's method suggests the approximation

$$\boldsymbol{B}^{(k)} = \boldsymbol{B}^{(k-1)} + \frac{\boldsymbol{y}^{(k)} - \boldsymbol{B}^{(k-1)}\boldsymbol{p}^{(k-1)}}{\langle \boldsymbol{p}^{(k-1)}, \boldsymbol{p}^{(k-1)} \rangle}[\boldsymbol{p}^{(k-1)}]^T ,$$

where $\boldsymbol{y}^{(k)} = \nabla f(\boldsymbol{x}^{(k)}) - \nabla f(\boldsymbol{x}^{(k-1)})$.

Because the Sherman-Morrison formula appeared to be useful for the solution of linear systems of some special form, we proceed with the complete derivation of Broyden's method. First, we remind that the outer product of two vectors $\boldsymbol{u} = (u_1, u_2, \ldots, u_n)^T$ and $\boldsymbol{v} = (v_1, v_2, \ldots, v_n)^T$, is defined as

$$\boldsymbol{u} \otimes \boldsymbol{v} = \boldsymbol{u}\boldsymbol{v}^T = \begin{pmatrix} u_1v_1 & u_1v_2 & \cdots & u_1v_n \\ u_2v_1 & u_2v_2 & \cdots & u_2v_n \\ \vdots & \vdots & \ddots & \vdots \\ u_nv_1 & u_nv_2 & \cdots & u_nv_n \end{pmatrix} .$$

The Sherman-Morrison formula provides a way to compute any $n \times n$ matrix $\boldsymbol{B}$ that can be written in the form

$$\boldsymbol{B} = \boldsymbol{A} + \boldsymbol{u} \otimes \boldsymbol{v} .$$

In such cases, the matrix $\boldsymbol{B}$ is called *rank-one update*, since the columns of the matrix $\boldsymbol{u} \otimes \boldsymbol{v}$ are constant multiples of the vector $\boldsymbol{u}$, and thus it is of rank one.

Let $\boldsymbol{x}$ be the solution of the system $\boldsymbol{A}\boldsymbol{x} = \boldsymbol{u}$. Then,

$$\boldsymbol{B}\boldsymbol{x} = (\boldsymbol{A} + \boldsymbol{u} \otimes \boldsymbol{v})\boldsymbol{x} = \boldsymbol{A}\boldsymbol{x} + \boldsymbol{u} \otimes \boldsymbol{v}^T\boldsymbol{x} = \boldsymbol{u} + \boldsymbol{u}(\boldsymbol{v}\boldsymbol{x}^T) = (1 + \boldsymbol{v}\boldsymbol{x}^T)\boldsymbol{u} ,$$

and

$$\boldsymbol{x} = (1 + \boldsymbol{v}\boldsymbol{x}^T)\boldsymbol{B}^{-1}\boldsymbol{u} .$$

Solving for $\boldsymbol{B}^{-1}\boldsymbol{u}$ and using the formula $\boldsymbol{x} = \boldsymbol{A}^{-1}\boldsymbol{u}$ we obtain

$$\boldsymbol{B}^{-1}\boldsymbol{u} = \frac{1}{1 + \boldsymbol{v}(\boldsymbol{A}^{-1}\boldsymbol{u})^T}\boldsymbol{A}^{-1}\boldsymbol{u} .$$

In case $\langle \boldsymbol{v}, \boldsymbol{A}^{-1}\boldsymbol{x} \rangle = 0$, the situation can be simplified even more. For example,

$$\begin{aligned} \boldsymbol{B}\boldsymbol{A}^{-1}\boldsymbol{x} &= (\boldsymbol{A} + \boldsymbol{u} \otimes \boldsymbol{v})\boldsymbol{A}^{-1}\boldsymbol{x} = \boldsymbol{A}\boldsymbol{A}^{-1}\boldsymbol{x} + \boldsymbol{u}\boldsymbol{v}^T\boldsymbol{A}^{-1}\boldsymbol{x} \\ &= \boldsymbol{x} + \boldsymbol{u}[\boldsymbol{v}(\boldsymbol{A}^{-1}\boldsymbol{x})^T] = \boldsymbol{x} , \end{aligned}$$

which is simplified to $\boldsymbol{B}^{-1}\boldsymbol{x} = \boldsymbol{A}^{-1}\boldsymbol{x}$.

This formula can lead to a formula for the computation of the inverse for more general matrices. For example, let $\boldsymbol{C}$ be an $n \times n$ matrix, and suppose we want to compute the matrix $\boldsymbol{D}$ such that $\boldsymbol{D}\boldsymbol{w} = \boldsymbol{z}$ for some given vectors $\boldsymbol{w}, \boldsymbol{z}$, and $\boldsymbol{D}\boldsymbol{y} = \boldsymbol{C}\boldsymbol{y}$ for $\boldsymbol{y}$ orthogonal to a given vector $\boldsymbol{g}$. Here $\boldsymbol{C} = \boldsymbol{A}^{-1}$, $\boldsymbol{D} = \boldsymbol{B}^{-1}$, $\boldsymbol{w} = \boldsymbol{u}$, $\boldsymbol{z} = 1/[1 + \boldsymbol{v}(\boldsymbol{A}^{-1}\boldsymbol{u})^T]\boldsymbol{A}^{-1}\boldsymbol{u}$, and $\boldsymbol{g} = \boldsymbol{A}^{-T}\boldsymbol{v}$. Setting

$$\boldsymbol{D} = \boldsymbol{C} + \frac{(\boldsymbol{z} - \boldsymbol{C}\boldsymbol{w}) \otimes \boldsymbol{g}}{\boldsymbol{g}\boldsymbol{w}^T} ,$$

then, if $\boldsymbol{g}\boldsymbol{y}^T = 0$, from the analysis above we have $\boldsymbol{D}\boldsymbol{y} = \boldsymbol{C}\boldsymbol{y}$. Moreover, we observe that

$$\boldsymbol{D}\boldsymbol{w} = \boldsymbol{C}\boldsymbol{w} + (\boldsymbol{z} - \boldsymbol{C}\boldsymbol{w}) = \boldsymbol{z} .$$

Applying all these in the definition of $\boldsymbol{D}$, we obtain

$$\begin{aligned}\boldsymbol{B}^{-1} &= \boldsymbol{A}^{-1} + \frac{\left[\left(\frac{1}{1+\boldsymbol{v}\cdot\boldsymbol{A}^{-1}\boldsymbol{u}}\boldsymbol{A}^{-1}\boldsymbol{u} - \boldsymbol{A}^{-1}\boldsymbol{u}\right)\otimes\boldsymbol{v}\right]\boldsymbol{A}^{-1}}{\boldsymbol{v}\cdot\boldsymbol{A}^{-1}\boldsymbol{u}} \\ &= \boldsymbol{A}^{-1} - \frac{\boldsymbol{A}^{-1}(\boldsymbol{u}\otimes\boldsymbol{v})\boldsymbol{A}^{-1}}{1+\boldsymbol{v}\cdot\boldsymbol{A}^{-1}\boldsymbol{u}} .\end{aligned}$$

This is the Sherman-Morrison formula, which can lead to an approximation of $[\boldsymbol{B}^{(k)}]^{-1}$ using the previous matrix $[\boldsymbol{B}^{(k-1)}]^{-1}$. More precisely, given $\boldsymbol{B}^{(k-1)}$ we have

$$\boldsymbol{B}^{(k)}(\boldsymbol{x}^{(k)} - \boldsymbol{x}^{(k-1)}) = \nabla f(\boldsymbol{x}^{(k)}) - \nabla f(\boldsymbol{x}^{(k-1)}) ,$$

and if $\boldsymbol{z}^T(\boldsymbol{x}^{(k)} - \boldsymbol{x}^{(k-1)}) = 0$ then $\boldsymbol{B}^{(k)}\boldsymbol{z} = \boldsymbol{B}^{(k-1)}\boldsymbol{z}$. The Sherman-Morrison formula then yields

$$\boldsymbol{B}^{(k)} = \boldsymbol{B}^{(k-1)} + \frac{\boldsymbol{y}^{(k-1)} - \boldsymbol{B}^{(k-1)}\boldsymbol{p}^{(k-1)}}{[\boldsymbol{p}^{(k-1)}]^T\boldsymbol{p}^{(k-1)}} \otimes \boldsymbol{p}^{(k)} ,$$

where $\boldsymbol{p}^{(k-1)} = \boldsymbol{x}^{(k)} - \boldsymbol{x}^{(k-1)}$, and $\boldsymbol{y}^{(k-1)} = \nabla f(\boldsymbol{x}^{(k)}) - \nabla f(\boldsymbol{x}^{(k-1)})$. Moreover, since

$$\boldsymbol{y}^{(k-1)} - \boldsymbol{B}^{(k-1)}\boldsymbol{p}^{(k-1)} = \nabla f(\boldsymbol{x}^{(k-1)}) ,$$

then

$$\boldsymbol{B}^{(k)} = \boldsymbol{B}^{(k-1)} + \frac{\nabla f(\boldsymbol{x}^{(k)})^T\boldsymbol{p}^{(k-1)}}{[\boldsymbol{p}^{(k-1)}]^T\boldsymbol{p}^{(k-1)}} .$$

Using the Sherman-Morrison formula we obtain

$$\begin{aligned}[\boldsymbol{B}^{(k)}]^{-1} &= [\boldsymbol{B}^{(k-1)}]^{-1} - \frac{[\boldsymbol{B}^{(k-1)}]^{-1}\left(\frac{1}{[\boldsymbol{p}^{(k-1)}]^T\boldsymbol{p}^{(k-1)}}\nabla f(\boldsymbol{x}^{(k)})\otimes\boldsymbol{p}^{(k-1)}\right)[\boldsymbol{B}^{(k-1)}]^{-1}}{1+[\boldsymbol{p}^{(k-1)}]^T[\boldsymbol{B}^{(k-1)}]^{-1}\left(\frac{1}{[\boldsymbol{p}^{(k-1)}]^T\boldsymbol{p}^{(k-1)}}\nabla f(\boldsymbol{x}^{(k)})\right)} \\ &= [\boldsymbol{B}^{(k-1)}]^{-1} - \frac{[\boldsymbol{B}^{(k-1)}]^{-1}(\nabla f(\boldsymbol{x}^{(k)})\otimes\boldsymbol{p}^{(k-1)})[\boldsymbol{B}^{(k-1)}]^{-1}}{[\boldsymbol{p}^{(k-1)}]^T\boldsymbol{p}^{(k-1)} + [\boldsymbol{p}^{(k-1)}]^T[\boldsymbol{B}^{(k-1)}]^{-1}\nabla f(\boldsymbol{x}^{(k)})} \\ &= [\boldsymbol{B}^{(k-1)}]^{-1} - \frac{(\boldsymbol{u}^{(k-1)}\otimes\boldsymbol{p}^{(k-1)})[\boldsymbol{B}^{(k-1)}]^{-1}}{[\boldsymbol{p}^{(k-1)}]^T(\boldsymbol{p}^{(k-1)}+\boldsymbol{u}^{(k-1)})} ,\end{aligned}$$

where $\boldsymbol{u}^{(k-1)} = [\boldsymbol{B}^{(k-1)}]^{-1}\nabla f(\boldsymbol{x}^{(k)})$.

11.3.3 Nonlinear optimization and the module `scipy.optimize`

In the module `scipy.optimize` the general purpose function `minimize` is provided for the minimization of a scalar function of one or more variables. One can use this function to employ any of the previously described methods, which is among at least 15 different methods! The general call of `minimize` is the following:

```
minimize(fun,x0,args,method,jac,hess,hessp,
         bounds,constraints,tol,callback,options)
```

where some of the arguments are the following:

`fun`: Is the objective function we want to minimize, which can be given as `fun(x,args)`

`x0`: Is the initial guess of the solution

`args`: Is a tuple containing any extra arguments needed by the function `fun`. (Optional with default value the empty tuple `()`)

`method`: Is a string defining the method we prefer to use for the minimization process. We can choose among many the methods we have described such as `'CG'` and `'Newton-CG'`. The complete list of all possible choices can be found online. (Optional with default value `method=None` in which case the method is chosen internally)

`jac`: Can be a function that returns the gradient vector. Alternatively, can be one of the strings `2-point`, `3-point` or `cs` for the numerical estimation of the gradient using finite differences. It can also be a boolean. If it is `True` then the function `fun` is assumed to return the gradient along with the objective function, otherwise, the gradient is estimated using the `2-point` finite difference estimation. (Optional with default value `jac=None`)

`tol`: Is the tolerance for the termination of the method. (Optional with default value `tol=None`)

`options`: Is a dictionary with the integer `maxiter` maximum number of iterations and the boolean value `disp` which determines whether messages related to convergence will be print on the screen. (Optional with default value `options=None`)

The function returns the `OptimizeResult` object in a variable `res`, which includes the solution `x`, and boolean variable `success` and the string `message` which describes the reason of the termination. As you may have observed, here we do not provide information related to parameters `hess`, `hessp`, `bounds`, `constraints` and `callback` since we didn't mention anything about these in the theory and they are optional arguments. Information about these arguments can be found in the official documentation of the module SciPy.

11.3.4 The Gauss-Newton approximation

In the linear least squares problem we seek for a minimizer of the error function

$$E(\boldsymbol{x}) = \frac{1}{2}\|\boldsymbol{A}\boldsymbol{x} - \boldsymbol{b}\|^2 .$$

In such a case Newton's method takes the form

$$\boldsymbol{H}(\boldsymbol{x}^{(k+1)} - \boldsymbol{x}^{(k)}) = -\nabla E(\boldsymbol{x}^{(k)}) ,$$

where the gradient ∇E and the Hessian matrix of the second derivatives of E can be computed explicitly by differentiating the function $E(\boldsymbol{x})$ as

$$\nabla E(\boldsymbol{x}) = \boldsymbol{A}^T(\boldsymbol{A}\boldsymbol{x} - \boldsymbol{b}) \quad \text{and} \quad \boldsymbol{H} = \boldsymbol{A}^T\boldsymbol{A} .$$

Thus, Newton's iteration can be written as

$$\boldsymbol{A}^T\boldsymbol{A}(\boldsymbol{x}^{(k+1)} - \boldsymbol{x}^{(k)}) = -\boldsymbol{A}^T(\boldsymbol{A}\boldsymbol{x}^{(k)} - \boldsymbol{b}) ,$$

which can be simplified to

$$\boldsymbol{A}^T\boldsymbol{A}\boldsymbol{x}^{(k+1)} = \boldsymbol{A}^T\boldsymbol{b} .$$

The Hessian $\boldsymbol{H} = \boldsymbol{A}^T\boldsymbol{A}$ is the idea behind the use of the Gauss-Newton method for the solution of the nonlinear least squares problem. In the nonlinear least squares problem the Hessian matrix $\boldsymbol{H}$ is being approximated by the product of two matrices $\boldsymbol{H} \approx 2\boldsymbol{J}^T\boldsymbol{J}$, where in practice the matrix $\boldsymbol{J}$ is the Jacobian matrix of the nonlinear fit function $\hat{y}(\boldsymbol{x})$.

More precisely, assume that we have a set of data (t_i, y_i) for $i = 0, 1, \ldots, N$ that we want to approximate by a fitting function $\hat{\boldsymbol{y}}(t; \boldsymbol{x})$ that depends perhaps in a nonlinear way on the n parameters $\boldsymbol{x} = (x_1, \ldots, x_n)$. Sometimes, for the sake of simplicity, we will write $\hat{\boldsymbol{y}}(\boldsymbol{x})$ ignoring the dependence on t. For example, if $n = 2$ and we consider a linear function $\hat{\boldsymbol{y}}(t; \boldsymbol{x}) = x_0 t + x_1$, then the problem is the linear least squares problem of Section 10.2. The general least squares problem is to find the minimizer of residual function

$$f(\boldsymbol{x}) = \sum_{i=0}^{N} (y_i - \hat{y}_i(\boldsymbol{x}))^2 \ ,$$

where $\hat{y}_i = \hat{\boldsymbol{y}}(t_i; \boldsymbol{x})$. Note that we changed the notation slightly from Section 10.2 so as to be in agreement with the previous analysis.

If $\boldsymbol{H}$ denotes the Jacobian matrix of ∇f then the Newton iteration is

$$\boldsymbol{H}(\boldsymbol{x}^{(k+1)} - \boldsymbol{x}^{(k)}) = -\nabla f(\boldsymbol{x}^{(k)}) \ .$$

The idea of Gauss-Newton's method is to avoid the computation of the second derivatives based on the following observation

$$f(\boldsymbol{x}) = \sum_{i=0}^{N} (y_i - \hat{y}_i)^2 = (\boldsymbol{y} - \hat{\boldsymbol{y}}(\boldsymbol{x}))^T (\boldsymbol{y} - \hat{\boldsymbol{y}}(\boldsymbol{x})) = \boldsymbol{y}^T \boldsymbol{y} - 2\boldsymbol{y}^T \hat{\boldsymbol{y}}(\boldsymbol{x}) + \hat{\boldsymbol{y}}(\boldsymbol{x})^T \hat{\boldsymbol{y}}(\boldsymbol{x}) \ ,$$

which yields to

$$\nabla f(\boldsymbol{x}) = -2\boldsymbol{J}^T (\boldsymbol{y} - \hat{\boldsymbol{y}}(\boldsymbol{x})) \ ,$$

where J is the Jacobian of the function $\hat{y}$ with

$$J_{ij} = \frac{\partial}{\partial x_j} \hat{y}_i \ .$$

Next, we need to estimate the Hessian matrix in order to derive the Gauss-Newton method. Taking the Taylor's expansion $\hat{\boldsymbol{y}}(\boldsymbol{x} + \delta \boldsymbol{x}) \approx \hat{\boldsymbol{y}}(\boldsymbol{x}) + \boldsymbol{J}\delta \boldsymbol{x}$, substitution into the expression of the function f yields

$$\begin{aligned} f(\boldsymbol{x} + \delta\boldsymbol{x}) &= \boldsymbol{y}^T \boldsymbol{y} - 2\boldsymbol{y}^T \hat{\boldsymbol{y}}(\boldsymbol{x} + \delta\boldsymbol{x}) + \hat{\boldsymbol{y}}(\boldsymbol{x} + \delta\boldsymbol{x})^T \hat{\boldsymbol{y}}(\boldsymbol{x} + \delta\boldsymbol{x}) \\ &\approx \boldsymbol{y}^T \boldsymbol{y} - 2\boldsymbol{y}^T (\hat{\boldsymbol{y}}(\boldsymbol{x}) + \boldsymbol{J}\delta\boldsymbol{x}) + (\hat{\boldsymbol{y}}(\boldsymbol{x}) + \boldsymbol{J}\delta\boldsymbol{x})^T (\hat{\boldsymbol{y}}(\boldsymbol{x}) + \boldsymbol{J}\delta\boldsymbol{x}) \\ &= (\boldsymbol{y} - \hat{\boldsymbol{y}}(\boldsymbol{x}))^T (\boldsymbol{y} - \hat{\boldsymbol{y}}(\boldsymbol{x})) - 2(\boldsymbol{y} - \hat{\boldsymbol{y}}(\boldsymbol{x}))^T \boldsymbol{J}\delta\boldsymbol{x} + \delta\boldsymbol{x}^T \boldsymbol{J}^T \boldsymbol{J}\delta\boldsymbol{x} \ . \end{aligned}$$

Comparing the last relation with the Taylor expansion of $f(\boldsymbol{x} + \delta\boldsymbol{x})$ we conclude that $\boldsymbol{J}^T \boldsymbol{J} \approx \frac{1}{2}\boldsymbol{H}$. Here, the Hessian cannot be equal to the matrix $2\boldsymbol{J}^T\boldsymbol{J}$ but based on the approximation $\boldsymbol{H} \approx 2\boldsymbol{J}^T\boldsymbol{J}$ we modify the Newton iteration to be

$$\boldsymbol{J}^T \boldsymbol{J}(\boldsymbol{x}^{(k+1)} - \boldsymbol{x}^{(k)}) = \boldsymbol{J}^T \left(\boldsymbol{y} - \hat{\boldsymbol{y}}(\boldsymbol{x}^{(k)})\right) \ . \tag{11.24}$$

The iteration (11.24) is known to as the Gauss-Newton iteration.

11.3.5 The Levenberg-Marquardt modification

Using the previous approximations for the $\nabla f(\boldsymbol{x})$ we can write the general method of steepest descent in the form

$$\boldsymbol{x}^{(k+1)} - \boldsymbol{x}^{(k)} = -\lambda_k \boldsymbol{J}^T \left(\boldsymbol{y} - \hat{\boldsymbol{y}}(\boldsymbol{x}^{(k)})\right) \ .$$

The Levenberg-Marquardt method is a modification of the Gauss-Newton method that combines both the method of steepest descent and the Gauss-Newton method

$$\boldsymbol{J}^T\boldsymbol{J}(\boldsymbol{x}^{(k+1)}-\boldsymbol{x}^{(k)})=\boldsymbol{J}^T\left(\boldsymbol{y}-\hat{\boldsymbol{y}}(\boldsymbol{x}^{(k)})\right) .$$

In particular, the Levenberg-Marguardt method for nonlinear least squares problems is formulated as

$$(\boldsymbol{J}^T\boldsymbol{J}+\lambda\boldsymbol{I})(\boldsymbol{x}^{(k+1)}-\boldsymbol{x}^{(k)})=\boldsymbol{J}^T\left(\boldsymbol{y}-\hat{\boldsymbol{y}}(\boldsymbol{x}^{(k)})\right) , \qquad (11.25)$$

where λ is usually chosen to be fairly large. Many attempts to determine the parameter λ lead to the following simple algorithm in adapting the value of λ depending on the value of the error $E(\boldsymbol{x})$: Choose a large initial value $\lambda=\lambda_0$ and a parameter $\nu>1$ for the first iteration. After computing $\boldsymbol{x}^{(1)}$, compute the value $E(\boldsymbol{x}^{(1)})$. Continue using $\lambda=\lambda_1=\lambda_0/\nu$ and compute again the value $E(\boldsymbol{x}^{(2)})$. If the values of the error functions are decreasing in a satisfactory way continue, otherwise use the value $\lambda=\lambda_k\nu^j$ for some j until better convergence is achieved.

The Levenberg-Marquardt method is an enhanced first order method, which is very popular in optimization and also in the training of neural networks of moderate size. Minimization algorithms in Python are implemented mainly in the module `scipy.optimize`. Gauss-Newton and Levenberg-Marquardt methods are implemented in the function `least_squares` of the previously mentioned module, and finds the local minimum of the cost function

$$F(x)=\frac{1}{2}\sum_{i=0}^{m-1}\rho(f_i(x)^2),\quad x\in[a,b] .$$

11.4 Introduction to Neural Networks

Methods of optimization are perhaps the most useful tools of computational mathematics in the study of artificial neural networks of machine learning. The training of a computer system, and in particular a machine learning procedure, is based on the minimization of appropriate functionals. Artificial neural networks are inspired by biological neural networks. These can be thought of as algorithms, that "learn" to perform tasks by considering examples. Artificial neural networks find application in fields such as computer science, data science, statistics and mathematics. Examples include computer vision, speech and face recognition, social networks, computer games, and even the solution of differential equations. In this section we introduce the notion of artificial neural networks, which we will call for simplicity neural networks from now on, and we explain the most basic algorithm we use to train them, the back-propagation algorithm.

11.4.1 A simple neural network

Think of an artificial neural network as a mathematical function $\boldsymbol{y}=\boldsymbol{f}(\boldsymbol{x};\boldsymbol{w})$ which depends on some parameters $\boldsymbol{w}$ and requires an input $\boldsymbol{x}$ to return the output $\boldsymbol{y}$. The function $\boldsymbol{f}$ is adjusted to approximate a function $\hat{\boldsymbol{f}}$ which corresponds to the real problem. We usually achieve this by minimizing the difference $\boldsymbol{f}-\hat{\boldsymbol{f}}$. In artificial neural networks we can adjust the function $\boldsymbol{f}$ in order to return desired values $\boldsymbol{y}$ for certain input $\boldsymbol{x}$. This is usually done by finding appropriate parameters $\boldsymbol{w}$. Artificial neural networks work similarly to biological neural networks. Neural networks consist of neurons. Neurons receive signals (input) from the environment and they react appropriately by sending (firing) signals to the brain, which

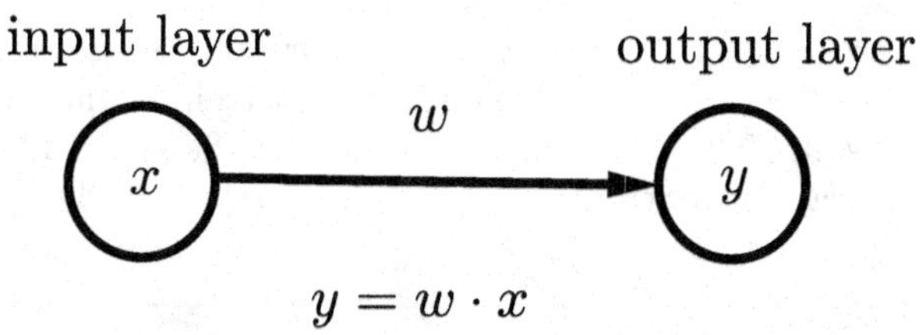

FIGURE 11.3
A perceptron: The simplest neural network with only two layers.

is responsible for the output $\boldsymbol{y}$. The output $\boldsymbol{y}$ in biological neural networks depends on the experiences we have or in other words on our training. Think what happens when we drive. If the input we receive from our eyes is a red traffic light, then our brain signals to stop since this is what we have learnt to do.

Artificial neural networks, like biological neural networks, have nodes called neurons that are responsible for receiving input, processing it and propagating information inside the network. Usually, because there are more than one input data, there are more than one neurons in a neural network. There are many different ways that neurons are connected. The connections between neurons are called synapses (singular synapsis). There are also many different strategies to process the input data. Each way determines a different architecture in artificial neural networks. In this brief introduction we consider only *feed-forward* networks where the connections (synapses) between neurons are direct and acyclic in the sense that the information can propagate only in one direction.

The perceptron

The simplest feed-forward neural network was introduced in 1957 by Frank Rosenblatt (1928–1971) [110] and is called *perceptron*. This neural network consists of two nodes (neurons). The perceptron receives the input x at the input node, and returns the output y at the output node. The process that this simple perceptron can perform for example could be to take the input x, multiply it with a *synaptic weight* w and turn in the output result y. This simple neural network can be represented schematically in Figure 11.3. We say in general that the input nodes consist of the input layer and the output nodes the output layer. So in this simple example we have only two nodes (the input and the output layers) and a weight w.

The output y is the product of the weight w with the input x such that $y = w \cdot x$. Given the input x, we need to estimate the unknown weight w so as to obtain the correct output y. This is actually the main task of a feed-forward neural network. Initially, we give to w any arbitrary value $w^{(0)}$. In this simple case we can *train* the algorithm to find (or in other words learn) what the weight should be so as to solve the problem in a trivial way. In order to train the algorithm to compute the correct output value y for general x we need to provide the algorithm with an example. It is worth mentioning that in general neural networks there might be more layers between the input and output layers. These layers are called *hidden layers*. In this simple case we just presented there are no hidden layers.

As an example, assume that the desired, weighted output for input 2 is 0.5. These data serve as *training data*. Now the determination of the neural network is complete. Using the equation $y = w \cdot x$ we can solve for w and we can use the specific value of w to compute the output y for any given input x. Obviously $w = y/x = 1/4$.

☞ A neural network is determined by its structure and the given training data. A neural network with the same structure but different training data will provide with totally different results. The procedure of estimating the synaptic weights is called training of the neural network. Sometimes we also say that the neural network is learning the synaptic weights.

Instead of using this simple technique and dividing to find the optimal weight, we present a more sophisticated way to estimate weights. This general methodology is used for more complicated neural networks and is called *back-propagation.* We introduce the error function borrowed from the problem of linear least squares

$$E(w) = |y - w \cdot x|^2 ,$$

which in the language of machine learning is called *cost function.* In order to estimate the optimal value of w we need to minimize the error $E(w)$ and find an approximation of the value $w^* = \text{argmin} E(w)$. We can approximate the $\text{argmin} E(w)$, using the gradient method (11.3), which in this case is

$$w^{(k+1)} = w^{(k)} - \lambda E'(w^{(k)}) .$$

It is noted that since the only independent variable in $E(w)$ is w, then the gradient $\nabla E(w)$ is reduced to the standard scalar derivative $E'(w)$. In the language of neural networks the parameter λ is called *learning rate.* We have discussed how to compute the value of this parameter in each step using the method of steepest descent. Here we take arbitrary learning rate and we focus only on neural network techniques.

Using the values $x = 2$ and $y = 0.5$ of our example we have $E(w) = |0.5 - 2w|^2$. Computing the derivative $E'(w) = -2 + 8w$ we can write the steepest descent method as

$$w^{(k+1)} = w^{(k)} - \lambda(-2 + 8w^{(k)}) .$$

Using an initial guess $w^{(0)} = 0.8$ and learning rate $\lambda = 0.1$ we compute the first five approximations $w^{(k)}$ for $k = 1, 2, \ldots, 5$ with the following simple code.

```
w = 0.8; lam = 0.1
for i in range(5):
    w = w - lam*(-2+8*w)
    print(w)
```

```
0.36
0.272
0.2544
0.25088
0.250176
```

We observe that the code converges to the optimal value $w = 0.25$, which is the same with the value we computed with the direct division. If we are given the input $x = 3$ and asked about the output y, we can immediately respond with $y = 0.75$ since $y = 0.25 \cdot 3$.

In the previous example we chose the learning rate $\lambda = 0.1$. Typically, this can be chosen using the techniques described in the previous sections. The method of determining the weight w using optimization techniques is called *back-propagation.* Also the specific form of learning we just described is called *supervised learning.*

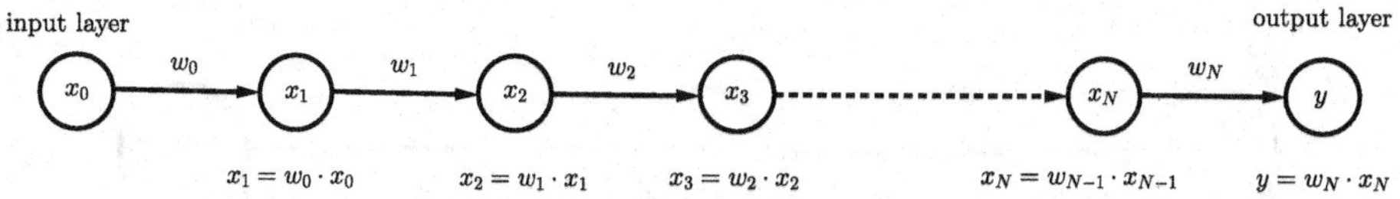

FIGURE 11.4
A serial neural network with hidden layers.

☞ When we say that a neural network is learning we mean that it is just minimizing a cost function to find the optimal synaptic weights involved. The computation of the optimal weights using the previous methodology is called back-propagation.

A serial neural network

In the perceptron model we had a neural network with only two layers (neurons), the input and output layers. A more complicated neural network consists of more neurons and more layers. As a second example we consider a neural network consisting of $N+2$ layers connected in a serial way as shown in Figure 11.4. Since we have $N+1$ neurons x_i, $i = 0, 1, \ldots, N$ we consider also $N+1$ weights w_i, for $i = 0, 1, \ldots, N$ for the computation of the values x_i for each of the hidden layers. The layers for $i = 1, \ldots, N$ we call them *hidden* as their values are computed internally in order to find the output value y. Also the number of hidden layers in a neural network determines the *depth* of the neural network, and that is why the training of neural networks is called *deep learning.*

In such topology each value $x_i = w_{i-1} \cdot x_{i-1}$, thus the output value is

$$\begin{aligned} y &= w_N \cdot x_N \\ &= w_N w_{N-1} \cdot x_{N-1} \\ &\vdots \\ &= w_N w_{N-1} \cdots w_0 \cdot x_0 \ . \end{aligned}$$

Such a neural network can be considered as perceptron where the output $y = w \cdot x$ with $x = x_0$ and $w = w_N w_{N-1} \cdots w_0$. All the layers between the input and the output are considered as hidden layers. If it is required to compute the synaptic weights w_i, $i = 0, 1, \ldots, N$, then by setting arbitrary initial guess $\boldsymbol{w}^{(0)}$ for the vector $\boldsymbol{w} = (w_0, w_1, \ldots, w_N)$ and considering the minimization of the cost function

$$E(\boldsymbol{w}) = |y - w_0 w_1 \cdots w_N x|^2 \ ,$$

with the help of the steepest descent method we have

$$\boldsymbol{w}^{(k+1)} = \boldsymbol{w}^{(k)} - \lambda \, \nabla E(\boldsymbol{w}^{(k)}) \ .$$

This time the gradient $\nabla E(\boldsymbol{w})$ is a vector defined as

$$\nabla E(\boldsymbol{w}) = \left(\frac{\partial}{\partial w_0} E(\boldsymbol{w}), \frac{\partial}{\partial w_1} E(\boldsymbol{w}), \ldots, \frac{\partial}{\partial w_N} E(\boldsymbol{w}) \right)^T \ .$$

☞ Synaptic weights are initialized usually with guesses, and then optimal values are computed using optimization techniques.

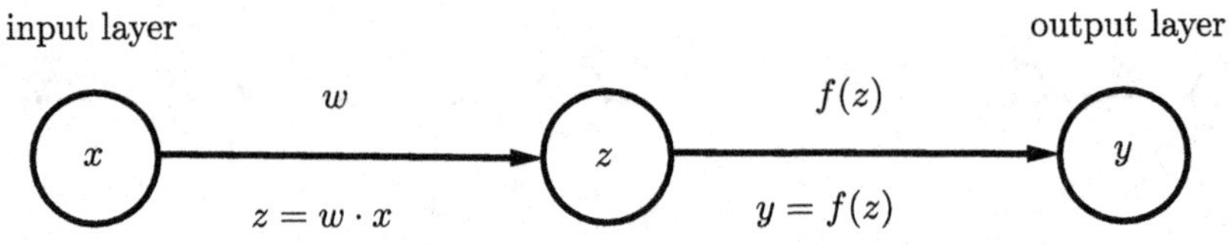

FIGURE 11.5
Perceptron with activation function.

11.4.2 Activation functions

Let a perceptron simulate one of our fingers that checks the water temperature (input) and returns a response (output). If the temperature is less than a certain value we might not want to do anything. On the other hand, if the water temperature is greater than a threshold, then we want the response output to represent an intense reaction. This can be done by allowing the neuron to apply an *activation* (or *learning*) function to the original output and return $y = f(w \cdot x)$ instead of just $w \cdot x$. The architecture of the perceptron is shown in Figure 11.5.

Examples of activation functions include the logistic (or step) function

$$f_{\text{step}}(z) = \begin{cases} 0, & \text{if } z < 0 \\ 1, & \text{if } z \geq 0 \end{cases},$$

which in practical terms returns a non-zero output only if $w \cdot x \geq 0$. The step function is not the most popular choice compared to other activation functions due to its discontinuity. The main feature of an activation function is that it returns small or zero values for arguments that are smaller than a certain value, and values close to one for arguments of larger value. Another candidate for activation function is the sigmoid function

$$f_{\text{sigmoid}}(z) = \frac{1}{1 + e^{-r \cdot z}},$$

for appropriate value $r \in \mathbb{R}$, $r > 0$. This function is called sigmoid due to its graph (Figure 11.6).

Other activation functions is the hyperbolic tangent $f_{\tanh}(z) = \tanh(r \cdot z)$ the inverse tangent $f_{\arctan}(z) = \arctan(z)$, the rectified linear unit (ReLU) function

$$f_{ReLU}(z) = (z)_+ = \max\{0, z\} = \begin{cases} 0, & \text{if } z < 0 \\ z, & \text{if } z \geq 0 \end{cases},$$

and others.

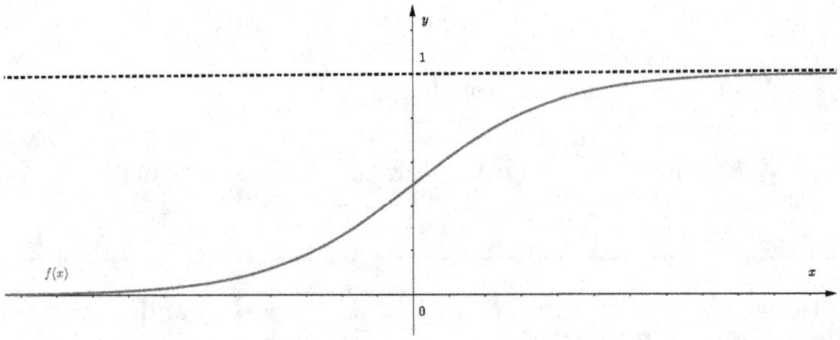

FIGURE 11.6
Sketch of the sigmoid function.

Since the output of the activation function will be the output of our simple neural network, we can only have output in the range of the activation function. For example, in the case of the sigmoid function we can have output $y \in (0, 1)$ only. If we need the output to be in a different interval, then we need to modify the activation function, for example, by multiplying it with an appropriate constant.

Looking back at the example with the single neuron and with training data $x = 2$, $y = 0.5$ using the sigmoid function (with $r = 1$) as activation function, we need to estimate a new optimal value of the weight w. This time the situation is more complicated. The optimal value can be found as $w^* = \text{argmin} E(w)$ where $E(w) = |y - f(w \cdot x)|^2$. Trying to find the gradient $E'(w)$ we obtain

$$E'(w) = -2(y - f(w \cdot x))xf'(w \cdot x) \ .$$

Since $f'(z) = e^{-z}/(1 + e^{-z})^2$ we have

$$E'(w) = -2xe^{-2 \cdot w} \frac{0.5(1 + e^{-2 \cdot w}) - 1}{(1 + e^{-2 \cdot w})^3} \ .$$

The steepest descent method then becomes

$$w^{(k+1)} = w^{(k)} - \lambda E'(w^{(k)}) \ ,$$

with given $w^{(0)} = 0.8$. Modifying the code for the perceptron appropriately we can check that in the case where the activation function is the sigmoid the convergence to the optimal weight $w = 0$ can be achieved by using large amount of iterations k. This is because the sigmoid function is very smooth with small derivative. The convergence can be improved by using larger values of learning rate λ.

11.4.3 General feedforward neural networks

General feedforward neural networks, can contain layers with multiple neurons. These neurons can be interconnected with other neurons of other layers with different synapses. A general neuron receives multiple input values x_i, processes them and computes their sum

$$s = \sum_{i=1}^{N} w_i \cdot x_i \ ,$$

adds a bias b to the sum s and then uses an activation function f to decide whether to fire a positive response or return a zero value. This can be summarized by considering the i-th neuron's output defined as

$$y = f\left(\sum_{i=1}^{N} w_i \cdot x_i + b\right) \ ,$$

and can be represented by the diagram of Figure 11.7.

In this general neuron we consider a *bias* term b, which is added to the linear combination of the synaptic weights and the input data. The bias term can be considered though as a weight $w_0 = b$ of given input $x_0 = 1$. Thus, we could write the summation as

$$z = \sum_{i=0}^{N} w_i \cdot x_i \ .$$

The role of the bias is to increase or decrease the chances of neuron's reaction after the value z passes through the activation function $f(z)$.

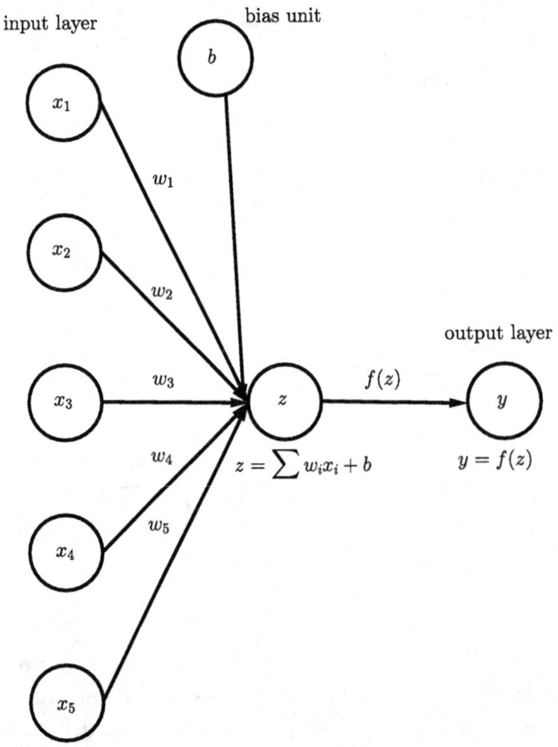

FIGURE 11.7
General perceptron with multiple input, bias unit and activation function.

The synaptic weights w_i and the bias value b are set initially to arbitrary values, and in order to find their optimal values, we consider a cost function of the form

$$E(\boldsymbol{w}) = \frac{1}{2}\left| y - f(\sum_{i=0}^{N} w_i x_i)\right|^2 . \tag{11.26}$$

As an example, consider the case $N = 2$ and the training data $x_1 = 0.7$, $x_2 = 0.3$, $y = 0.5$. For simplicity, we consider the simplest activation function $f(z) = z$ and bias $w_0 = b$ with $x_0 = 1$. For the cost function 11.26) with $N = 2$, we have

$$\nabla E(\boldsymbol{w}) = \left(-x_0\left(y - \sum_{i=0}^{2} w_i x_i\right),\ -x_1\left(y - \sum_{i=0}^{2} w_i x_i\right),\ -x_2\left(y - \sum_{i=0}^{2} w_i x_i\right)\right)^T .$$

The optimal synaptic weights (including the bias) can be found with the help of the gradient method

$$\boldsymbol{w}^{(k+1)} = \boldsymbol{w}^{(k)} - \lambda\, \nabla E(\boldsymbol{w}) ,$$

with λ an arbitrary small value for the learning rate. Using the steepest descent method we obtain the values $b = w_0 \approx 0.05$, $w_1 \approx 0.33$ and $w_2 \approx 0.71$.

It is worth noting that avoiding the use of activation function, then the cost function can be written as

$$E(\boldsymbol{w}) = \frac{1}{2}\|\boldsymbol{y} - \boldsymbol{A}\boldsymbol{x}\|_2^2 ,$$

where $\boldsymbol{x} = (x_0, x_1, x_2)^T$, $\boldsymbol{y} = y$ and $\boldsymbol{A} = (w_0, w_1, w_2)$, and thus the problem of training a neural network can be seen as not very different from solving a least squares problem.

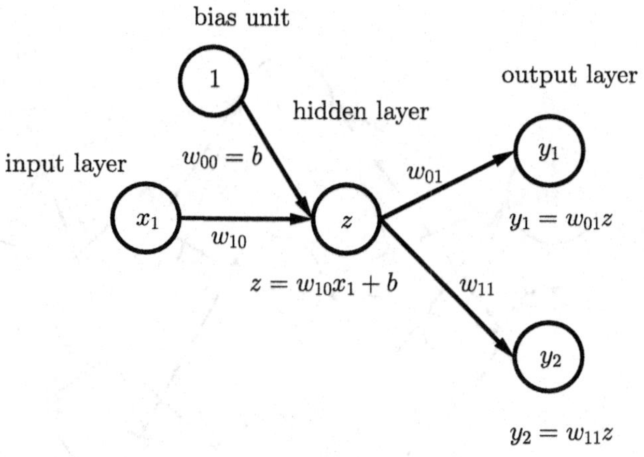

FIGURE 11.8
Perceptron with a hidden layer and multiple input and output neurons.

In general, a neural network consists of multiple input nodes x_i and multiple output nodes y_i. In such case the cost function can be defined as $E(\boldsymbol{w}) = \frac{1}{2}\sum_i |y_i - \sum_j w_{ij}x_j|^2$. This cost function is again of the same kind as before. To demonstrate the optimization of a neural network with multiple output we consider the simple case with two nodes y_1 and y_2 in the output layer and one node x_1 in the input layer. Furthermore, we assume a bias $b = w_0$ and also the respective given input $x_0 = 1$. To keep the discussion simple we consider the activation function $f(z) = z$. Such a simple neural network is represented in Figure 11.8.

In this simple example, the cost function is

$$E(\boldsymbol{w}) = \frac{1}{2}\sum_{i=1}^{2} |y_i - w_{i1}z|^2 = \frac{1}{2}\sum_{i=1}^{2} |y_i - w_{i1}(w_{10}x_1 + b)|^2$$
$$= \frac{1}{2}\sum_{i=1}^{2} |y_i - w_{i1}(w_{10}x_1 + w_{00}x_0)|^2 \ .$$

The problem is four-dimensional and we need to find optimal values for

$$\boldsymbol{w} = (w_{00}, w_{10}, w_{11}, w_{21})^T \ .$$

With the help of chain rule of differentiation we obtain

$$\nabla E(\boldsymbol{w}) = \begin{pmatrix} -\sum_{i=1}^{2}(y_i - w_{i1}z)w_{i1}x_0 \\ -\sum_{i=1}^{2}(y_i - w_{i1}z)w_{i1}x_1 \\ -(y_1 - w_{11}z)z \\ -(y_2 - w_{21}z)z \end{pmatrix} ,$$

with $z = w_{10}x_1 + w_{00}x_0$. Taking arbitrary initial guess $\boldsymbol{w} = (1, 1, 1, 1)^T$ and using the method of steepest descent with training data $x_1 = 0.4$, $y_1 = 0.3$, $y_2 = 0.4$ we obtain the optimal solution $\boldsymbol{w} \approx (0.35, 0.74, 0.46, 0.62)^T$. Note that the term $\frac{1}{2}$ in the definition of cost functions could be avoided without any difference in the results but it is included for convenience in the calculations.

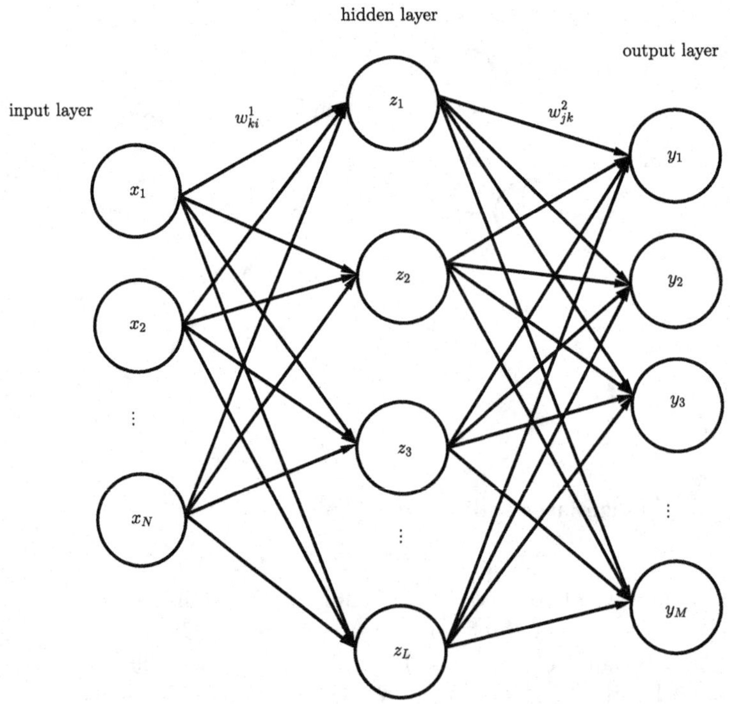

FIGURE 11.9
A neural network with multiple input and output layers, and one hidden layer.

More general neural networks

In most of the applications we prefer to use multiple hidden layers with multiple neurons. Consider a neural net with N inputs x_i, $i = 1, 2, \ldots, N$, M outputs y_j, $j = 1, 2, \ldots, M$ and one hidden layer with L neurons. Denote the activation function of each neuron in the hidden layer by f^1_k, $k = 1, 2, \ldots, L$, and that of the output layer by f^2_j, $j = 1, 2, \ldots, M$. Let w^1_{ki}, $k = 1, 2, \ldots, L$, $i = 1, 2, \ldots, N$ be the synaptic weights for inputs into the hidden layer, and w^2_{jk}, $j = 1, 2, \ldots, M$, $k = 1, 2, \ldots, L$ the synaptic weights for the output layer. For simplicity, we ignore biases that anyway can be incorporated into the input data. Such a neural network is depicted in Figure 11.9.

Denoting the entries of the hidden layer by

$$z_k = f^1_k \left(\sum_{i=1}^{N} w^1_{ki} x_i \right) ,$$

then the j-th output in the output layer will be

$$y_j = f^2_j \left(\sum_{k=1}^{L} w^2_{jk} z_k \right) = f^2_j \left(\sum_{k=1}^{L} w^2_{jk} f^1_k \left(\sum_{i=1}^{N} w^1_{ki} x_i \right) \right) .$$

Let's now denote the unknown vector of synaptic weights by

$$\boldsymbol{w} = \{ w^1_{ki}, w^2_{jk} \ : \ i = 1, 2, \ldots, N, \ k = 1, 2, \ldots, L, \ j = 1, 2, \ldots, M \} .$$

Given the training data $\boldsymbol{x} = (x_1, x_2, \ldots, x_N)^T$ and $\boldsymbol{y} = (y_1, y_2, \ldots, y_M)^T$, the training of the neural network requires the estimation of the optimal synaptic weights that minimize the cost function

$$E(\boldsymbol{w}) = \frac{1}{2}\sum_{j=1}^{M}\left|y_j - f_j^2\left(\sum_{k=1}^{L} w_{jk}^2 z_k\right)\right|^2 = \frac{1}{2}\sum_{j=1}^{M}\left|y_j - f_j^2\left(\sum_{k=1}^{L} w_{jk}^2 f_k^1\left(\sum_{i=1}^{N} w_{ki}^1 x_i\right)\right)\right|^2 .$$

To solve this minimization problem we employ a gradient descent method as before. This requires though the computation of the partial derivatives of the cost function E which using the chain rule of differentiation we can write them as

$$\frac{\partial E}{\partial w_{jk}^2}(\boldsymbol{w}) = \left(y_j - f_j^2\left(\sum_{k=1}^{L} w_{jk}^2 z_k\right)\right) f_j^{2'}\left(\sum_{s=1}^{L} w_{js}^2 z_s\right) z_k ,$$

where the symbol $'$ denotes the ordinary derivative with respect to the independent variable. To simplify notation, we write

$$\delta_j = \left(y_j - f_j^2\left(\sum_{k=1}^{L} w_{jk}^2 z_k\right)\right) f_j^{2'}\left(\sum_{s=1}^{L} w_{js}^2 z_s\right) ,$$

so the partial derivatives can be expressed in the form

$$\frac{\partial E}{\partial w_{jk}^2}(w) = -\delta_j z_k .$$

Similarly, we have

$$\frac{\partial E}{\partial w_{ki}^2}(\boldsymbol{w}) = -\left(\sum_{i=1}^{L} \delta_i w_{ik}^1\right) f_k^{1'}\left(\sum_{s=1}^{N} w_{ks}^1 x_i\right) x_i .$$

Having computed the gradient of the cost function we can perform the minimization algorithm of our choice to compute the synaptic weights. After computing the synaptic weights we can use them anytime we want to compute unknown values $\boldsymbol{y}$ for given input $\boldsymbol{x}$.

Instead of presenting theoretical examples of the backpropagation algorithm, we present the way it works in practice and with Python. There are several libraries dedicated to neural networks. Such libraries are the modules TensorFlow, PyTorch and others. In this book we present briefly the module Scikit-Learn. For a detailed discussion of artificial neural networks, we refer to the references and exercises at the end of this chapter.

11.4.4 Machine learning and the module `sklearn`

The module `sklearn`[5] is a simple and efficient tool for basic machine learning problems such as *classification*, *regression*, *clustering*, *dimensionality reduction*, and others. In this introductory text we consider only the classification and the regression problems. The regression problem is the problem of approximating data and predicting the unknown from the known, like what we achieved with interpolation and least-squares approximation. The methods used for solving the regression problem are called *regressors*. The classification problem is quite different. In particular, the classification problem is the problem of identifying the category of an object, given some specified categories. An example of a classification problem

[5] `sklearn` is an acronym for the library Scikit-Learn

is the problem of identifying spam emails from regular emails. In the next section we will present another classification problem, which is the recognition of handwritten numbers. The methods we use to solve classification problems are called *classifiers*.

We begin with a brief presentation of the module `sklearn` and in particular a class that can solve regression problems using a few commands. The module `sklearn` contains two public attributes named `coefs_` and `intercepts_`. The first one is a list of weight matrices where its i-th column contains the weights between the i and $i+1$ layers, while the second is a list of bias vectors where the i-th vector represents the bias values added to layer $i+1$.

The module that we care more about here is the submodule `sklearn.neural_network` and its classes `MLPRegressor` and `MLPClassifier`. These classes implement *backpropagation* algorithms to train a general perceptron such as the one in Figure 11.7. The `MLP` in the names of these classes is an abbreviation of the words multi-layer perceptron, which is mainly setup by these classes. The optimization is based on either the so-called L-BFGS method which is a quasi-Newton-type method or the `sgd` and `adam` methods, which are both stochastic gradient descent methods. Let's make a parenthesis here to discuss the stochastic gradient descent method.

Convergence of gradient descent methods based on the iteration $\boldsymbol{w}^{(k+1)} = \boldsymbol{w}^{(k)} - \lambda\nabla E(\boldsymbol{w}^{(k)})$ can be time consuming, especially, when the computation of the gradient is considered. In applications of neural networks we often use a modification of the gradient descent methods called *stochastic gradient descent method* where at every step only small sets of data is used for training the network. These small sets are called *minibatches* and are chosen randomly.

Using M minibatches leads to the modification of the function $E(\boldsymbol{x}) = \frac{1}{M}\sum E_i(\boldsymbol{x})$, which takes into account the contribution of the M minibatches only. An additional advantage of the stochastic gradient descent is that using minibatches we avoid a phenomenon called *overfitting*. Overfitting happens when we approximate noisy data with great accuracy, like the noise is part of the actual solution, leading to inaccurate predictions. Adaptation in the various choices of the learning rate result in a variety of stochastic gradient descent methods, one of which is the so-called Adam method[6]. For more information related to the stochastic gradient method and its variants, we refer to the references at the end of this chapter.

We proceed with the general call of the class `MLPRegressor`, which is the following:

```
MLPRegressor(hidden_layer_sizes, activation, solver, alpha, batch_size,
learning_rate, learning_rate_init, power_t, max_iter, other_options)
```

where

`hidden_layer_sizes`: Is a tuple where the i-th element represents the number of neurons in the i-th hidden layer. Its default value is the tuple `(100,)` indicating that there is only one hidden layer with 100 neurons

`activation`: Is a string defining the activation function and can take one of the following values: `'indentity'`, `'logistic'`, `'tanh'`, `'relu'`. (Optional with default value `activation='relu'`)

`solver`: Is a string defining the optimization method and can take one of the following values `'lbfgs'`, `'sgd'`, `'adam'`. (Optional with default value `solver='adam'`)

`alpha`: Is a floating point number identifying an L^2 penalty parameter. (Optional with default value `alpha=0.0001`)

[6]The name Adam was derived from the name of the method "adaptive moment estimation"

`batch_size`: Is an integer identifying the size of minibatches for the stochastic optimization method and is ignored if `lbfgs` is the choice for the `solver` parameter. (Optional with default value `batch_size=auto`)

`learning_rate`: Is a string defining the learning rate schedule for weight updates. It can take the value `constant` for constant learning rate given by the parameter `learning_rate_init`, `'invscaling'` when the learning rate is gradually decreasing at each time step `t` using an inverse scaling exponent of `power_t` and `'adaptive'` when the learning rate changes adaptively. (Optional with default value `learning_-rate='constant'`)

`learning_rate_init`: Is the constant value of the learning rate. (Optional with default value `learning_rate_init=0.001`)

`power_t`: Is the exponent for the inverse scaling learning rate and is only used with the `'sgd'` solver. (Optional with default value `power_t=0.5`)

`max_iter`: Is the maximum number of iterations we allow the optimization method to perform. (Optional with default value `max_iter=200`)

Other input parameters include the *optional* parameters `shuffle`, `random_state`, `tol`, `verbose`, `momentum`, `nesterovs_momentum`, `early_stopping`, `validation_fraction`, `beta_1`, `beta_2`, `epsilon`, `n_iter_no_change` and `max_fun`. We refer to the online manual of the Scikit-Learn module for a description of these values.

Similarly the general call of the class `MLPClassifier` is the following

```
MLPClassifier(hidden_layer_sizes, activation, solver, alpha, batch_size,
learning_rate, learning_rate_init, power_t, max_iter, other_options)
```

and the various arguments coincide with those of the `MLPRegressor`, for which we refer to the previous description and the online manual of Scikit-Learn.

After we have setup a neural network we can train it using the member function `fit`. If we have created a neural net in the variable `mlp`, then we can train it by calling

```
mlp.fit(x_train,y_train)
```

As an example we consider the approximation of the function $f(x) = \sin(2\pi x) + \sin(7\pi x)$ in the interval $[-1, 1]$ using neural networks methodology. The idea is to use some points of the function f as given training data and the rest as testing data. This problem is analogous to regression we studied before.

First, we setup our data by generating the data set `x`, `y` with `x` consisting of 2000 uniformly distributed points in $[-1, 1]$ and `y` containing the corresponding values of f.

```
def f(x):
    y = np.sin(2.0*np.pi*x) + np.sin(7.0*np.pi*x)
    return y
h = 0.001
x = np.arange(-1.0,1.0+h,h)
y = f(x)
plt.plot(x,y)
plt.xlabel('$x$')
plt.ylabel('$y$',rotation=0)
plt.show()
```

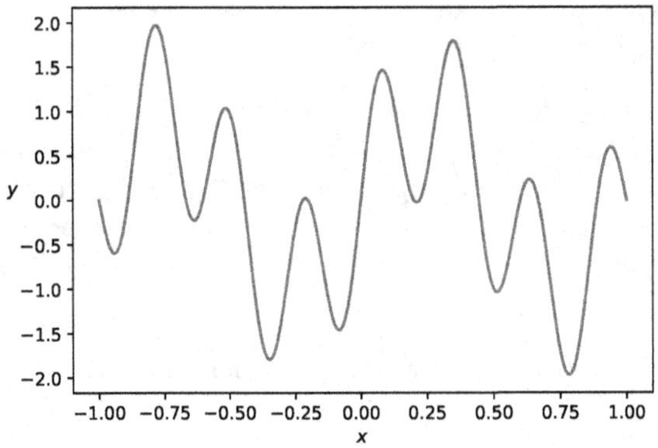

The next step is to select randomly data points to use them as training data. Usually, we select 70-80% of the data as training data, and in this example we choose 75%. The separation between testing and training data can be done with the function `train_test_split` of the module `sklearn.model_selection`.

```
from sklearn.model_selection import train_test_split
x_train, x_test, y_train, y_test = train_test_split(x, y, test_size=0.25)
plt.plot(x_train, y_train, '.', x_test, y_test,'x')
plt.xlabel('$x$')
plt.ylabel('$y$',rotation=0)
plt.show()
```

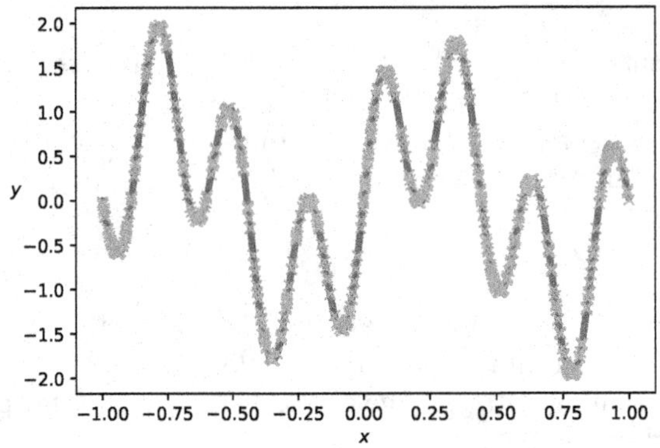

In this figure we can see that most of the points (dots) are training data. Then using the testing data, we will try to reconstruct the shape of the function f.

The back-propagation algorithm with one hidden layer of 100 neurons in total for this problem can be implemented as easy as the following code:

```
# reshape the date into (N,1)-dimensional array
x_train = x_train.reshape(-1,1)
x_test = x_test.reshape(-1,1)
from sklearn.neural_network import MLPRegressor
# setup the network
mlp = MLPRegressor(hidden_layer_sizes=[100])
# train the network
mlp.fit(x_train,y_train)
```

```
# use the trained network to obtain results
predictions = mlp.predict(x_test)
plt.plot(x,y,x_test, predictions, 'r.')
plt.xlabel('$x$')
plt.ylabel('$y$',rotation=0)
plt.show()
```

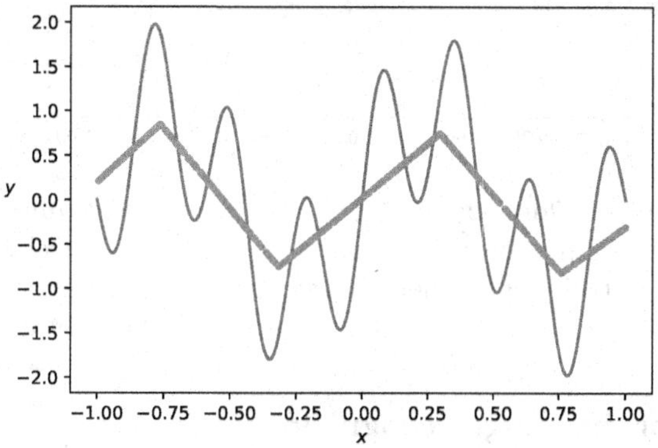

In this code we obtain in the output `mlp` the setup of the neural network, then with the `fit` function we train the network and finally we receive the predictions with the function `predict`.

The predicted values are not very accurate, although the result is better than what we would have gotten in the case of a linear least squares approximation. On the other hand, it demonstrates the ease of use of the Scikit-Learn with its default parameters. To obtain better results we increase the number of hidden layers and their neurons to six hidden layers with `[20, 50, 50, 50, 50, 20]` neurons.

```
x_train = x_train.reshape(-1,1)
x_test = x_test.reshape(-1,1)
from sklearn.neural_network import MLPRegressor
mlp = MLPRegressor(hidden_layer_sizes=[20,50,50,50,50,20])
mlp.fit(x_train,y_train)
predictions = mlp.predict(x_test)
plt.plot(x,y,x_test, predictions, 'r.')
plt.xlabel('$x$')
plt.ylabel('$y$',rotation=0)
plt.show()
```

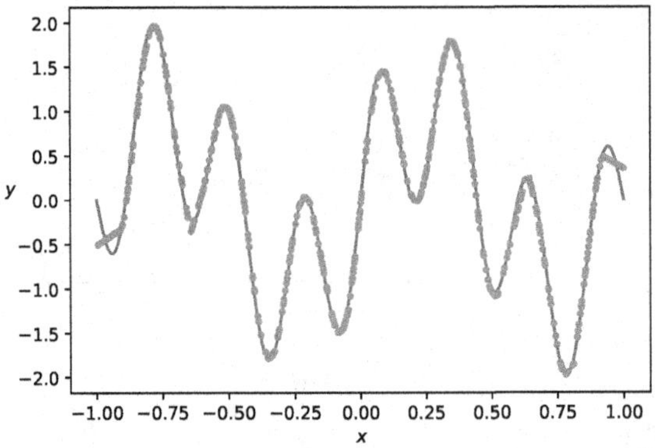

In addition to the `MLPRegressor` the Scikit-Learn is equipped with the class `MLPClassifier` which has a number of attributes and methods specialized for solving classification problems. After training the network we use the method `predict` to predict new samples. We will see all these in action in the following application example.

11.4.5 Application in image recognition

Perhaps one of the most important classification problems that has been solved with machine learning is the digits recognition and the recognition of car registration plates by machines. In this section we show how to train and test a simple neural network to recognize handwritten numbers from 0 to 9. For training and testing the network we need a set of images with handwritten digits. For this reason, we will use a dataset provided by the `sklearn` module. This dataset can be loaded into the memory using the command

```
from sklearn import datasets
digits = datasets.load_digits()
```

which will load all the images and their attributes into the Python class `digits`. The images are stored into the array `digits.images`, which is a three-dimensional array with shape `(1797, 8, 8)` meaning that we have 1797 images of 8×8 pixels. Using `subplots` we plot the first 10 digits with the commands

```
figure, axes = plt.subplots(nrows=1, ncols=10, figsize=(10, 3))
for i in range(10):
    axes[i].set_axis_off()
    axes[i].imshow(digits.images[i],cmap='binary')
plt.show()
```

Each of these digits is an image in grey scale. Each pixel is an entry of an array and its value represents the intensity of grey. The rest of the images are similar but all different. You can try plotting more of such images by modifying the previous code appropriately. For more information about digital image representation, we refer to Section 10.5.6.

As a first step we consider a simple neural network with 64 neurons for the input layer. Each neuron corresponds to a pixel. We will also consider a hidden layer of 15 neurons and an output layer with 10 neurons for the digits 0 to 9.

Since the input layer is a vector of 64 entries, we need to reshape the input images and transform them into `(64,)` vectors, and store them as columns of an array `x` instead of square 8×8 arrays.

```
x = digits.images.reshape((len(digits.images), -1))
print(x.shape)
```

We also keep a record of the digits using the attribute `target` of the class `digits` with the command `y = digits.target`.

Next, we split our data into training and testing data by separating the first 1000 images to be used for training and the rest 797 for testing. We do the same for their labels and we store them into two different arrays with names `x_train`, `y_train`, and `x_text`, `y_test`.

```
x_train = x[:1000]; y_train = y[:1000]
x_test = x[1000:]; y_test = y[1000:]
```

After having all the data ready, we build a neural network with 15 hidden layers using the class `MLPClassifier`. We store the neural network in a variable called `NN`. For this neural network we consider the activation function `logistic` and learning rate 0.1, while for solver we use the `sgd` which implements a stochastic gradient descent method. The complete call should look like

```
from sklearn.neural_network import MLPClassifier
NN = MLPClassifier(hidden_layer_sizes=[15], activation='logistic',
                   solver='sgd', learning_rate_init=.1, verbose=True)
```

We give the input parameter `verbose=True` so as to observe the progress during the training of the neural network and receive important information. The training of the network is completed with the method `fit`

```
NN.fit(x_train,y_train)
```

Since we have computed all the weights we can let our neural network recognize our test images using the method `predict`. For the sake of convenience we show the first 50 digits from the set `x_test`. In addition to these predictions we present the exact values stored in the array `y_test`.

```
predictions = NN.predict(x_test)
print('predicted values=', predictions[:50])
print('exact values=', y_test[:50])
```

`predicted values=[1 4 0 5 3 6 9 6 1 7 5 4 4 7 2 8 2 2 5 7 9 5 4 4 9 0 8 9 8 0 1 2 3 4 5 6 7 8 3 0 1 2 3 4 5 6 7 8 5 0]`

`exact values=[1 4 0 5 3 6 9 6 1 7 5 4 4 7 2 8 2 2 5 7 9 5 4 4 9 0 8 9 8 0 1 2 3 4 5 6 7 8 9 0 1 2 3 4 5 6 7 8 9 0]`

Observe that the 49-th digit is 9 but our network predicted 5. We can avoid such errors by choosing different parameters for the neural network. This is not the only error in this set of data but as you can see the mixed numbers are quite the same. We can easily compare them using the command `sum(predictions==y_test)`. This will return the total number of correct predictions, while subtracting from `len(predictions)` we estimate the wrong predictions. In our test case we got 67 wrong predictions and 730 correct. How many wrong predictions did you get?

11.5 Further Reading

Suggested books for further reading in optimization include the books [13, 23, 93, 98]. These are also the books that we have followed closely. Optimization methods for the numerical solution of linear systems, such as the method of steepest descent and conjugate gradient are analyzed in books of numerical linear algebra such as [112, 29, 53, 48, 24]. An important review on methods of optimization is the review paper [33]. Applications of optimization in artificial neural networks can be found in [59]. For a complete introduction to neural network techniques, refer to [49, 86, 92]. The paper [62] consists of an introduction to deep learning and the backpropagation algorithm and is highly recommended.

Chapter Highlights

- Unconstrained optimization is the branch of optimization that focuses on the minimization of functions.
- The solution $\boldsymbol{x}^*$ minimizing a function $f(\boldsymbol{x})$ is called minimizer.
- Gradient methods are iterative methods where at every iteration the new approximation is computed along the direction of the gradient $-\nabla f(\boldsymbol{x})$, the direction of the steepest descent.
- General gradient descent methods converge linearly.
- The method of steepest descent is a particular gradient method where the step of the minimization is defined so as to minimize the value $f(\boldsymbol{x}^{(k+1)})$.
- The steepest descent method finds applications in the solution of linear systems of equations by minimizing the function

$$f(\boldsymbol{x}) = \frac{1}{2}\langle \boldsymbol{A}\boldsymbol{x}, \boldsymbol{x}\rangle - \langle \boldsymbol{b}, \boldsymbol{x}\rangle \ .$$

- The conjugate gradient method is an alternative of the steepest descent method in which the directions are more general and can be derived using the notion of conjugate directions.
- The conjugate gradient method for linear systems converges if $\boldsymbol{A}$ is symmetric and positive definite. It also requires at most n iterations to converge to the exact solution with exact arithmetic.
- The convergence of iterative methods for linear systems can be accelerated using preconditioning techniques.
- Other popular methods for the minimization of functions are Newton-type methods such as the classical Newton method and other approximations such as quasi-Newton and Levenberg-Marquardt methods.
- One of the pillars of modern machine learning is the branch of optimization.
- Neural networks represented by connected nodes (neurons) are characterized by undetermined weights.
- The estimation of weights (known as backpropagation) can be done using optimization techniques to minimize errors on given training data.
- The error in neural network processes is usually expressed as a least squares error and is known as cost function.
- SciKit-Learn is a Python library for machine learning problems. Other libraries are the TensorFlow and PyTorch.

Exercises

1. **(Golden section search algorithm)** The golden section search algorithm is an iterative algorithm for the approximation of the value $x^* = \text{argmin} f(x)$. Let $x^* \in [a, b]$ be the unique minimizer of the objective function f. Like the bisection method, the initial interval will be replaced by a smaller one in the next step of the method. The main difference is that in order to locate an interval with the minimizer we need two points to detect whether the function f is increasing or decreasing. For this reason, in each step we compute the values

$$\begin{array}{ll} x = a + r(b - a), & u = f(x) \\ y = a + r^2(b - a), & v = f(y) \end{array} ,$$

where $r = 1/\rho$ and ρ is the golden section ratio

$$\rho = \frac{1 + \sqrt{5}}{2} \approx 1.62 \ .$$

If $u > v$ then the minimum of f lies in the interval $[a, x]$, which will be the input interval for the next step. If $u \leq v$ then the minimum lies in the interval $[y, b]$. We repeat the previous steps in the new interval until the interval is small enough to meet appropriate tolerance.

   (a) Show that r is a root of the equation $r^2 + r - 1 = 0$.

   (b) Implement the previous method in a Python function.

   (c) Use the method to find the minimum of the function $f(x) = x^2 + x + 1$.

   (d) Compare your results with the results of the function `golden` of the module `scipy.optimize`.

2. Prove that $\|\cdot\|_1$, $\|\cdot\|_2$ and $\|\cdot\|_\infty$ as defined in Chapter 9 are vector norms. Let $\boldsymbol{A}$ be a symmetric and positive definite matrix. Prove that the map $\|\cdot\|_A : \mathbb{R}^n \to \mathbb{R}$ with

$$\|\boldsymbol{x}\|_A = \boldsymbol{x}^T \boldsymbol{A} \boldsymbol{x} \ ,$$

is a vector norm. This norm is known as A-norm.

3. Consider the minimization of the linear least squares error $E(\boldsymbol{x}) = \frac{1}{2}\|\boldsymbol{A}\boldsymbol{x} - \boldsymbol{b}\|^2$, where $\boldsymbol{A}$ is a real $m \times n$ matrix, and b an n-dimensional column vector.

   (a) Write and implement the steepest descent method for solving the specific linear least squares problem.

   (b) Write and implement the conjugate gradient method for solving the same problem.

   (c) Test your codes for the problem with

$$\boldsymbol{A} = \begin{pmatrix} 0 & 1.1 \\ 1 & 0 \\ 0 & -0.2 \end{pmatrix}, \qquad \boldsymbol{b} = \begin{pmatrix} 1 \\ -1.1 \\ -0.2 \end{pmatrix},$$

   which has optimal solution the vector $\boldsymbol{x} = (-1.1, 1)^T$.

   (d) Compare the results with those obtained using the *SVD* of $\boldsymbol{A}$.

4. Consider the function $f(\boldsymbol{x}) = x_1^3 - x_1x_2 - x_1 + x_1x_2^3 - x_2^4$ with $\boldsymbol{x} = (x_1, x_2)^T$.

(a) Find all critical points for which $\nabla f(\boldsymbol{x}) = 0$.

(b) Verify which are maxima and minima.

(c) Use the steepest descent method to compute approximations of all these points.

5. Consider the function $f(\boldsymbol{x}) = 100(x_1^2 - x_2^2) + (1 - x_1)^2$, with $\boldsymbol{x} = (x_1, x_2)^T$.

   (a) Use the steepest descent method to approximate the minima of f.

   (b) Use the conjugate gradient method to approximate the minima of f.

6. Consider the linear system $\boldsymbol{A}\boldsymbol{x} = \boldsymbol{b}$ with

$$\boldsymbol{A} = \begin{pmatrix} 1 & 0 \\ 0 & \lambda \end{pmatrix} ,$$

   with $\lambda > 0$ and $\boldsymbol{x} = \boldsymbol{b} = \boldsymbol{0}$.

   (a) If $\boldsymbol{x}^{(k)} = (x_1^{(k)}, x_2^{(k)})^T$, show that the method of steepest descent is

$$\boldsymbol{x}^{(k+1)} = \frac{x_1^{(k)} x_2^{(k)} (\lambda - 1)}{(x_1^{(k)})^2 + \lambda^3 (x_2^{(k)})^2} \begin{pmatrix} \lambda^2 x_2^{(k)} \\ -x_1^{(k)} \end{pmatrix} .$$

   Sketch a diagram with the isolines of the functional $f(\boldsymbol{x})$ as defined in (11.12) and some approximations $\boldsymbol{x}^{(k)}$, $k = 1, 2, 3, \ldots$ when $\boldsymbol{x}^{(0)} = (5, 1)^T$.

   (b) Show that if $x^{(k)} = c(\lambda, \pm 1)^T$, then the inequality (11.17) holds as an equality.

   (c) If $\lambda = 100$, how many steps of the method are needed until the error $\langle \boldsymbol{A}\boldsymbol{e}^{(k)}, \boldsymbol{e}^{(k)} \rangle^{1/2}$ becomes less than $\varepsilon \langle \boldsymbol{A}\boldsymbol{e}^{(0)}, \boldsymbol{e}^{(0)} \rangle^{1/2}$ for given $\varepsilon > 0$?

7. Consider the linear system $\boldsymbol{A}\boldsymbol{x} = \boldsymbol{b}$ with

$$\boldsymbol{A} = \begin{pmatrix} 2 & 1 & 1 \\ 1 & 2 & 1 \\ 1 & 1 & 2 \end{pmatrix} \quad \text{and} \quad \boldsymbol{b} = \begin{pmatrix} 2 \\ 0 \\ 2 \end{pmatrix} .$$

   (a) Solve the system with the conjugate gradient method using as initial guess the vector $\boldsymbol{x}_0 = (0, 0, 0)^T$.

   (b) Solve the system with the Jacobi and Gauss-Seidel methods.

   (c) Compare the methods in terms of convergence, speed and accuracy.

8. Consider an artificial neural network with three neurons x_1, x_2, x_3 as input layer, a hidden layer with two neurons z_1, z_2 and the output layer comprised two neurons y_1, y_2. The neural network has a bias unit connected to both z_1, z_2 and a different bias connected to y_1, y_2. The specific neural network can be described with the graph diagram 11.10. Given the training data $\hat{\boldsymbol{x}} = ((\hat{y}_1, \hat{y}_2, \hat{x}_3)^T = (1, 4, 5)^T$ with desired output $\hat{\boldsymbol{y}} = (\hat{y}_1, \hat{y}_2)^T = (0.1, 0.05)^T$:

   (a) Write a Python function that will implement the f_{sigmoid} activation function.

   (b) Write a Python function that given the synaptic weights and input data will compute the values

$$f_{\text{sigmoid}}(w_1 x_1 + w_3 x_2 + w_5 x_3 + b_1) = z_1 ,$$
$$f_{\text{sigmoid}}(w_2 x_1 + w_4 x_2 + w_6 x_3 + b_1) = z_2 ,$$

   and

$$f_{\text{sigmoid}}(w_7 z_1 + w_9 z_2 + b_2) = y_1 ,$$
$$f_{\text{sigmoid}}(w_8 z_1 + w_{10} z_2 + b_2) = y_2 .$$

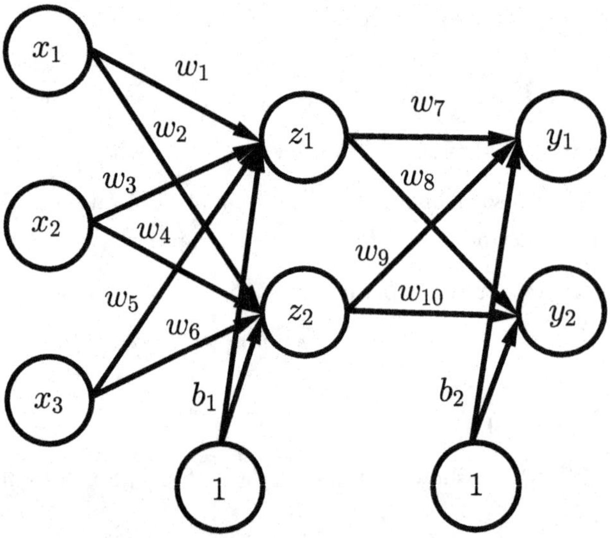

FIGURE 11.10
General neural network.

(c) Initialize the synaptic weights with the values

$$\begin{array}{lllll} w_1 = 0.1, & w_2 = 0.2, & w_3 = 0.3, & w_4 = 0.4, & w_5 = 0.5, \\ w_6 = 0.6, & w_7 = 0.7, & w_8 = 0.8, & w_9 = 0.9, & w_{10} = 0.1, \end{array}$$

and biases $b_1 = b_2 = 0.5$. Forward propagate the network to compute the output $\boldsymbol{y} = (y_1, y_2)^T$. Estimate the error in the forward propagation using the cost function

$$E(\boldsymbol{w}) = \frac{1}{2}\|\boldsymbol{y} - \hat{\boldsymbol{y}}\|_2^2 \ .$$

Estimate also the default synaptic weights and biases.

(d) Compute the gradient

$$\nabla E = \left(\frac{\partial E}{\partial w_1}, \frac{\partial E}{\partial w_2}, \ldots, \frac{\partial E}{\partial w_{10}}\right)^T .$$

For the last one you will need to use the chain rule. Then write a Python function that will return the gradient of the cost function.

(e) Write a Python program to back propagate the network using the given training data and compute the optimal synaptic weights and biases.

(f) Use Python to compute the output of the specific neural network for $\boldsymbol{x} = (x_1, x_2, x_3)^T = (1, 6, 9)^T$.

9. In this problem we analyze the housing market of the Boston state of the USA and we predict the value of a house.

(a) Load the `sklearn` dataset with name `load_boston`. The particular dataset returns a dictionary with five variables: `data`, `target`, `feature_names`, and `DESCR`. The variable `feature_names` is explained in the variable `DESCR`.

(b) From this dataset extract the number of rooms and the price of the houses in two variables names as `X_rooms` and `y_price`.

(c) Split the data using the function `train_test_split` of `sklearn` into 80% training and 20% testing data.

(d) Use the `MLPRegressor` to construct a neural network with six hidden layers of $10, 50, 50, 50, 50$ and 10 perceptrons respectively.

(e) Use the appropriate `sklearn` function to train and test the neural network for this problem.

(f) Explore the predictions and compute a mean error using vector norms.

12

Eigenvalue Problems

We have seen that computing the *SVD* of a matrix $\boldsymbol{A}$ requires the computation of the eigenvalues of the matrix $\boldsymbol{A}^T\boldsymbol{A}$. This is a simple example where the knowledge of eigenvalues can be very important. In addition to numerical methods that require the knowledge of eigenvalues, there are applications like the PageRank algorithm of Google, that rely on the fast computation of the principal eigenvalue of a large matrix. In this chapter, we discuss numerical methods for the computation of eigenvalues and eigenvectors of matrices. Initially we discuss algorithms for the estimation of selected eigenvalues, for example the maximum in magnitude eigenvalue. Although this algorithm is very simple, it has a wide range of applications. We close this chapter with the *QR* iteration, which is not limited to the computation of selective eigenvalues.

12.1 Eigenvalues and Eigenvectors

We begin with a revision of eigenvalues and eigenvectors. Some of their properties can be found useful in the development of efficient algorithms for their computation. We also include some practical information for computations with the so-called Jordan normal form. Readers interested in the numerical methods only, they can skip that part and continue with Section 12.2.

12.1.1 Basic properties

In Section 2.2.12 we defined the eigenvalues of an $n \times n$ matrix $\boldsymbol{A}$ to be the roots of the characteristic equation $\det(\boldsymbol{A} - \lambda \boldsymbol{I}) = 0$ where $\phi(\lambda) = \det(\boldsymbol{A} - \lambda \boldsymbol{I})$ is the characteristic polynomial. Because of this definition, if λ is an eigenvalue of $\boldsymbol{A}$, then the matrix $\boldsymbol{A} - \lambda \boldsymbol{I}$ is singular (not invertible).

As an immediate consequence of the previous definition is that $\boldsymbol{A}$ and its transpose share exactly the same eigenvalues. To see this suppose that λ is an eigenvalue of $\boldsymbol{A}$ such that $\det(\boldsymbol{A} - \lambda \boldsymbol{I}) = 0$. Then,

$$\det(\boldsymbol{A}^T - \lambda \boldsymbol{I}) = \det(\boldsymbol{A} - \lambda \boldsymbol{I})^T = \det(\boldsymbol{A} - \lambda \boldsymbol{I}) = 0 ,$$

which means that λ is also an eigenvalue of $\boldsymbol{A}^T$.

The set of all eigenvalues of $\boldsymbol{A}$ is called spectrum and is denoted by $\sigma(\boldsymbol{A})$. Since eigenvalues are zeros of a polynomial, they might not be simple roots. The *algebraic multiplicity* τ of an eigenvalue λ_j is the number of times the eigenvalue λ_j appears as a root of the characteristic polynomial $\phi(\lambda)$, that is the largest integer τ such that $(\lambda - \lambda_j)^\tau$ divides exactly the characteristic polynomial.

The linear system $(\boldsymbol{A} - \lambda \boldsymbol{I})\boldsymbol{x} = \boldsymbol{0}$ for an eigenvalue λ will have non-trivial solutions $\boldsymbol{x} = \boldsymbol{u} \neq \boldsymbol{0}$ (in addition to $\boldsymbol{x} = \boldsymbol{0}$) since the matrix $\boldsymbol{A} - \lambda \boldsymbol{I}$ is not invertible. The non-trivial solutions $\boldsymbol{u}$ are called eigenvectors associated with the eigenvalue λ. Suppose that

DOI: 10.1201/9781003287292-12

an eigenvalue λ has m linearly independent eigenvectors $\boldsymbol{u}_i$ such that $\boldsymbol{A}\boldsymbol{u}_i = \lambda\boldsymbol{u}_i$ for all $i = 1, 2, \ldots, m$. Define the space produced by the eigenvectors of $\boldsymbol{A}$ corresponding to the eigenvalue λ

$$V_\lambda = \{\boldsymbol{u} : \ (\boldsymbol{A} - \lambda\boldsymbol{I})\boldsymbol{u} = \boldsymbol{0}\} \ .$$

The space V_λ is called *eigenspace* or the *kernel* of the matrix $\boldsymbol{A} - \lambda\boldsymbol{I}$. For this reason, sometimes we write $V_\lambda = \ker(\boldsymbol{A} - \lambda\boldsymbol{I})$ while $m = \dim(V_\lambda)$.

The *geometric multiplicity* of an eigenvalue λ is the number $d = \dim(V_\lambda)$, the dimension of V_λ. The geometric multiplicity is actually the number of linearly independent eigenvectors associated with the eigenvalue λ. It can be shown that $d \leq \tau$. If $d = \tau$ for each eigenvalue of $\boldsymbol{A}$, then the matrix $\boldsymbol{A}$ is called *nondefective* (or *matrix of simple structure*). If $\boldsymbol{A}$ has an eigenvalue with $d < \tau$, then it is called *defective.*

Two matrices $\boldsymbol{A}$, $\boldsymbol{J}$ are called *similar* if there is an invertible matrix $\boldsymbol{P}$ such that $\boldsymbol{A} = \boldsymbol{P}\boldsymbol{J}\boldsymbol{P}^{-1}$. If λ is an eigenvalue of $\boldsymbol{J}$ with corresponding eigenvector $\boldsymbol{u}$, then the following calculations reveal that λ is also eigenvalue of $\boldsymbol{A}$ and $\boldsymbol{P}\boldsymbol{u}$ a corresponding eigenvector.

$$\begin{aligned}
\boldsymbol{J}\boldsymbol{u} &= \lambda\boldsymbol{u} \\
\boldsymbol{J}\boldsymbol{P}^{-1}\boldsymbol{P}\boldsymbol{u} &= \lambda\boldsymbol{u} \\
\boldsymbol{P}\boldsymbol{J}\boldsymbol{P}^{-1}\boldsymbol{P}\boldsymbol{u} &= \lambda\boldsymbol{P}\boldsymbol{u} \\
\boldsymbol{A}\boldsymbol{P}\boldsymbol{u} &= \lambda\boldsymbol{P}\boldsymbol{u}
\end{aligned}$$

In the theory of linear algebra it is known that any square matrix $\boldsymbol{A}$ is similar to its Jordan normal form $\boldsymbol{J}$, or else there is an upper triangular matrix $\boldsymbol{J}$ with diagonal entries the eigenvalues of $\boldsymbol{A}$ and an invertible and orthogonal matrix $\boldsymbol{P}$ such that $\boldsymbol{A} = \boldsymbol{P}\boldsymbol{J}\boldsymbol{P}^{-1}$.

12.1.2 Diagonalization of a matrix and the Jordan normal form

Suppose that an $n \times n$ matrix $\boldsymbol{A}$ has n distinct eigenvalues. This means that the spectrum of $\boldsymbol{A}$ is $\sigma(\boldsymbol{A}) = \{\lambda_1, \lambda_2, \ldots, \lambda_n\}$ with $\lambda_i \neq \lambda_j$, $i \neq j$. If $\boldsymbol{A}$ is nondefective and $\boldsymbol{u}_1, \boldsymbol{u}_2, \ldots, \boldsymbol{u}_n$ are eigenvectors associated with the eigenvalues $\lambda_1, \lambda_2, \ldots, \lambda_n$ of $\sigma(\boldsymbol{A})$, then they will be linearly independent (since they correspond to different eigenvalues), [105].

Let

$$\boldsymbol{P} = \begin{pmatrix} \boldsymbol{u}_1 & \boldsymbol{u}_2 & \cdots & \boldsymbol{u}_n \end{pmatrix} \ ,$$

the matrix with columns the eigenvectors of $\boldsymbol{A}$. Since the columns of $\boldsymbol{P}$ are linearly independent, then $\det(\boldsymbol{P}) \neq 0$, i.e. $\boldsymbol{P}$ is invertible. Moreover, $\boldsymbol{A}\boldsymbol{u}_i = \lambda_i\boldsymbol{u}_i$, $i = 1, 2, \ldots, n$. Then,

$$\boldsymbol{A}\,\boldsymbol{P} = \boldsymbol{A}\begin{pmatrix} \boldsymbol{u}_1 & \boldsymbol{u}_2 & \cdots & \boldsymbol{u}_n \end{pmatrix} = \begin{pmatrix} \lambda_1\boldsymbol{u}_1 & \lambda_2\boldsymbol{u}_2 & \cdots & \lambda_n\boldsymbol{u}_n \end{pmatrix} \ ,$$

which in matrix notation can be written as

$$\boldsymbol{A}\,\boldsymbol{P} = \begin{pmatrix} \boldsymbol{u}_1 & \boldsymbol{u}_2 & \cdots & \boldsymbol{u}_n \end{pmatrix} \begin{pmatrix} \lambda_1 & 0 & \cdots & 0 \\ 0 & \lambda_2 & \ddots & 0 \\ 0 & \ddots & \ddots & 0 \\ 0 & \cdots & 0 & \lambda_n \end{pmatrix} = \boldsymbol{P}\,\boldsymbol{D} \ ,$$

where

$$\boldsymbol{D} = \begin{pmatrix} \lambda_1 & 0 & \cdots & 0 \\ 0 & \lambda_2 & \ddots & 0 \\ 0 & \ddots & \ddots & 0 \\ 0 & \cdots & 0 & \lambda_n \end{pmatrix} = \mathrm{diag}\{\lambda_1, \lambda_2, \ldots, \lambda_n\} \ .$$

The relation $\boldsymbol{AP} = \boldsymbol{PD}$ implies that $\boldsymbol{A} = \boldsymbol{PDP}^{-1}$. We call $\boldsymbol{A}$ *diagonalizable.*

> ☞ If an $n \times n$ matrix $\boldsymbol{A}$ is nondefective, then the n linearly independent eigenvectors of $\boldsymbol{A}$ form a similarity transform $\boldsymbol{P}$ such that $\boldsymbol{A} = \boldsymbol{PDP}^{-1}$ where $\boldsymbol{D} = \text{diag}\{\lambda_1, \lambda_2, \ldots, \lambda_n\}$.

Diagonalizable matrices is a special case but in general not every matrix is similar to a diagonal matrix. On the other hand, a square (complex) matrix $\boldsymbol{A}$ is similar to a block diagonal matrix

$$\boldsymbol{J} = \begin{pmatrix} \boldsymbol{J}_1 & & \\ & \ddots & \\ & & \boldsymbol{J}_p \end{pmatrix},$$

where the blocks $\boldsymbol{J}_i$ are called Jordan blocks. A Jordan block is upper-triangular

$$\boldsymbol{J}_i = \begin{pmatrix} \lambda & 1 & & & \\ & \lambda & 1 & & \\ & & \ddots & \ddots & \\ & & & \lambda & 1 \\ & & & & \lambda \end{pmatrix},$$

where λ is an eigenvalue of $\boldsymbol{A}$. In the special case of a diagonalizable matrix, the Jordan blocks are 1×1 blocks consisting of the eigenvalues of $\boldsymbol{A}$. More information about the computation of the Jordan canonical form can be found in the Appendix.

12.1.3 Other properties

Let $\boldsymbol{A} = \boldsymbol{PJP}^{-1}$, then any power of $\boldsymbol{A}$ can be computed as

$$\boldsymbol{A}^k = \boldsymbol{PJ}^k\boldsymbol{P}^{-1} .$$

To see this, take $k = 2$ and write

$$\boldsymbol{A}^2 = \boldsymbol{A} \cdot \boldsymbol{A} = \boldsymbol{PJP}^{-1} \cdot \boldsymbol{PJP}^{-1} = \boldsymbol{PJ}^2\boldsymbol{P}^{-1} .$$

The same applies to more general functions $f(\boldsymbol{A})$ of a matrix $\boldsymbol{A}$.

The trace of a matrix $\boldsymbol{A}$ is the sum of all its diagonal elements

$$\text{trace}(\boldsymbol{A}) = \sum_{i=1}^{n} a_{ii} .$$

Since by the Jordan normal form theorem any $n \times n$ matrix $\boldsymbol{A}$ is similar to an upper triangular matrix $\boldsymbol{J}$ where the diagonal entries of $\boldsymbol{J}$ are the eigenvalues of $\boldsymbol{A}$ repeated depending on their multiplicity, then the trace of $\boldsymbol{A}$ is the same with the sum of all the eigenvalues of $\boldsymbol{A}$

$$\text{trace}(\boldsymbol{A}) = \sum_{i=1}^{n} a_{ii} = \sum_{i=1}^{n} \lambda_i ,$$

where λ_i are the eigenvalues of $\boldsymbol{A}$, even with non-trivial multiplicity. This follows from the fact that $\text{trace}(\boldsymbol{AB}) = \text{trace}(\boldsymbol{BA})$ for any square matrices $\boldsymbol{A}$ and $\boldsymbol{B}$.

Using again the Jordan normal form of any square matrix $\boldsymbol{A}$ we have that the determinant of $\boldsymbol{A}$ is the product of the eigenvalues of $\boldsymbol{A}$ (counting their multiplicity)

$$\det(\boldsymbol{A}) = \det(\boldsymbol{PJP}^{-1}) = \det(\boldsymbol{P})\det(\boldsymbol{J})\det(\boldsymbol{P}^{-1}) = \prod_{i=1}^{n} \lambda_i \ .$$

Thus, the knowledge of the eigenvalues of a matrix $\boldsymbol{A}$ can solve the computationally expensive problem of computing its determinant.

Consider the shifted matrix $\boldsymbol{A} + s\boldsymbol{I}$ for some $s \in \mathbb{R}$. If $\boldsymbol{u}$ is an eigenvector of $\boldsymbol{A}$ that corresponds to an eigenvalue λ, then

$$(\boldsymbol{A} + s\boldsymbol{I})\boldsymbol{u} = \boldsymbol{Au} + s\boldsymbol{u} = \lambda\boldsymbol{u} + s\boldsymbol{u} = (\lambda + s)\boldsymbol{u} \ .$$

This means that the shifted matrix $\boldsymbol{A} + s\boldsymbol{I}$ has the shifted eigenvalues $\lambda + s$ and shares the same eigenvectors with $\boldsymbol{A}$.

In general, the eigenvalues of a real matrix $\boldsymbol{A}$ do not need to be real. Some of them can be complex, unless the matrix is symmetric $\boldsymbol{A} = \boldsymbol{A}^T$. For example, suppose $\boldsymbol{A}$ is symmetric with eigenvalue λ and eigenvector $\boldsymbol{u} \neq \boldsymbol{0}$, then $\boldsymbol{Au} = \lambda\boldsymbol{u}$. Taking the complex conjugate on both sides of the previous relation and the transpose of the result we have

$$(\overline{\boldsymbol{Au}})^T = (\overline{\lambda\boldsymbol{u}})^T \ ,$$

which can be written as

$$\overline{\boldsymbol{u}}^T\boldsymbol{A}^T = \overline{\lambda}\overline{\boldsymbol{u}}^T \ .$$

If we multiply from the right both sides of the last relation with $\boldsymbol{u}$ we get

$$\overline{\boldsymbol{u}}^T\boldsymbol{A}^T\boldsymbol{u} = \overline{\lambda}\overline{\boldsymbol{u}}^T\boldsymbol{u} \ , \tag{12.1}$$

Similarly, if we multiply the relation $\boldsymbol{Au} = \lambda\boldsymbol{u}$ with $\overline{\boldsymbol{u}}^T$ from the left we have

$$\overline{\boldsymbol{u}}^T\boldsymbol{Au} = \lambda\overline{\boldsymbol{u}}^T\boldsymbol{u} \ . \tag{12.2}$$

If we subtract (12.1) from (12.2) and use the fact that $\boldsymbol{A}^T = \boldsymbol{A}$ we have that

$$(\overline{\lambda} - \lambda)\overline{\boldsymbol{u}}^T\boldsymbol{u} = 0 \ ,$$

or better

$$\|\boldsymbol{u}\|^2(\overline{\lambda} - \lambda) = 0 \ .$$

Since $\boldsymbol{u} \neq \boldsymbol{0}$ we have that $\overline{\lambda} - \lambda = 0$ which implies that $\overline{\lambda} = \lambda$ and thus $\lambda \in \mathbb{R}$.

Interesting are also the properties of the eigenvectors of a symmetric matrix. Two different eigenvectors $\boldsymbol{u}_i$, $\boldsymbol{u}_j$ of a real symmetric matrix $\boldsymbol{A}$ corresponding to the eigenvalues $\lambda_i \neq \lambda_j$, are orthogonal, in the sense that $\boldsymbol{u}_i^T\boldsymbol{u}_j = 0$ if $i \neq j$. To see this, take the transpose of $\boldsymbol{Au}_i = \lambda_i\boldsymbol{u}_i$ to obtain $\boldsymbol{u}_i^T\boldsymbol{A} = \lambda_i\boldsymbol{u}_i^T$. Multiplication with $\boldsymbol{u}_j$ gives

$$\lambda_i\boldsymbol{u}_i^T\boldsymbol{u}_j = \boldsymbol{u}_i^T\boldsymbol{Au}_j = \boldsymbol{u}_i^T\lambda_j\boldsymbol{u}_j = \lambda_j\boldsymbol{u}_i^T\boldsymbol{u}_j \ .$$

Transfering everything to the left side yields

$$(\lambda_i - \lambda_j)\boldsymbol{u}_i^T\boldsymbol{u}_j = 0 \ .$$

Since $\lambda_i \neq \lambda_j$ implies $\boldsymbol{u}_i^T\boldsymbol{u}_j = 0$, which means that $\boldsymbol{u}_i \perp \boldsymbol{u}_j$.

It is worth mentioning that solving (12.2) for the eigenvalue λ we obtain the formula

$$\lambda = \frac{\overline{\boldsymbol{u}}^T\boldsymbol{Au}}{\overline{\boldsymbol{u}}^T\boldsymbol{u}} = \frac{\overline{\boldsymbol{u}}^T\boldsymbol{Au}}{\|\boldsymbol{u}\|_2^2} \ , \tag{12.3}$$

which gives the value of the eigenvalue λ given its eigenvector $\boldsymbol{u}$.

12.1.4 First estimation of eigenvalues using Gerschgorin disks

A first estimation of the location of the eigenvalues of a matrix $\boldsymbol{A}$ can be found using Gerschgorin's theorem.

Theorem 12.1 (Gerschgorin). *Any eigenvalue λ of an $n \times n$ matrix $\boldsymbol{A}$ is located in at least one of the discs with center the diagonal entry a_{ii} of $\boldsymbol{A}$ and radius the trace*

$$r_i = \sum_{\substack{j=1\\ j\neq i}}^{n} |a_{ij}| \ .$$

These disks are known as Gerschgorin disks. An immediate consequence is that the spectrum of $\boldsymbol{A}$ is contained in the union of Gerschgorin disks of $\boldsymbol{A}$,

$$\sigma(\boldsymbol{A}) \subset \cup_{i=1}^{n}\{z : |z - a_{ii}| \leq r_i\} \subset \mathbb{C} \ .$$

Proof. Let λ be an eigenvalue of $\boldsymbol{A}$ with corresponding eigenvector $\boldsymbol{x}$. We write $\boldsymbol{A} = \boldsymbol{D} - \boldsymbol{L} - \boldsymbol{U}$ where $\boldsymbol{D}$ is the diagonal matrix with the entries a_{ii}, $i = 1, 2, \ldots, n$, while the rest of the entries of $\boldsymbol{A}$ are in the matrix $-(\boldsymbol{L} + \boldsymbol{U})$, as in Section 9.7.2. Then we have that

$$\begin{aligned}(\boldsymbol{A} - \lambda\boldsymbol{I})\boldsymbol{x} &= 0\\ (\boldsymbol{D} - \lambda\boldsymbol{I})\boldsymbol{x} - (\boldsymbol{L} + \boldsymbol{U})\boldsymbol{x} &= 0\\ (\lambda\boldsymbol{I} - \boldsymbol{D})\boldsymbol{x} &= -(\boldsymbol{L} + \boldsymbol{U})\boldsymbol{x} \ .\end{aligned}$$

This implies that the i-th entry of the vector $(\lambda\boldsymbol{I} - \boldsymbol{D})\boldsymbol{x}$ is

$$(\lambda - a_{ii})x_i = \sum_{\substack{j=1\\ j\neq i}}^{n} a_{ij}x_j, \quad \text{for} \quad i = 1, 2, \ldots, n \ .$$

Taking modulus on both sides we get

$$|\lambda - a_{ii}|\ |x_i| \leq \sum_{\substack{j=1\\ j\neq i}}^{n} |a_{ij}|\ |x_j| \ ,$$

where we made use of the triangles inequality $|a + b| \leq |a| + |b|$.

Let x_k be the largest in magnitude entry of the eigenvector $\boldsymbol{x}$. Then after scaling $\boldsymbol{x}$ with x_k we have

$$|\lambda - a_{ii}| \leq \sum_{\substack{j=1\\ j\neq i}}^{n} |a_{ij}|\frac{|x_j|}{|x_k|} \leq \sum_{\substack{j=1\\ j\neq i}}^{n} |a_{ij}| \ ,$$

since $|x_j|/|x_k| \leq 1$. Thus, any eigenvalue λ is contained in at least one of the Gerschgorin disks, which proves the theorem. □

For example, consider the matrix

$$\boldsymbol{A} = \begin{pmatrix} -2 & -4 & 2\\ -2 & 1 & 2\\ 4 & 2 & 5 \end{pmatrix} \ ,$$

with spectrum $\sigma(\boldsymbol{A}) = \{-5, 3, 6\}$. The Gerschgorin disks in this case are the sets

$$D_1 = \{z : |z + 2| \leq 6\}, \quad D_2 = \{z : |z - 1| \leq 4\}, \quad D_3 = \{z : |z - 5| \leq 6\} \ .$$

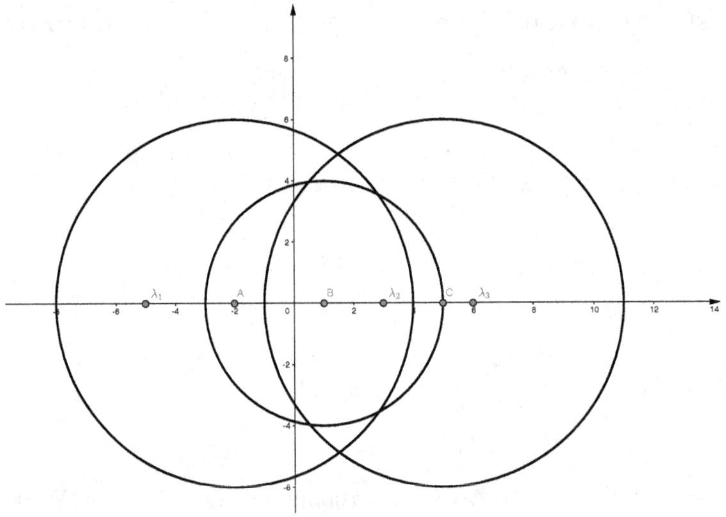

FIGURE 12.1
The Gerschgorin disks and the eigenvalues of a matrix $\boldsymbol{A}$.

The Gerschgorin disks and with the locations of the eigenvalues are depicted in Figure 12.1. Apart from its beauty and theoretical use, the particular theorem can only provide a rough estimation of the location of the eigenvalues of a matrix, especially for the extreme eigenvalues. For more accurate estimations we need to use more sophisticated numerical algorithms.

12.2 Numerical Approximation of Eigenvalues and Eigenvectors

There are several numerical methods for the approximation of eigenvalues of matrices. Namely, the power method, inverse iteration, Raleigh quotient iteration, Arnoldi iteration, Lanczos algorithm, QR algorithm and many more. In this section we present some numerical methods for the computation of selected eigenvalues such as the one with maximum magnitude. We will call this eigenvalue *principal eigenvalue.* We start with the simplest method known as the power method or power iteration.

12.2.1 Power method

The *power method* or *power iteration* is an iterative method for the estimation of the *principal* eigenvalue of an $n \times n$ matrix $\boldsymbol{A}$. This is the largest in magnitude eigenvalue of $\boldsymbol{A}$. The eigenvector that corresponds to the eigenvalue with the largest magnitude is called the *principal* eigenvector. Assume that $\boldsymbol{A}$ has N eigenvalues with

$$|\lambda_1| > |\lambda_2| \geq \cdots \geq |\lambda_N| \ .$$

It is important for the convergence of this method to notice that the eigenvalue λ_1 is not repeated. Assume also that each eigenvalue λ_i has an eigenvector $\boldsymbol{u}_i$ for $i = 1, 2, \ldots, N$ that altogether form a linearly independent set of eigenvectors

$$E = \{\boldsymbol{u}_1, \boldsymbol{u}_2, \ldots, \boldsymbol{u}_N\} \ .$$

Since the N eigenvectors $\boldsymbol{u}_i$, $i = 1, 2, \ldots, N$ are linearly independent, they form a basis of $\mathbb{R}^N$. And so any vector $\boldsymbol{x}^{(0)} \in \mathbb{R}^N$ can be expressed as a linear combination of all these $\boldsymbol{u}_i$

$$\boldsymbol{x}^{(0)} = c_1\boldsymbol{u}_1 + c_2\boldsymbol{u}_2 + \cdots + c_N\boldsymbol{u}_N \ .$$

We choose arbitrarily such a vector $\boldsymbol{x}^{(0)} \neq \boldsymbol{0}$ to be an initial approximation of $\boldsymbol{u}_1$. Then we define the iterative method

$$\boldsymbol{x}^{(k+1)} = \boldsymbol{A}\boldsymbol{x}^{(k)} \qquad \text{for } k = 1, 2, \ldots \ ,$$

for the approximation of the eigenvector $\boldsymbol{u}_1$.

Note that

$$\boldsymbol{x}^{(k)} = \boldsymbol{A}\boldsymbol{x}^{(k-1)} = \boldsymbol{A}\boldsymbol{A}\boldsymbol{x}^{(k-2)} = \boldsymbol{A}^2\boldsymbol{x}^{(k-2)} = \cdots = \boldsymbol{A}^k\boldsymbol{x}^{(0)} \ .$$

Because in every iteration we implicitly construct powers of $\boldsymbol{A}$, we call this particular numerical method *power method.*

Moreover,

$$\begin{aligned}
\boldsymbol{A}^k\boldsymbol{x}^{(0)} &= \boldsymbol{A}^k(c_1\boldsymbol{u}_1 + c_2\boldsymbol{u}_2 + \cdots + c_N\boldsymbol{u}_N) \\
&= c_1\boldsymbol{A}^k\boldsymbol{u}_1 + c_2\boldsymbol{A}^k\boldsymbol{u}_2 + \cdots + c_N\boldsymbol{A}^k\boldsymbol{u}_N \\
&= c_1\lambda_1^k\boldsymbol{u}_1 + c_2\lambda_2^k\boldsymbol{u}_2 + \cdots + c_N\lambda_N^k\boldsymbol{u}_N \\
&= \lambda_1^k\left[c_1\boldsymbol{u}_1 + \left(\frac{\lambda_2}{\lambda_1}\right)^k c_2\boldsymbol{u}_2 + \cdots + \left(\frac{\lambda_N}{\lambda_1}\right)^k c_N\boldsymbol{u}_N\right] \ ,
\end{aligned}$$

which implies

$$\frac{\boldsymbol{A}^k\boldsymbol{x}^{(0)}}{\lambda_1^k} = c_1\boldsymbol{u}_1 + \left(\frac{\lambda_2}{\lambda_1}\right)^k c_2\boldsymbol{u}_2 + \cdots + \left(\frac{\lambda_N}{\lambda_1}\right)^k c_N\boldsymbol{u}_N \ . \tag{12.4}$$

Since $\lambda_1 > \lambda_j$ for all $j \geq 2$ we have that

$$\frac{\lambda_j}{\lambda_1} < 1 \qquad \text{for } j = 2, 3, \ldots, N \ .$$

Taking the limit $k \to \infty$ in (12.4), we have

$$\lim_{k\to\infty} \frac{\boldsymbol{A}^k\boldsymbol{x}^{(0)}}{\lambda_1^k} = c_1\boldsymbol{u}_1 \ .$$

Thus, if we could estimate the eigenvalue λ_1, then we can choose the vector

$$\boldsymbol{u}_1 \approx \frac{\boldsymbol{x}^{(k)}/\lambda_1^k}{\|\boldsymbol{x}^{(k)}/\lambda_1^k\|} \ ,$$

for large value of k to be an approximation of the eigenvector $\boldsymbol{u}_1$.

Suppose that the maximum entry of the vector $\boldsymbol{x}^{(k)}$ is its j-th entry $x_j^{(k)}$ for some j. In order to find the eigenvalue λ_1, we first define the function $f_j(\boldsymbol{x}) = x_j$ where x_j is the j-th entry of the vector $\boldsymbol{x}$ and $j = \text{argmax}(\boldsymbol{x}^{(k)})$ fixed. Then we have that

$$f_j(\boldsymbol{x}^{(k)}) = \lambda_1^k c_1 f_j(\boldsymbol{u}_1) + \left(\frac{\lambda_2}{\lambda_1}\right)^k c_2 f_j(\boldsymbol{u}_2) + \cdots + \left(\frac{\lambda_N}{\lambda_1}\right)^k c_N f_j(\boldsymbol{u}_N) \ .$$

For convenience denote

$$E^{(k)} = \left(\frac{\lambda_2}{\lambda_1}\right)^k c_2 f_j(\boldsymbol{u}_2) + \cdots + \left(\frac{\lambda_N}{\lambda_1}\right)^k c_N f_j(\boldsymbol{u}_N) ,$$

which has

$$\lim_{k\to\infty} E^{(k)} = 0 .$$

Computing the ratio

$$\frac{f_j(\boldsymbol{x}^{(k+1)})}{f_j(\boldsymbol{x}^{(k)})} = \lambda_1 \frac{c_1 f_j(\boldsymbol{u}_1) + E^{(k+1)}}{c_1 f_j(\boldsymbol{u}_1) + E^{(k)}} ,$$

and taking the limit $k \to \infty$ we obtain the maximum in magnitude eigenvalue λ_1 to be

$$\lambda_1 = \lim_{k\to\infty} \frac{f_j(\boldsymbol{x}^{(k+1)})}{f_j(\boldsymbol{x}^{(k)})} .$$

Therefore, we approximate this eigenvalue with the number

$$\lambda_1 \approx \frac{f_j(\boldsymbol{x}^{(k+1)})}{f_j(\boldsymbol{x}^{(k)})} ,$$

for a large value of k.

☞ The convergence of the power iteration is geometric and depends on the ratio $\lambda_2/\lambda_1 < 1$. If $\lambda_2 \approx \lambda_1$ then the convergence is very slow. On the other hand, if $\lambda_2 \ll \lambda_1$ then the convergence will be fast.

The algorithm of the power method is summarized in Algorithm 48.

Algorithm 48 Power method

Set a tolerance TOL for the accuracy
Set the maximum number of iterations $MAXIT$
Initialize the vector $\boldsymbol{x}^{(k)} \neq \boldsymbol{0}$ and the eigenvalue $\lambda_1 = 0$, and set $k = 0$
Set $Error = TOL + 1$
Compute $j = \text{argmax}_j(x_j^{(k)})$
while $Error > TOL$ and $k < MAXIT$ **do**
  Compute new $\boldsymbol{x}^{(k+1)} = \boldsymbol{A}\boldsymbol{x}^{(k)}$
  Set $\lambda_0 = x_j^{(k+1)}/x_j^{(k)}$
  Set $Error = |\lambda_1 - \lambda_0|$
  Increase the counter $k = k + 1$
  Set $\lambda_1 = \lambda_0$
end while
Return the approximate eigenvalue λ_1 and the eigenvector $\boldsymbol{u}_1 = (\boldsymbol{x}^{(k)}/\lambda_1^k)/\|\boldsymbol{x}^{(k)}/\lambda_1^k\|$

Algorithm 48 is implemented in the Python function `powermethod` below:

```
def powerMethod(A, x0, tol = 1.e-6, maxit = 100):
    # initialize the parameters
    iteration = 0
    error = tol+1.0
```

```
    lambda0 = 0.0
    # find the index for the maximum entry
    j = np.argmax(x0)
    while (error > tol and iteration < maxit):
        x1 = np.dot(A,x0)
        # find the approximation of the eigenvalue
        lambda1 = x1[j]/x0[j]
        error = np.abs(lambda0-lambda1)
        iteration = iteration + 1
        lambda0 = lambda1
        x0 = x1.copy()
    k = iteration
    u1 = (x1/lambda1**k)/npl.norm(x1/lambda1**k,np.inf)
    print('number of iterations required = ',iteration)
    # return eigenvalue and corresponding eigenvector
    return lambda1, u1
```

Note that for the computation of the eigenvector $\boldsymbol{u}_1$, we normalize its value using the maximum norm. As testing example we consider the matrix

$$\boldsymbol{A} = \begin{pmatrix} -2 & -4 & 2 \\ -2 & 1 & 2 \\ 4 & 2 & 5 \end{pmatrix}$$

with spectrum $\sigma(\boldsymbol{A}) = \{-5, 3, 6\}$. Using our function `powerMethod` we compute as expected the maximum in magnitude eigenvalue while we verify that $\|\boldsymbol{A}\boldsymbol{u}_1 - \lambda_1 \boldsymbol{u}_1\| \approx 0$.

```
# define matrix and initial guess x0
A = np.array([[-2.,-4.,2.],[-2.,1.,2.],[4.,2.,5.]])
x0 = np.array([1,1,1])
# compute eigenvalue and eigenvector
lambda1,u1 = powerMethod(A,x0, tol=1.e-10, maxit=1000)
# print eigenvalue and eigenvector
print(lambda1)
print(u1)
# compute and print the residual ||Au1 - λ1u1||
error = npl.norm(np.dot(A,u1) - lambda1*u1)
print('actual error = ', error)
```

```
number of iterations required =  158
6.000000000040729
[0.0625 0.375  1.    ]
actual error =  4.315242285689494e-11
```

12.2.2 Inverse power method

The inverse power method (or else the inverse iteration) is a modification of the power method for the computation of the minimum in magnitude eigenvalue of $\boldsymbol{A}$. Assume that $\boldsymbol{A}$ has eigenvalues

$$|\lambda_1| \geq |\lambda_2| \geq \cdots \geq |\lambda_{N-1}| > |\lambda_N| > 0\ .$$

If $\boldsymbol{u}_N$ is the eigenvector that corresponds to the eigenvalue λ_N, then we have

$$\boldsymbol{A}\boldsymbol{u}_N = \lambda_N \boldsymbol{u}_N ,$$

which can be rearranged to the formula

$$\boldsymbol{A}^{-1}\boldsymbol{u}_N = \frac{1}{\lambda_N}\boldsymbol{u}_N .$$

Thus, the eigenvalue λ_N^{-1} is the dominant eigenvalue of $\boldsymbol{A}^{-1}$. To approximate the dominant eigenvalue of $\boldsymbol{A}^{-1}$ we can employ the power method, which can be expressed as

$$\boldsymbol{x}^{(k+1)} = \boldsymbol{A}^{-1}\boldsymbol{x}^{(k)} .$$

Since the direct computation of the inverse $\boldsymbol{A}^{-1}$ is not a good practice, we prefer to compute the vector $\boldsymbol{x}^{(k+1)}$ by solving the system

$$\boldsymbol{A}\boldsymbol{x}^{(k+1)} = \boldsymbol{x}^{(k)} .$$

This can be done efficaciously using the *LU* decomposition, and the function `lu_factor` and `lu_solve` as described in Sections 9.1.2 and 9.6. The modification of the function `powermethod` is very simple and we call it `inversePowerMethod`

```
def inversePowerMethod(A, x0, tol = 1.e-6, maxit = 100):
    # initialize the parameters
    iteration = 0
    error = tol+1.0
    lambda0 = 0.0
    lu, piv = spl.lu_factor(A)
    # find the index for the maximum entry
    j = np.argmax(x0)
    while (error > tol and iteration < maxit):
        x1 = spl.lu_solve((lu, piv), x0)
        # find the approximation of the eigenvalue
        lambda1 = x1[j]/x0[j]
        error = np.abs(lambda0-lambda1)
        iteration = iteration + 1
        lambda0 = lambda1
        x0 = x1.copy()
    k = iteration
    u1 = (x1/lambda1**k)/npl.norm(x1/lambda1**k,np.inf)
    print('number of iterations required = ',iteration)
    # return eigenvalue and corresponding eigenvector
    return 1/lambda1, u1
```

Using the matrix from the example with the power method and the following code we obtain the eigenvalue 3 with the smallest magnitude.

```
A = np.array([[-2.,-4.,2.],[-2.,1.,2.],[4.,2.,5.]])
x0 = np.array([-1,1,1])
lambda1,u1 = inversePowerMethod(A,x0, tol=1.e-10, maxit=1000)
print(lambda1)
```

```
number of iterations required =  44
3.0000000002520206
```

The approximation of the rest of the eigenvalues can be done with more sophisticated numerical methods or the simple shifted power method. The last method is the subject of the next paragraph.

12.2.3 Shifted inverse power method

So far we have seen how to approximate the largest and smallest in magnitude eigenvalues of a matrix $\boldsymbol{A}$. As example we used the matrix

$$\boldsymbol{A} = \begin{pmatrix} -2 & -4 & 2 \\ -2 & 1 & 2 \\ 4 & 2 & 5 \end{pmatrix} ,$$

that has spectrum $\sigma(\boldsymbol{A}) = \{-5, 3, 6\}$. The smallest in magnitude eigenvalue is 3 and the largest one is 6. The question is if we could use the power method to approximate the other eigenvalue as well. The answer is positive. We can compute the eigenvalue -5 by using a modified power method known to as the *shifted power method.* The name shifted is because we shift the spectrum of $\boldsymbol{A}$ so as the desired eigenvalue becomes closer to 0.

The shifted power method relies on the fact that if λ is an eigenvalue of $\boldsymbol{A}$ and $\boldsymbol{u}$ its corresponding eigenvector, then $\lambda - \alpha$ is an eigenvalue of the matrix $\boldsymbol{A} - \alpha\boldsymbol{I}$ with the same corresponding eigenvector $\boldsymbol{u}$. Similarly, the matrix $(\boldsymbol{A} - \alpha\boldsymbol{I})^{-1}$ has the eigenvalue $1/(\lambda - \alpha)$. To compute an eigenvalue λ, which is not any of the extreme in magnitude, we choose α such that $\lambda - \alpha$ is the closest to 0 eigenvalue of the matrix $\boldsymbol{A} - \alpha\boldsymbol{I}$, and then we use the inverse power method to find it. The choice of α can rely on experimentation or intuition. In our example, from Gerschgorin's theorem we know that there is an eigenvalue in the disk with center -2 and radius 6. Using the center of the Gerschgorin disk $\alpha = -2$ to shift the matrix we obtain

$$\boldsymbol{A} + 2\boldsymbol{I} = \begin{pmatrix} 0 & -4 & 2 \\ -2 & 3 & 2 \\ 4 & 2 & 7 \end{pmatrix} .$$

Using the inverse power method we approximate the eigenvalue $\lambda - \alpha = -3$ for the matrix $\boldsymbol{A} + 2\boldsymbol{I}$. Solving for λ with $\alpha = -2$, we obtain $\lambda = -5$. We leave the implementation of the shifted power method as an exercise for the reader.

12.2.4 Basic QR iteration

Power methods are very simple and easy to implement. Their major drawback is that they return only one eigenvalue at a time. On the other hand, more sophisticated methods can lead to the approximation of all the eigenvalues of a matrix at once. One such example is the QR iteration.

The QR iteration is an iterative method based on a similarity transformation of $\boldsymbol{A}$ to a matrix $\boldsymbol{T}$ for which it is easier to compute its eigenvalues. In what follows we describe briefly the basic QR iteration, where for the sake of simplicity, assume that $\boldsymbol{A}$ is a real $n \times n$ matrix with n distinct eigenvalues

$$|\lambda_1| > |\lambda_2| > \cdots > |\lambda_n| .$$

Choosing an orthogonal matrix $\boldsymbol{Q}^{(0)}$ we define $\boldsymbol{T}^{(0)} = [\boldsymbol{Q}^{(0)}]^T \boldsymbol{A} \boldsymbol{Q}^{(0)}$. Usually, we take $\boldsymbol{Q}^{(0)} = \boldsymbol{I}$, and thus $\boldsymbol{T}^{(0)} = \boldsymbol{A}$. Then we perform the QR factorization

$$\boldsymbol{Q}^{(k+1)} \boldsymbol{R}^{(k+1)} = \boldsymbol{T}^{(k)}, \quad \text{for } k = 0, 1, \ldots ,$$

and we compute the new matrix

$$T^{(k+1)} = R^{(k+1)}Q^{(k+1)} ,$$

for all $k = 0, 1, 2, \ldots$.

In this way we produce every time the matrix

$$\begin{aligned} T^{(k+1)} &= R^{(k+1)}Q^{(k+1)} = [Q^{(k+1)}]^T[Q^{(k+1)}R^{(k+1)}]Q^{(k+1)} = [Q^{(k+1)}]^T T^{(k)}Q^{(k+1)} \\ &= [Q^{(0)}Q^{(1)}\cdots Q^{(k+1)}]^T A[Q^{(0)}Q^{(1)}\cdots Q^{(k+1)}] . \end{aligned}$$

Thus, the matrix $T^{(k)}$ is always similar to A, and thus they share the same eigenvalues. One can prove that the limit

$$\lim_{k\to\infty} T^{(k)} = T ,$$

where T contains the n eigenvalues of A in its main diagonal. Based on the last assertion, the QR iteration returns all the eigenvalues of A on the main diagonal of T. The algorithm of the basic QR iteration is presented in Algorithm 49.

Algorithm 49 Basic QR iteration

Set a tolerance TOL for the accuracy
Set the maximum number of iterations $MAXIT$
Initialize the matrices $Q^{(0)} = I$, $T^{(0)} = A$ and set $k = 0$
Set $Error = TOL + 1$
while $Error > TOL$ and $k < MAXIT$ **do**
  Compute the QR decomposition $Q^{(k+1)}R^{(k+1)} = T^{(k)}$
  Compute the new $T^{(k+1)} = R^{(k+1)}Q^{(k+1)}$
  Compute $Error = \|\text{diag}(T^{(k+1)}) - \text{diag}(T^{(k)})\|$
  Increase the counter $k = k + 1$
end while
Return the approximate eigenvalues $\text{diag}(T^{(k+1)})$

For the implementation of the basic QR iteration algorithm 49 we will use the QR factorization of Python as it was described in Section 10.4.4.

```
def basicQRiteration(A, tol = 1.e-6, maxit = 100):
    # initialize the parameters
    iteration = 0
    error = tol+1.0
    T = A.copy()
    #set first approximation of eigenvalues
    eigs = np.diag(T)
    while (error > tol and iteration < maxit):
        #extract the diagonal entries of T
        eig0=eigs
        Q, R = npl.qr(T)
        T = np.dot(R,Q)
        #extract the new diagonal entries
        eigs = np.diag(T)
        error = npl.norm(eigs-eig0)
        iteration = iteration + 1
    print('number of iterations required = ',iteration)
```

```
    # return eigenvalue and corresponding eigenvector
    return eigs
```

In the implementation of the basic QR iteration in the function `basicQRiteration` after computing the array `T`, the code returns the approximate eigenvalues in the vector `eigs`. For example, in the case of the matrix

$$\boldsymbol{A} = \begin{pmatrix} -2 & -4 & 2 \\ -2 & 1 & 2 \\ 4 & 2 & 5 \end{pmatrix} .$$

with spectrum $\sigma(\boldsymbol{A}) = \{-5, 3, 6\}$ we get

```
A = np.array([[-2.,-4.,2.],[-2.,1.,2.],[4.,2.,5.]])
eigs = basicQRiteration(A, tol = 1.e-6, maxit = 100)
print(eigs)
```

```
number of iterations required =  89
[ 5.99999972 -4.99999972  3.        ]
```

It is worth mentioning that the basic QR iteration will not order the eigenvalues but they appear in a rather arbitrary order. The justification of this simple method along with more sophisticated methods can be found in the references at the end of this chapter.

12.2.5 Application in computer networks

In this section we present the idea behind a very successful algorithm used by Google Search to rank webpages based on the likelihood to be useful to the user. This algorithm is called PageRank, and it was named after the term "webpage" and the co-founder of Google, Larry Page. Assume that there is an internet *surfer* who has submitted a search query to a search engine. The search engine identifies and gathers information of all possible webpages that contain the keywords of the search. The question we will discuss below is the following: How can we order the search results to present first the most relevant webpages? For example, when I searched on Google for the phrase "numerical analysis" I got 523,000,000 results as one can see in Figure 12.2. How a search engine can decide which one is the most useful for me and present it first? The answer to this question is to transform the problem in an eigenvalue problem and compute the eigenvector of the dominant eigenvalue. The idea of transforming ranking problems into eigenvalue problems goes back to the work [102]. Many works followed until 1996 when Larry Page and Sergey Brin developed the PageRank algorithm as a new kind of search engine, [17].

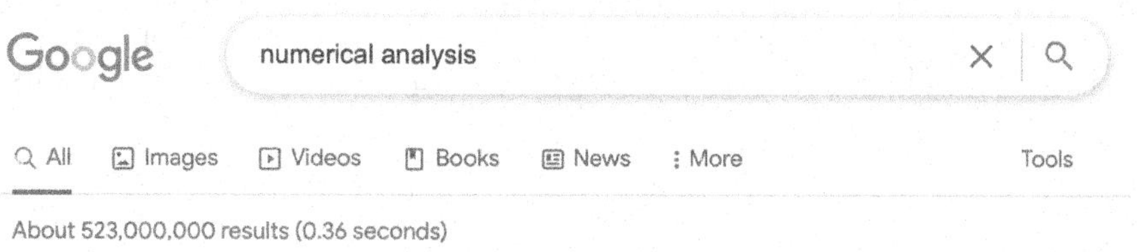

FIGURE 12.2
Google search.

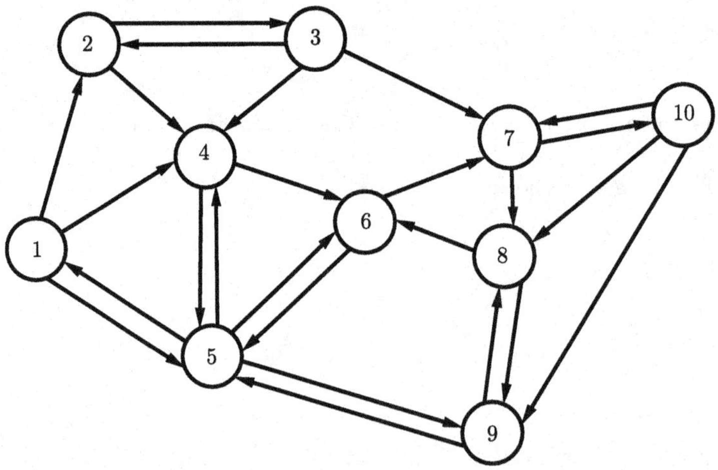

FIGURE 12.3
Connectivity graph of webpages and their links.

TABLE 12.1
Number of departing links per node

i	1	2	3	4	5	6	7	8	9	10
d_i	3	2	3	2	4	2	2	2	2	3

For the sake of simplicity assume that the search identifies 10 webpages with the term 'numerical analysis'. Using these 10 webpages we create a directed graph. The 10 nodes of this graph represent the webpages and the directed edges of the graph the links between them. This graph is shown in Figure 12.3. In this way we can refer to the webpages with the numbers $\{1, 2, 3, 4, 5, 6, 7, 8, 9, 10\}$, and the problem has been transformed into ordering this set based on the importance of the website.

In order to define the importance of the j-th webpage we assume that the i-th node has d_i departing links to other nodes. Table 12.1 shows the number of links departing from node i. Naturally, the importance of the j-th node must depend on the links it receives from other nodes and also on the importance of these nodes. Suppose that the links between the nodes are all equally important. We define the importance of a webpage to be the number

$$r_j = \sum_{i \to j} \frac{r_i}{d_i} , \tag{12.5}$$

where $i \to j$ means that the sum is over the nodes i that have a departing link to the j-th node, r_i is the importance of the i-th node, and d_i is the number of links departing from the i-th node. This number is also called the score of the j-th node. For example, the 4th node in the graph of Figure 12.3 receives links from the nodes 1, 2, 3, 5. Then its score is

$$r_4 = \frac{r_1}{d_1} + \frac{r_2}{d_2} + \frac{r_3}{d_3} + \frac{r_5}{d_5} .$$

Using the formula (12.5) for the scores of all the nodes $i = 1, 2, \dots, 10$, we obtain 10 equations with 10 unknowns:

$$
\begin{aligned}
r_1 &= \tfrac{r_5}{4} & r_6 &= \tfrac{r_4}{2} + \tfrac{r_5}{4} + \tfrac{r_8}{2} \\
r_2 &= \tfrac{r_1}{3} + \tfrac{r_3}{3} & r_7 &= \tfrac{r_3}{3} + \tfrac{r_6}{2} + \tfrac{r_{10}}{3} \\
r_3 &= \tfrac{r_2}{2} & r_8 &= \tfrac{r_7}{2} + \tfrac{r_9}{2} + \tfrac{r_{10}}{3} \\
r_4 &= \tfrac{r_1}{3} + \tfrac{r_2}{2} + \tfrac{r_3}{3} + \tfrac{r_5}{4} & r_9 &= \tfrac{r_5}{4} + \tfrac{r_8}{2} + \tfrac{r_{10}}{3} \\
r_5 &= \tfrac{r_1}{3} + \tfrac{r_4}{2} + \tfrac{r_6}{2} + \tfrac{r_9}{2} & r_{10} &= \tfrac{r_7}{2}
\end{aligned}
$$

If we denote by $\boldsymbol{r}$ the vector with entries r_i, then the previous linear system can be written as

$$\boldsymbol{r} = \boldsymbol{G}\boldsymbol{r} \tag{12.6}$$

where

$$
\boldsymbol{G} = \begin{pmatrix}
0 & 0 & 0 & 0 & 1/4 & 0 & 0 & 0 & 0 & 0 \\
1/3 & 0 & 1/3 & 0 & 0 & 0 & 0 & 0 & 0 & 0 \\
0 & 1/2 & 0 & 0 & 0 & 0 & 0 & 0 & 0 & 0 \\
1/3 & 1/2 & 1/3 & 0 & 1/4 & 0 & 0 & 0 & 0 & 0 \\
1/3 & 0 & 0 & 1/2 & 0 & 1/2 & 0 & 0 & 1/2 & 0 \\
0 & 0 & 0 & 1/2 & 1/4 & 0 & 0 & 1/2 & 0 & 0 \\
0 & 0 & 1/3 & 0 & 0 & 1/2 & 0 & 0 & 0 & 1/3 \\
0 & 0 & 0 & 0 & 0 & 0 & 1/2 & 0 & 1/2 & 1/3 \\
0 & 0 & 0 & 0 & 1/4 & 0 & 0 & 1/2 & 0 & 1/3 \\
0 & 0 & 0 & 0 & 0 & 0 & 1/2 & 0 & 0 & 0
\end{pmatrix} .
$$

In realistic situations this matrix can contain billions of entries and it might not be even possible to be stored. Solving a linear system such as (12.6) using the methods of Chapter 9, it will not be efficient. On the other hand, $\boldsymbol{G}$ has some nice properties. If we see the system (12.6) as an eigensystem with $\lambda_1 = 1$ an eigenvalue of $\boldsymbol{G}$ with corresponding eigenvector $\boldsymbol{r}$, then we could use the power method of Section 12.2.1. First we note that the connectivity graph such as the graph in Figure 12.3 is a *Markov chain* with transition matrix $\boldsymbol{G}$. Moreover, the sum of the entries of each column of $\boldsymbol{G}$ equals 1. Such a matrix is called *stochastic* or *column stochastic*. The following lemma states that the spectral radius of $\boldsymbol{G}$ is $\sigma(\boldsymbol{G}) = 1$.

Lemma 12.2. *The maximum in magnitude eigenvalue of a stochastic matrix $\boldsymbol{G}$ is the eigenvalue $\lambda_1 = 1$.*

Proof. Let λ be an eigenvalue of $\boldsymbol{G}$ with corresponding eigenvector $\boldsymbol{x}$. Then, $\boldsymbol{G}\boldsymbol{x} = \lambda\boldsymbol{x}$. Since $\boldsymbol{G}$ is a stochastic matrix then $\|\boldsymbol{G}\|_1 = 1$. On the other hand, $\|\boldsymbol{G}\boldsymbol{x}\|_1 = \|\lambda\boldsymbol{x}\|_1 = |\lambda|\|\boldsymbol{x}\|_1$, which gives

$$|\lambda| = \frac{\|\boldsymbol{G}\boldsymbol{x}\|_1}{\|\boldsymbol{x}\|_1} \le \frac{\|\boldsymbol{G}\|_1\|\boldsymbol{x}\|_1}{\|\boldsymbol{x}\|_1} = \|\boldsymbol{G}\|_1 = 1 .$$

Thus, any eigenvalue of $\boldsymbol{G}$ has magnitude less than or equal to 1. Also the eigenvalue $\lambda_1 = 1$ is an eigenvalue of $\boldsymbol{G}$. To see this first observe that $\boldsymbol{G}^T$ has the eigenvalue $\lambda_1 = 1$ with corresponding eigenvector the vector $(1, 1, \dots, 1)^T$. From Section 12.1.1 we know that $\boldsymbol{G}$ and $\boldsymbol{G}^T$ share the same eigenvalues, and thus λ_1 is also an eigenvalue of $\boldsymbol{G}$. We conclude that λ_1 is the eigenvalue with maximum magnitude since all others have $|\lambda| \le 1$. □

The last question is if the eigenvalue λ_1 has multiplicity 1, that is $\lambda_1 > |\lambda|$ for all $\lambda \in \sigma(\boldsymbol{G})$ with $\lambda \ne \lambda_1$. The answer is given by Perron-Frobenius theorem. Its proof goes beyond the scope of this book, and we just mention the main result of this theorem which is the following:

Theorem 12.3 (Perron-Frobenius). *If* $\boldsymbol{G}$ *is a positive* $n \times n$ *matrix, then* $|\lambda| < \rho(\boldsymbol{G})$ *for any eigenvalue* λ *with* $\lambda \neq \rho(\boldsymbol{G})$.

Now that we know that $\lambda_1 = 1$ is a simple eigenvalue of $\boldsymbol{G}$ and all the rest are strictly less than 1 in magnitude, we can use the Algorithm 48 of the power method from Section 12.2.1. This method relies only on matrix vector multiplication, and thus it is very efficient for the computation of the eigenvector $\boldsymbol{r}$. Recall, that the eigenvector contains the scores of the webpages.

Using the code of Section 12.2.1 we compute the following vector $\boldsymbol{r}$ which we normalize so as the sum of its entries to be 1 as its entry is a probability.

```
A = np.array([[0,0,0,0,1/4,0,0,0,0,0],
              [1/3,0,1/3,0,0,0,0,0,0,0],
              [0,1/2,0,0,0,0,0,0,0,0],
              [1/3,1/2,1/3,0,1/4,0,0,0,0,0],
              [1/3,0,0,1/2,0,1/2,0,0,1/2,0],
              [0,0,0,1/2,1/4,0,0,1/2,0,0],
              [0,0,1/3,0,0,1/2,0,0,0,1/3],
              [0,0,0,0,0,0,1/2,0,1/2,1/3],
              [0,0,0,0,1/4,0,0,1/2,0,1/3],
              [0,0,0,0,0,0,1/2,0,0,0]] )
x0 = np.ones(10)/10.0
lambda1,u1 = powermethod(A,x0, tol=1.e-10, maxit=1000)
r1 = u1/npl.norm(u1,1)
print(r1)
```

```
number of iterations required =  52
[0.05417957 0.02167183 0.01083591 0.08668731 0.21671827 0.16821465
 0.10526316 0.14138287 0.14241486 0.05263158]
```

Based on the principal eigenvector $\boldsymbol{r}$, the webpage with the highest score is the one corresponding to node 5 of the graph. Its score is `0.21671827`. This webpage is more likely to be useful to the user, and it will appear on the top of the search results. Based on the output of the PageRank algorithm, the webpages will appear in the following order $\{5, 6, 9, 8, 7, 4, 1, 10, 2, 3\}$. This is the main idea behind an algorithm that we use almost every day.

12.3 Further Reading

Most classical books of numerical analysis such as [18, 22, 108, 115, 90, 100] dedicate part of their exposition to eigenvalue problems. Specialized books to linear algebra problems such as [29, 135, 48, 105] contain extensive analysis of eigenvalue problems. For more information on the Jordan normal form, we refer to the books [128, 67, 103] and the references therein. Some specialized books dedicated only to the numerical eigenvalue problem include the books [113, 80, 97, 140]. A simple introduction to the PageRank algorithm can be found in [105] and [115] while more information can be found in the review article [63], and the references therein.

Chapter Highlights

- The matrices $\boldsymbol{A}$ and $\boldsymbol{A}^T$ have the same eigenvalues.
- Similar matrices share the same eigenvalues.
- If a matrix has an eigenvalue with geometric multiplicity less than algebraic multiplicity, then it is called defective.
- If each eigenvalue of a matrix has geometric multiplicity equal to algebraic multiplicity, then the matrix is called nondefective.
- A nondefective matrix $\boldsymbol{A}$ with n linearly independent eigenvectors is similar to a diagonal matrix with diagonal entries the eigenvalues of $\boldsymbol{A}$.
- The Jordan normal form of a matrix $\boldsymbol{A}$ is an upper-triangular matrix $\boldsymbol{J}$ with diagonal elements all the eigenvalues of the matrix $\boldsymbol{A}$. The matrix $\boldsymbol{J}$ is similar to $\boldsymbol{A}$. If $\boldsymbol{A}$ is nondefective, then $\boldsymbol{J}$ is diagonal.
- Real, symmetric matrices have real eigenvalues and orthogonal eigenvectors.
- The eigenvalues of a matrix $\boldsymbol{A}$ are contained in the union of the Gerschgorin disks of $\boldsymbol{A}$.
- The i-th Gerschgorin disk of $\boldsymbol{A}$ has center at a_{ii} and radius

$$r_i = \sum_{j=1, j\neq i}^{n} |a_{ij}| .$$

- The power iteration is used for the approximation of the eigenvalue of $\boldsymbol{A}$ with the maximum magnitude and its corresponding eigenvector.
- The convergence of the power iteration depends on the ratio $\lambda_2/\lambda_1 < 1$. If $\lambda_2 \approx \lambda_1$, then the convergence is very slow. On the other hand, if $\lambda_2 \ll \lambda_1$, then the convergence will be fast.
- For the approximation of the eigenvalue with the minimum magnitude of $\boldsymbol{A}$ we use the inverse power method, which approximates the maximum in magnitude eigenvalue of $\boldsymbol{A}^{-1}$.
- Other eigenvalues of a matrix $\boldsymbol{A}$ can be computed with the shifted inverse power method. This method relies on shifting the spectrum of $\boldsymbol{A}$ in such a way that the desired eigenvalue becomes close to 0.
- The QR is a more sophisticated algorithm for the estimation of the eigenvalues of a matrix $\boldsymbol{A}$.
- The PageRank algorithm relies on the computation of the eigenvector of the principal eigenvalue of a stochastic matrix.

Exercises

1. Show that for any 2×2 matrix $\boldsymbol{A}$ the characteristic polynomial is

$$\phi(\lambda) = \lambda^2 - \text{trace}(\boldsymbol{A})\lambda + \det(\boldsymbol{A}) \ .$$

2. For any $n \times m$ matrix $\boldsymbol{A}$ show that $\boldsymbol{T} = \frac{1}{2}(\boldsymbol{A} + \boldsymbol{A}^T)$ is symmetric.

3. Consider the $n \times n$ tridiagonal matrix

$$\boldsymbol{A} = \text{tridiag}(b, a, c) = \begin{pmatrix} a & c & & & \\ b & a & c & & \\ & b & a & \ddots & \\ & & \ddots & \ddots & c \\ & & & b & a \end{pmatrix} ,$$

with $bc > 0$.

(a) Prove that $\boldsymbol{A}$ has eigenvalues

$$\lambda_k = a + 2\sqrt{bc}\cos\frac{\pi k}{n+1}, \qquad k = 1, 2, \ldots, n \ .$$

(b) Why A is diagonalizable?

(c) Show that if $a = 2$ and $b = c = -1$, then $\lambda_k = 4\sin^2(\pi k/2(n+1))$.

(d) Use this fact to prove that $\boldsymbol{A}$ is positive definite.

4. Consider the matrix

$$\boldsymbol{A} = \begin{pmatrix} 1 & 1 \\ 1 & 0 \end{pmatrix} .$$

(a) Compute the eigenvalues and the corresponding eigenvectors of the matrix $\boldsymbol{A}$.

(b) Find the sequence of numbers $\{x_n\}$ with $x_0 = 0$ and $x_1 = 1$ and the rest of the terms are defined recursively by the formula

$$\boldsymbol{x}^{(n)} = \boldsymbol{A}\boldsymbol{x}^{(n-1)}, \qquad \text{for} \qquad n = 1, 2, \ldots \ ,$$

where $\boldsymbol{x}^{(n)} = (x_n, x_{n-1})^T$.

(c) Determine a recursive formula $F(x_n, x_{n-1}, x_{n-2}) = 0$ for the terms of this sequence.

(d) Compute the ratio x_n/x_{n-1} for $n = 1, 2, \ldots, 10^6$. What do you observe?

(e) If $x_n = 832,040$ find the term x_{n+1} without using the term x_{n-1}.

[Hint: The sequence x_n of this problem is known to as the Fibonacci sequence and the golden ratio is $(1 + \sqrt{5})/2$.]

5. Let $\boldsymbol{x} = (x, y)^T$. Consider the line described by the equation

$$Q : 5x^2 + 8xy + 5y^2 = 1 \ .$$

(a) Identify the shape of the line Q

(b) Find a symmetric matrix $\boldsymbol{A}$ such that $Q(\boldsymbol{x}) = \boldsymbol{x}^T\boldsymbol{A}\boldsymbol{x}$.

(c) Find the eigenvalues and eigenvectors of $\boldsymbol{A}$, and show that $\boldsymbol{A}$ is diagonalizable, i.e. write

$$\boldsymbol{A} = \boldsymbol{P}\boldsymbol{J}\boldsymbol{P}^T ,$$

where $\boldsymbol{P}$ has the eigenvectors of $\boldsymbol{A}$ as columns and $\boldsymbol{J}$ is diagonal with the eigenvalues in its main diagonal.

(d) Use the equation $\boldsymbol{x}^T\boldsymbol{A}\boldsymbol{x} = 1$ and the previous question to write the equation of the line Q in its canonical form.

(e) Use the power method and its modifications for the computation of the eigenvalues and eigenvectors of $\boldsymbol{A}$, and validate your theoretical calculations.

6. Consider the matrix

$$\boldsymbol{A} = \begin{pmatrix} 3 & -2 & -1 \\ 1 & -2 & 1 \\ 1 & -1 & 0 \end{pmatrix} .$$

(a) Find analytically all the eigenvalues and eigenvectors of $\boldsymbol{A}$.

(b) Confirm your results using the function `eig` of the module `numpy.linalg`.

(c) Estimate the principal eigenvalue of the matrix $\boldsymbol{A}$ using the power method using $\boldsymbol{x}_0 = (-1, 1, 1)^T$. Explain why the Python code provided in the book fails when $\boldsymbol{x}_0 = (1, 1, 1)^T$.

(d) Explain why we cannot use the inverse power method for the computation of the minimum in magnitude eigenvalue of $\boldsymbol{A}$.

(e) Use the shifted inverse power method for the approximation of the minimum in magnitude eigenvalue of $\boldsymbol{A}$.

(f) Compare the solutions of the QR iteration with the previous acquired results.

7. Consider the generalized eigenvalue problem

$$\boldsymbol{C}\boldsymbol{x} = \lambda \boldsymbol{B}\boldsymbol{x} , \tag{12.7}$$

where $\boldsymbol{C}$ and $\boldsymbol{B}$ are symmetric $n \times n$ matrices, while $\boldsymbol{B}$ is also positive definite.

(a) Using the Cholesky decomposition of $\boldsymbol{B}$ transform (12.7) into an equation of the form

$$\boldsymbol{A}\boldsymbol{u} = \lambda \boldsymbol{u} .$$

This is called the standard form of an eigenvalue problem.

(b) Consider the matrices

$$\boldsymbol{C} = \begin{pmatrix} 4 & -1 & 0 \\ -1 & 4 & -1 \\ 0 & -1 & 4 \end{pmatrix} \quad \text{and} \quad \boldsymbol{B} = \begin{pmatrix} 2 & -1 & 0 \\ -1 & 2 & -1 \\ 0 & -1 & 1 \end{pmatrix} .$$

Find the standard form of the generalized eigenvalue problem (12.7).

(c) Use the power method and its variants to approximate all the eigenvalues λ of (12.7).

(d) Use the QR iteration to approximate all the eigenvalues λ of (12.7).

(e) Compare your results with the function `eig` of the module `numpy.linalg`.

[Hint: For more information, we refer to [99] and [78]]

8. The quotient

$$\frac{\boldsymbol{x}^T\boldsymbol{A}\boldsymbol{x}}{\boldsymbol{x}^T\boldsymbol{x}} ,$$

is called the Rayleigh quotient.

(a) Show that if $\boldsymbol{u}$ is an eigenvector of $\boldsymbol{A}$, then the corresponding eigenvalue λ of $\boldsymbol{A}$ is

$$\lambda = \frac{\boldsymbol{u}^T\boldsymbol{A}\boldsymbol{u}}{\boldsymbol{u}^T\boldsymbol{u}} .$$

(b) If all the eigenvalues of $\boldsymbol{A}$ are positive, then what can you say about the matrix $\boldsymbol{A}$?

(c) Modify the power method accordingly so as to use the Rayleigh quotient formula for the computation of the principal eigenvalue.

(d) Modify the inverse power iteration in the same way for the computation of the eigenvalue with the minimum magnitude.

(e) Test your codes with the matrix

$$\boldsymbol{A} = \begin{pmatrix} -2 & -4 & 2 \\ -2 & 1 & 2 \\ 4 & 2 & 5 \end{pmatrix} .$$

9. Consider a 3×3 matrix $\boldsymbol{A}$ with eigenvalues

$$|\lambda_1| > |\lambda_2| > |\lambda_3| .$$

(a) Show that if we apply the power method with an initial condition

$$\boldsymbol{x}^{(0)} = c_2\boldsymbol{u}_2 + c_3\boldsymbol{u}_3 ,$$

then we obtain an approximation to the eigenvalue λ_2.

(b) Show that for any vector in the form $\boldsymbol{x} = a_1\boldsymbol{u}_1 + a_2\boldsymbol{u}_2 + a_3\boldsymbol{u}_3$, $a_1, a_2, a_3 \in \mathbb{R}$, the vector

$$\boldsymbol{x}^{(0)} = (\boldsymbol{A} - \lambda_1\boldsymbol{I})\boldsymbol{x} ,$$

is of the form $\boldsymbol{x}^{(0)} = c_2\boldsymbol{u}_2 + c_3\boldsymbol{u}_3$.

(c) Show that the power method with initial condition

$$\boldsymbol{x}^{(0)} = (\boldsymbol{A} - \lambda_2\boldsymbol{I})(\boldsymbol{A} - \lambda_1\boldsymbol{I})\boldsymbol{x} ,$$

will approximate the eigenvalue λ_3.

(d) Write a Python code to implement this technique. Use your code to approximate all the eigenvalues of the matrix

$$\boldsymbol{A} = \begin{pmatrix} -2 & -4 & 2 \\ -2 & 1 & 2 \\ 4 & 2 & 5 \end{pmatrix} .$$

[This technique is known as *annihilation technique*]

A

Computing Jordan Normal Forms

In this Appendix, we review briefly a very important matrix factorization known to as the Jordan normal form or Jordan canonical form, a subject that is not covered usually in basic linear algebra courses. As we have already seen, this plays a significant role in some of the proofs of convergence in linear algebra, and it is also a very important tool for the study of eigenvalues. We will see how we can write any matrix $\boldsymbol{A}$ as a product $\boldsymbol{PJP}^{-1}$, where $\boldsymbol{J}$ is a sparse, upper-diagonal matrix associated with the eigenvalues of $\boldsymbol{A}$, and $\boldsymbol{P}$ invertible matrix associated with the eigenvectors of $\boldsymbol{A}$. First we consider the easiest case where $\boldsymbol{J}$ is a diagonal matrix. For more information and a more *in-depth* presentation of the subject, we refer to [67].

Generalized Eigenvectors and the Jordan Normal Form

Let $\boldsymbol{A}$ be an $n \times n$ (complex) matrix and $\lambda \in \sigma(\boldsymbol{A})$ an eigenvalue with algebraic multiplicity τ. Let $d = \dim \ker(\boldsymbol{A} - \lambda \boldsymbol{I})$ be the geometric multiplicity of λ (the number of linearly independent eigenvectors), with $d < \tau$. In such case $\boldsymbol{A}$ is called *defective.* (Recall that the space $\ker(\boldsymbol{A} - \lambda\boldsymbol{I}) = V_\lambda$ is the kernel (or null) of $\boldsymbol{A} - \lambda\boldsymbol{I}$ and contains vectors $\boldsymbol{u}$ such that $(\boldsymbol{A} - \lambda\boldsymbol{I})\boldsymbol{u} = \boldsymbol{0}$). This means that $\boldsymbol{A}$ is a defective matrix (not of simple structure), and therefore it is not diagonalizable using a similarity transformation. This is because there are not enough linearly independent eigenvectors for the construction of the matrix $\boldsymbol{P}$. For each eigenvalue λ with $d < \tau$, the d linearly independent eigenvectors can be completed with $(\tau - d)$ *generalized eigenvectors* to form a set of τ linearly independent eigenvectors (simple and generalized together).

A nonzero vector $\boldsymbol{u}_k$ is a *generalized eigenvector* of $\boldsymbol{A}$ corresponding to the eigenvalue λ if

$$(\boldsymbol{A} - \lambda\boldsymbol{I})^k \boldsymbol{u}_k = \boldsymbol{0} \quad \text{for some } k \in \mathbb{N} ,$$

which is equivalent to say that $\boldsymbol{u}_k \in \ker(\boldsymbol{A} - \lambda\boldsymbol{I})^k$. A generalized eigenvector $\boldsymbol{u}_k$ is called of rank k if

$$(\boldsymbol{A} - \lambda\boldsymbol{I})^k \boldsymbol{u}_k = \boldsymbol{0} \ \text{ and } \ (\boldsymbol{A} - \lambda\boldsymbol{I})^{k-1} \boldsymbol{u}_k \neq \boldsymbol{0} .$$

According to this definition, the (classical) eigenvectors of $\boldsymbol{A}$ are generalized eigenvector of rank 1.

If $\boldsymbol{u}_k$ is a rank k generalized eigenvector of $\boldsymbol{A}$ with respect to the eigenvalue λ then

$$(\boldsymbol{A} - \lambda\boldsymbol{I})\boldsymbol{u}_k = \boldsymbol{u}_{k-1} ,$$

where $\boldsymbol{u}_{k-1}$ is a rank $(k-1)$ generalized eigenvector of $\boldsymbol{A}$. Therefore, for the generalized eigenvector $\boldsymbol{u}_k$ there is a chain

$$\{\boldsymbol{u}_k, \boldsymbol{u}_{k-1}, \ldots, \boldsymbol{u}_1\} ,$$

DOI: 10.1201/9781003287292-A

of generalized eigenvectors associated with the eigenvalue λ of rank $k, k-1, \ldots, 1$ respectively with

$$\begin{aligned}
&\boldsymbol{u}_k \ , \\
&\boldsymbol{u}_{k-1} = (\boldsymbol{A} - \lambda \boldsymbol{I})\boldsymbol{u}_k \Rightarrow \boldsymbol{A}\boldsymbol{u}_k = \lambda \boldsymbol{u}_k + \boldsymbol{u}_{k-1} \ , \\
&\boldsymbol{u}_{k-2} = (\boldsymbol{A} - \lambda I)^2 \boldsymbol{u}_k = (\boldsymbol{A} - \lambda \boldsymbol{I})\boldsymbol{u}_{k-1} \Rightarrow \boldsymbol{A}\boldsymbol{u}_{k-1} = \lambda \boldsymbol{u}_{k-1} + \boldsymbol{u}_{k-2} \ , \\
&\vdots \\
&\boldsymbol{u}_1 = (\boldsymbol{A} - \lambda \boldsymbol{I})^{k-1}\boldsymbol{u}_k = \cdots = (\boldsymbol{A} - \lambda I)\boldsymbol{u}_1 = \boldsymbol{0} \Rightarrow \boldsymbol{A}\boldsymbol{u}_1 = \lambda \boldsymbol{u}_1 \ .
\end{aligned}$$

The number of these chains is defined by the geometric multiplicity d, while the maximal possible length p is called the *index of annihilation.* These eigenvectors are also linearly independent. This can be seen by taking a linear combination of these vectors with $a_k \boldsymbol{u}_k + a_{k-1}\boldsymbol{u}_{k-1} + \cdots a_1 \boldsymbol{u}_1 = \boldsymbol{0}$, and then substituting $\boldsymbol{u}_{k-1}, \ldots, \boldsymbol{u}_1$ with the previous formulas and multiplying with $(\boldsymbol{A} - \lambda \boldsymbol{I})^{k-1}$. Since $\boldsymbol{u}_k$ is a generalized eigenvector, then $a_k = 0$. It is noted that any chain of generalized eigenvectors is also called *Jordan chain* or string ending at $\boldsymbol{u}_k$.

If $\{\boldsymbol{u}_k, \boldsymbol{u}_{k-1}, \cdots, \boldsymbol{u}_1\}$ is a chain of generalized eigenvectors corresponding to the eigenvalue λ, then as we have seen, we have

$$\boldsymbol{A}\boldsymbol{u}_k = \lambda \boldsymbol{u}_k + \boldsymbol{u}_{k-1}, \ \boldsymbol{A}\boldsymbol{u}_{k-1} = \lambda \boldsymbol{u}_{k-1} + \boldsymbol{u}_{k-2} \ \cdots \ \boldsymbol{A}\boldsymbol{u}_1 = \lambda \boldsymbol{u}_1 \ .$$

If $\boldsymbol{P}$ is the matrix with columns the linear independent generalized eigenvectors $\begin{pmatrix}\boldsymbol{u}_1 & \boldsymbol{u}_2 & \cdots & \boldsymbol{u}_k\end{pmatrix}$ then

$$\begin{aligned}
\boldsymbol{A}\,\boldsymbol{P} &= \boldsymbol{A}\begin{pmatrix}\boldsymbol{u}_1 & \boldsymbol{u}_2 & \cdots & \boldsymbol{u}_k\end{pmatrix} \\
&= \begin{pmatrix}\boldsymbol{A}\boldsymbol{u}_1 & \boldsymbol{A}\boldsymbol{u}_2 & \cdots & \boldsymbol{A}\boldsymbol{u}_k\end{pmatrix} \\
&= \begin{pmatrix}\lambda\boldsymbol{u}_1 & \lambda\boldsymbol{u}_2 + \boldsymbol{u}_1 & \cdots & \lambda\boldsymbol{u}_k + \boldsymbol{u}_{k-1}\end{pmatrix} \\
&= \begin{pmatrix}\boldsymbol{u}_1 & \boldsymbol{u}_2 & \cdots & \boldsymbol{u}_k\end{pmatrix}\begin{pmatrix}\lambda & 1 & 0 & \cdots & 0 \\ 0 & \lambda & 1 & \cdots & 0 \\ \vdots & & & & \vdots \\ 0 & 0 & 0 & \cdots & 1 \\ 0 & 0 & 0 & \cdots & \lambda\end{pmatrix} \\
&= \boldsymbol{P}\boldsymbol{J}_k \ .
\end{aligned}$$

The $k \times k$ matrix $\boldsymbol{J}_k$ is called *Jordan block.*

The trivial 1×1 Jordan block $\boldsymbol{J}_1(\lambda) = (\lambda)$ implies that the main diagonal of a diagonal matrix consists of Jordan blocks $\boldsymbol{J}_1$. Other examples of Jordan blocks are the following 2×2 and 3×3 Jordan blocks

$$\boldsymbol{J}_2(\lambda) = \begin{pmatrix}\lambda & 1 \\ 0 & \lambda\end{pmatrix} \quad \textit{and} \quad \boldsymbol{J}_3(\lambda) = \begin{pmatrix}\lambda & 1 & 0 \\ 0 & \lambda & 1 \\ 0 & 0 & \lambda\end{pmatrix} \ .$$

In general, if $\boldsymbol{P}$ is the $n \times n$ matrix with all the chains of generalized eigenvectors of matrix $\boldsymbol{A}$, then $\boldsymbol{P}$ is invertible (with linearly independent columns, the generalized eigenvectors). Therefore, we can write $\boldsymbol{A}$ as $\boldsymbol{A} = \boldsymbol{P}\boldsymbol{J}\boldsymbol{P}^{-1}$ where $\boldsymbol{J}$ is the so-called *Jordan normal form* of $\boldsymbol{A}$. Moreover, $\boldsymbol{J}$ is a block-diagonal matrix where each diagonal block is a Jordan block.

Computation of Jordan normal form

Let $d_i = \dim\ker(\boldsymbol{A} - \lambda \boldsymbol{I})^i$ and $r_i = \operatorname{rank}(\boldsymbol{A} - \lambda \boldsymbol{I})^i$, $i = 1, 2, \cdots$, then

$$d_i = n - r_i \ .$$

It is now obvious tha:

- the number of simple eigenvectors is $d_1 = n - r_1$.
- the number of generalized eigenvectors of rank 2 is $d_2 - d_1 = (n - r_2) - (n - r_1) = r_1 - r_2$.
- the number of generalized eigenvectors of rank p is $d_p - d_{p-1} = r_{p-1} - r_p$.

We form the vector $\boldsymbol{s}$ with these numbers

$$\boldsymbol{s} = \begin{pmatrix} n - r_1 & r_1 - r_2 & \cdots & r_{p-1} - r_p \end{pmatrix} ,$$

known as the Segré characteristic. Using the Segré characteristic we construct a diagram that reveals the Jordan blocks. This diagram is called the Ferrer diagram and can be constructed using the following steps:

Step 1: In the first row we draw $n - r_1$ asterisks ★ to represent the generalized eigenvectors of rank 1 (simple eigenvectors).

Step 2: In the second row we draw $r_1 - r_2$ ★ to represent the generalized eigenvectors of rank 2.

Step 3: We continue until the p-th row where we draw $r_{p-1} - r_p$ ★ corresponding to the generalized eigenvectors of rank p.

The number of the asterisks appearing in each column of this diagram shows the rank of the corresponding Jordan blocks which we use to formulate the matrix $\boldsymbol{J}$.

For example, consider a 10×10 matrix $\boldsymbol{A}$ with $\lambda \in \sigma(\boldsymbol{A})$, an eigenvalue of $\boldsymbol{A}$ with $r_1 = 5$, $r_2 = 3$, $r_3 = 2$, $r_4 = 1$, $r_5 = 1$ (where $r_i = \text{rank}(\boldsymbol{A} - \lambda \boldsymbol{I})^i$, $i = 1, 2, \ldots, 5$). The minimum number $p \in \mathbb{N}$ such that $r_p = r_{p+1}$ is $p = 4$ is the index of annihilation. Consider the differences $n - r_1 = 10 - 5 = 5$, $r_1 - r_2 = 2$, $r_2 - r_3 = 1$, $r_3 - r_4 = 1$. Then the characteristic Segré is $(5, 2, 1, 1)$. The Ferrer diagram is:

$$\begin{array}{rcccccc}
n - r_1 = 5 & \longrightarrow & \bigstar & \bigstar & \bigstar & \bigstar & \bigstar \\
r_1 - r_2 = 2 & \longrightarrow & \bigstar & \bigstar & & & \\
r_2 - r_3 = 1 & \longrightarrow & \bigstar & & & & \\
r_3 - r_4 = 1 & \longrightarrow & \bigstar & & & & \\
\hline
 & & 4 & 2 & 1 & 1 & 1
\end{array}$$

The numbers $4, 2, 1, 1, 1$ are called the Weyr characteristics, and show the rank of the Jordan blocks corresponding to the eigenvalue λ. We repeat the same procedure for each eigenvalue λ, and we write the Jordan normal form using the computed Jordan blocks.

☞ To compute the Jordan normal form of a matrix $\boldsymbol{A}$ we follow the steps:

1. Compute all the eigenvalues of matrix $\boldsymbol{A}$
2. For each eigenvalue compute all the eigenvectors of rank 1
3. For each eigenvalue compute the Ferrer diagram
4. Construct the Jordan normal form

Examples

As a first example consider the matrix

$$\boldsymbol{A} = \begin{pmatrix} 4 & 0 & 1 \\ 2 & 3 & 2 \\ 1 & 0 & 4 \end{pmatrix} .$$

First we compute the eigenvalues by solving the characteristic equation $\det(\boldsymbol{A} - \lambda \boldsymbol{I}) = 0$. The characteristic polynomial is

$$\det(A - \lambda I) = \begin{vmatrix} 4-\lambda & 0 & 1 \\ 2 & 3-\lambda & 2 \\ 1 & 0 & 4-\lambda \end{vmatrix} = (3-\lambda)^2(5-\lambda) .$$

The characteristic equation $(3-\lambda)^2(5-\lambda) = 0$ has only two roots $\lambda_1 = 3$ and $\lambda_2 = 5$. For $\lambda_1 = 3$ we have $\tau_1 = 2$ and we compute two eigenvectors of rank 1 (since $d_1 = n - r_1 = 3 - 1 = 2$)

$$\boldsymbol{u}_1 = \begin{pmatrix} 0 & 1 & 0 \end{pmatrix}^T \quad \text{and} \quad \boldsymbol{u}_2 = \begin{pmatrix} -1 & 0 & 1 \end{pmatrix}^T .$$

Therefore, λ_1 has the same algebraic and geometric multiplicity $\tau_1 = d_1 = 2$. For λ_2 we compute only one eigenvector $\boldsymbol{u}_3 = \begin{pmatrix} 1 & 2 & 1 \end{pmatrix}^T$ and so this eigenvalue has algebraic multiplicity $\tau_2 = 1$ and geometric multiplicity $d_2 = 1$. Since all eigenvalues have the same algebraic and geometric multiplicity we conclude that the matrix $\boldsymbol{A}$ is diagonalizable. Although we can formulate directly the Jordan normal form by using the eigenvalues for the main diagonal, we continue using our methodology.

Computing the matrices $(\boldsymbol{A} - \lambda_i \boldsymbol{I})^2$ for $i = 1, 2$ we observe that they have ranks $r_2 = r_1$ and therefore $r1 - r2 = 0$ in both cases. This means that there are no generalized eigenvectors. So the Ferrer diagram consists of only one row with the simple eigenvector (of rank 1). For example, for $\lambda_1 = 3$ we have

$n - r_1 = 2$	$\longrightarrow$	★	★
$r_1 - r_2 = 0$	$\longrightarrow$		
		1	1

which means that there will be two Jordan blocks of dimension 1. More precisely the blocks [3] and [3]. For $\lambda_2 = 5$ again we have

$n - r_1 = 1$	$\longrightarrow$	★
$r_1 - r_2 = 0$	$\longrightarrow$	
		1

which means that there will be one more Jordan block of dimension 1, that is [5]. Therefore, the Jordan normal form in this case is

$$\boldsymbol{J} = \begin{pmatrix} 3 & 0 & 0 \\ 0 & 3 & 0 \\ 0 & 0 & 5 \end{pmatrix} .$$

Moreover, defining $\boldsymbol{P}$ to be the matrix with columns the eigenvectors of matrix $\boldsymbol{A}$

$$\boldsymbol{P} = \begin{pmatrix} 0 & -1 & 1 \\ 1 & 0 & 2 \\ 0 & 1 & 1 \end{pmatrix} ,$$

then easily we can verify that $\boldsymbol{A} = \boldsymbol{P}\boldsymbol{J}\boldsymbol{P}^{-1}$.

As another example we consider the problem of finding the Jordan normal form of the matrix

$$\boldsymbol{A} = \begin{pmatrix} 1 & 1 & 1 \\ 0 & 1 & 0 \\ 0 & 0 & 1 \end{pmatrix} .$$

In order to find the eigenvalues we first compute the matrix

$$\boldsymbol{A} - \lambda \boldsymbol{I} = \begin{pmatrix} 1-\lambda & 1 & 1 \\ 0 & 1-\lambda & 0 \\ 0 & 0 & 1-\lambda \end{pmatrix} .$$

In this case, the matrix $\boldsymbol{A}$ has only one eigenvalue, $\lambda = 1$ with algebraic multiplicity $\tau = 3$, and obviously, because the last two rows for $\lambda = 1$ are zero, yields $r_1 = \text{rank}(\boldsymbol{A} - \boldsymbol{I}) = 1$. Thus, for $\lambda = 1$ the matrix $\boldsymbol{A}$ has $d_1 = n - r_1 = 3 - 1 = 2$ eigenvectors of rank 1, namely

$$\boldsymbol{u}_1 = \begin{pmatrix} 1 & 0 & 0 \end{pmatrix}^T \quad \text{and} \quad \boldsymbol{u}_2 = \begin{pmatrix} 0 & 1 & -1 \end{pmatrix}^T .$$

To find the generalized eigenvector in this case, the matrix $(\boldsymbol{A} - \boldsymbol{I})^2$ is not helpful since it coincides with the 3×3 zero matrix. For this reason, we solve the system

$$(\boldsymbol{A} - \boldsymbol{I})\boldsymbol{u}_3 = \boldsymbol{u}_1 ,$$

which gives $\boldsymbol{u}_3 = \begin{pmatrix} 0 & 0 & 1 \end{pmatrix}^T$. In this case, the transformation matrix $\boldsymbol{P} = \begin{pmatrix} \boldsymbol{u}_1 & \boldsymbol{u}_2 & \boldsymbol{u}_3 \end{pmatrix}$. The Ferrer diagram is

$n - r_1 = 2$	$\longrightarrow$	★	★
$r_1 - r_2 = 1$	$\longrightarrow$	★	
		2	1

which means that $\boldsymbol{J}$ consists of a 2×2 and 1×1 Jordan blocks. The Jordan normal form is then

$$\boldsymbol{J} = \begin{pmatrix} 1 & 1 & 0 \\ 0 & 1 & 0 \\ 0 & 0 & 1 \end{pmatrix} .$$

Finally, we consider the problem of finding the Jordan normal form of the matrix

$$\boldsymbol{A} = \begin{pmatrix} -1 & -1 & 0 \\ 0 & -1 & -2 \\ 0 & 0 & -1 \end{pmatrix} .$$

In this example there is only one eigenvalue $\lambda = -1$ (easy to see since matrix $\boldsymbol{A}$ is upper-triangular). We observe that the matrix

$$\boldsymbol{A} - \lambda \boldsymbol{I} = \boldsymbol{A} + \boldsymbol{I} = \begin{pmatrix} 0 & -1 & 0 \\ 0 & 0 & -2 \\ 0 & 0 & 0 \end{pmatrix} ,$$

has $r_1 = \text{rank}(\boldsymbol{A} - \lambda \boldsymbol{I}) = 2$. Thus, $d_1 = n - r_1 = 3 - 2 = 1$ and there is only one eigenvector, namely $\boldsymbol{u}_1 = \begin{pmatrix} 1 & 0 & 0 \end{pmatrix}^T$. Solving the system $(\boldsymbol{A}+\boldsymbol{I})\boldsymbol{u}_2 = \boldsymbol{u}_1$ we compute only one generalized eigenvector of rank 2, the eigenvector $\boldsymbol{u}_2 = \begin{pmatrix} 0 & -1 & 0 \end{pmatrix}^T$, which implies that $d_2 = 1$. Similarly, the system $(\boldsymbol{A} + \boldsymbol{I})\boldsymbol{u}_3 = \boldsymbol{u}_2$ leads to the generalized eigenvector of rank 3, $\boldsymbol{u}_3 = \begin{pmatrix} 0 & 0 & 1/2 \end{pmatrix}^T$, and thus $d_3 = 1$.

The Ferrer diagram in this case is

$$\begin{array}{rcc} n - r_1 = 1 & \longrightarrow & \bigstar \\ r_1 - r_2 = 1 & \longrightarrow & \bigstar \\ r_2 - r_3 = 1 & \longrightarrow & \bigstar \\ \hline & & 3 \end{array}$$

and therefore there is only one 3×3 Jordan block forming the Jordan normal form of $\boldsymbol{A}$

$$\boldsymbol{J} = \begin{pmatrix} -1 & 1 & 0 \\ 0 & -1 & 1 \\ 0 & 0 & -1 \end{pmatrix} .$$

Note that the previous methodology can be used even when there are complex eigenvalues. For more information, we refer to the books [128, 67, 103, 21, 139] and the references therein.

Bibliography

[1] G. Akrivis and V. Dougalis. *Introduction to numerical analysis.* Cretan University Press, 2004.

[2] W. Ames. *Numerical methods for partial differential equations.* Academic Press, New York, 1977.

[3] T. Apostol. *Calculus, Volume 1, One-variable calculus, with an introduction to linear algebra*, volume 1. Wiley, 1991.

[4] T. Apostol. *Calculus, Volume 2, Multi-variable calculus and linear algebra with applications to differential equations and probability*, volume 2. Wiley, 1991.

[5] I. Argyros, Y. Cho, and S. Hilout. *Numerical methods for equations and its applications.* CRC Press, Boca Raton, FL, 2012.

[6] I. Argyros and F. Szidarovszky. *The theory and applications of iteration methods.* CRC Press, Boca Raton, FL, 2018.

[7] K. Atkinson. *An introduction to numerical analysis.* Wiley, New York, 2008.

[8] O. Axelsson. *Iterative solution methods.* Cambridge University Press, Cambridge, 1994.

[9] J. Barzilai and J. Borwein. Two-point step size gradient methods. *IMA Journal of Numerical Analysis*, 8:141—148, 1988.

[10] G. Birkhoff. General mean value and remainder theorems with applications to mechanical differentiation and quadrature. *Trans. Amer. Math. Soc.*, 7:107–136, 1906.

[11] G. Birkhoff and G.-C. Rota. *Ordinary differential equations.* John Wiley & Sons, 1978.

[12] Å. Björck. *Numerical methods for least squares problems.* SIAM, Philadelphia, 1996.

[13] S. Boyd and L. Vandenberghe. *Convex optimization.* Cambridge University Press, Cambridge, 2004.

[14] H. Brass and K. Petras. *Quadrature theory, The theory of numerical integration on a compact interval.* American Mathematical Society, Providence, Rhode Island, 2011.

[15] F. Brauer and C. Castillo-Chavez. *Mathematical models for communicable diseases.* SIAM, Philadelphia, 2012.

[16] E. Bressert. *SciPy and NumPy.* O'Reilly, 1st edition, 2013.

[17] S. Brin and L. Page. The anatomy of a large-scale hypertextual web search engine. *Computer networks and ISDN systems*, 30:107–117, 1998.

[18] R. Burden and J. Faires. *Numerical analysis.* Brooks/Cole, Cengage Learning, 2011.

[19] J. Butcher. *Numerical methods for ordinary differential equations.* Johm Wiley & Sons, UK, 2016.

[20] F. Chaitin-Chatelin and V. Frayssé. *Lectures on finite precision computations.* SIAM, 1996.

[21] C.-T. Chen. *Linear system theory and design.* Oxford university press, New York, Oxford, 1999.

[22] E. Cheney and D. Kincaid. *Numerical mathematics and computing.* Brooks/Cole: Cengage Learning, 2013.

[23] E. Chong and S. Żak. *An introduction to optimization.* John Wiley & Sons, Inc., Hoboken, New Jersey, 2013.

[24] P. Ciarlet. *Introduction to numerical linear algebra and optimisation.* Cambridge University Press, Cambridge, (English translation) 1989.

[25] S. Conte and C. De Boor. *Elementary numerical analysis: An algorithmic approach.* SIAM, Philadelphia, 2018.

[26] G. Dahlquist and Å. Björck. *Numerical methods.* Dover Publications, Inc. New York, NY, USA, 1974.

[27] G. Dahlquist and Å. Björck. *Numerical methods in scientific computing.* SIAM, Philadelphia, 2008.

[28] Y.-H. Dai and Y. Yuan. A nonlinear conjugate gradient method with a strong global convergence property. *SIAM J. Optim.*, 10:177–182, 1999.

[29] B.N. Datta. *Numerical linear algebra and applications.* SIAM, Philadelphia, 2010.

[30] P. Davis and P. Rabinowitz. *Methods of numerical integration.* Academic Press Inc., San Diego, 1984.

[31] C. De Boor. *A practical guide to splines.* Springer-Verlag New York, 1978.

[32] J. Demmel. *Applied numerical linear algebra.* SIAM, Philadelphia, 1997.

[33] J. Dennis and J. Moré. Quasi-newton methods, motivation and theory. *SIAM review*, 19:46–89, 1977.

[34] J. Dennis and R. Schnabel. *Numerical methods for unconstrained optimization and nonlinear equations.* SIAM, Philadelphia, 1996.

[35] J. Dongarra, C. Moler, J. Bunch, and G. Stewart. *LINPACK users' guide.* SIAM, Philadelphia, 1979.

[36] J. Dormand. *Numerical methods for differential equations: a computational approach.* CRC Press, Boca Raton, FL, 2018.

[37] V. Dougalis, D. Noutsos, and A. Hadjidimos. Numerical linear algebra. University of Thessaly, 2012.

[38] G. Evans. *Practical numerical integration.* Wiley, New York, 1993.

[39] K. Feng and M. Qin. *Symplectic geometric algorithms for Hamiltonian systems.* Springer-Verlag Berlin Heidelberg, 2010.

[40] R. Feynman. The best short works of Richard P. Feynman. In J. Robbins, editor, *The pleasure of finding things out.* Perseus books, 1999.

[41] R. Fletcher and C. Reeves. Function minimization by conjugate gradients. *The Computer Journal*, 7:149–154, 1964.

[42] G. Forsythe. Pitfalls in computation, or why a math book isn't enough. *The American Mathematical Monthly*, 77:931–956, 1970.

[43] G. Forsythe, M. Malcolm, and C. Moler. Computer methods for mathematical computations. *Prentice-Hall, Inc. Englewood Cliffs, N.J.*, 1977.

[44] W. Gear. *Numerical initial value problems in ordinary differential equations.* Prentice-Hall, Englewood Cliffs, N.J., 1971.

[45] A. George and J. Liu. *Computer solution of large sparse positive definite systems.* Prentice-Hall Englewood Cliffs, New Jersey, 1981.

[46] D. Goldberg. What every computer scientist should know about floating-point arithmetic. *ACM computing surveys*, 23:5–48, 1991.

[47] G. Golub and J. Ortega. *Scientific computing: an introduction with parallel computing.* Academic Press Inc., San Diego, 1993.

[48] G. Golub and C. Van Loan. *Matrix computations.* Johns Hopkins University Press, 2012.

[49] I. Goodfellow, Y. Bengio, and A. Courville. *Deep learning.* MIT press, Cambridge, 2016.

[50] Steven I Gordon and Brian Guilfoos. *Introduction to Modeling and Simulation with MATLAB® and Python.* CRC Press, Boca Raton, FL, 2017.

[51] D. Griffiths and D. Higham. *Numerical methods for ordinary differential equations.* Springer-Verlag London, 2010.

[52] C. Groetsch. Lanczo's generalized derivative. *The American Mathematical Monthly*, 105:320–326, 1998.

[53] W. Hackbusch. *Iterative solution of large sparse systems of equations.* Springer Switzerland, 2016.

[54] E. Hairer, C. Lubich, and G. Wanner. *Geometric numerical integration: Structure-preserving algorithms for ordinary differential equations.* Springer-Verlag Berlin Heidelberg, 2006.

[55] E. Hairer, S. Nørsett, and G. Wanner. *Solving ordinary differential equations I: Nonstiff problems.* Springer-Verlag Berlin Heidelberg, 1993.

[56] E. Hairer and G. Wanner. *Solving ordinary differential equations II: Stiff and differential-algebraic problems.* Springer-Verlag Berlin Heidelberg, 1996.

[57] G. Hämmerlin and K.-H. Hoffmann. *Numerical mathematics.* Springer-Verlag New York, 2012.

[58] C. Harris, J. Millman, S. van der Walt, R. Gommers, et al. Array programming with NumPy. *Nature*, 585:357–362, 2020.

[59] S. Haykin. *Neural networks and learning machines.* Pearson, 2009.

[60] M. Hestenes and E. Stiefel. Methods of conjugate gradients for solving linear systems. *Journal of research of the National Bureau of Standards*, 49:409–436, 1952.

[61] H. Hethcote. The mathematics of infectious diseases. *SIAM review*, 42:599–653, 2000.

[62] C. Higham and D. Higham. Deep learning: An introduction for applied mathematicians. *SIAM Review*, 61:860–891, 2019.

[63] D. Higham and A. Taylor. The sleekest link algorithm. *Mathematics Today*, 39:192–197, 2003.

[64] N. Higham. *Accuracy and stability of numerical algorithms.* SIAM, Philadelphia, 2002.

[65] N. Higham, editor. *Princeton companion to applied mathematics.* Princeton University Press, 2015.

[66] F. Hildebrand. *Introduction to numerical analysis.* Dover Publications, Inc. New York, USA, 1987.

[67] R. Horn and C. Johnson. *Matrix analysis.* Cambridge University Press, 2012.

[68] A. Householder. Unitary triangularization of a nonsymmetric matrix. *Journal of the ACM*, 5:339–342, 1958.

[69] A. Householder. *The theory of matrices in numerical analysis.* Dover Publications, Inc. New York, USA, 2013.

[70] E. Isaacson and H.B. Keller. *Analysis of numerical methods.* Dover Publications, Inc. New York, USA, 1994.

[71] A. Iserles. *A first course in the numerical analysis of differential equations.* Cambridge University Press, Cambridge, 2008.

[72] R. Johansson. *Numerical Python: Scientific Computing and Data Science Applications with Numpy, SciPy and Matplotlib.* Apress, Berkeley, CA, 2019.

[73] D. Kahaner, C. Moler, and S. Nash. *Numerical methods and software.* Prentice-Hall Englewood Cliffs, New Jersey, 1989.

[74] H. Karimi, J. Nutini, and M. Schmidt. Linear convergence of gradient and proximal-gradient methods under the Polyak-Łojasiewicz condition. In P. Frasconi, N. Landwehr, G. Manco, and J. Vreeken, editors, *Machine Learning and Knowledge Discovery in Databases*, pages 795–811. Springer International Publishing, 2016.

[75] C. Kelley. *Iterative method for linear and nonlinear equations.* SIAM, Philadelphia, 1995.

[76] D. Kincaid and E. Cheney. *Numerical analysis: mathematics of scientific computing.* American Mathematical Society, Providence, Rhode Island, 2009.

[77] D. Kincaid, J. Respess, D. Young, and R. Grimes. Algorithm 586: Itpack 2c: A fortran package for solving large sparse linear systems by adaptive accelerated iterative methods. *ACM Transactions on Mathematical Software (TOMS)*, 8:302–322, 1982.

[78] J. Kiusalaas. *Numerical methods in engineering with Python 3.* Cambridge University Press, 2013.

[79] Q. Kong, T. Siauw, and A. Bayen. *Python programming and numerical methods: A guide for engineers and scientists.* Elsevier Inc., 2020.

[80] D. Kressner. *Numerical methods for general and structured eigenvalue problems.* Springer-Verlag Berlin Heidelberg, 2005.

[81] E. Kreyszig. *Introductory functional analysis with applications.* Wiley, New York, 1978.

[82] H. Langtangen and S. Linge. *Finite difference compute with PDEs: A modern software approach.* Springer Nature, 2010.

[83] B. Leimkuhler and S. Reich. *Simulating Hamiltonian dynamics.* Cambridge University Press, 2004.

[84] M. Lutz. *Learning Python.* O'Reilly, 5th edition, 2013.

[85] J. Lyness and C. Moler. Numerical differentiation of analytic functions. *SIAM Journal on Numerical Analysis*, 4:202–210, 1967.

[86] S. Marsland. *Machine learning: An algorithmic perspective.* Chapman and Hall/CRC, New York, 2009.

[87] E. Matthes. *Python crash course: A hands-on, project-based introduction to programming.* No Starch Press, US, 2nd edition, 2019.

[88] W. McKinney. *Python for data analysis: Data wrangling with Pandas, NumPy, and IPython.* O'Reilly, 2nd edition, 2017.

[89] C. Moler. *Experiments with MATLAB.* Mathworks, 2001.

[90] C. Moler. *Numerical computing with MATLAB.* SIAM, 2004.

[91] J. Moré, B. Garbow, and K. Hillstrom. User Guide for MINPACK-1. Technical Report III, Argonne, 1980.

[92] K. Murphy. *Probabilistic Machine Learning: An introduction.* MIT Press, 2022.

[93] J. Nocedal and S. Wright. *Numerical optimization.* Springer-Verlag New York, 2006.

[94] J. Ortega. *Numerical analysis: a second course.* SIAM, Philadelphia, 1990.

[95] J. Ortega and W. Rheinboldt. *Iterative solution of nonlinear equations in several variables.* SIAM, Philadelphia, 2000.

[96] A. Ostrowski. *Solution of Equations and Systems of Equations: Pure and Applied Mathematics: A Series of Monographs and Textbooks, Vol. 9.* Academic Press Inc., New York, 2016.

[97] B. Parlett. *The symmetric eigenvalue problem.* SIAM, Philadelphia, 1998.

[98] A. Peressini, F. Sullivan, and J. Uhl Jr. *The mathematics of nonlinear programming.* Springer-Verlag, 1988.

[99] G. Peters and J. Wilkinson. $ax = \lambda bx$ and the generalized eigenproblem. *SIAM Journal on Numerical Analysis*, 7:479–492, 1970.

[100] G. Phillips and P. Taylor. *Theory and applications of numerical analysis.* Academic press, London, 1996.

[101] R. Piessens, E. de Doncker-Kapenga, C. Überhuber, and D. Kahaner. *Quadpack, A subroutine package for automatic integration.* Springer-Verlag Berlin Heidelberg, 1983.

[102] G. Pinski and F. Narin. Citation influence for journal aggregates of scientific publications: Theory, with application to the literature of physics. *Information Processing & Management*, 12:297–312, 1976.

[103] R. Piziak and P. Odell. *Matrix theory: from generalized inverses to Jordan form.* Chapman and Hall/CRC, 2007.

[104] E. Polak and G. Ribiere. Note sur la convergence de méthodes de directions conjuguées. *ESAIM: Mathematical Modelling and Numerical Analysis*, 3:35–43, 1969.

[105] D. Poole. *Linear algebra: A modern introduction.* Cengage Learning, 4th edition, 2015.

[106] G. Poore. PythonTeX: reproducible documents with LaTeX, Python, and more. *Computational Science & Discovery*, 8:014010, 2015.

[107] M. Powell. *Approximation theory and methods.* Cambridge University Press, 1981.

[108] A. Quarteroni, R. Sacco, and F. Saleri. *Numerical mathematics.* Springer-Verlag Berlin Heidelberg, 2007.

[109] A. Ralston and P. Rabinowitz. *A first course in numerical analysis.* Dover Publications, Inc. New York, USA, 2001.

[110] F. Rosenblatt. *Principles of neurodynamics; perecptrons and the theory of brain mechanisms.* Spartan books, Washington D.C., 1962.

[111] C. Runge. Über empirische funktionen und die interpolation zwischen äquidistanten ordinaten. *Zeitschrift für Mathematik und Physik*, 46:224–243, 1901.

[112] Y. Saad. *Iterative methods for sparse linear systems.* SIAM, Philadelphia, 2003.

[113] Y. Saad. *Numerical methods for large eigenvalue problems.* SIAM, Philadelphia, 2011.

[114] J.M. Sanz-Serna and M.P. Calvo. *Numerical Hamiltonian problems.* Chapman and Hall/CRC Press, London, 1994.

[115] T. Sauer. *Numerical analysis.* Pearson, 2018.

[116] M. Schatzman. *Numerical analysis: a mathematical introduction.* Oxford University Press, 2002.

[117] M. Schultz. *Spline analysis.* Prentice-Hall Englewood Cliffs, New Jersey, 1973.

[118] L. Schumaker. *Spline functions: basic theory.* Cambridge University Press, Cambridge, 2007.

[119] L. Schumaker. *Spline functions: Computational methods.* SIAM, Philadelphia, 2015.

[120] R. Scott. *Numerical Analysis.* Princeton University Press, 2011.

[121] L. Shampine. *Numerical solution of ordinary differential equations.* CRC Press, Boca Raton, FL, 1994.

[122] M. Spivak. *Caluclus*, volume 3rd. Cambridge University Press, 2006.

[123] W. Squire and G. Trapp. Using complex variables to estimate derivatives of real functions. *SIAM review*, 40:110–112, 1998.

[124] G. Stewart. *Afternotes on numerical analysis.* SIAM, Philadelphia, 1996.

[125] G. Stewart. *Afternotes goes to graduate school, lectures on advanced numerical analysis.* SIAM, Philadelphia, 1998.

[126] J. Stewart. *Calculus.* Cengage Learning, Inc, 2011.

[127] J. Stoer and R. Bulirsch. *Introduction to numerical analysis.* Springer-Verlag New York, 2013.

[128] G. Strang. *Linear algebra and its applications.* Thomson Brooks/Cole, 2006.

[129] G. Strang. *Introduction to linear algebra.* Wellesley-Cambridge Press, 4th edition, 2009.

[130] G. Strang. *Linear algebra and learning from data.* Wellesley-Cambridge Press, 1st edition, 2019.

[131] A. Stroud and D. Secrest. *Gaussian quadrature formulas.* Prentice-Hall, Englewood Cliffs, N.J., 1966.

[132] E. Süli and D. Mayers. *An introduction to numerical analysis.* Cambridge, UK, 2003.

[133] T. Tao. *Analysis I.* Springer, Singapore, 2016.

[134] T. Tao. *Analysis II.* Springer, Singapore, 2016.

[135] N. Trefethen and D. III Bau. *Numerical linear algebra.* SIAM, Philadelphia, 1997.

[136] J. VanderPlas. *Python data science handbook: Essential tools for working with data.* O'Reilly, 1st edition, 2016.

[137] R. Varga. *Matrix iterative analysis.* Springer-Verlag Berlin Heidelberg, 2000.

[138] P. Virtanen, R. Gommers, T. Oliphant, et al. SciPy 1.0: Fundamental Algorithms for Scientific Computing in Python. *Nature Methods*, 17:261–272, 2020.

[139] S. Weintraub. *Jordan canonical form: theory and practice.* Morgan & Claypool Publishers, 2009.

[140] J. Wilkinson. *The algebraic eigenvalue problem.* Clarendon Press, Oxford, 1965.

[141] J. Wilkinson. *Rounding errors in algebraic processes.* Dover Publications, Inc. New York, USA, 1994.

[142] D. Young. *Iterative solution of large linear systems.* Academic Press, New York, 1971.

Index

Printed in the United States
by Baker & Taylor Publisher Services